The Bird

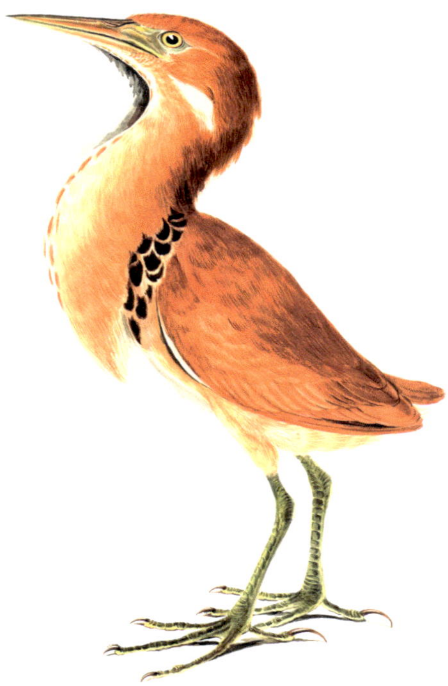

LAURENCE KING
First published in Great Britain in 2021
by Laurence King Publishing
an imprint of The Orion Publishing Group Ltd
Carmelite House, 50 Victoria Embankment
London EC4Y 0DZ

An Hachette UK Company

1 3 5 7 9 10 8 6 4 2

A CIP catalogue record for this book is available from the British Library.

ISBN 978 1 786 277 312

Commissioning Editor: Zara Larcombe
Senior Editors: Sara Goldsmith, Felicity Maunder
Designer: Alexandre Coco
Production Manager: Simon Walsh

Printed in China by C&C Offset Printing Co., Ltd.

BOOK CHAIN PROJECT

www.laurenceking.com
www.orionbooks.co.uk

The Bird

The Great Age of Avian Illustration

PHILIP KENNEDY

Splendours of Nature:

A brief history of ornithological illustration

For centuries the bird has sparked our creativity, our curiosity and our imagination. Traces of this fascination can be found in our earliest artworks, and their cultural presence in our everyday lives continues to be seen even to this day. Throughout the world, our curiosity for birds has been explored and expressed in many distinctive forms. An eagerness to understand them has led to many incredible discoveries and creations. It has been through the act of making pictures that we have gained some of our most valuable knowledge about these magnificent creatures.

This book is about birds, but it is also inescapably a book about pictures. Every image in this book has been created by somebody who saw value in this act of creation. Drawing is a way for us to communicate our understanding of the world, but it is also a process that leads to a greater comprehension of it. The illustrations in this book exist because drawing can express things in a way that words often cannot. The act of drawing is the act of looking, and our understanding of birds has been shaped fundamentally by this simple act.

While ornithology – the scientific study of birds – only truly emerged as a discipline in the mid-eighteenth century, we have always been curious about birds. Indeed, people have been drawing them since they began to make marks. Our prehistoric ancestors first took to rocks and caves to depict many of the creatures that lived alongside them. In the vast wilderness of Arnhem Land in Australia, we can find a red ochre depiction of two giant emu-like birds. This is the work of Aboriginal Australians; an image that some have estimated may have been created around 40,000 years ago. On the other side of the world, people had also been drawing birds. In the Lascaux caves in France, we can find the depiction of a small bird that perches beside the figure of a man. Created roughly 15,000 years ago, this man is somewhat enigmatic as its head is that of a bird. What this means, we do not know. But we can deduce that it had some importance or some greater meaning. The figure was designed to be a permanent message that could be pushed forwards into the future. These are drawings as a means of communication and transmission of knowledge.

Some could suggest that the bird-headed figure was something other than simply an image. While the cave is filled with many depictions of animals, this figure could be read as a symbol: a series of marks that represent a more complex, and generally more abstract, idea. Indeed, our relationship with birds has continued to have this duality. The bird is something that we wish to understand, but it is also a symbol that can help us to understand ourselves. The early stages of our history are filled with such dichotomies. Here, images

HOOPOE
Upupa epops
Lucas Schan, *Historia animalium*, 1551–8

The woodcut illustrations in Gessner's *Historia animalium* were made by the natural history artist and wildfowler Lucas Schan. In addition to providing these illustrations, Schan also added descriptions of the behaviour of some of these birds.

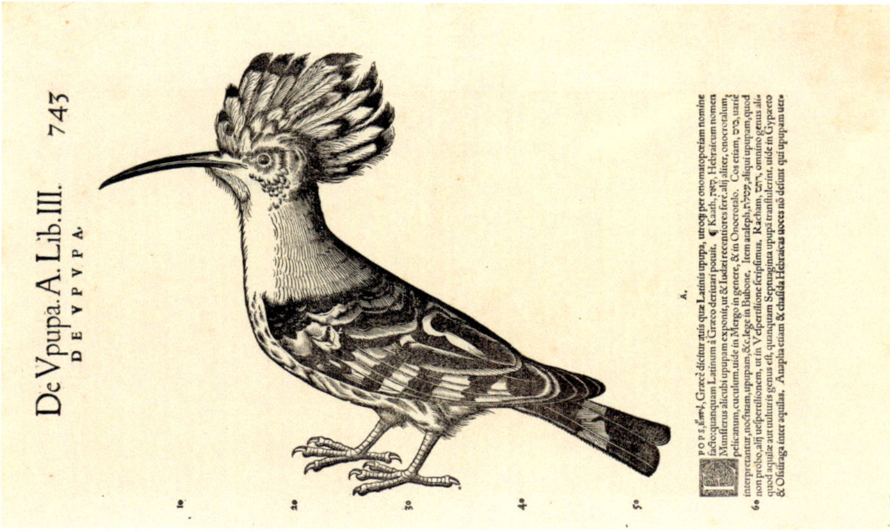

get mixed with symbols and the objective often blurs with the subjective. In our early history, the boundaries between art and science did not exist, and so our pursuit of a deeper understanding of birds took a variety of forms.

Throughout the world, birds have been central to countless myths and legends. They have been represented as gods and deities in many religions and their flight can often signify a link between heaven and earth. Often this link is represented physically: the Egyptian deity of Horus, for example, is depicted as a falcon-headed man; in Greek mythology, Pegasus is a divine horse that has the wings of a bird; in the Aztec religion, Huītzilōpōchtli is often represented as either a hummingbird or an eagle. While some religions may not choose to depict their god with the characteristics of a bird, many will still use the bird as a sacred symbol – an example of the great reverence that we have always had for these creatures.

While it is easy to point to the mythical and spiritual role that birds have played in the lives of early humans, our more physical and secular relationship with them is arguably somewhat less prevalent – or, at least, less visible. Birds and their eggs have been sources of food for humans since their origin and – if knowledge of birds was to be transmitted – it is safe to assume that we did so as a means to know which ones to avoid and which ones to hunt. Yet as time moved forward, so too did our curiosity.

Ancient Authorities & Mythical Beasts

The classification of birds goes back at least as far as ancient Greece, where Aristotle (384–322 BCE) studied the natural world and attempted to group animals according to their similarities. His work *Historia animalium* [*History of Animals*] (*ca.* 350 BCE) is the first known system of classification and in it, Aristotle worked to define the characteristics that separate different groups of animals – exploring their anatomy, their behaviour and their habitat. This was one of the most influential zoological works ever written and it continued to be a primary source of scientific knowledge until at least the sixteenth century.

In the intervening years, knowledge of birds and other animals continued to be shared. Perhaps most notably in the work of the Roman naturalist Pliny the Elder (*ca.* 24–79 CE), whose *Historia naturalis* (77 CE) collected much of the knowledge of the time and can be credited as being the first scientific encyclopaedia. Ambitious in its scope, this work's zoological sections used Aristotle's division of nature. Yet despite Pliny being an authority on scientific knowledge at the time, today much of the information in *Historia Naturalis* is known to be inaccurate. While the work does little to explore new avenues of thought, it remains an impressive compilation of the knowledge of the time.

Indeed, accuracy is something that tends to be lacking in the majority of these early scholarly works. This is most prevalent during the Middle Ages, when

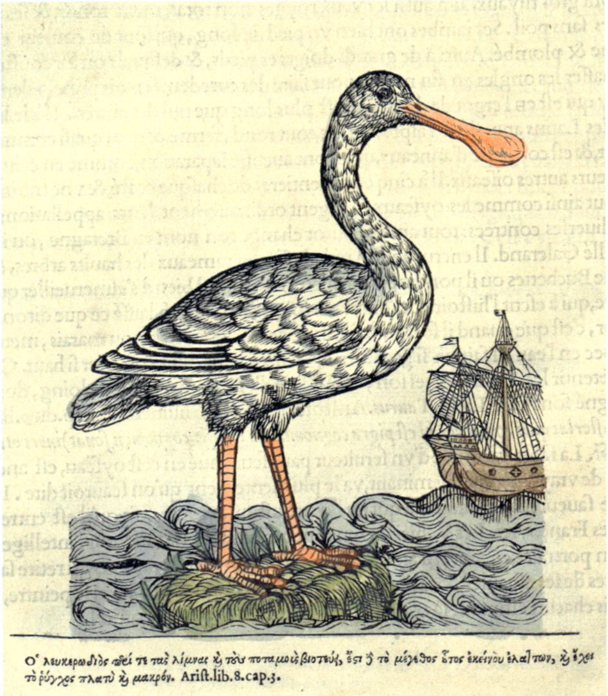

'WHITE HERON',
ACTUALLY
EUROPEAN SPOONBILL
Platalea leucorodia

Uncredited, *L'Histoire de la nature des oyseaux*, 1555

Published in 1555, Pierre Belon's *L'Histoire de la nature des oyseaux* was one of the first ornithological works illustrated with original drawings. Featuring 160 woodcuts, Belon based his descriptions of birds on Aristotelian principles. A French naturalist and writer, Belon was a pioneer of natural history in the sixteenth century. As a Renaissance scholar he studied and wrote on a broad range of topics including anatomy, botany and architecture.

PASSERINES

Uncredited, *The Ornithology of Francis Willughby*, 1678

Various artists contributed to Ray and Willughby's work. Identification is the primary objective of this work, with compositions laid out clearly and the names of each bird written next to their image.

medieval animal encyclopedias – known as bestiaries – were produced in Europe. Building upon the early zoology seen in the works of Aristotle and Pliny, these works took a more Christian understanding of creation. Produced in the belief that the natural world had been created by God, these works looked to extract important symbolic meaning from the animals around us. Typically pieced together from earlier texts, bestiaries explained the religious significance that each creature held and attached a Christian moral to each entry. They were made by hand and were among the most richly illuminated manuscripts of their era. Words and pictures worked alongside one another to portray both real and imaginary creatures, with pictures drawing readers in and helping them to remember the names of each entry as well as the morals that were attached to it. Birds and animals were once again used as symbols and pictures of them became tools for teaching.

It was the Renaissance that eventually saw a break away from these traditional symbolic depictions and started to bring about a revival in the ancient practice of studying the natural world through observation. At the time, Aristotle's work was being rediscovered and people began to look at nature again with fresh eyes. Not only did they see a new beauty within it, but they also saw beauty in attempting to understand it. In art, people such as Pisanello (*ca.* 1395–*ca.* 1455), Leonardo da Vinci (1452–1519) and Albrecht Dürer (1471–1528) were all discovering that to depict the natural world accurately you need to have a deep understanding of it.

Despite these new ways of thinking, a modern understanding of the natural world did not arrive instantaneously. It was a slow and gradual process that happened over centuries and was informed by many factors. The invention of the printing press in the sixteenth century was one such factor, as it allowed for scientific minds to share their understandings of the world with others. The printed book was – and is – a vessel for knowledge. It can be built upon, shared and expanded. In the early days of printing, we can see some examples of this starting to emerge, with early naturalists using the power of the press to collate, conserve and expand on existing understandings of nature.

In 1544, the English naturalist William Turner (*ca.* 1508–68) published *Avium praecipuarum*, a work credited with being the first-ever printed book devoted solely to birds. It was written in Latin – the scientific language of the day – and each bird's name has its origins in this classical language. Subtitled *A Short and Succinct History of the Principal Birds Noticed by Pliny and Aristotle*, Turner's book presented the work of the ancient authorities, who at the time were considered to be the central thinkers on science. In addition to their observations he also included notes and descriptions from his own extensive knowledge. This was a real innovation and an important step forward in the study of nature.

Turner was a friend of the great Swiss naturalist Conrad Gessner (1516–65), whose *Historia animalium* (1551–8) was the first truly modern zoological work.

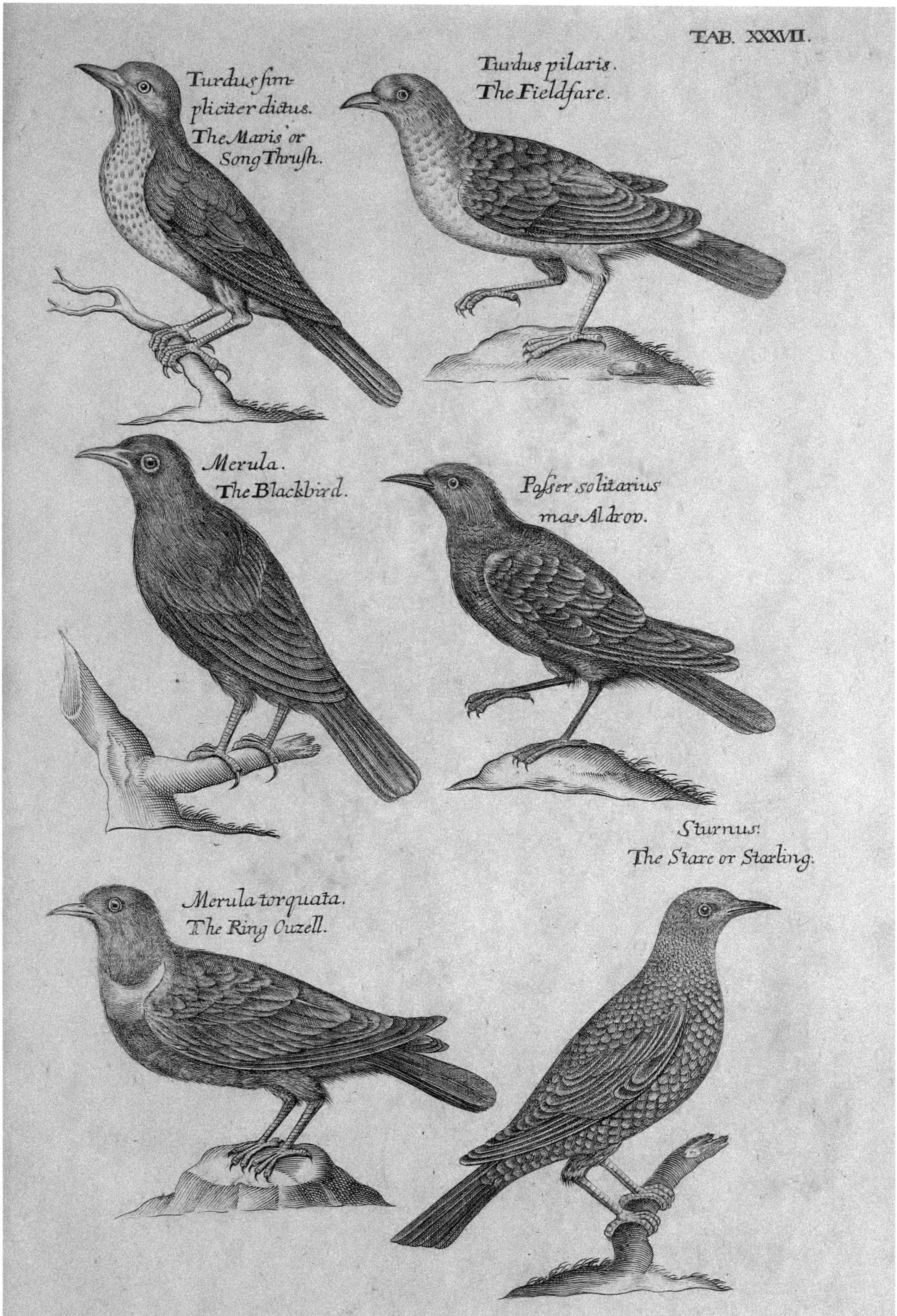
TAB. XXXVII.
Turdus simpliciter dictus.
The Mavis or Song Thrush.
Turdus pilaris.
The Fieldfare.
Merula.
The Blackbird.
Passer solitarius mas Aldrov.
Sturnus.
The Stare or Starling.
Merula torquata.
The Ring Ouzell.

L'AIGRETTE (EGRET)

François Nicolas Martinet, *Histoire Naturelle Des Oiseaux*, 1770–86

In addition to engraving plates for Buffon, Martinet also provided plates for other titles including Mathurin Jacques Brisson's *Ornithologie* (1760) and the *Encyclopedia* of Diderot and d'Alembert.

Sharing the same name as Aristotle's seminal work and spanning over 4,500 pages, this series of books was a repository of knowledge that collected everything from the classical works and medieval texts to biblical tales and personal observations. The book was rare for its time in that it was richly illustrated. This heralded the birth of zoological illustration and set the standard for the books that would follow. Of the five volumes, book three centres on birds and Gessner chose to ignore Aristotle's system of classification, instead simply presenting his subjects in alphabetical order.

His aim for the project was not to redefine classification – it was to present as much information as possible and, over the five books, he delivered on this ambition. *Historia animalium* is a masterwork of its time and is the defining work of Renaissance natural history. It would take a little over 100 years for a new system of categorization to appear, but when it did, it heralded what many would consider to be the beginning of scientific ornithology.

What's In A Name?

Published in 1676, *The Ornithology of Francis Willughby* aimed to describe all the then-known birds of the world. Written by the English naturalist John Ray (1627–1705), it contained a number of his own observations alongside a great deal of work researched by his friend and colleague Francis Willughby (1635–72). The book provided the first truly useful system of classification. Rather than arranging birds by how they function, as Aristotle did, Willughby and Ray arranged them by their characteristics of form – their morphology – with birds grouped by features such as the shape of their beak, the size of their body and the structure of their feet. Their intention was to show the order that they believed God had defined when He was creating all life. The system of classification that they outlined revolutionized how we thought about birds and their work provided the first working definition of a species.

Willughby and Ray's work influenced many other people, but the most celebrated of these is undoubtedly Carl Linnaeus (1707–78). Born in Sweden, he implemented a clear rank-based system of classification and then used that system for developing new names for the plants and animals that he wrote about. Linnaeus ranked things according to class, order, genera and species and created scientific names by combining the Latin name of each species with the genus that they belonged to. So, for example, humans belong to the genus Homo and within that genus, they belong to the species Homo sapiens. This two-part naming system is known as the binomial nomenclature.

Linnaeus first began assigning binomial names to birds in the tenth edition of his *Systema Naturae* (1758–9). This was an important step in the history of ornithology as it allowed scientists to accurately identify individual species. For example, the American robin and the European robin may share a common name, but they are not that closely related. In fact, the American robin, *Turdus migratorius*, belongs to the thrush family; while the smaller

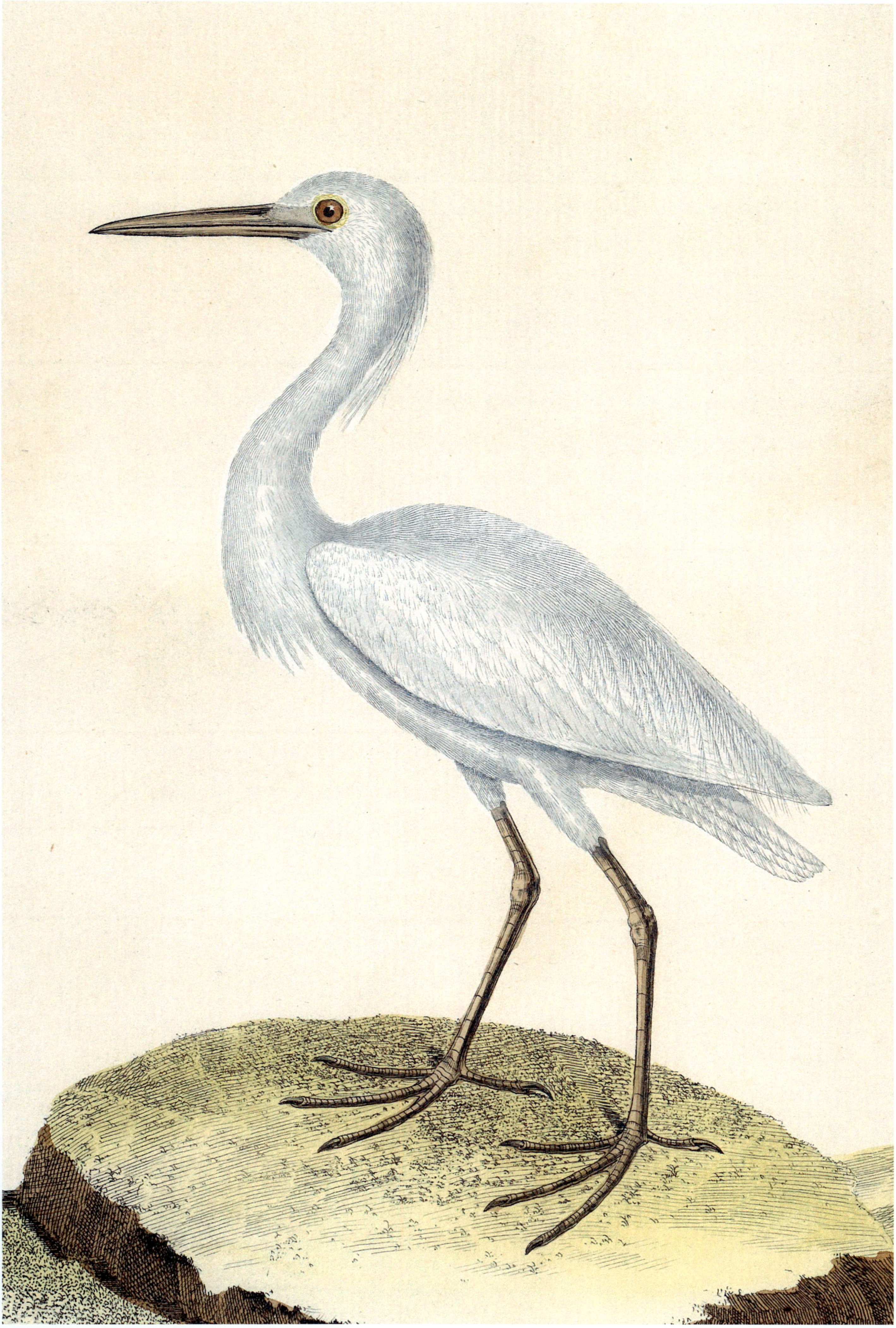

European robin, *Erithacus rubecula*, belongs to the Old World flycatcher family. Linnaeus' system was accurate, precise and rational. Even today, his hierarchical rank-based system is the one that we continue to use and, while some changes have occurred over the years, the system has helped us shape our understanding of the natural world.

Indeed, Linnaeus' work is emblematic of its time. The eighteenth century was the Age of Enlightenment and, throughout much of Europe, we can see examples of people actively looking at the world and eagerly trying to pin it down. If we take the 1750s alone, we not only see Linnaeus' work, but also events such as the establishment of the British Museum (1753), the publication of Samuel Johnson's *Dictionary of the English Language* (1755), and the beginning of Diderot and d'Alembert's hugely important *Encyclopédie* (1751–72). It was this pursuit of knowledge and understanding that brought about an end to many old ways of thinking and allowed for a modern new world to emerge.

But perhaps this is jumping ahead a little too quickly. While knowledge, printing and classification were all instrumental in the emergence of this new world, there was another factor that we have not yet addressed. That is the emergence of the New World itself.

Discoveries

It is commonly agreed that the New World was discovered by the Europeans in the year 1492. This discovery came during a point in history that has been dubbed both the Age of Exploration or the Age of Discovery. This era saw Western Europe push itself outwards across the globe in search of trade routes, new land and raw materials. Beginning in the fifteenth century and lasting roughly until the seventeenth century, the names for this period are misnomers as they imply that humans only began to explore and discover things at this point in time. The truth is that for millennia humans have been exploring and discovering the world. Terms like 'Age of Exploration' and 'Age of Discovery' sound noble and adventurous, but their true aim is to whitewash an era that was defined by European greed and savage imperialism.

The motivation for exploration was not the expansion of knowledge; it was about the acquisition of resources, power and wealth. The 'discoveries' that emerged from this time were a byproduct of ruthless expansion and it would be remiss not to mention that they came at the expense of both indigenous people and indigenous knowledge. It was conquest and colonization that introduced Europe to new plants, animals, culture, food, technology and ideas. Even when unique flora and fauna were discovered, they were valued predominantly for their rarity and exoticism, rather than their inherent worth. Yet despite how these discoveries were valued, it is necessary to stress that they were valued. Collected by royalty and nobility, birds and animals from distant lands were seen as objects for display. Sometimes they were stuffed, other times they were kept in menageries. These foreign creatures became trophies of colonial competition. To display a creature or specimen from a foreign land was to show your wealth and your global reach. As more

THE RAVEN

Thomas Bewick, *A History of British Birds*, 1797–1804 (this edition: 1885)

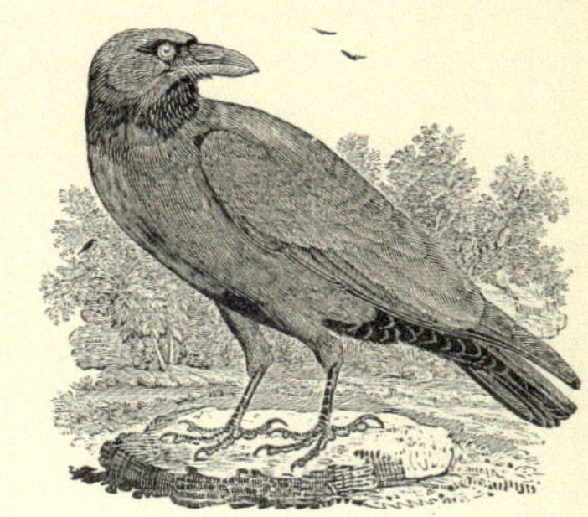

72 BRITISH BIRDS.

THE RAVEN.

GREAT CORBIE CROW.

(*Corvus Corax*, Linn.—*Corbeau noir*, Temm.)

Is the largest of this kind; the length is above two feet: breadth four. The female is somewhat less. The bill is strong, and thick at the base, measuring rather more than two inches and a half in length, and covered with strong hairs and bristles, which extend above half its length, covering the nostrils: the general colour of the upper parts is a fine glossy black, reflecting a blue tint in particular lights; the under parts are duller, and of a dusky hue.

ships returned, more cultures were exploited and so wealth trickled down through European society. The brutal extraction of resources from around the globe created a new class within Europe, and these people were eager to demonstrate their wealth and social standing. To have something rare was to have something of value, and to have something of value was to show your importance.

Even the earliest European voyages returned with new animals and plants that had never been seen before in Europe. It heralded a new curiosity and enthusiasm for nature. Up until this point, it had been assumed that Europeans had encountered every species on the planet; there was an assumption that knowledge of the natural world was complete. Before this, there was a deeply-ingrained understanding that nature was a direct manifestation of God and that each animal had allegorical importance, but suddenly Europeans were confronted by a host of new species. This revelation meant that their understanding of the world was forced to shift. These were species that were not contained within any bestiary and so they could not be quickly categorised by some pre-existing symbolic significance. Suddenly there was a massive increase in the number of known species in the world and they were appearing at an unprecedented rate. The only efficient way that all of these species could be categorized was simply by describing them.

This is when a new way of studying the world began to emerge and we start to see European thought make its first steps towards being truly scientific in its outlook. But, this is not to say that a new secular understanding of the world suddenly arrived. Linnaeus himself had believed that he was chosen by God to reveal the precise ways in which Creation was ordered, and European collectors no doubt took inspiration in the story of Noah's Ark when building their menageries. A central part of European expansion was about the rise of European Christian culture. This was grounded in the concept of 'one true faith' and an implicit part of this was the destruction of other cultures. What the Age of Exploration did was strip away the allegorical and symbolic role of the animal and allow people to focus on how the world was ordered. The whimsical drawings of the bestiaries were replaced by more refined illustrations in the seventeenth century, and by the eighteenth century – they had become even more sophisticated. At the centre of this progression was the growing popularity of the printed book.

The Beauty of Books

The publishing revolution of the eighteenth century was tightly rooted in the Age of Enlightenment. During this period, antiquity came to an end and science worked tirelessly to dispel old superstitions. Reason began to replace faith as the cornerstone of knowledge and those eager to spread their thoughts and understanding typically did so through books.

Factual books had previously only been read by specialists but, by the eighteenth century, learning had become more democratic. The exploitation and the extraction of resources from abroad also meant that there was more wealth in Europe, and this brought about a more affluent society that demanded more works to read and to own. People were becoming increasingly eager to understand the world around them, and publishers realized that there was a market for all types of books – as well as a market for all types of readers. If you had an enlightened mind it was important to demonstrate this through the books that you owned.

Meanwhile, with more opportunities to print, there was a far greater exploration into what a book could do and what value it held. Certainly, it was a tool for knowledge, but it was also capable of being incredibly beautiful and immersive. If you had a specific idea, the book was a way you could share it with the world. Inversely, a book was also a way to share the world with a single reader. Only a select few people were afforded the opportunity to travel to far-off places and those that did would bring back specimens. Yet specimens could only be shared with a small number of people. If you published a series of drawings of that specimen, you could share it with a far greater number of people. For those interested in sharing the beauty and importance of nature, publishing was the perfect way to do so.

One of the first truly great examples of this is seen in the work of Mark Catesby (1683–1749). An English naturalist, he undertook two extensive voyages to North America in the early eighteenth century. Not only was his work the first bird book to be printed in colour, but it was also the first account of the continent's flora and fauna. Published between 1729 and 1747, it introduced the wonders of Britain's American colonies to Europe and helped natural history scientists identify the birds, animals and flora that existed there. Plants were of particular interest to many because they were an important source for food and medicine. Books on botany were far more numerous at this time than books on birds or animals.

Catesby never saw himself as an artist, and while his work does have a primitive edge, it is nonetheless attractive. Yet it is important to remember that these are primarily illustrations and are not intended to be works of art – their aim is to convey information. Catesby was producing his work at a time when science was emerging as the dominant way to view the world, and his illustrations are there to aid identification. His *Natural History of Carolina, Florida and the Bahama Islands* was hugely influential, particularly in science, and for over 100 years, he remained the authority on American natural history. The work came several years before Linnaeus' system of taxonomy and Linnaeus even included information from this work in the tenth edition of his *Systema Naturae*.

When Linnaeus' framework was put in place, later works in ornithology became even more systematic in their approach, with creators focusing primarily on defining and identifying birds as precisely as possible. While some ornithologists drew in the field, they were typically the exception to the rule. Instead, most worked from materials they got from explorers or collectors. They established networks of correspondents throughout Europe where they exchanged specimens from other collectors and fellow scientists. The studio became a central part in the study of birds, with illustrators producing work that could inform and communicate. Drawing and science were simply ways to understand the world, and they worked together harmoniously during the creation of these works.

Collaboration & Support

European Enlightenment philosophy tended to emphasize the individual and their creative self-expression. The image of the lone genius who has rare artistic skill or possesses a unique way of thinking is rooted in the philosophy of this age. In the West, we still tend to gravitate towards this narrative. This very book is guilty of perpetuating it: we remain drawn to stories of people who have demonstrated exceptional intellect or unparalleled creative ability. Catesby, for example, taught himself how to etch and then spent twenty years preparing and publishing his *Natural History*. It is a romantic image of extraordinary self-belief and determination. Yet the reality is that these works were very rarely produced with such self-sufficiency. There is no denying the importance of, say, the French naturalist Georges-Louis Leclerc (1707–88) who was Comte de Buffon. His multi-volume encyclopedia, *Histoire Naturelle* (1749–1804) was undoubtedly a pioneering work in the natural sciences, but such works were typically massive productions. The series' bird books – *Histoire Naturelle des Oiseaux* (1770–83) – credits the art and engraving to François-Nicolas Martinet (1731–1800), but Buffon himself makes it clear that more than eighty artists and workers produced the plates in this set of books.

Collaboration also existed between the creator and the collector. Take, for example, the English ornithologist George Edwards (1694–1773). Known as the father of British ornithology, part of his success must surely be attributed to the collector Sir Hans Sloane (1660–1753). A physician and president of the Royal Society, Sloane commissioned Edwards to produce illustrations of his collection. Sloane made much of his profits through the Atlantic slave trade, and this allowed him to collect tens of thousands of items which then became the basis for what is today known as the British Museum. Gaining access to such collections was an integral part in the production of these works.

This shared sense of collaborating for the greater good is something that was popular at the time and became even more prevalent as the nineteenth century emerged. Learned societies were run by gentlemen of science;

DARWIN'S FINCHES

John Gould, *Journal of researches into the geology and natural history of the various countries visited by H.M.S. Beagle*, 1840 (this edition: 1845)

Darwin enlisted the help of John Gould in identifying the specimens he had collected on the Galápagos Islands. In this illustration by Gould we see four of the species of finch that Darwin observed. Darwin postulated that the variation in the shape and size of their beaks was an example of adaptive radiation.

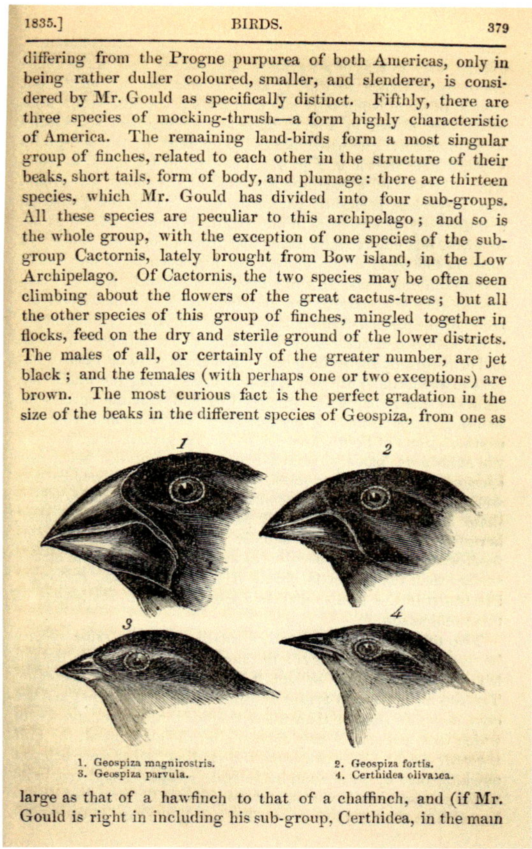
1835.] BIRDS. 379

differing from the Progne purpurea of both Americas, only in being rather duller coloured, smaller, and slenderer, is considered by Mr. Gould as specifically distinct. Fifthly, there are three species of mocking-thrush—a form highly characteristic of America. The remaining land-birds form a most singular group of finches, related to each other in the structure of their beaks, short tails, form of body, and plumage: there are thirteen species, which Mr. Gould has divided into four sub-groups. All these species are peculiar to this archipelago; and so is the whole group, with the exception of one species of the sub-group Cactornis, lately brought from Bow island, in the Low Archipelago. Of Cactornis, the two species may be often seen climbing about the flowers of the great cactus-trees; but all the other species of this group of finches, mingled together in flocks, feed on the dry and sterile ground of the lower districts. The males of all, or certainly of the greater number, are jet black; and the females (with perhaps one or two exceptions) are brown. The most curious fact is the perfect gradation in the size of the beaks in the different species of Geospiza, from one as

1. Geospiza magnirostris. 2. Geospiza fortis.
3. Geospiza parvula. 4. Certhidea olivasea.

large as that of a hawfinch to that of a chaffinch, and (if Mr. Gould is right in including his sub-group, Certhidea, in the main

these were wealthy men such as Sloane, who were keen to make grand gestures that could demonstrate the value they placed on knowledge. Supporting the production of expensive natural history prints was prestigious and it could also be profitable. Naturalists relied on subscribers who were eager to demonstrate their participation in the study of natural history. As production was costly, the number of subscribers to a work would typically dictate the number of copies that were produced. Printing was still an expensive process and there were additional expenses such as paper, binding and colourists. For this reason, these projects were hugely ambitious and the subscription model was the best way to fund them.

The Print Process

While we have mentioned that these illustrations were printed, it may be worth a little time exploring how this process worked. In fact, the illustrations reproduced in this book were not all produced in the same way. Manuscripts, such as bestiaries, were written and drawn by hand, but when the printing press was invented by Johannes Gutenberg (*ca.* 1400–1468) in the fifteenth century, it brought about an information revolution. His press used metal movable type, which allowed printers to reuse letters for each new page in their book. Known as letterpress printing, this technique suited the written word and broadly remained in use until the second half of the twentieth century. While these letters could be reused over several publications, illustrations were typically unique to a single project, so a whole range of methods to reproduce an image were developed over time.

Early illustrators tended to use woodcuts, a method which involved a draughtsman drawing their design onto a plank of wood and then using a knife or a graver to cut into it to remove the parts which they did not need to print. Like letterpress, woodcuts are a form of relief printing, so both can easily be printed together. In the 1790s, the British engraver and natural history author Thomas Bewick (1753–1828) refined this process by rotating the wood through ninety degrees and engraving into the wood grain. Using the tougher side of the wood allowed for more impressions to be made and woodblock engraving became one of the most widely used methods of reproduction during the nineteenth century.

Between the seventeenth and nineteenth century, copper plate engraving was the medium of choice for most natural history illustrators. Unlike relief printing – where the draughtsman creates a raised image on which ink is applied – engraving uses a process known as intaglio. Here, the image that is to be printed lies within recessed areas of the copper plate. Working from a drawing, a draughtsman will copy the image onto a plate using a cutting tool known as a burin. This technique allows for a more delicate line, but it does not allow for large blocks of ink to be used.

Etching was also a popular process at this time. Like engraving, it is a method of intaglio printmaking, but instead of cutting deep into the plate itself to reveal the image, an etcher will first coat their plate in wax and scrape lines into that wax to produce their image. Once complete, the plate is placed in a bath of acid which seeps into the grooves that have been drawn into the wax and dissolves the exposed metal below. If an engraving or etching was to be produced in colour, this would have to be added by hand.

By the mid-nineteenth century, lithography was starting to become more popular. In this process, the image is drawn with a wax crayon onto a polished piece of limestone. When the block is inked and wiped clean, the ink remains stuck to where the wax crayon had been applied. This allows for the drawing to be then transferred to paper. This process allowed for prints to have a much wider range of marks and tones than earlier relief and intaglio methods and allowed for more of the detail of an artist's drawing to be captured.

Despite advancements in printing, all of these methods required time, patience and skill. We mentioned earlier how work is rarely the output of a single individual and the process of developing an illustration often required one person to create the drawing, somebody else to transfer it to stone or plate, somebody to print it and finally somebody else to add colour. Some of these roles appear as credits on the prints themselves but the inner workings of a studio were often kept private, and frequently creators were left unacknowledged. Throughout this book, you might spot the occasional credit appearing on a reproduced plate. Typically, these credits use abbreviations of Latin words and these can provide us with a greater understanding of the work. Some of the more common words to appear on prints include: 'del' – which denotes the artist who produced the drawing that is reproduced, 'lith' – used to credit the draughtsman responsible for transferring the drawing to the lithographic stone, and 'imp', from *impressit*, crediting the printer who printed the work.

A Golden Age

Today, many of these prints remain just as impressive and as fascinating as they did when they were first printed. By the 1800s, ornithological illustration had reached its Golden Age. People like Alexander Wilson (1766–1813) and John James Audubon (1785–1851) were travelling America and developing important works that not only celebrated the birds of the country but also celebrated the idea of the country itself. Colonial expansion also continued, particularly after the end of the Napoleonic Wars (1803–15), and this increased the number of known species around the world. Expeditions would routinely include natural historians and amateur artists. This included the HMS Beagle expedition (1831–6) which led to Darwin's theory of evolution by natural selection, and while Darwin himself was not an ornithologist, his work provided explanations for numerous features in the lives of birds and encouraged a stronger scientific interest in the field.

Some books, such as Prideaux John Selby's *Illustrations of British Ornithology* (1819–34), looked inwards, yet the majority of them tended to look outwards. This is perhaps no more evident than in the work of John Gould (1804–81), who worked with a host of illustrators to produce multi-volume publications that focused chiefly on the birds of specific places. Titles featured surveys of the Himalayas, Europe, Australia and Asia, to name but four. For some, these works were a way to claim ownership over the native fauna and a means to demonstrate some control over nature and authority over the unknown. Yet their legacy paved the way for the future of ornithology and helped to inspire future generations who are now working to preserve and protect the natural world.

In this book, we will delve into the remarkable world of these creations and present some of the finest bird illustrations that were produced at this time. This carefully curated selection of images represents the diverse and ever-evolving world of ornithological illustration, reproducing work by such renowned figures as Audubon, Gould, Catesby, Wilson, and Selby. Through these chapters, we also briefly explore the lives of some of these key figures in the hope of gaining a greater understanding of their work. Despite the variety of illustrations found on these pages, the splendour of birds is a consistency that remains throughout. With an estimated 10,000 species on the planet, the bird remains endlessly captivating. It is through this this diversity of form that this book takes its structure. Here, you will encounter birds in all shapes and sizes, from tropical birds and seabirds to poultry and birds of prey. This book spans the globe, bringing together the familiar with the exotic and the common with the rare. The illustrations collected here demonstrate the sheer beauty of the avian world and stand to remind us why birds continue to be a central part of our creativity, our curiosity and our imagination.

OPPOSITE:
WESTERN MEADOWLARK
Sturnella neglecta

John James Audubon,
The Birds of America, 1827–38

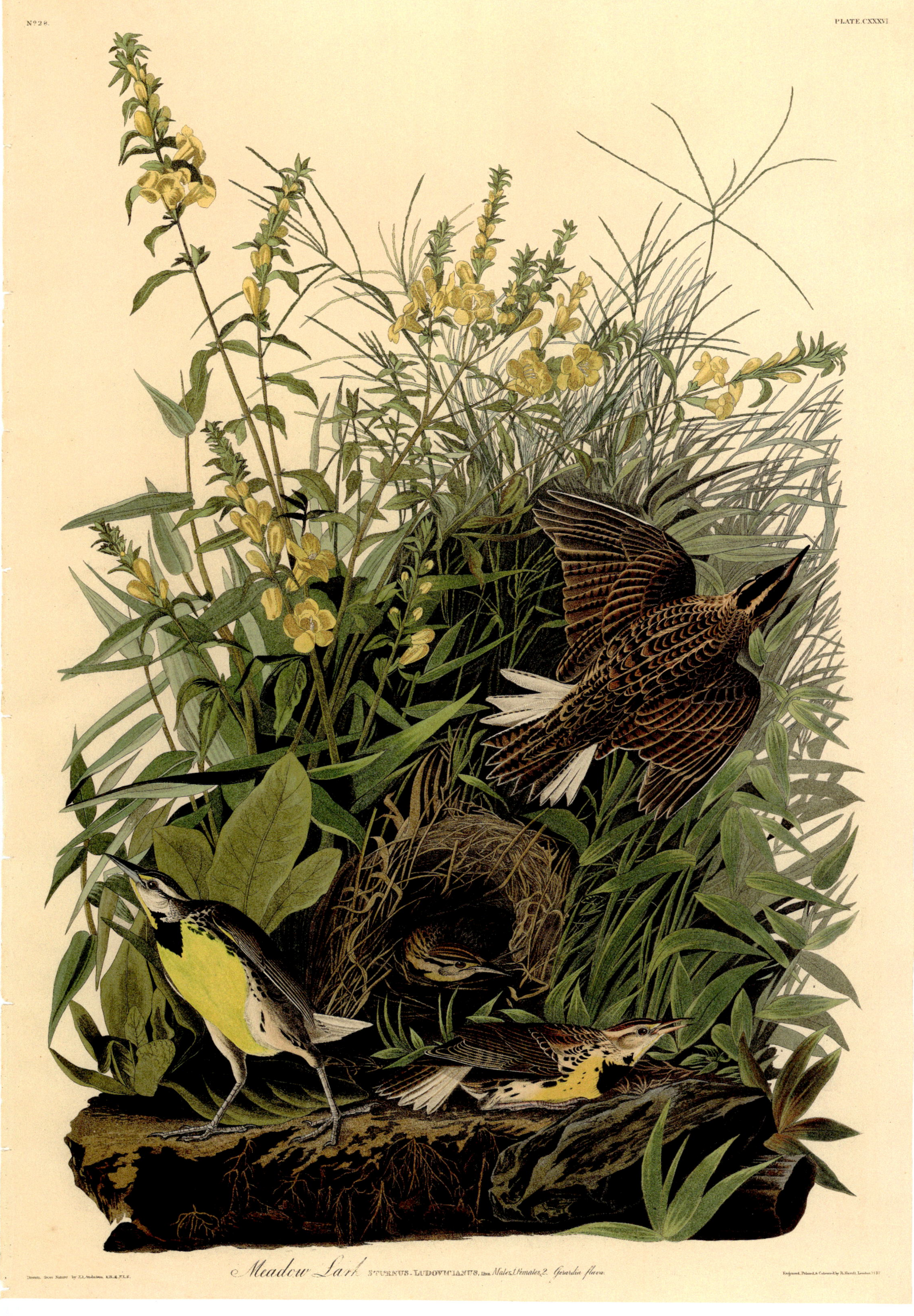

N° 28.
PLATE CXXXVI
Meadow Lark STURNUS LUDOVICIANUS, Linn. Males 1. Females 2. Gerardia flava.

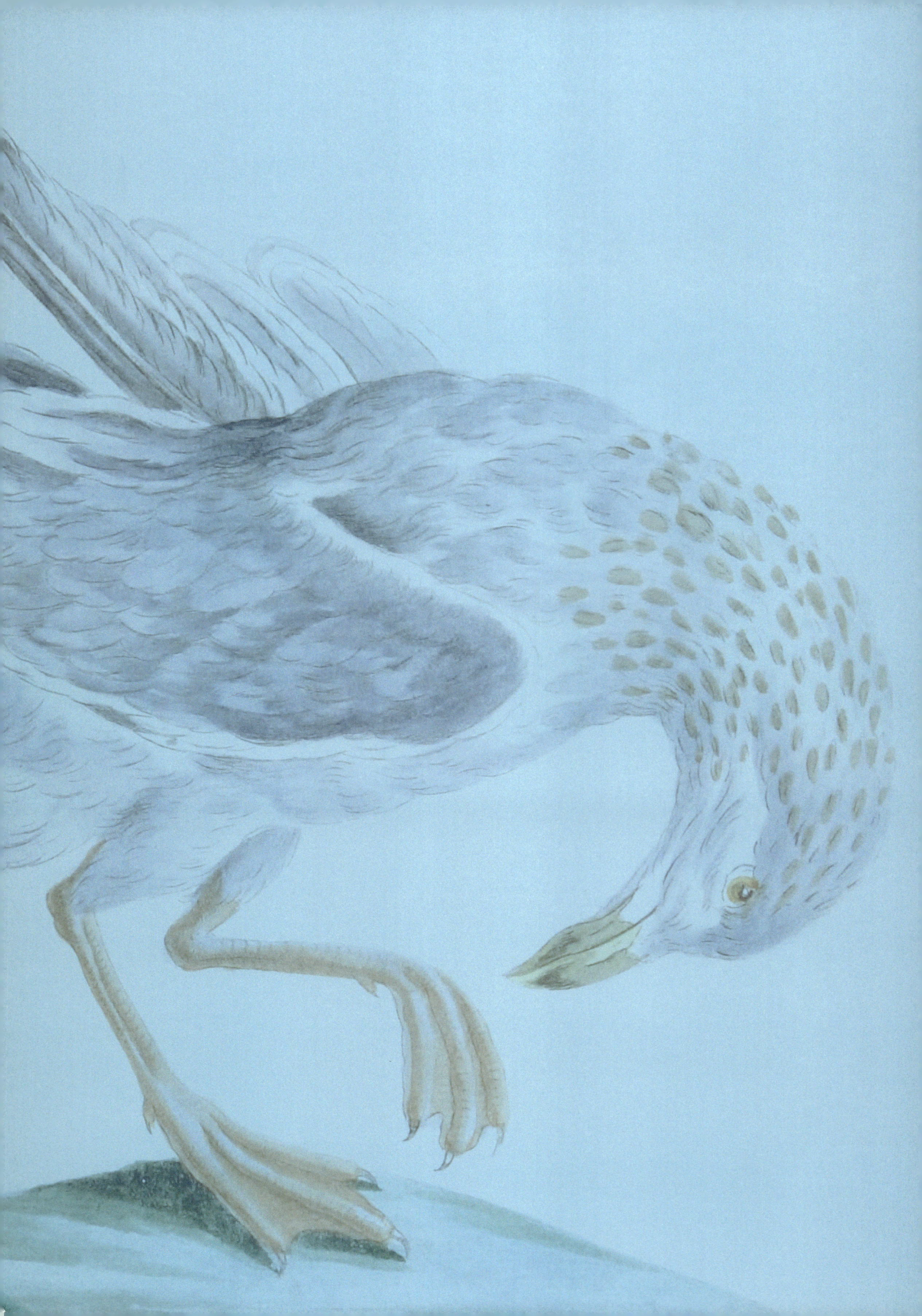

Seabirds

Flying, gliding, floating, swimming and diving: seabirds are beautifully adapted for ocean living. Some spend most of their lives out of sight of land, while others prefer to hug the coast. They often nest in large congregations known as colonies and, to us, they are mysterious and elusive.

Flight has made birds extremely adaptable. Not only have they mastered almost every habitat on land, but the power and reach of flight has also allowed them to spread all over the earth's coasts, seas and oceans, which make up more than two thirds of the surface of the planet.

Birds have developed a broad range of skills for living a life of harmony with the sea. Some swim at the ocean surface, some plunge in from a height, some dive into the depths chasing fish and squid, while others fly above the surface and drop down to pick up items. Some follow sea mammals and many are drawn to upwellings. Some birds are coastal, while others are pelagic, spending most of their lives at sea. From the giant albatross to the tiny storm petrel, seabirds come in all shapes and sizes.

Many seabirds are elusive, breeding in remote colonies and spending the majority of their lives far out on the ocean. For this reason, they are far more difficult to study than the birds we find on land. Yet, as the illustrations in this chapter will demonstrate, these birds were still documented and recorded. Indeed, it is important to remember that the sea played a significant role in the early history of ornithology. The so-called Age of Exploration had opened people's eyes to the incredible variety of form and colour that could be found in the living world; while the sea also brought curious minds like Mark Catesby (1683–1749; see page 44), Alexander Wilson (1766–1813; see page 102) and John Gould (1804–81) to new lands and new discoveries. These journeys often required months at sea and it is easy to imagine these voyaging ornithologists developing a strong attachment to seabirds.

The journeys that they undertook also draw a nice parallel with the journeys made each year by many birds. A large number of seabirds undertake an annual migration, with many

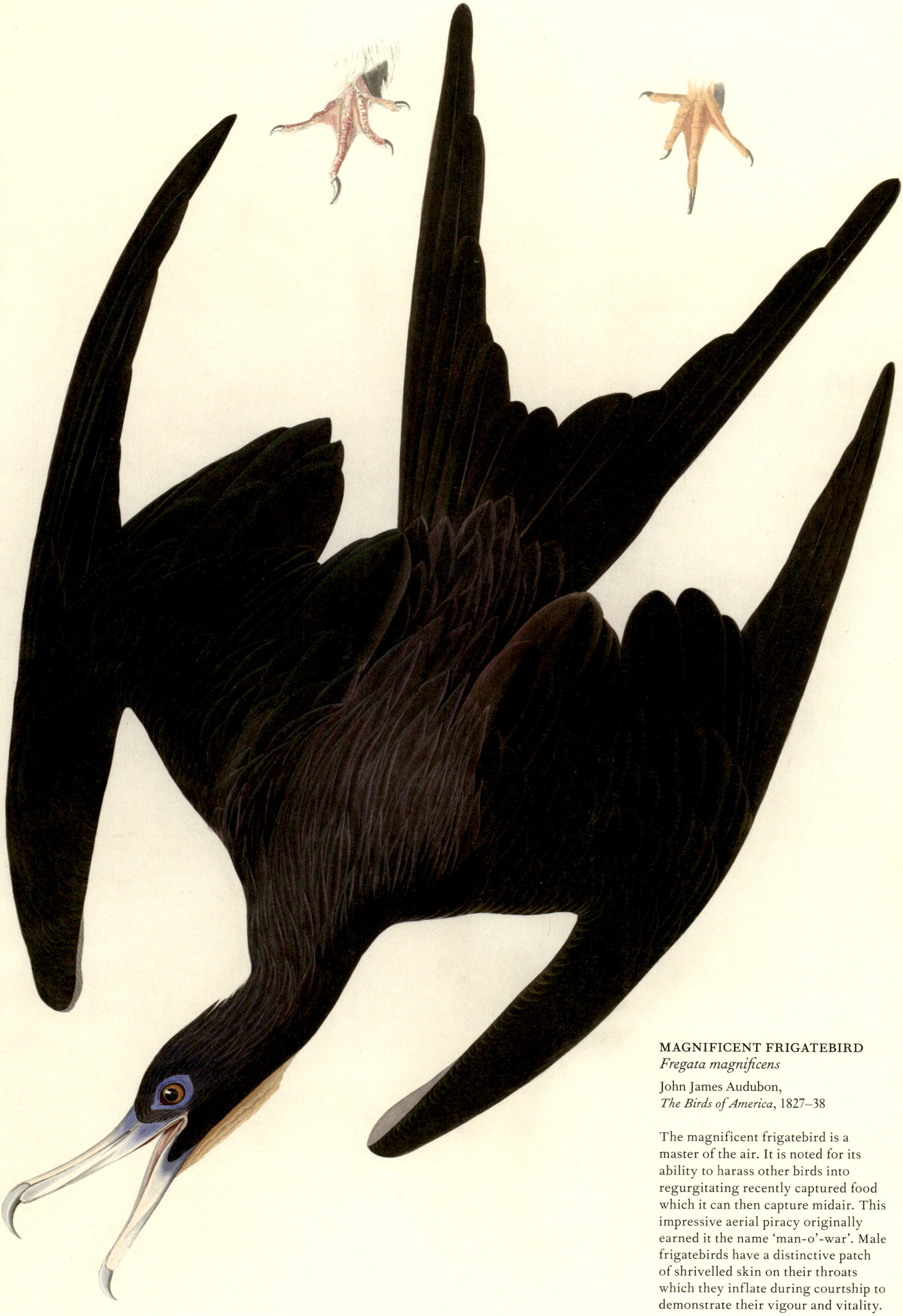

MAGNIFICENT FRIGATEBIRD
Fregata magnificens

John James Audubon,
The Birds of America, 1827–38

The magnificent frigatebird is a master of the air. It is noted for its ability to harass other birds into regurgitating recently captured food which it can then capture midair. This impressive aerial piracy originally earned it the name 'man-o'-war'. Male frigatebirds have a distinctive patch of shrivelled skin on their throats which they inflate during courtship to demonstrate their vigour and vitality.

LITTLE TERN
Sternula albifrons

John Gould & William Matthew Hart, *The Birds of New Guinea and the Adjacent Papuan Islands*, 1875–88

Terns are part of the *Laridae* family of seabirds that also includes gulls, noddies and skimmers. The little tern is – as its name might suggest – the smallest of the terns. These birds breed in large colonies on coastlines and are always alert for danger. Attacks can come from the ground and the air, and so little terns will rely on safety in numbers. This illustration shows the birds both in flight and resting. It was published originally in John Gould's *The Birds of New Guinea and the Adjacent Papuan Islands*.

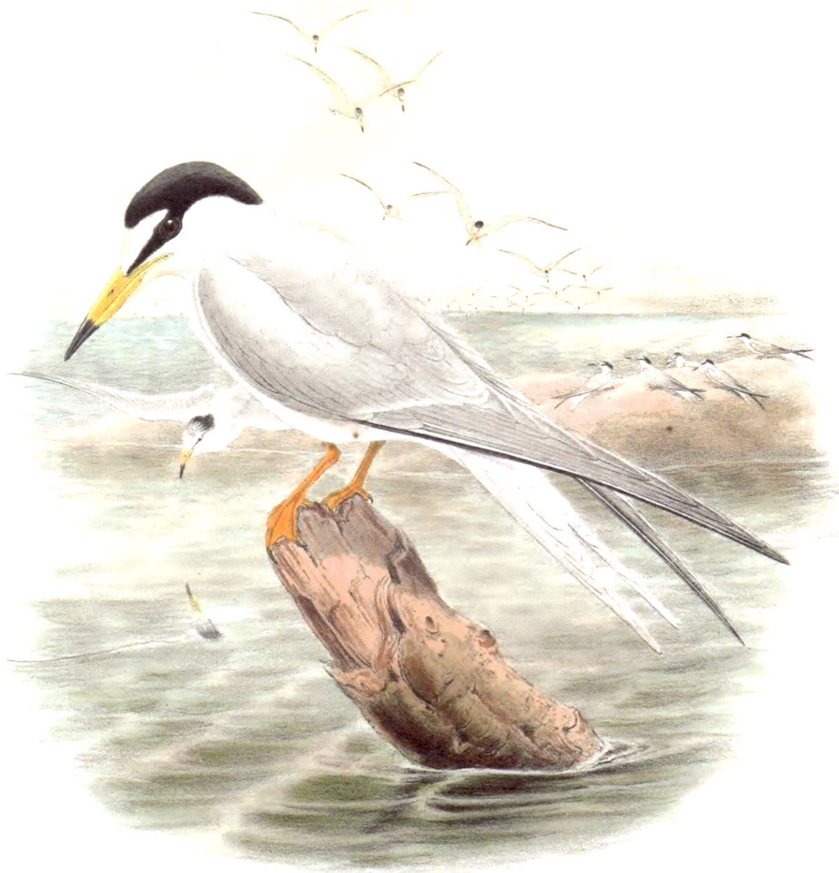

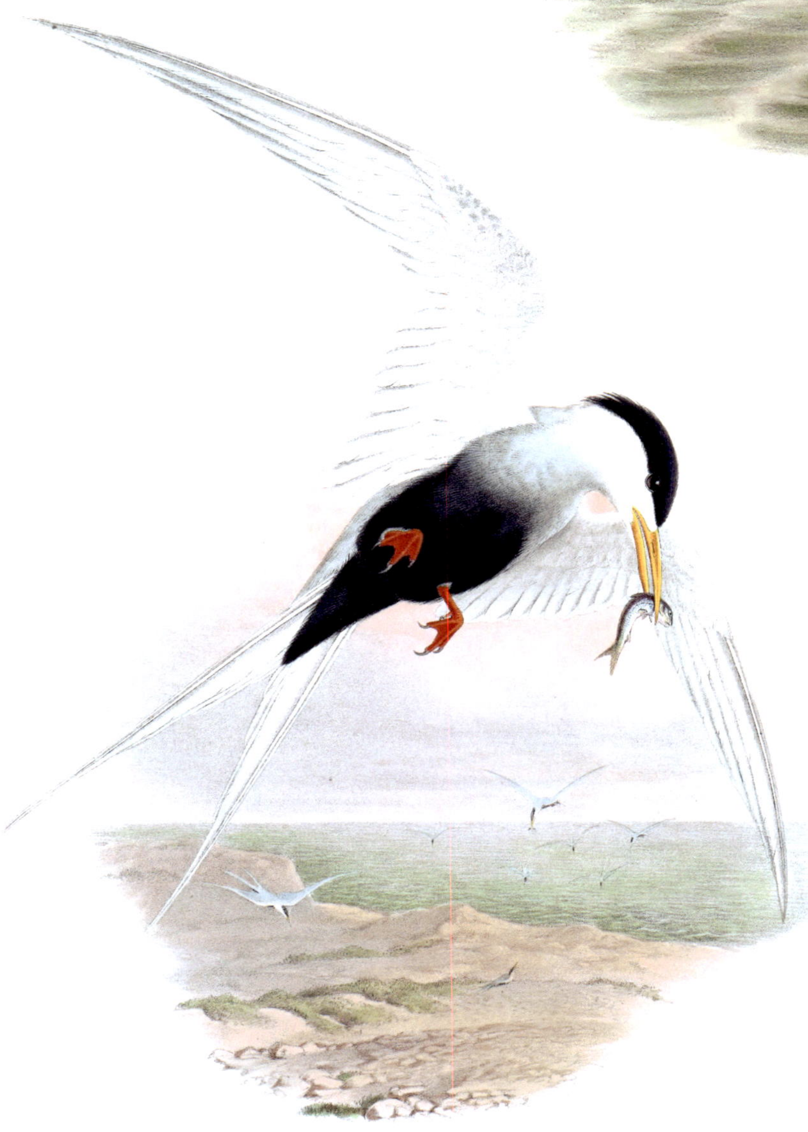

BLACK-BELLIED TERN
Sterna acuticauda

John Gould & Henry Constantine Richter, *The Birds of Asia*, 1850–83

Here we see another tern that also comes from a work by Gould yet this plate is taken from the book *The Birds of Asia*. Found near larger rivers in the Indian subcontinent, the black-bellied tern occurs mostly in India as well as parts of Pakistan, Nepal and Bangladesh. Its distinctive black belly is its breeding plumage and appears during summer. Outside of the breeding season, its belly is whitish. Its diet consists of small fish and insects, which it seeks out over land and water.

COMMON MURRE
Uria aalge

George Graves, *British Ornithology*, 1811–21

Graves' illustration of a common murre identifies this bird as a 'foolish guillemot'. This name was quite common for the species and comes from the bird's notoriously tame nature. Some accounts even describe the bird standing unmoved after several attempts at shooting it.

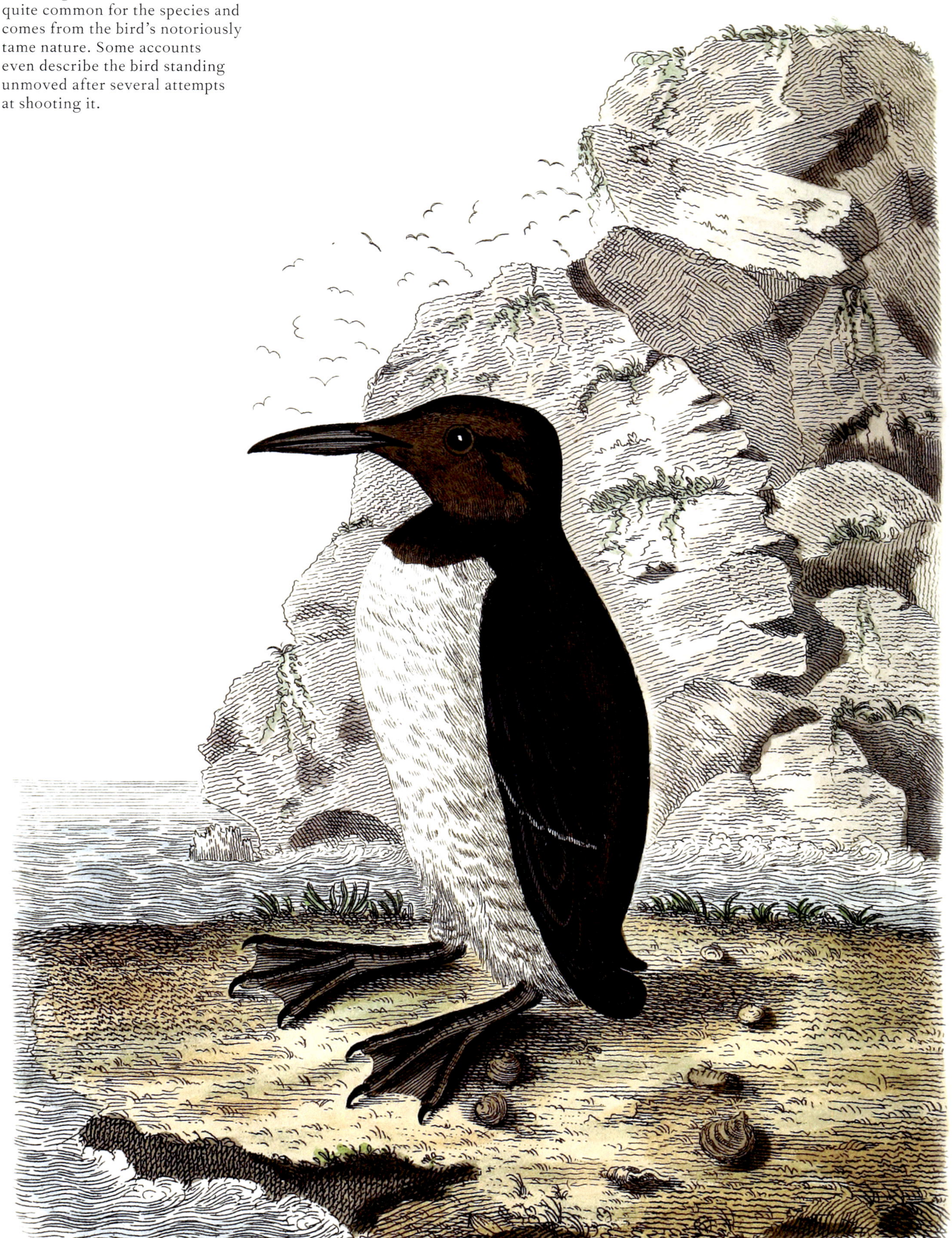

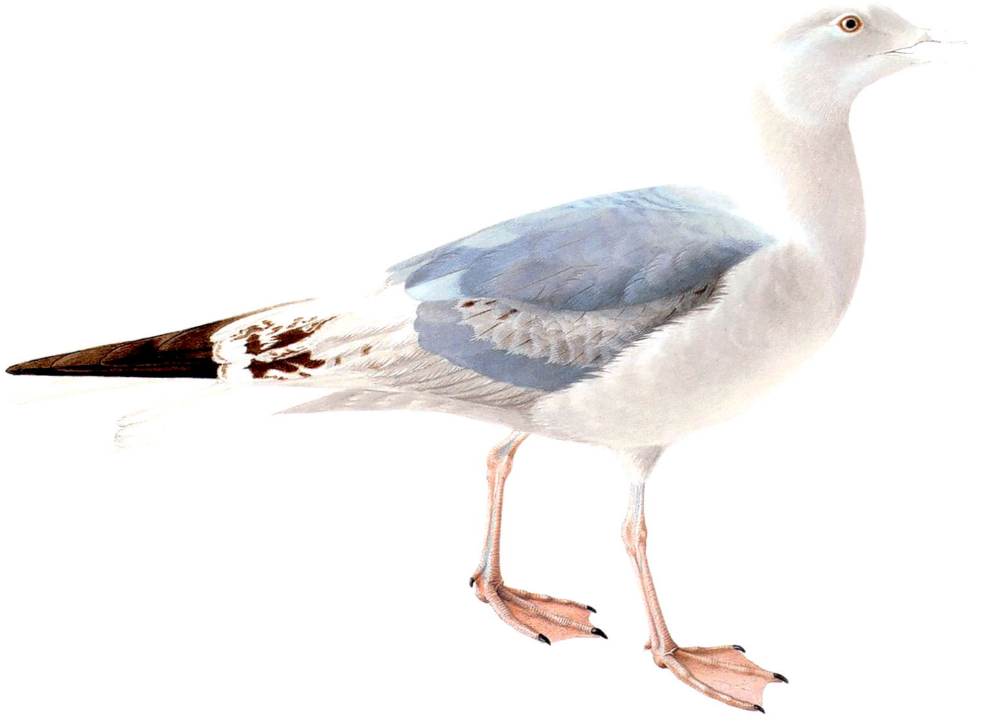

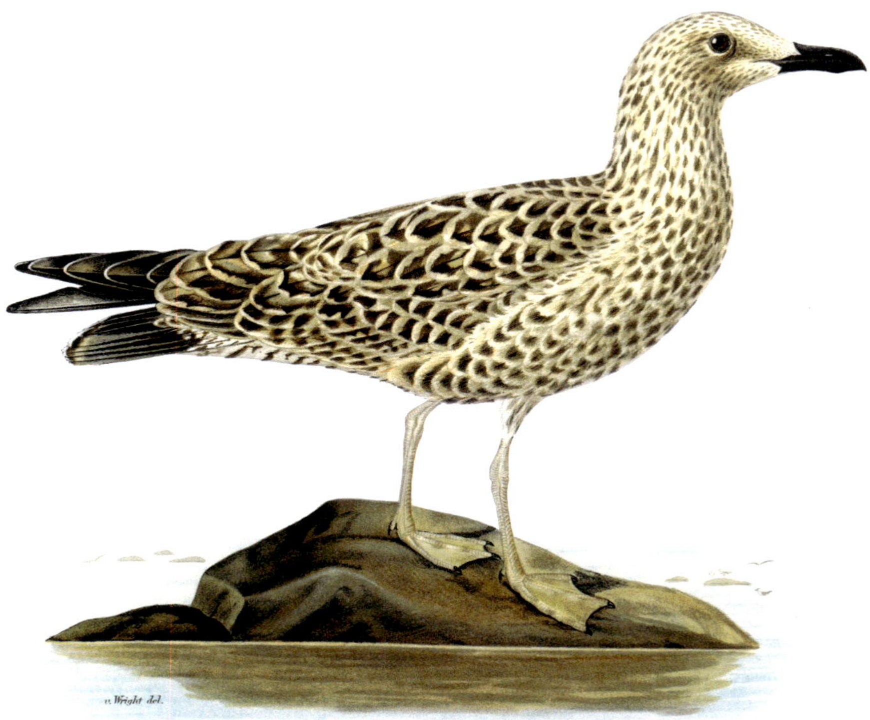

TOP:
EUROPEAN HERRING GULL
Larus argentatus

BOTTOM:
LESSER BLACK-BACKED GULL
Larus fuscus

Magnus von Wright, painting 1835, from Magnus von Wright & Wilhelm von Wright, *Svenska Fåglar*, 1828–38 (this edition: 1929)

Gulls are smart and adaptable birds. This painting of a European herring gull (top) was made in 1835 by Swedish-Finnish painter and ornithologist Magnus von Wright. Von Wright and his brother Wilhelm were skilled zoological illustrators who illustrated the work *Svenska Fåglar*. The image of the lesser black-backed gull (bottom) shows the bird in its first cycle plumage.

RED-BILLED TROPICBIRD
& EUROPEAN STORM-PETREL
Phaethon aethereus & Hydrobates pelagicus

Mark Catesby, *The Natural History of Carolina, Florida and the Bahama Islands*, 1729–47

This plate features a red-billed tropicbird and a European storm-petrel. It was made by English naturalist Mark Catesby for his book *Natural History of Carolina, Florida and the Bahama Islands*. Published between 1729 and 1747, this work was the first printed account of the flora and fauna of North America. Catesby's engravings were made from original watercolours that he produced predominantly in the field. His illustrations typically show birds in motion in an attempt to show them in a way that best exhibits their markings and other distinctive traits.

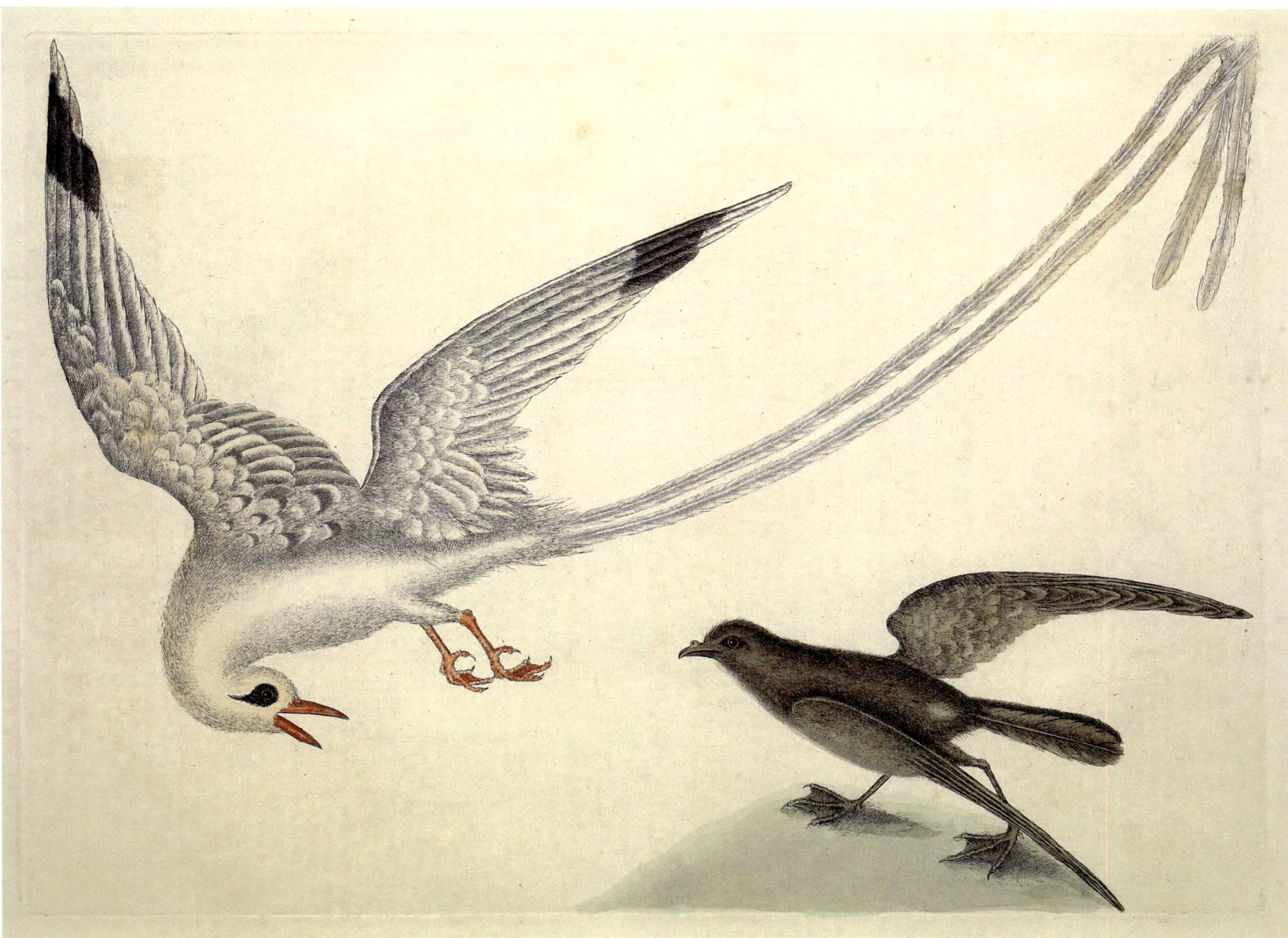

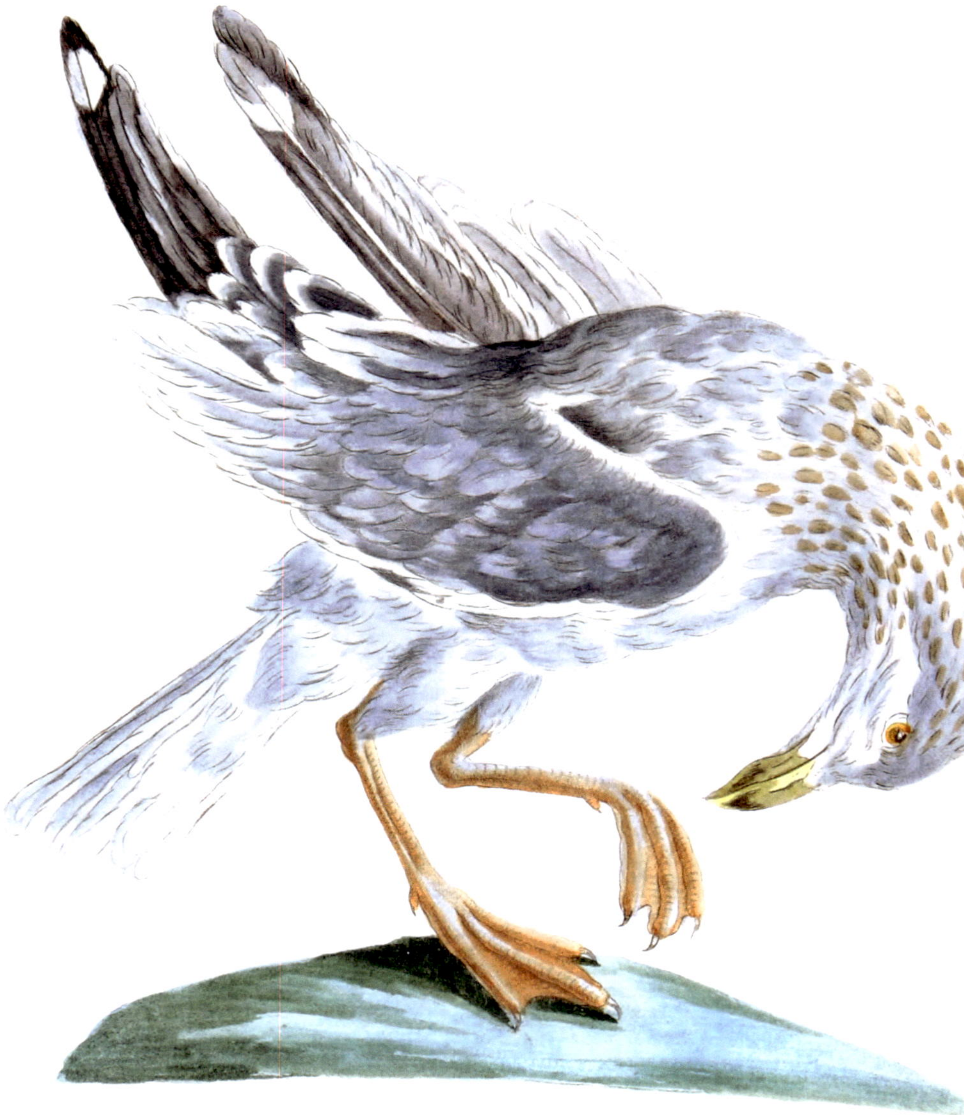

EUROPEAN HERRING GULL
Larus argentatus

Lorenzo Lorenzi & Violante Vanni,
Ornithologia methodice digesta, 1767–76

Italian artists Lorenzo Lorenzi and Violante Vanni produced this striking illustration for Saverio Manetti's five-volume work *Ornithologia methodice digesta*. Considered to be one of the greatest bird books of the eighteenth century, it is also arguably one of the most splendid books produced in Italy during that time. It contains 600 finely coloured copper plates, making it one of the largest surveys of ornithology of its era.

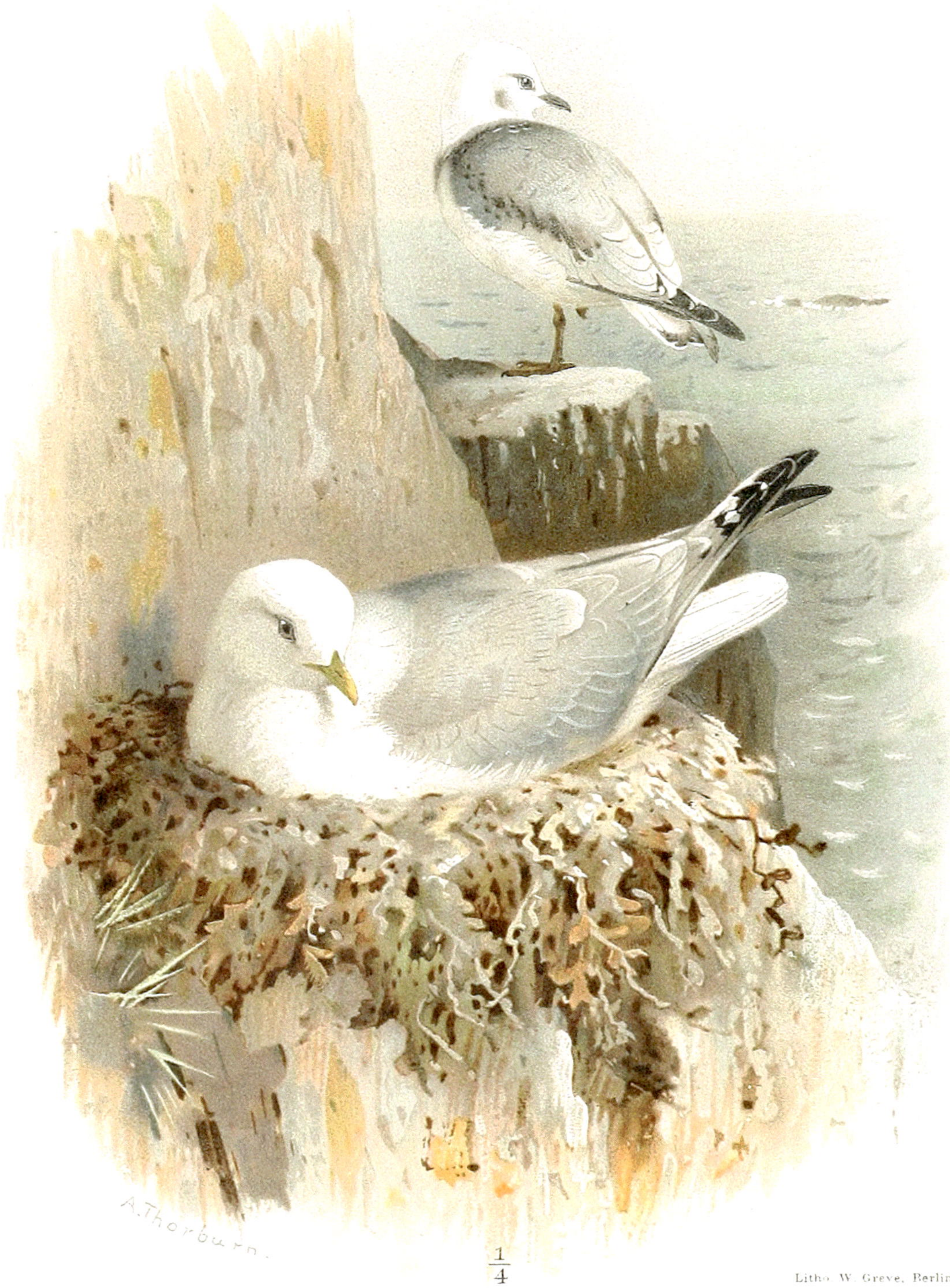

BLACK-LEGGED KITTIWAKES
Rissa tridactyla

Archibald Thorburn,
Coloured Figures of the Birds of the British Islands, 1885–97

The kittiwake takes its name from its distinctive call, which resembles the sound of the word 'kit-ti-wake'. Breeding season for the kittiwake occurs between May and September, when birds build colonies on sheer cliffs. Mud and grass will often be used for building their nest which they will later line with other materials such as moss and seaweed. Of all the birds in the *Laridae* family, kittiwakes are among the most pelagic and – outside of the breeding season – they spend the majority of their time at sea.

OVERLEAF:
COMMON MURRE (GUILLEMOT)
Uria aalge
Flamborough, June 5th 1877
John Gerrard Keulemans,
Ornithological Miscellany, 1876–8

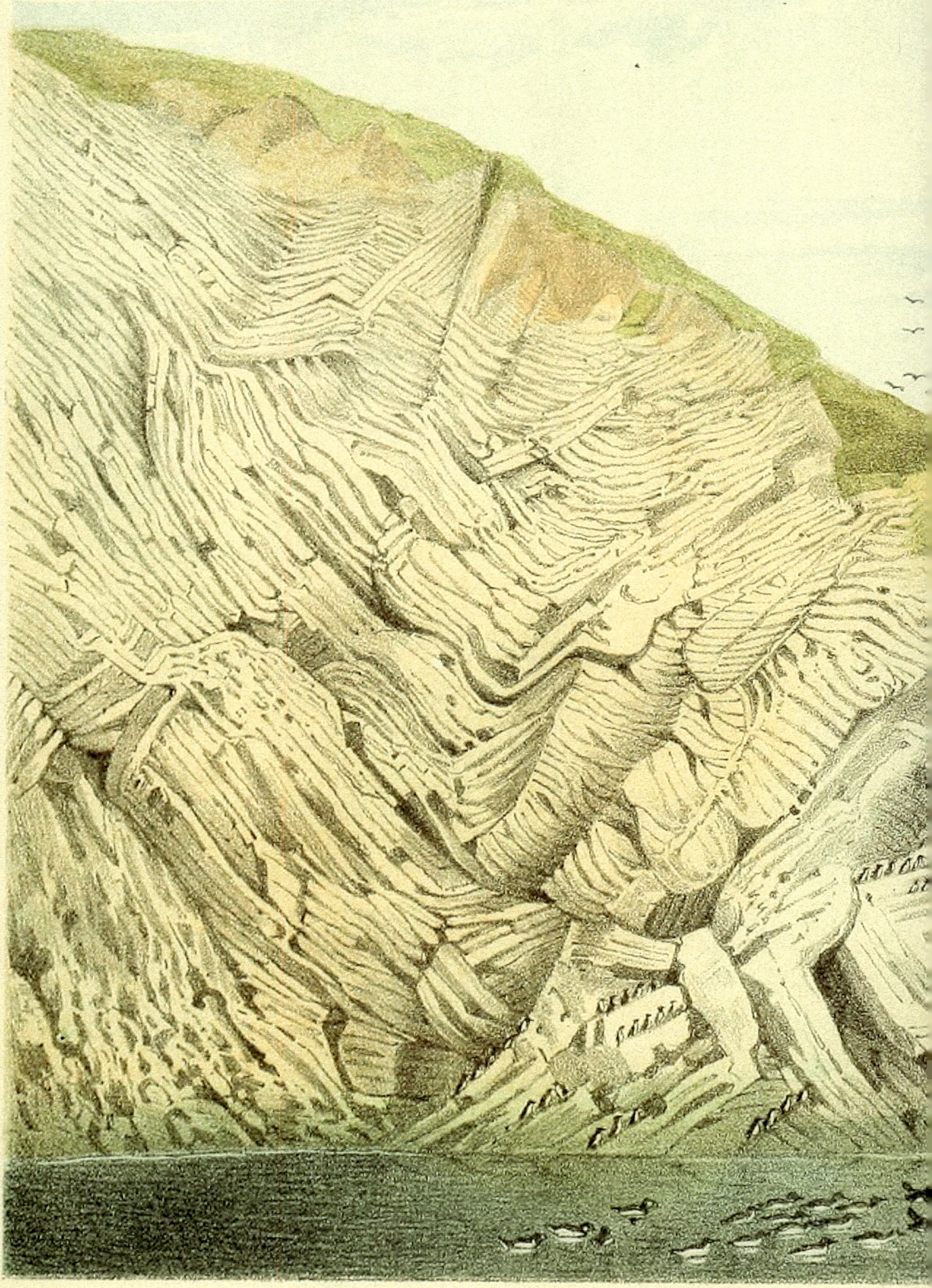

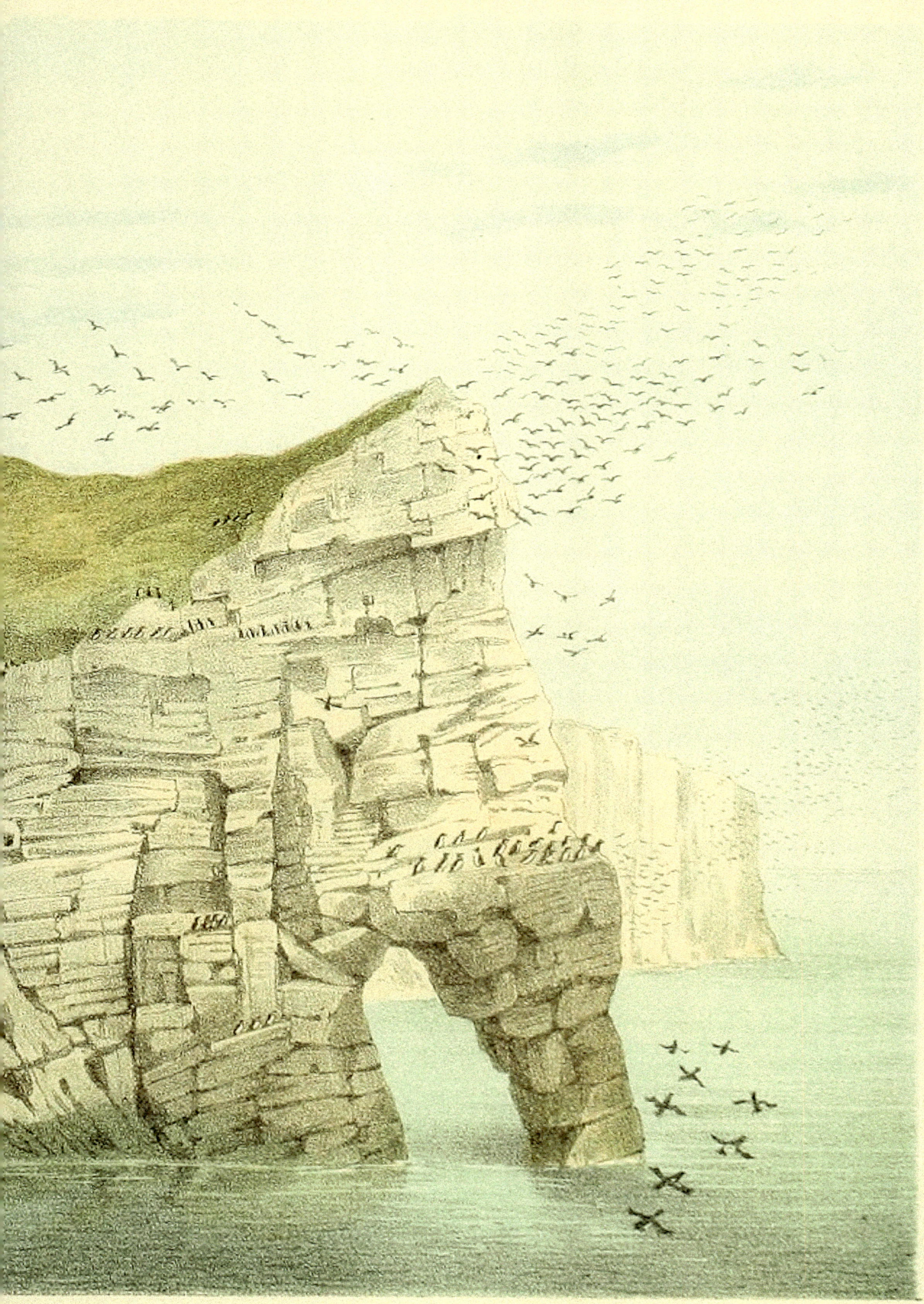

TOP:

LEACH'S STORM-PETREL & EUROPEAN STORM-PETREL
Oceanodroma leucorhoa & Hydrobates pelagicus

John Gould & Elizabeth Gould, *The Birds of Europe*, 1832–7

Few illustrators drew in the field. The common practice involved the illustrator receiving bird skins and drawing directly from these. Frequently an illustrator would not have even seen the bird that they were drawing in real life and that could occasionally lead to errors in their work. An example of this can be seen in this illustration by John and Elizabeth Gould that shows a storm-petrel standing in a rather improbable position – its left leg raised and pointed backwards.

BOTTOM:

LESSER NODDIES
Anous tenuirostris

John Gould & Henry Constantine Richter, *The Birds of Australia*, 1840–8

The lesser noddy tends to build its nest in tall mangrove trees. Nests are built with seaweed and excrement. The illustration seen here is taken from Gould's *The Birds of Australia* and shows both a male and female bird. Both sexes are physically similar and these birds show no seasonal variation in their plumage. The Australian lesser noddy is endemic to Australia and nests on the Houtman Abrolhos Islands off the country's west coast.

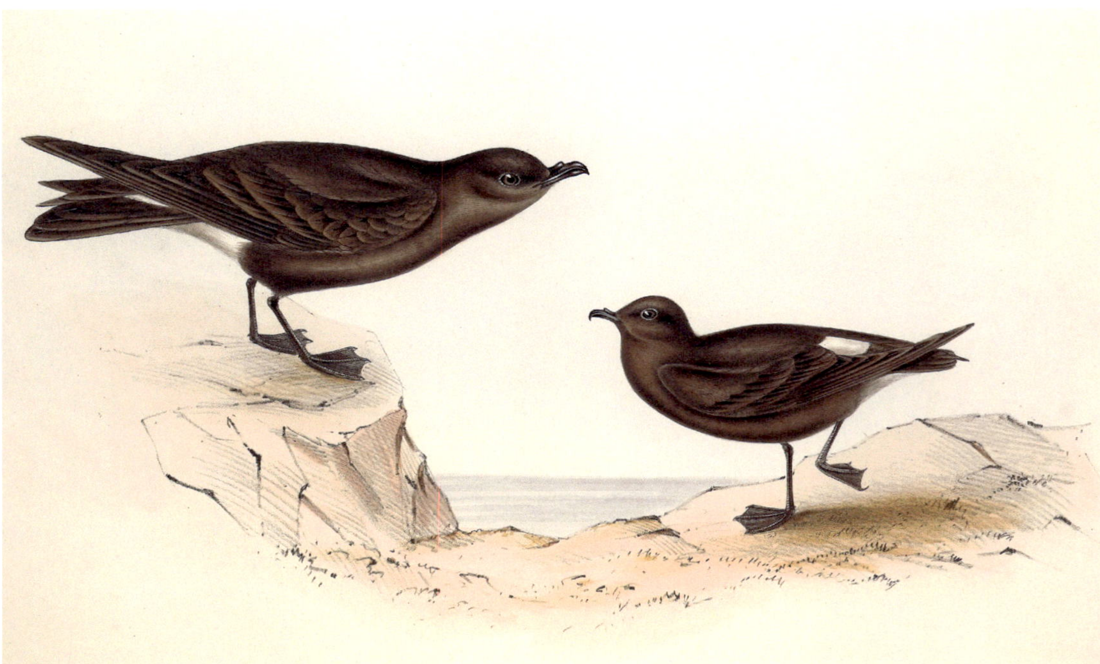

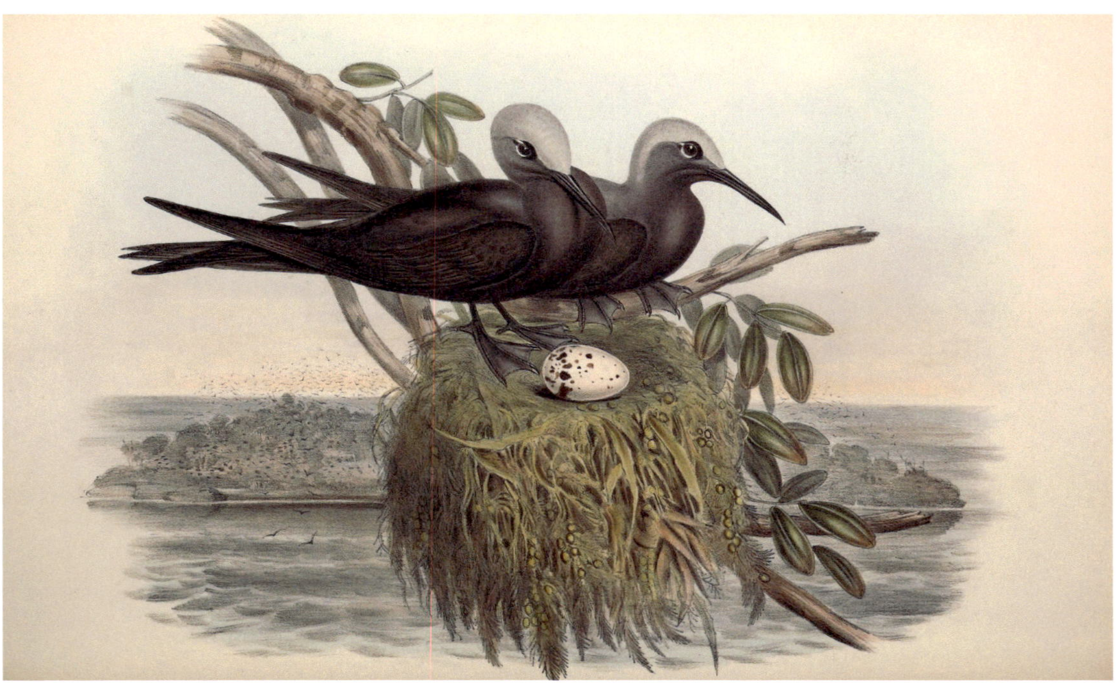

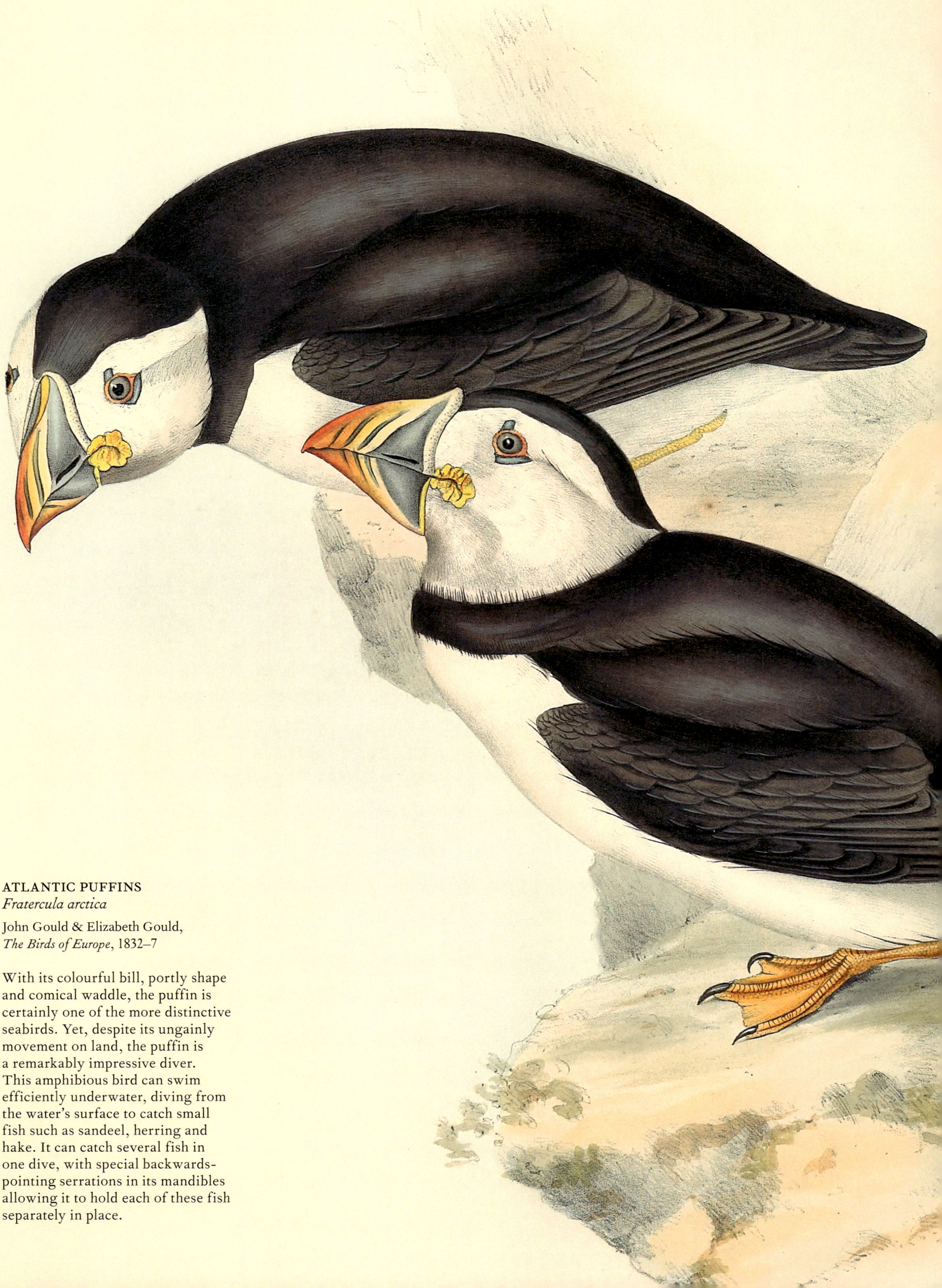

ATLANTIC PUFFINS
Fratercula arctica

John Gould & Elizabeth Gould,
The Birds of Europe, 1832–7

With its colourful bill, portly shape and comical waddle, the puffin is certainly one of the more distinctive seabirds. Yet, despite its ungainly movement on land, the puffin is a remarkably impressive diver. This amphibious bird can swim efficiently underwater, diving from the water's surface to catch small fish such as sandeel, herring and hake. It can catch several fish in one dive, with special backwards-pointing serrations in its mandibles allowing it to hold each of these fish separately in place.

ATLANTIC PUFFIN
Fratercula arctica

Eleazar Albin & Elizabeth Albin, *A Natural History of Birds*, 1731–8

This plate comes from *A Natural History of Birds*, a three-volume work that was written by English naturalist and watercolourist Eleazar Albin. Albin wrote and illustrated several natural history books during the eighteenth century, and his *Natural History of Birds* contains 101 copper plate illustrations that were drawn from life. These works were produced by Albin in collaboration with his daughter Elizabeth.

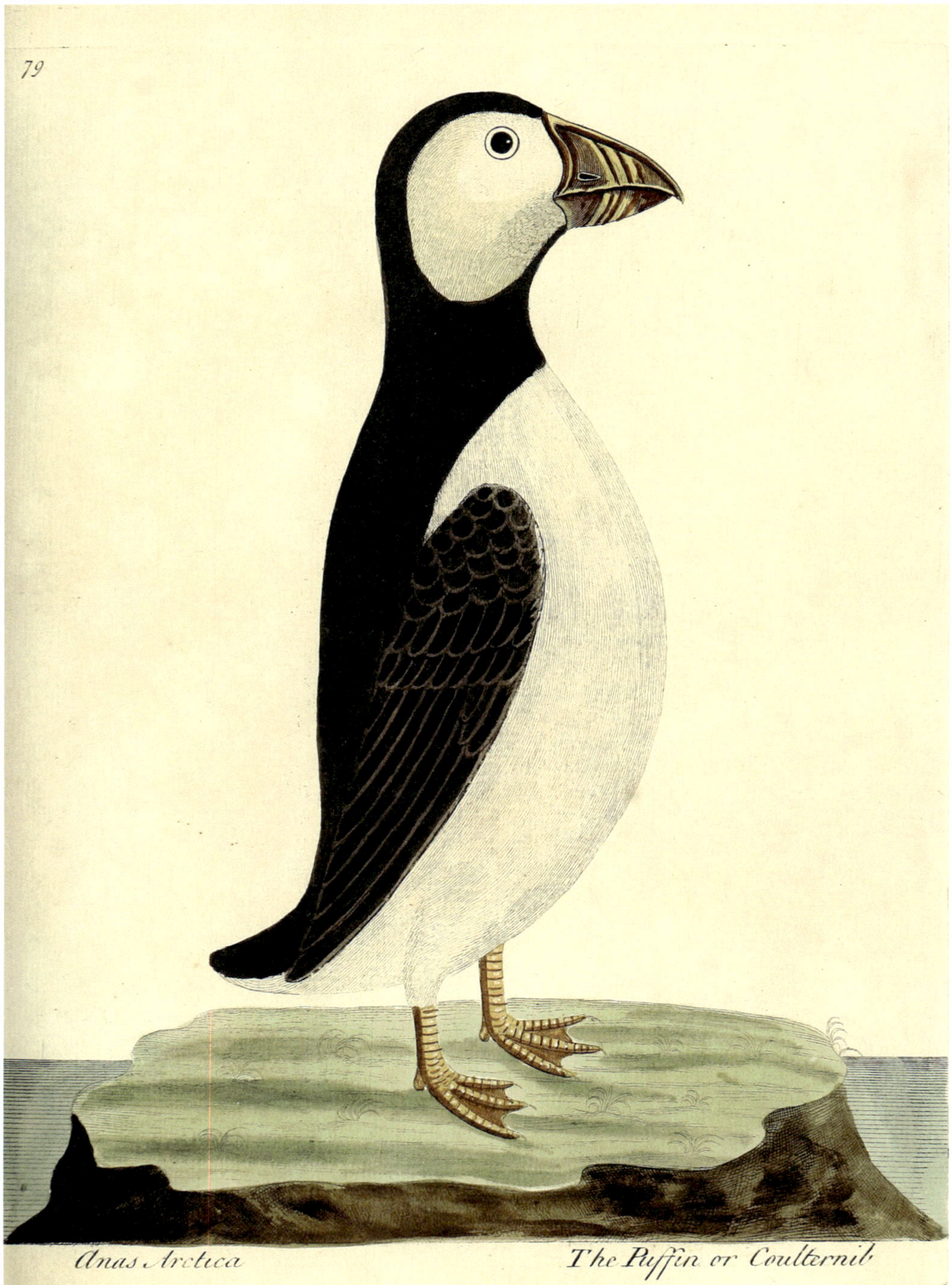

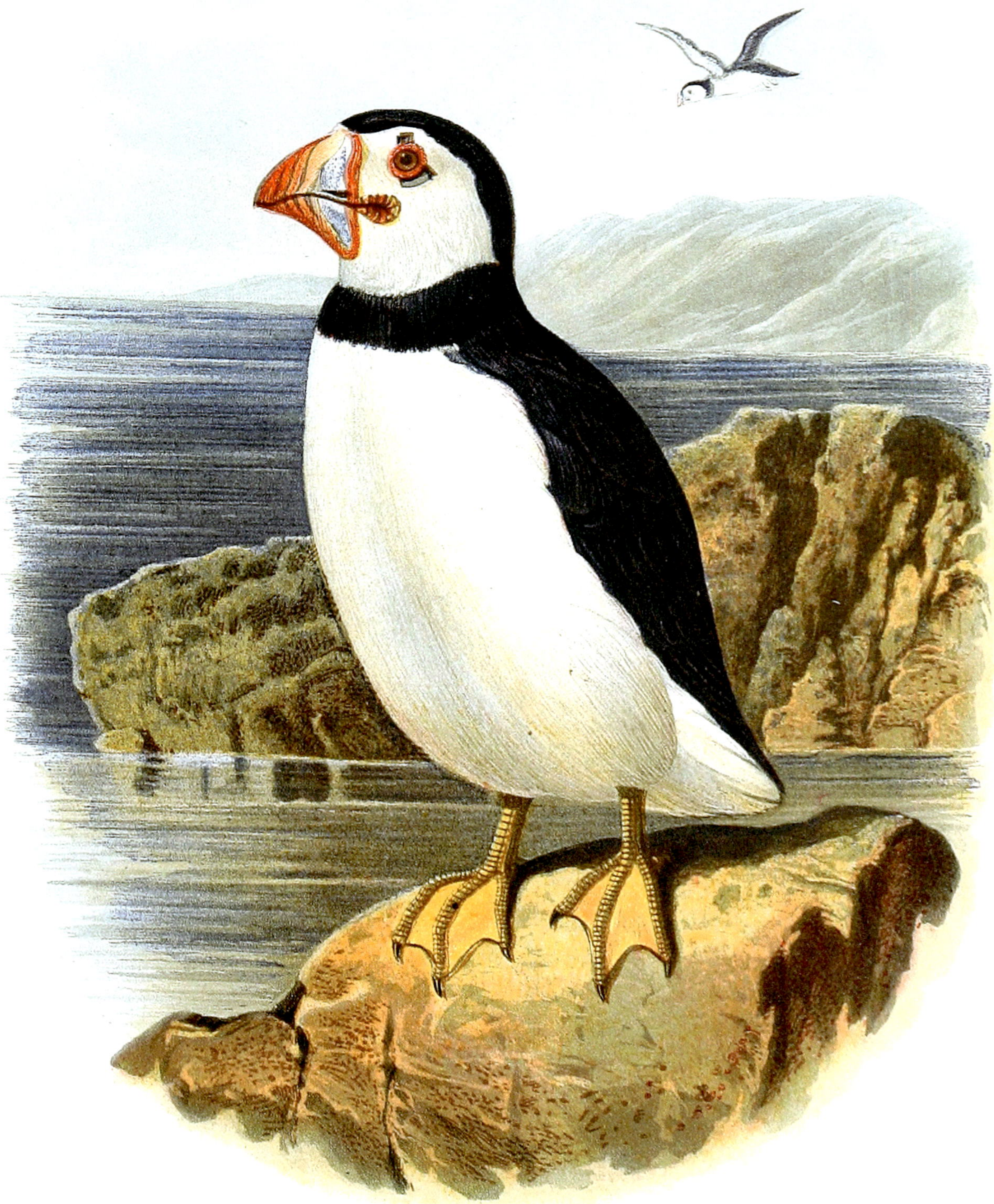

ATLANTIC PUFFIN
Fratercula arctica

Artist unknown,
Cassell's Book of Birds, *ca*. 1875

The scientific name for the Atlantic puffin is *Fratercula arctica*, which comes from the Medieval Latin words meaning 'arctic friar'. This is a reference to how the bird's black-and-white plumage resembles that of a monastic robe. Puffins breed in large colonies on cliffs and offshore islands. Some nest among cracks and crevices, while others make their home by burrowing in the soil. Despite being relatively small birds, puffins are incredibly resilient and will often endure severe winters at sea.

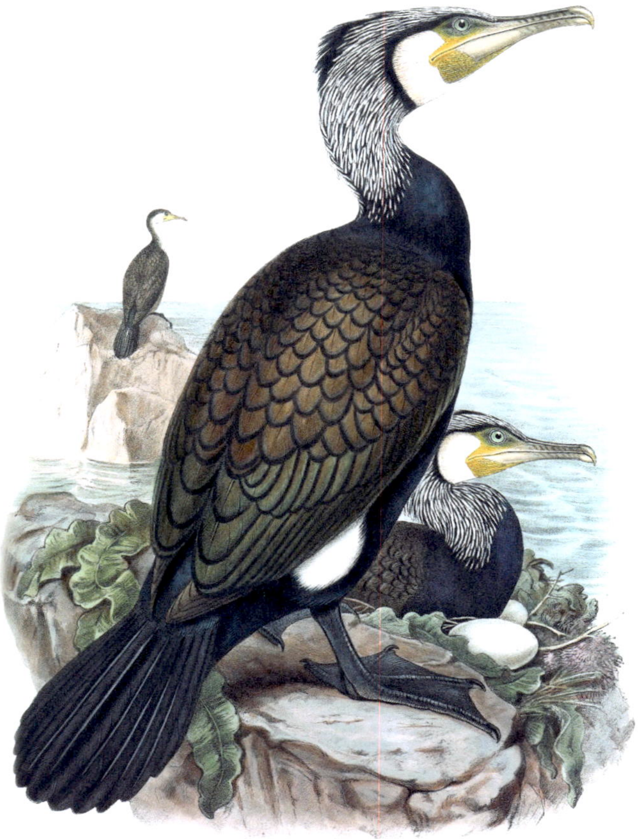

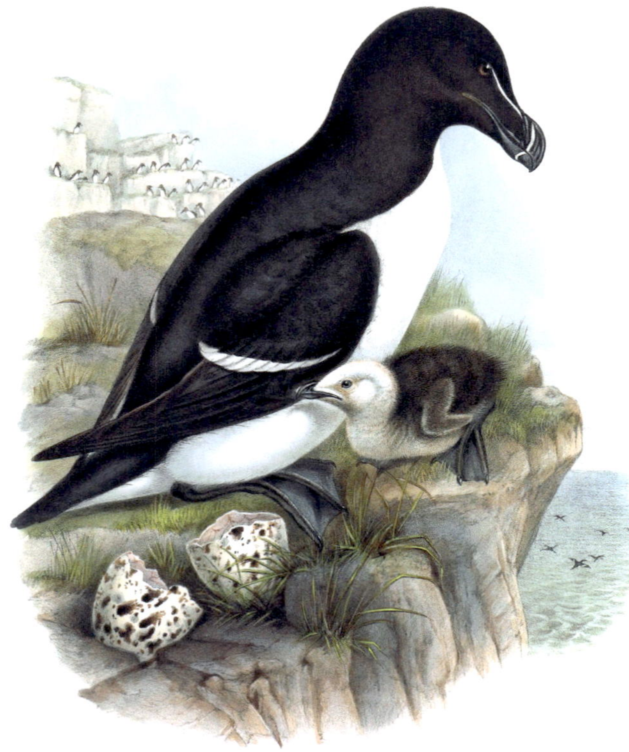

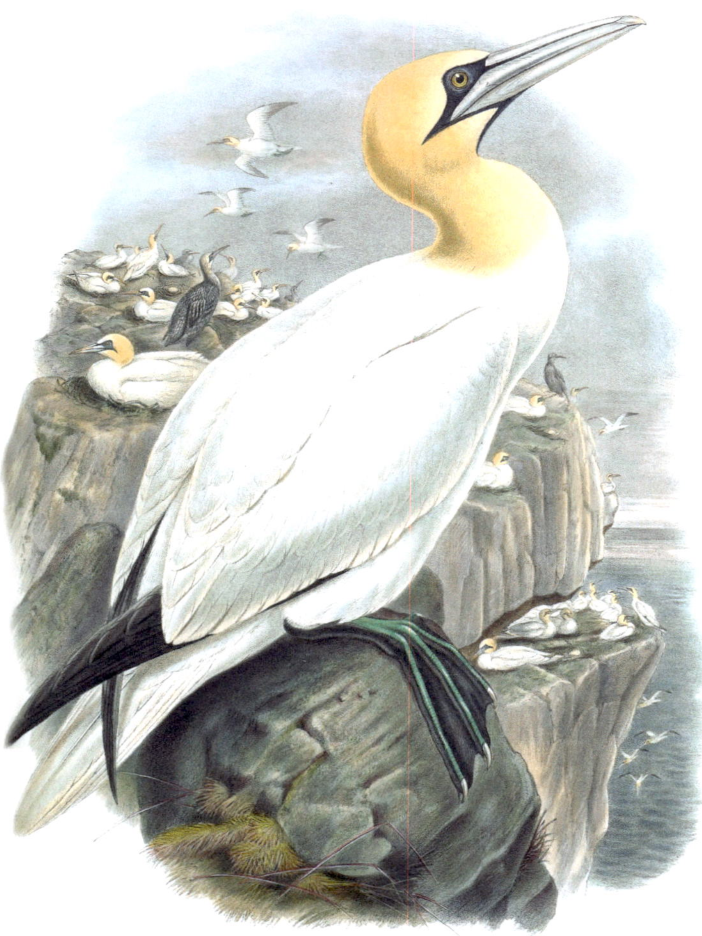

TOP LEFT:
GREAT CORMORANT
Phalacrocorax carbo

TOP RIGHT:
RAZORBILL
Alca torda

BOTTOM LEFT:
NORTHERN GANNET
Morus bassanus

OPPOSITE:
EUROPEAN SHAG
Gulosus aristotelis

John Gould & Henry Constantine Richter, *The Birds of Great Britain*, 1862–73

The Birds of Great Britain was published by John Gould between 1862 and 1873. The book focussed on birds from his native Britain, and for this reason it seems somewhat surprising that it arrived relatively late in his career. Yet this five-volume work demonstrates Gould's constant drive for improvement and the book shows a distinctive development in his aesthetic style. Many of its illustrations feature ambitious compositions that depict multiple birds, their nesting habits and occasionally even their young.

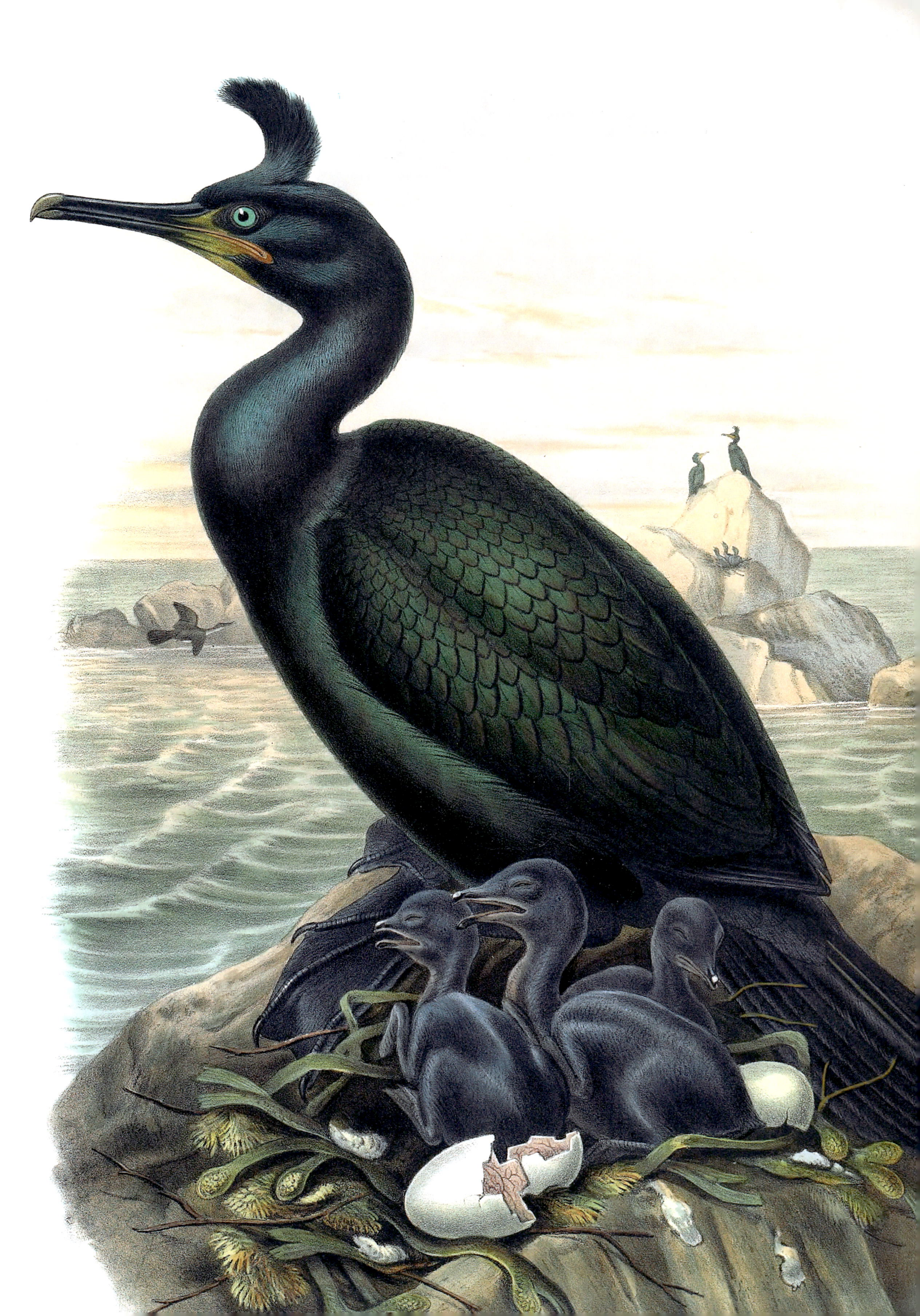

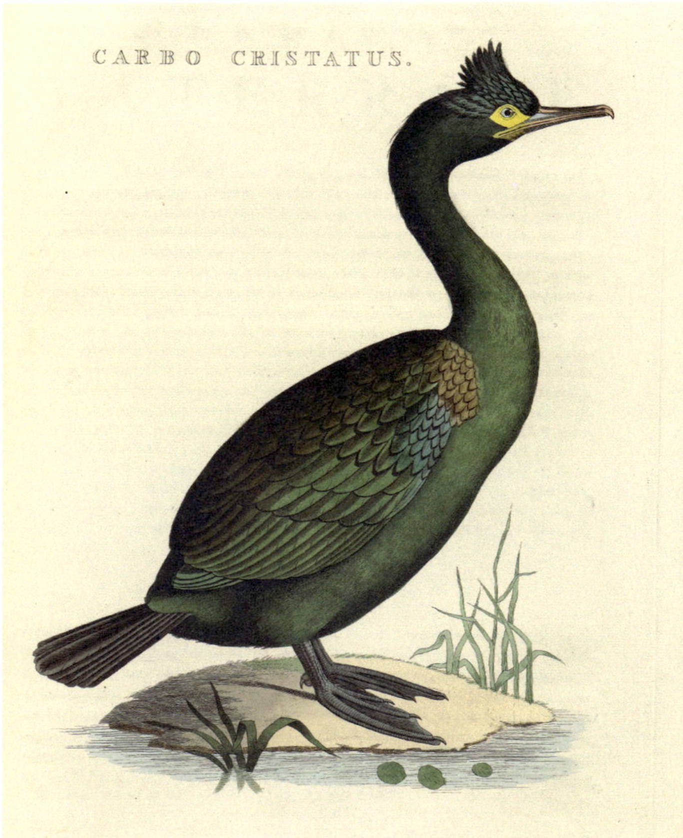

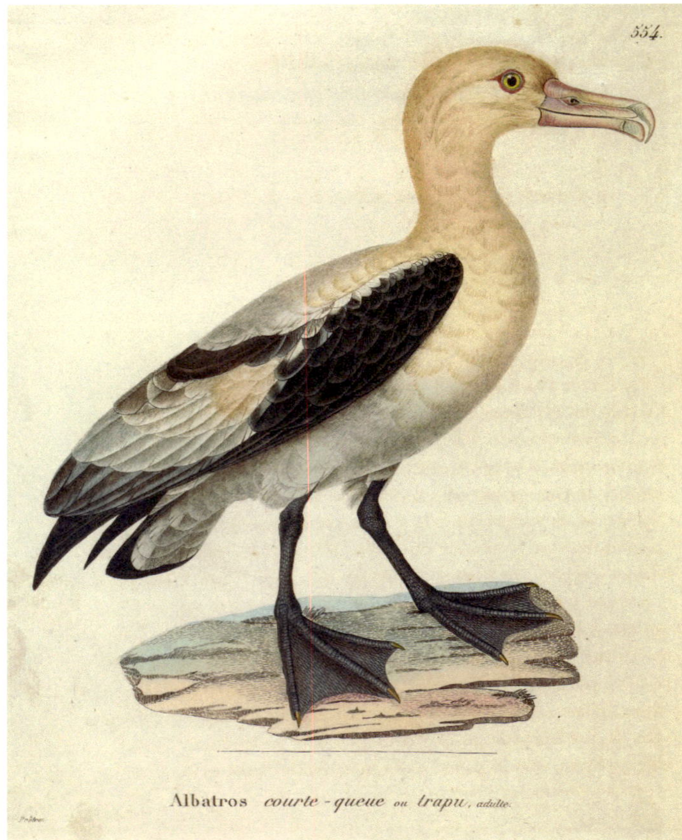

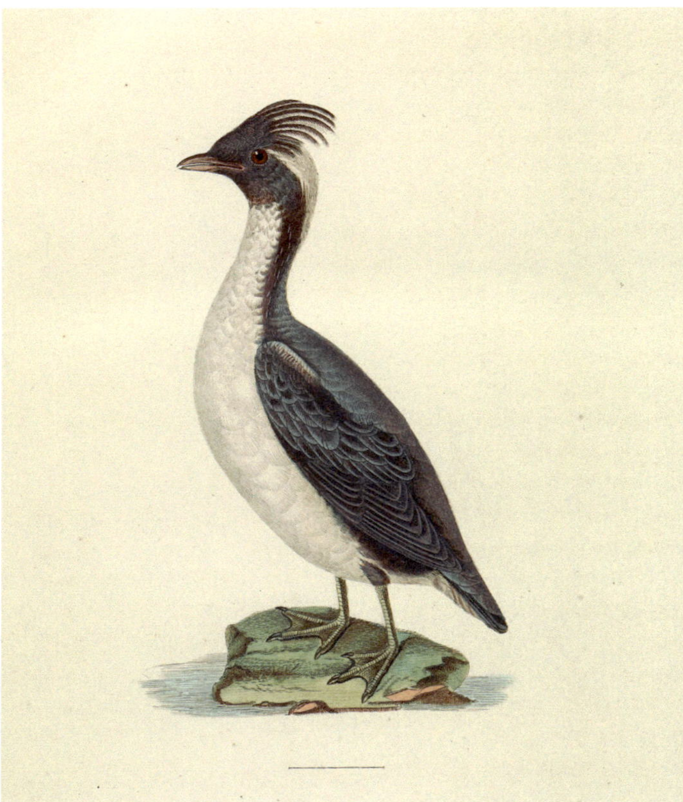

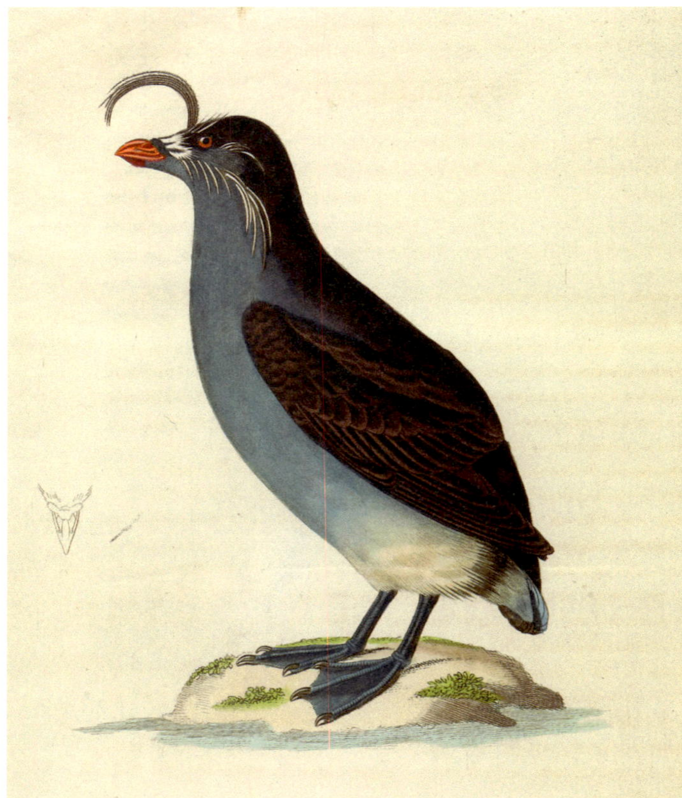

TOP LEFT:

EUROPEAN SHAG
Gulosus aristotelis

Jan Christiaan Sepp and son, *Nederlandsche Vogelen*, 1770–1829

TOP RIGHT:

SHORT-TAILED ALBATROSS
Phoebastria albatrus

Jean-Gabriel Prêtre, *Nouveau recueil de planches coloriées d'oiseaux*, 1820–39

BOTTOM LEFT:

JAPANESE MURRELET
Synthliboramphus wumizusume

Jean-Gabriel Prêtre, *Nouveau recueil de planches coloriées d'oiseaux*, 1820–39

BOTTOM RIGHT:

WHISKERED AUKLET
Aethia pygmaea

Nicolas Huet II, *Nouveau recueil de planches coloriées d'oiseaux*, 1820–39

ALBATROSS
(*perhaps* WANDERING ALBATROSS)
Diomedea exulans

George Edwards, *A Natural History of Uncommon Birds*, 1743–51

The immense bill of the albatross is the main focus in this plate by English naturalist and ornithologist George Edwards. Appearing in his hugely important book *A Natural History of Uncommon Birds*, this study highlights the impressive nasal gland that is found at the base of the bird's bill. This gland secretes a saline solution that helps ensure that the bird does not take in excess salt while living at sea.

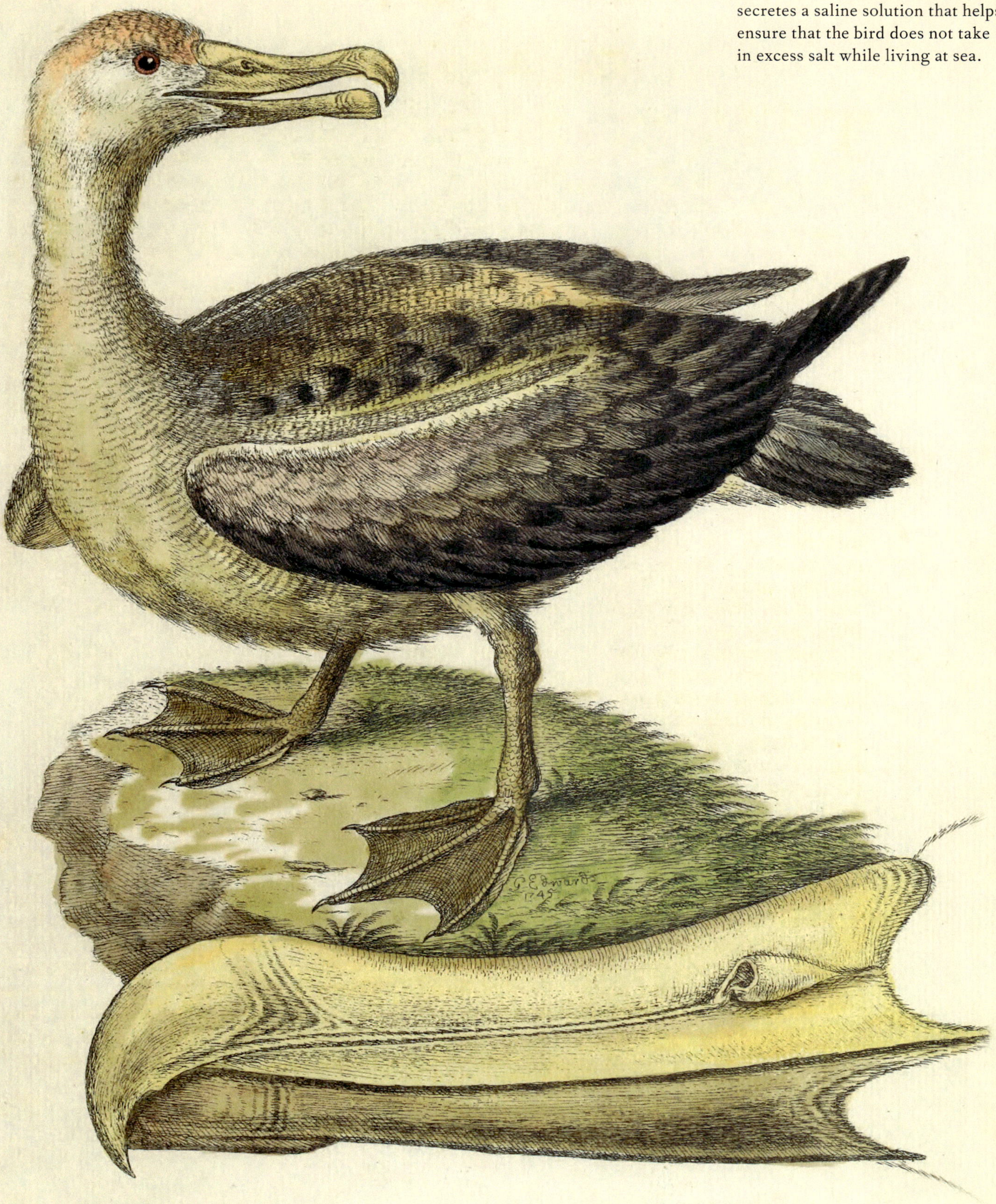

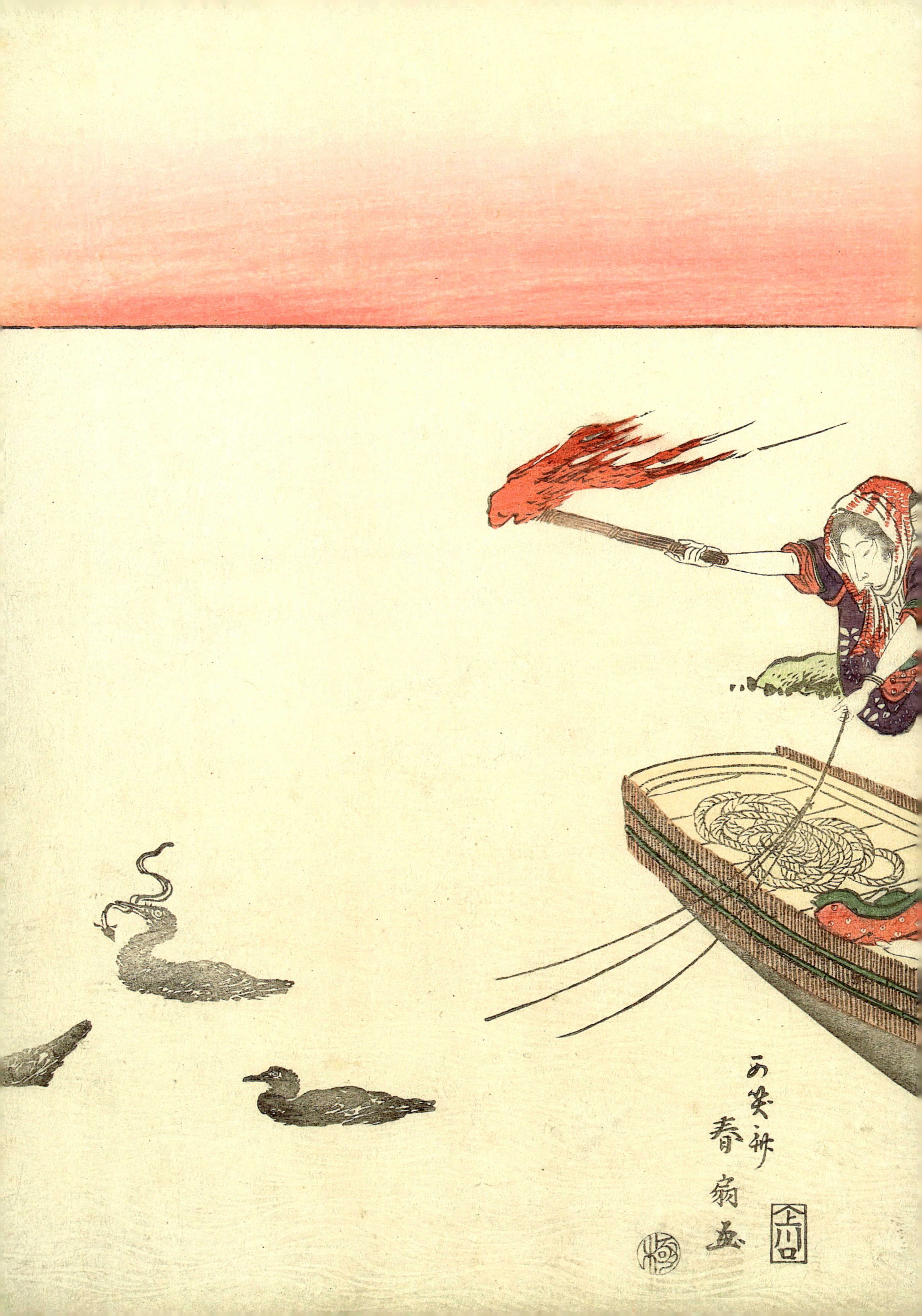
春扇画

GREAT OR JAPANESE CORMORANT
Phalacrocorax carbo or *P. capillatus*

Katsukawa Shunsen, *Cormorant Fishing at Night*, colour woodcut, 1800–10

Cormorants live entirely on fish and have become experts at catching them. With excellent diving skills, they can propel themselves underwater with their feet and their wings. Their talent for fishing saw their domestication as far back as 300 BCE, when captive birds were trained to fish and deliver their catch. A noose around their neck would prevent them from swallowing any large fish that they caught. Originally started in the East, the practice was documented in Venice in the sixteenth century and Britain and France in the seventeenth century. Today, this practice continues in China and Japan but is now primarily performed for tourists.

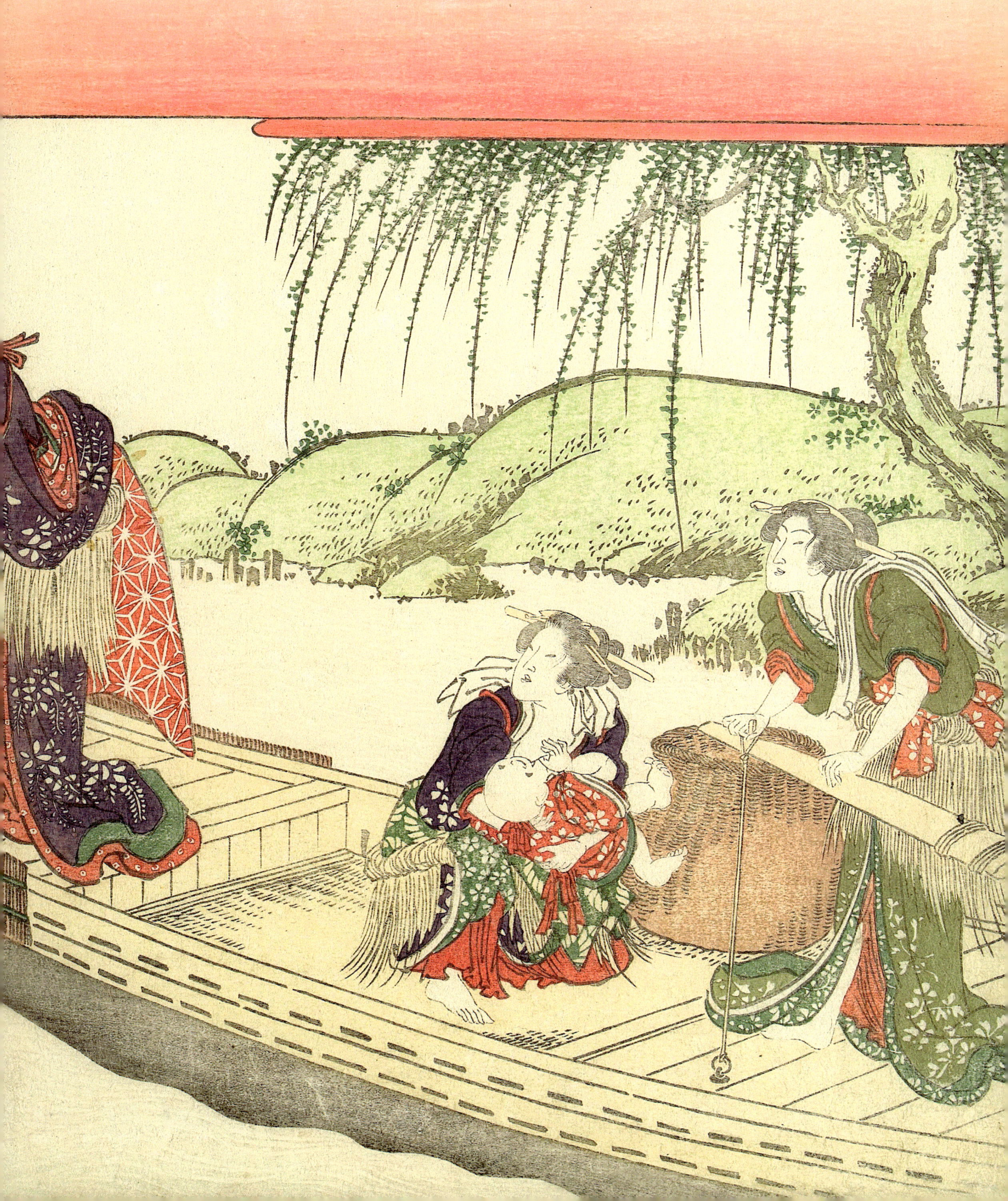

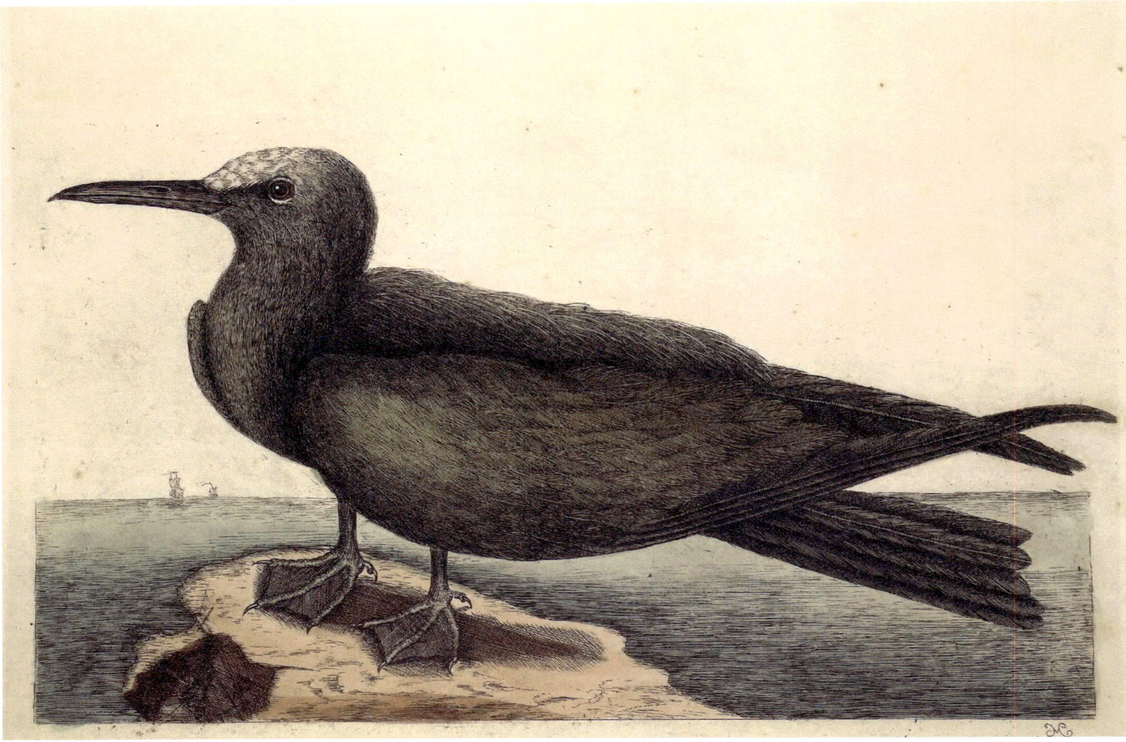

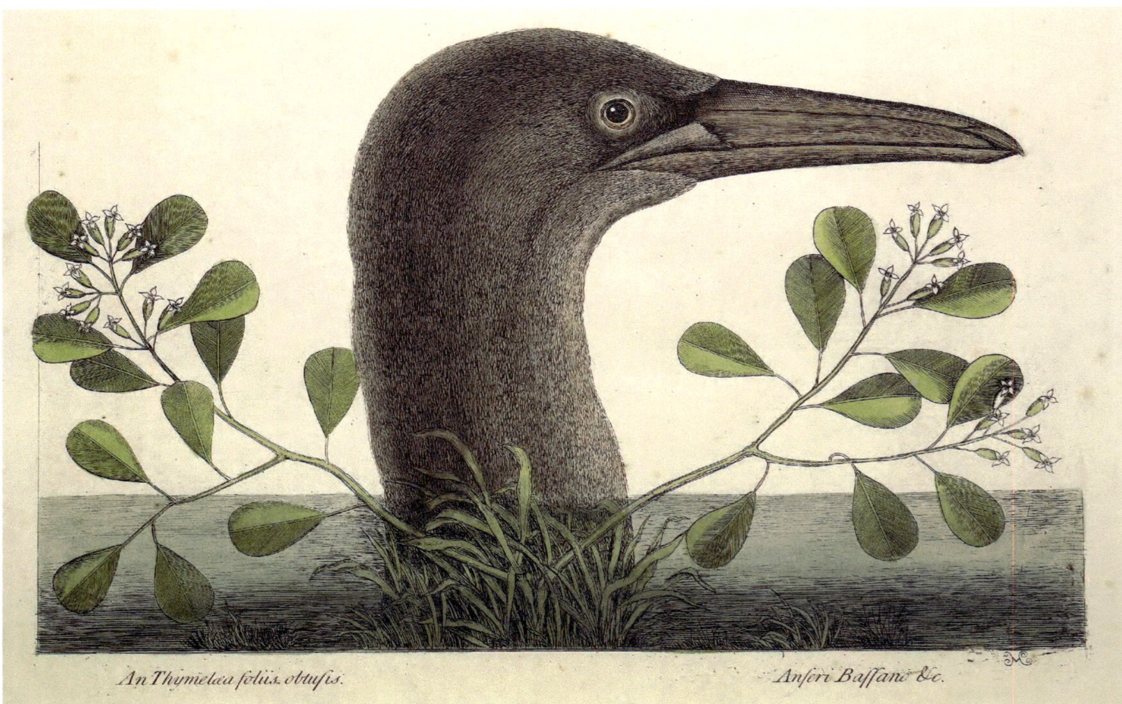

TOP:

BROWN NODDY

Anous stolidus

Mark Catesby, *The Natural History of Carolina, Florida and the Bahama Islands*, 1729–47

BOTTOM:

BROWN BOOBY

Sula leucogaster

Mark Catesby, *The Natural History of Carolina, Florida and the Bahama Islands*, 1729–47

Both of these illustrations come from Mark Catesby's *Natural History of Carolina, Florida and the Bahama Islands*. Catesby often illustrated his birds with information about the region in which he found them. His noddy was seen in the Bahama Islands and is shown alongside two sailing ships. Meanwhile, this brown booby is seen by a white mangrove.

BLACK-HEADED GULL
Chroicocephalus ridibundus

Prideaux John Selby, *Illustrations of British Ornithology*, 1819–34

The black-headed gull takes its name from the dark hood it develops during its summer plumage. This plumage is lost during winter, with only two dark spots remaining close to the bird's eyes. It is seen here in an illustration produced by English ornithologist, botanist and artist Prideaux John Selby. Often thought of as the 'Gentleman Naturalist', Selby's best-known work, *Illustrations of British Ornithology*, was the first book to feature illustrations of British birds at life size.

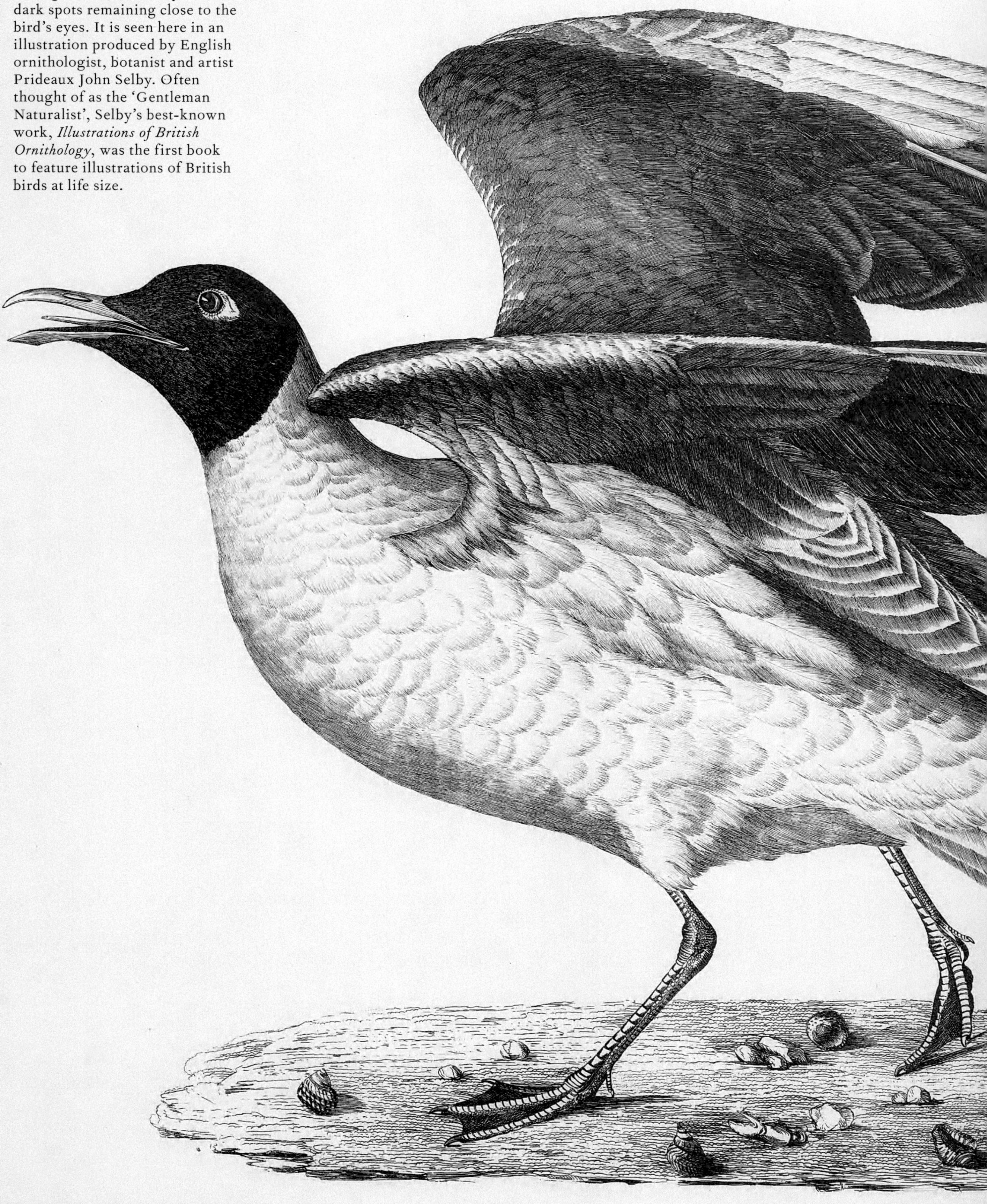

Mark Catesby
1683–1749

Catesby's *The Natural History of Carolina, Florida and the Bahama Islands* introduced the wonders of the New World to Europe and demonstrated the value of recording natural history as a means of understanding the world.

The earliest book to depict North American wildlife was produced by the English naturalist Mark Catesby. Published in parts between 1729 and 1747, *The Natural History of Carolina, Florida and the Bahama Islands* was a pioneering work that helped to steer natural history into an age of rationality and reason. At the time, Carolina, Florida and the Bahama Islands were all part of the Southern Colonies within British America. While the American continent had been known to Europeans for more than 200 years, most of it remained relatively unknown and Catesby's work was the first comprehensive survey of its flora and fauna.

Born in Essex in 1683, Catesby's first voyage to America came at the age of twenty-nine when an endowment from his father allowed him to pursue an interest in the natural world. This trip took him to Virginia and, from there, Jamaica. To the Europeans of the time, the New World was a vast land filled with uncertainty, opportunity and promise. There was no knowing what lived there, and so for Catesby it was the ideal place to feed his insatiable curiosity. The native plant life held particular interest for him and he gradually earned a reputation for his skills in collecting plants and seeds. For wealthy men, botany had become the science of choice. It held the promise of new spices, new dyes and new medicine. When Catesby eventually returned to England in 1719, he did so with an impressive collection of new specimens.

At the time, England was at the centre of science and at the heart of it was the Royal Society. Established in 1690, and still in operation today, it is the oldest national scientific institution in the world. Catesby's work in botany had brought him to the attention of a number of members in the society who encouraged him to return to America on an expedition to collect plants. The trip was endorsed by the society's president, Sir Isaac Newton (1643–1727), and had the financial backing of several important figures of the day, including the English botanist William Sherard (1659–1728) and the physician Sir Hans Sloane (1660–1753).

When Catesby returned to the Colonies in 1722 he did so with a new sharpness and specificity. Now travelling under the auspices of the Royal Society, he was no longer just a casual observer – he was now a 'Man of Science'. From 1722 to 1726, he made his way through several new territories, seeing places such as Carolina, Florida and Georgia. He adopted new systematic methods in his research, carefully collecting specimens and producing many paintings in the field. He took all of his observations directly from nature. These were wild and unsafe places. The American Revolution was still a century away, and threats such as malaria and attacks from pirates were all too real. Specimens he found were hard to ship back to England and larger findings would have needed to have been stripped and dried before being transported. This often posed a challenge in the severe wet weather of Carolina. Smaller specimens such as snakes could be preserved in bottles of rum and other spirits, but these could often be stolen by sailors who were looking to quench their boredom on the long journey home. Despite all this, Catesby managed to collect an enormous quantity of new birds, amphibians, fish, insects, mammals, reptiles, plants and seeds and bring them back to England.

Upon his return he was keen to translate these findings into a book, but his sponsors were not so eager on financing such an expensive publication. Undeterred by this setback, he took it upon himself to produce the book independently. This was a mammoth undertaking. Catesby was not an artist and to achieve his ambition he was forced to teach himself how to etch. With the help of French etcher Joseph Goupy (1689–1769), he transferred his watercolour and gouache paintings onto copper plates. The first instalment of volume one was published in 1729 and it took another fourteen years for the book to be completed. The appendix was finished and issued four years after that.

The Natural History of Carolina, Florida and the Bahama Islands was a remarkable achievement. The final publication spanned two folio volumes and contained 220 illustrated plates featuring depictions of birds, amphibians, reptiles, fish, insects, mammals and plants. Of the 109 birds contained in the work, many of Catesby's descriptions are the very first recordings we have of these species. For decades, the book remained the greatest work of its kind. It introduced the wonders of the New World to Europe and demonstrated the value of recording natural history as a means to better understand the world.

Catesby's book was easily the most expensive British book to be produced in the eighteenth century. Yet despite its importance – and its expense – it did not make him a rich man. He died in 1749 at the age of sixty-six with 'old age' being listed as his cause of death. He was buried in the churchyard of St Luke's in the London Borough of Islington, but his grave is now unknown. After his death science shifted its interest and taxonomy became the main focus for many. Despite this, Catesby played a foundational role in the story of ornithology – influencing not only Linnaeus but also the work of people like Alexander Wilson, John James Audubon and even Charles Darwin.

BROWN BOOBY
Sula leucogaster

Mark Catesby, *The Natural History of Carolina, Florida and the Bahama Islands*, 1729–47

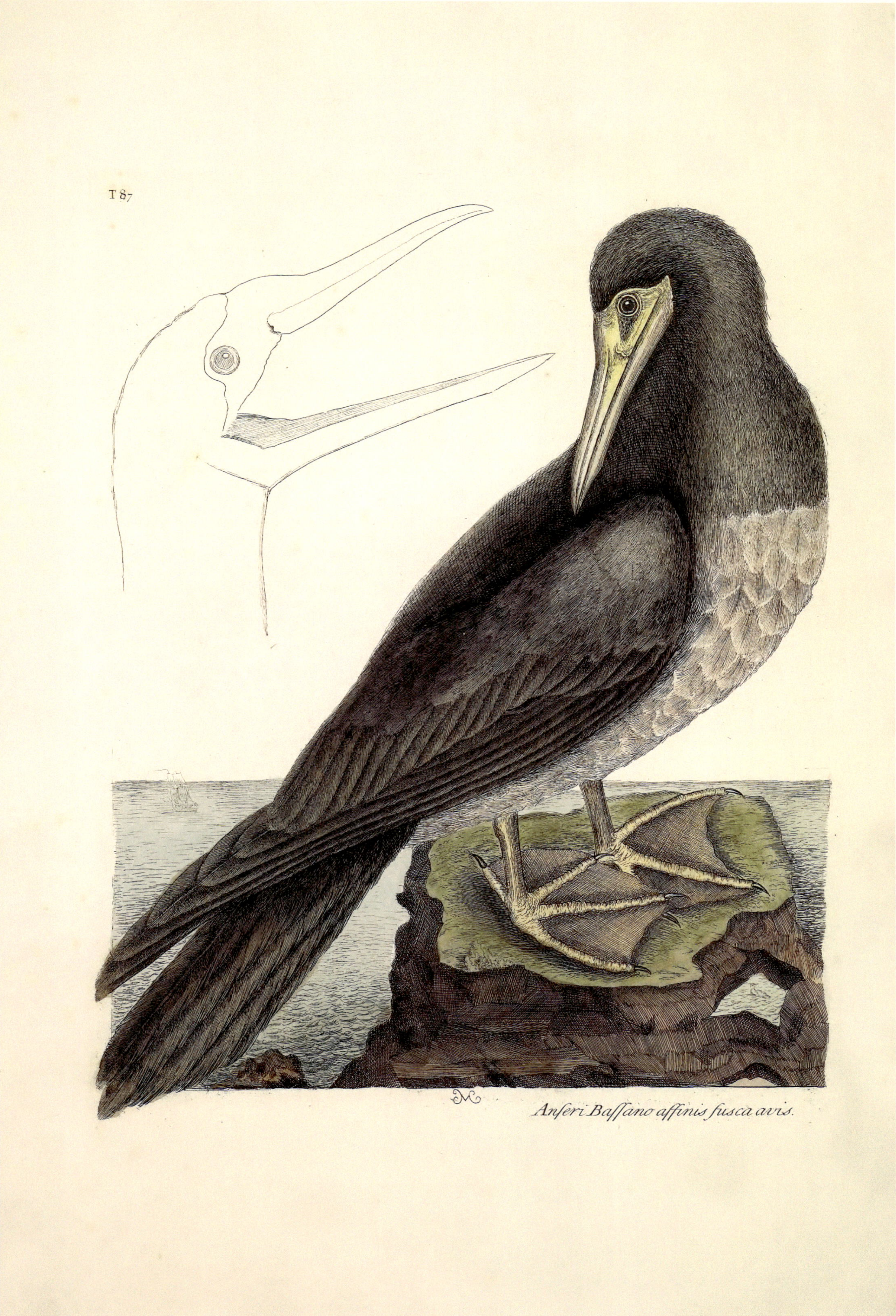
T 87
Anseri Bassano affinis fusca avis.

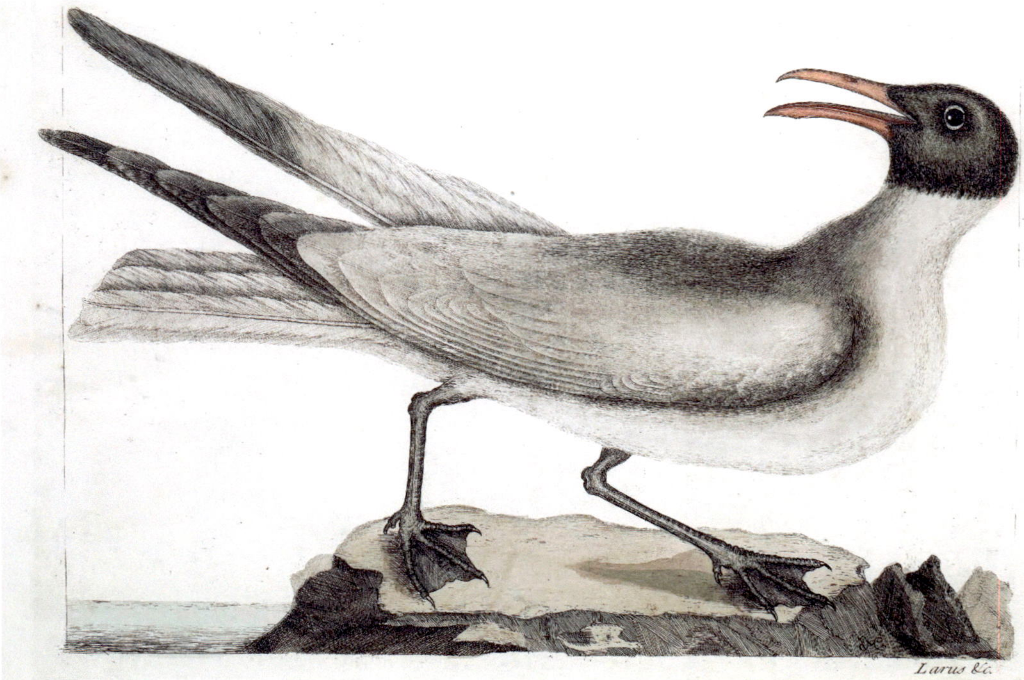

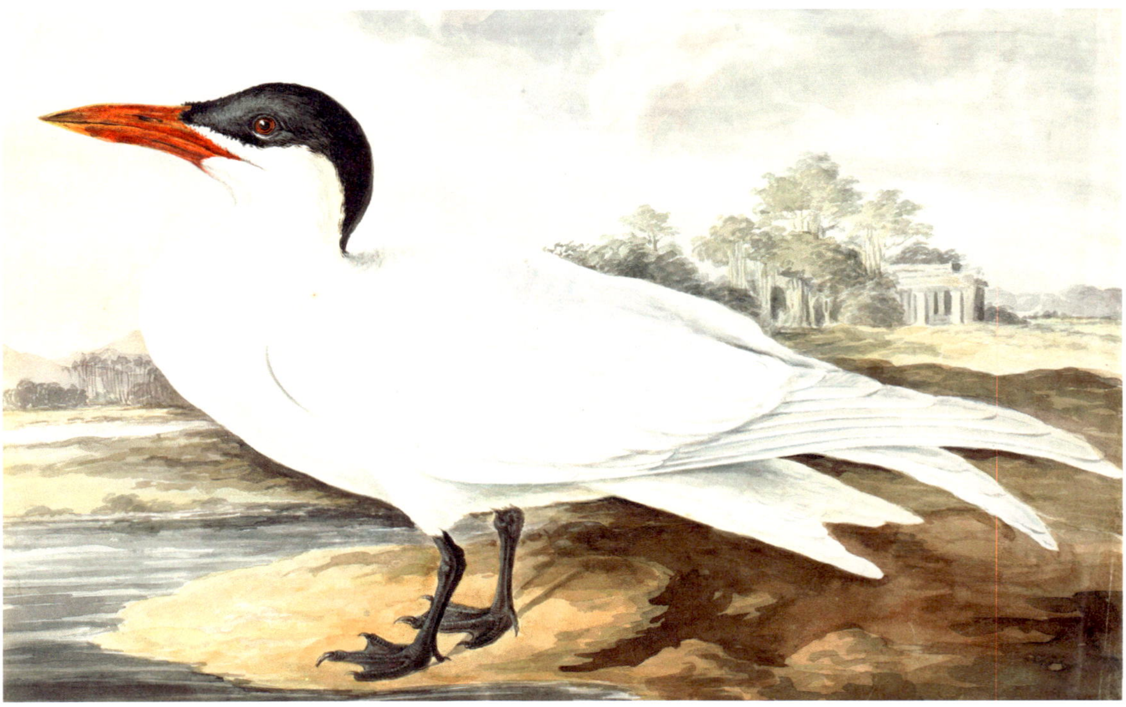

TOP:

LAUGHING GULL

Leucophaeus atricilla

Mark Catesby, *The Natural History of Carolina, Florida and the Bahama Islands*, 1729–47

Despite South Carolina having many gulls, this laughing gull is the only gull documented by Catesby in his book. It is named for its call, which resembles a raucous high-pitched laugh.

BOTTOM:

CASPIAN TERN

Hydroprogne caspia

Elizabeth Gwillim, watercolour, 1801

Elizabeth Gwillim was an English artist who moved to India in 1801 when her husband Henry Gwillim was asked to serve in the Supreme Court at Madras. There, she produced a series of roughly 200 watercolours of Indian birds.

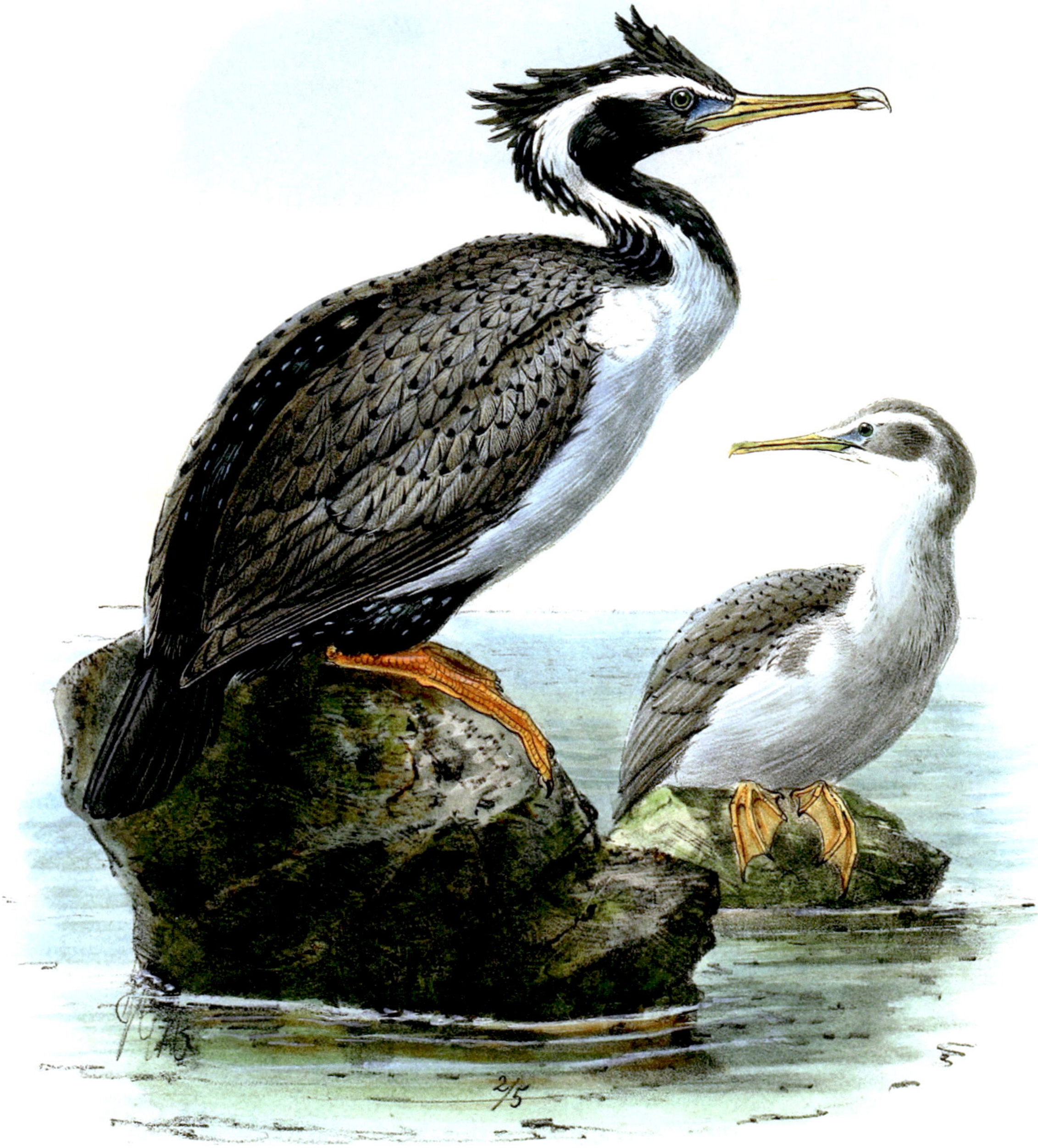

SPOTTED SHAG
Phalacrocorax punctatus

John Gerrard Keulemans, *A History of the Birds of New Zealand*, 1872–3

The spotted shag is a species of cormorant that is endemic to New Zealand. Shown here is an adult (left) and a juvenile (right). The illustration was created by Dutch bird illustrator John Gerrard Keulemans. Born in Rotterdam, Keulemans spent much of his life living and working in England, where he illustrated many important bird books. Among these was *A History of the Birds of New Zealand*, a seminal and popular title that was written by acclaimed New Zealand ornithologist Sir Walter Buller.

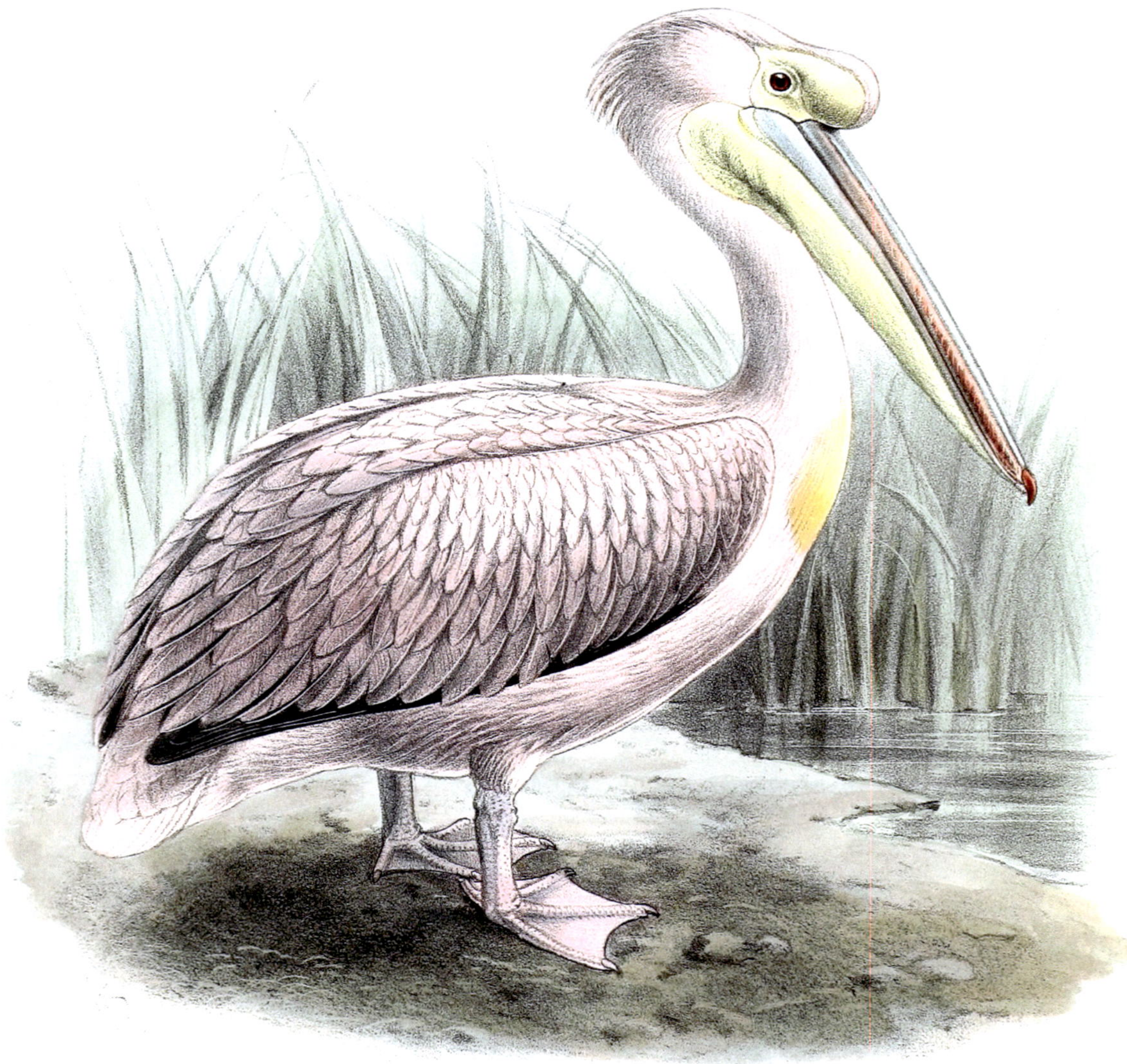

GREAT WHITE PELICAN
Pelecanus onocrotalus

John Gerrard Keulemans,
A History of the Birds of Europe, 1871–81

OPPOSITE:
AMERICAN WHITE PELICAN
Pelecanus erythrorhynchos

John James Audubon,
The Birds of America, 1827–38

The American white pelican is one of the biggest birds in North America. Its impressive size is captured brilliantly in this work by John James Audubon.

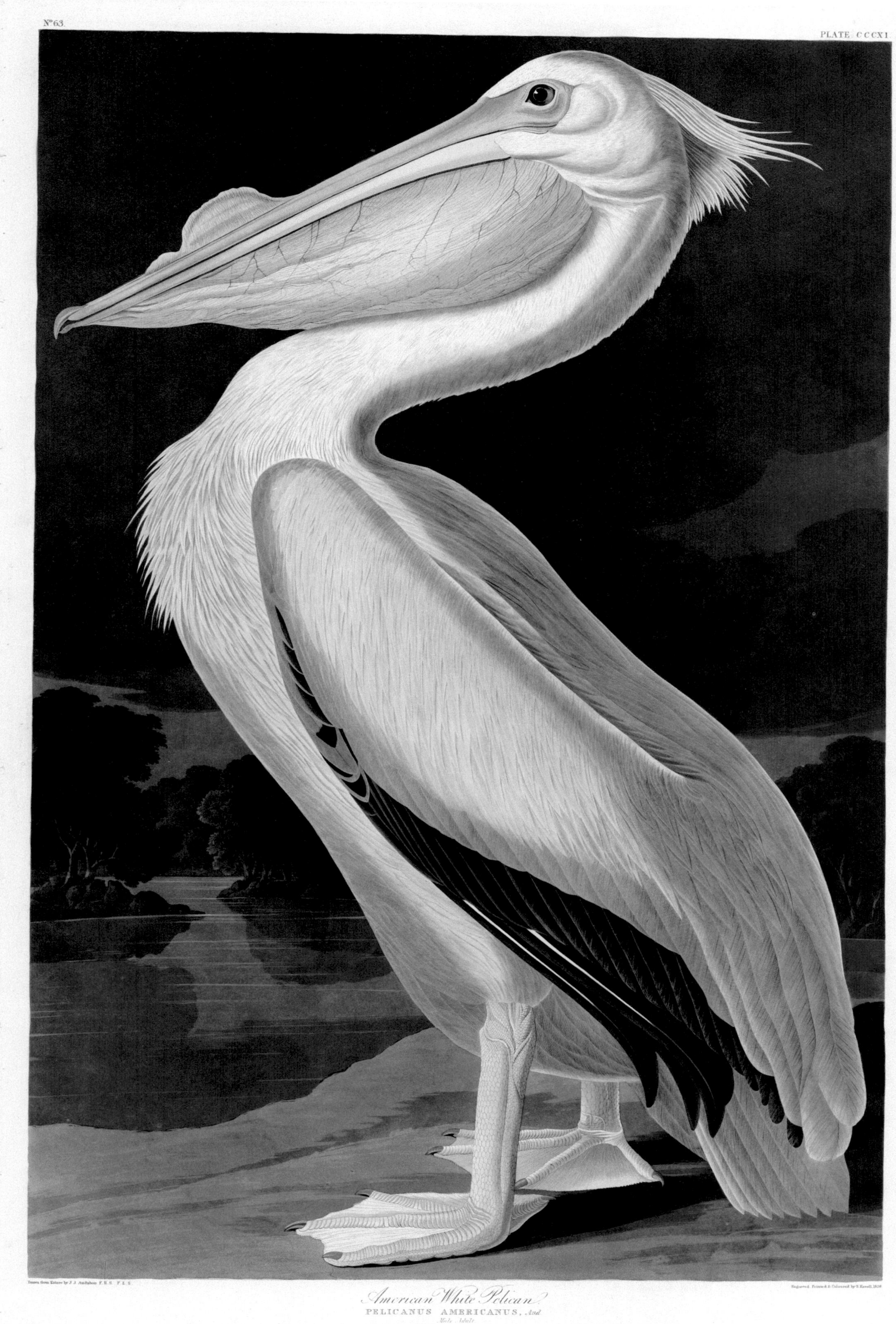
No. 63.
PLATE CCCXI
Drawn from Nature by J. J. Audubon F.R.S. F.L.S.
Engraved, Printed & Coloured by R. Havell, 1836
American White Pelican
PELICANUS AMERICANUS, Aud.
Male Adult.

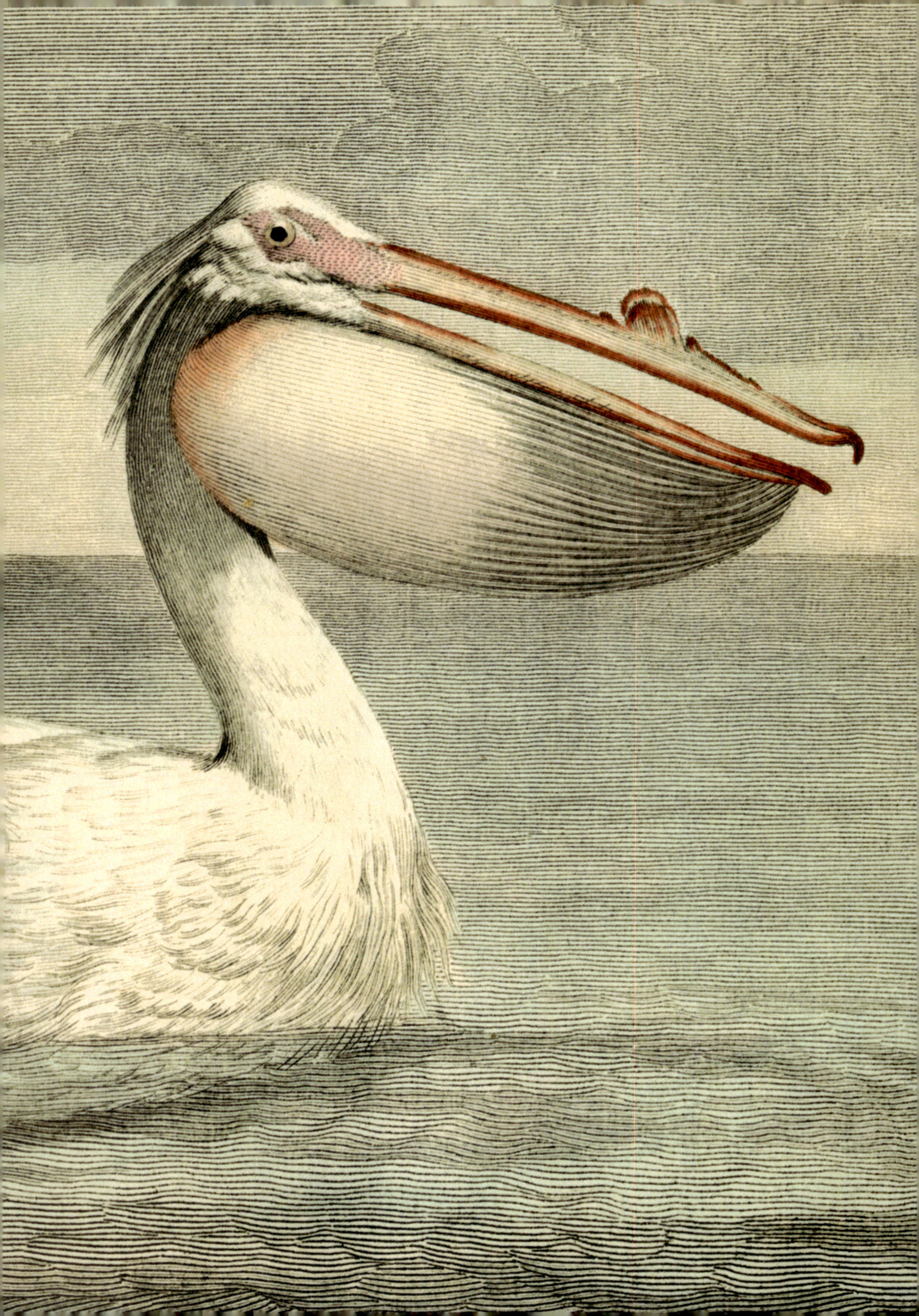

OPPOSITE:
AMERICAN WHITE PELICAN
Pelecanus erythrorhynchos

Charles Reuben Ryley,
Museum Leverianum, 1792–6

Pelicans boast an enormous bill. Their top mandible is flat and long while their lower one acts as a frame for an enormous pouch. This so-called 'gular pouch' is used for trapping and holding fish. The pelican can expand and contract it in size, enabling it to swoop in and catch fish as though it were using a large net.

DALMATIAN PELICANS
Pelecanus crispus

Edward Lear, *The Birds of Europe*, 1832–7

The Dalmatian pelican is the largest member of the pelican family. This plate shows both an adult and its young. Drawn by artist and author Edward Lear, these specimens had been sent to John Gould from the shores of Dalmatia – a historical region of Croatia from where this species' name originated.

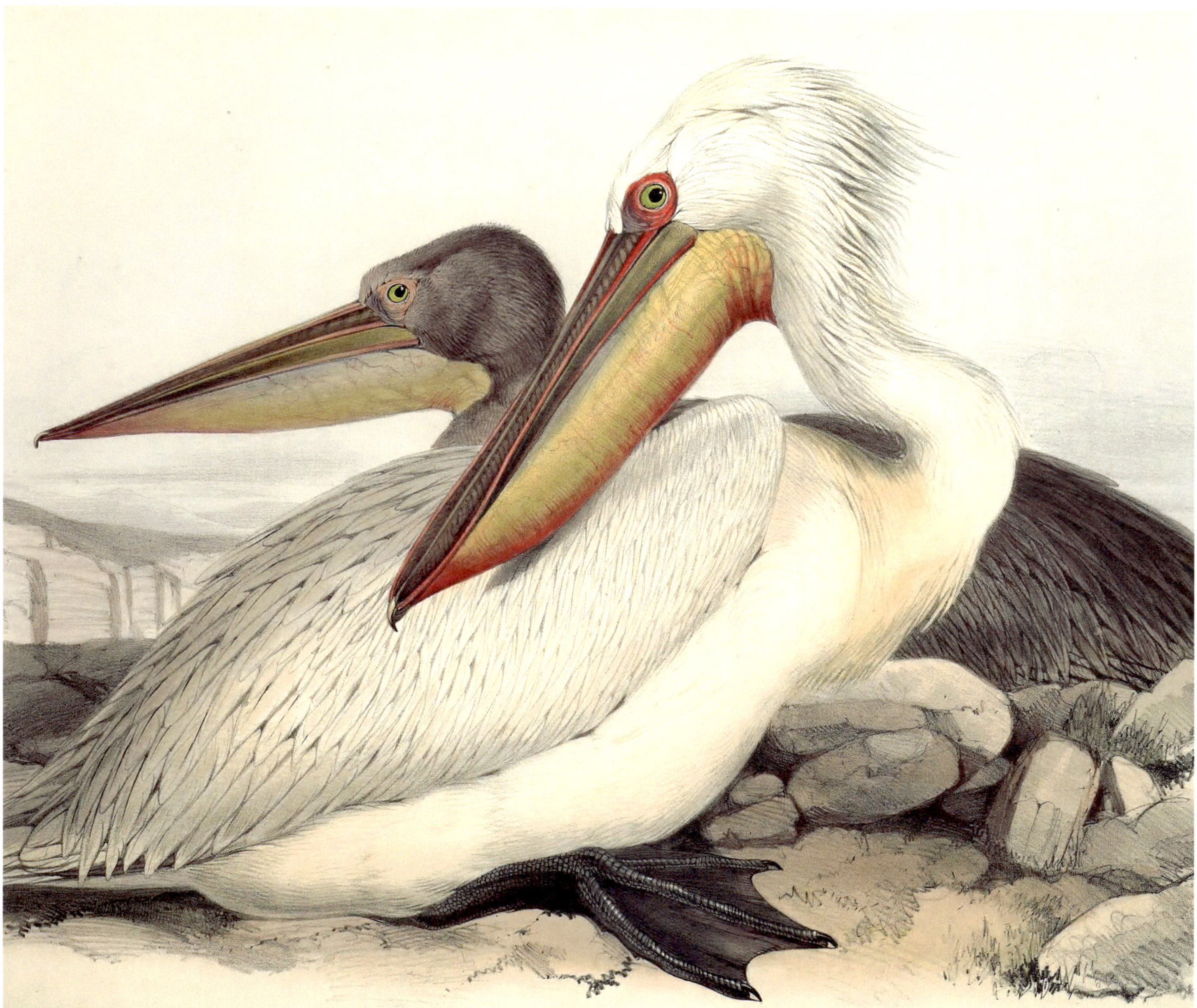

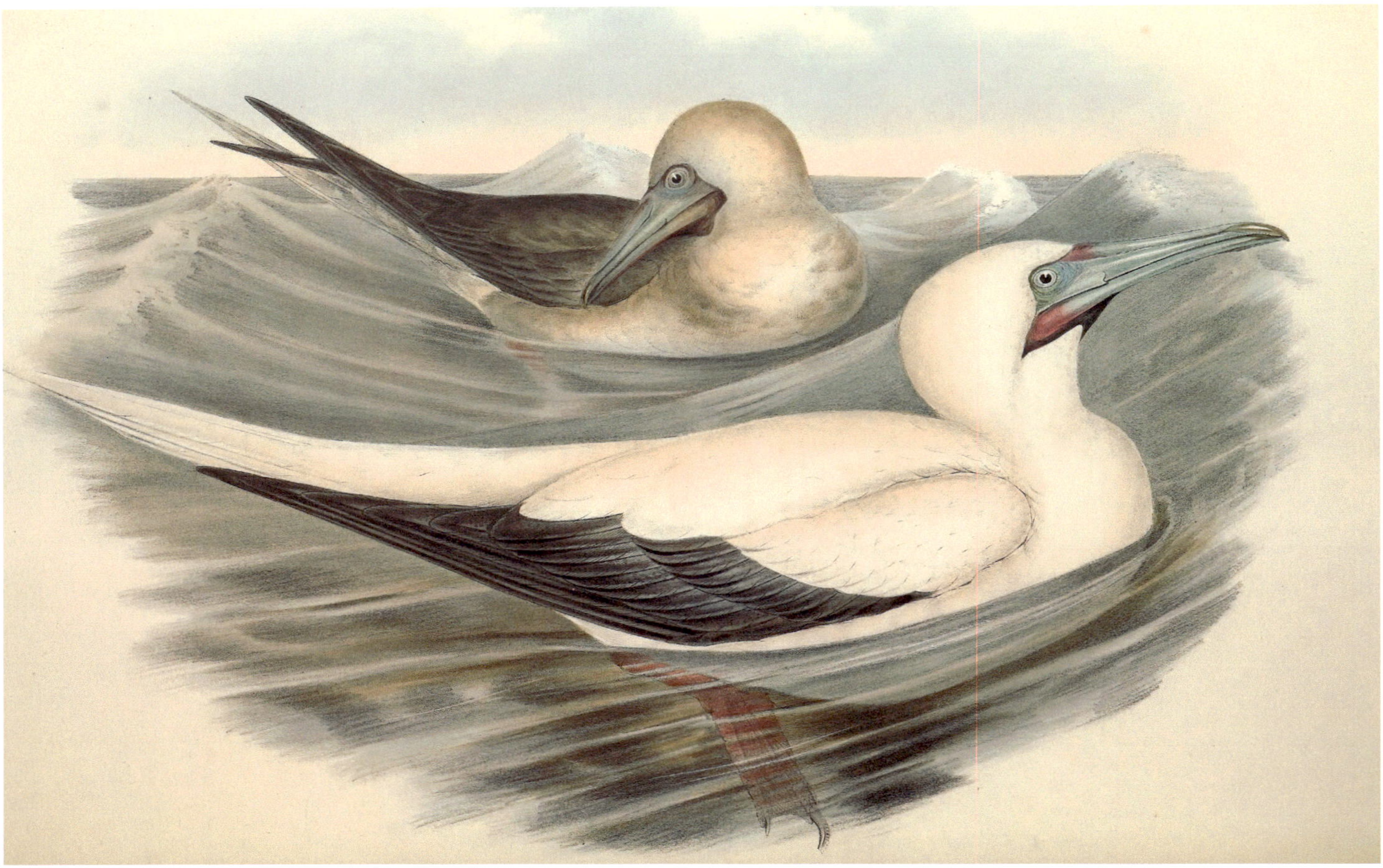

RED-FOOTED BOOBIES
Sula sula

John Gould & Henry Constantine Richter, *The Birds of Australia*, 1840–8

Boobies are impressive and agile fliers, but they can be somewhat clumsy on land. Their comical waddle earned them the name 'booby', which comes from the Spanish word *bobo*, meaning 'foolish' or 'silly'.

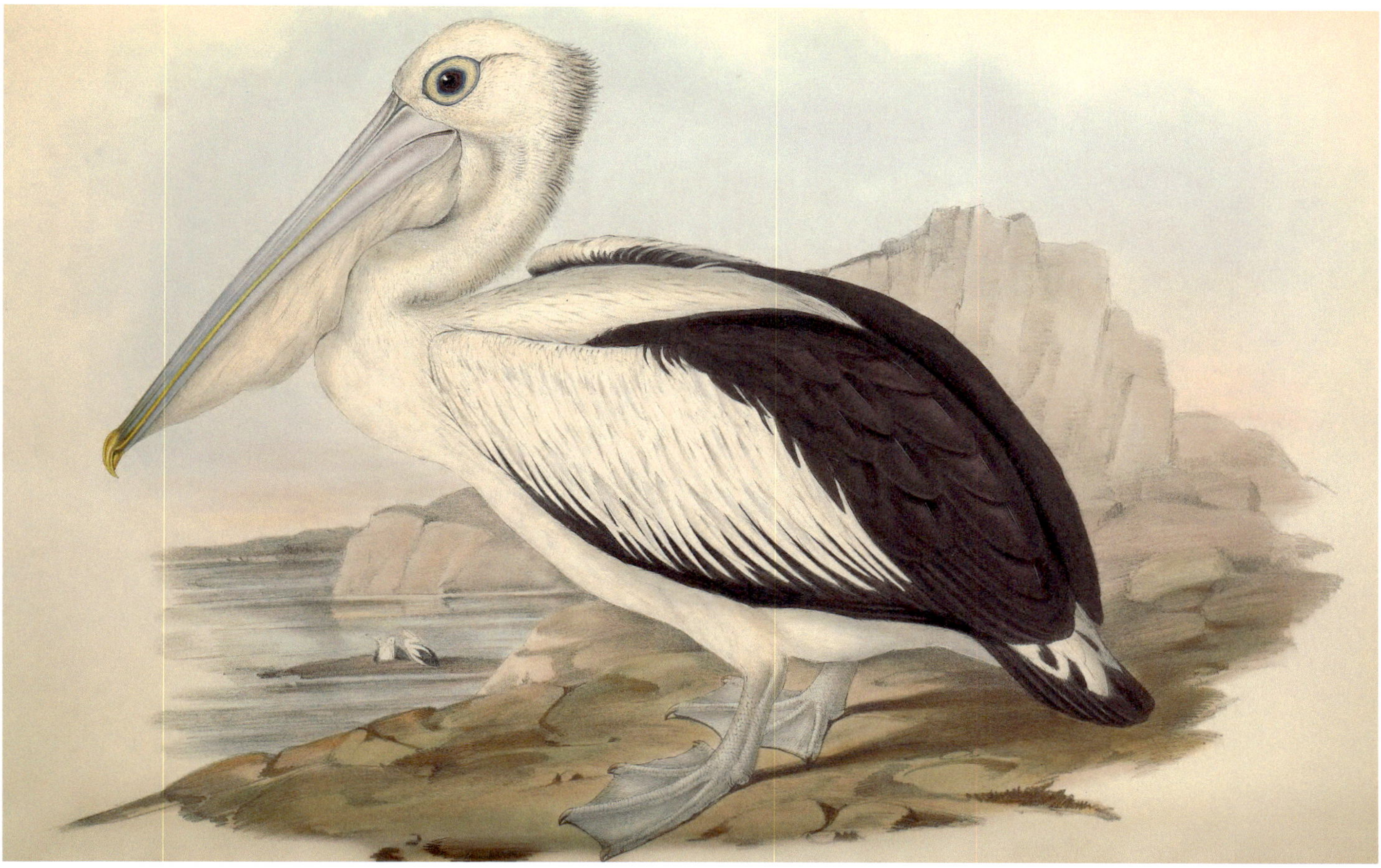

AUSTRALIAN PELICAN
Pelecanus conspicillatus

Elizabeth Gould,
The Birds of Australia, 1840–8

This Australian pelican is illustrated by Elizabeth Gould. In 1838 both Elizabeth and her husband John travelled to Australia to document the birds of the country. This pelican would have been a specimen that John collected on this trip and which Elizabeth would have drawn while based in Hobart.

OVERLEAF:
SCOULTON MERE
The Broad Water, May 15th 1877

John Gerrard Keulemans,
Ornithological Miscellany, 1876–8

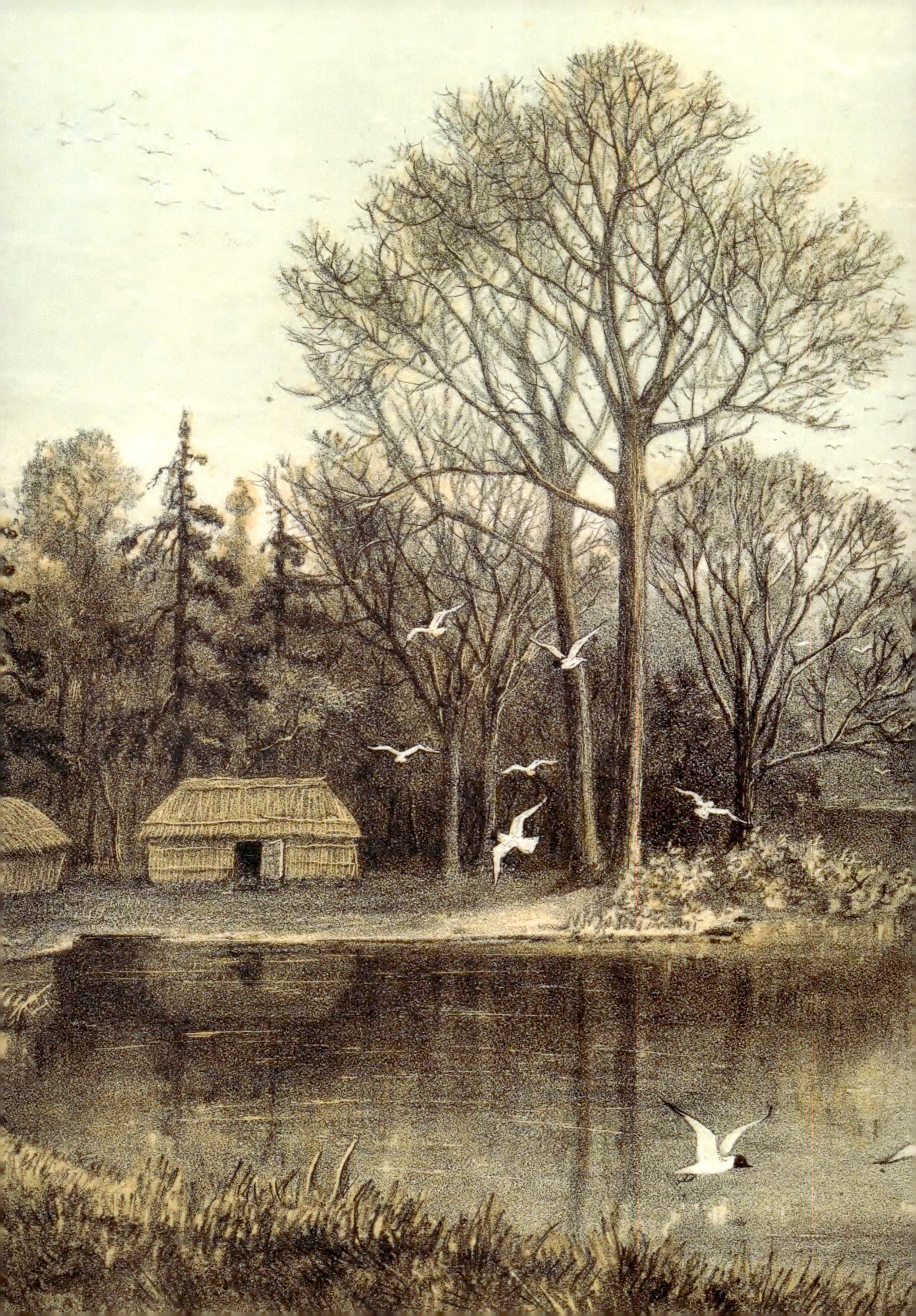

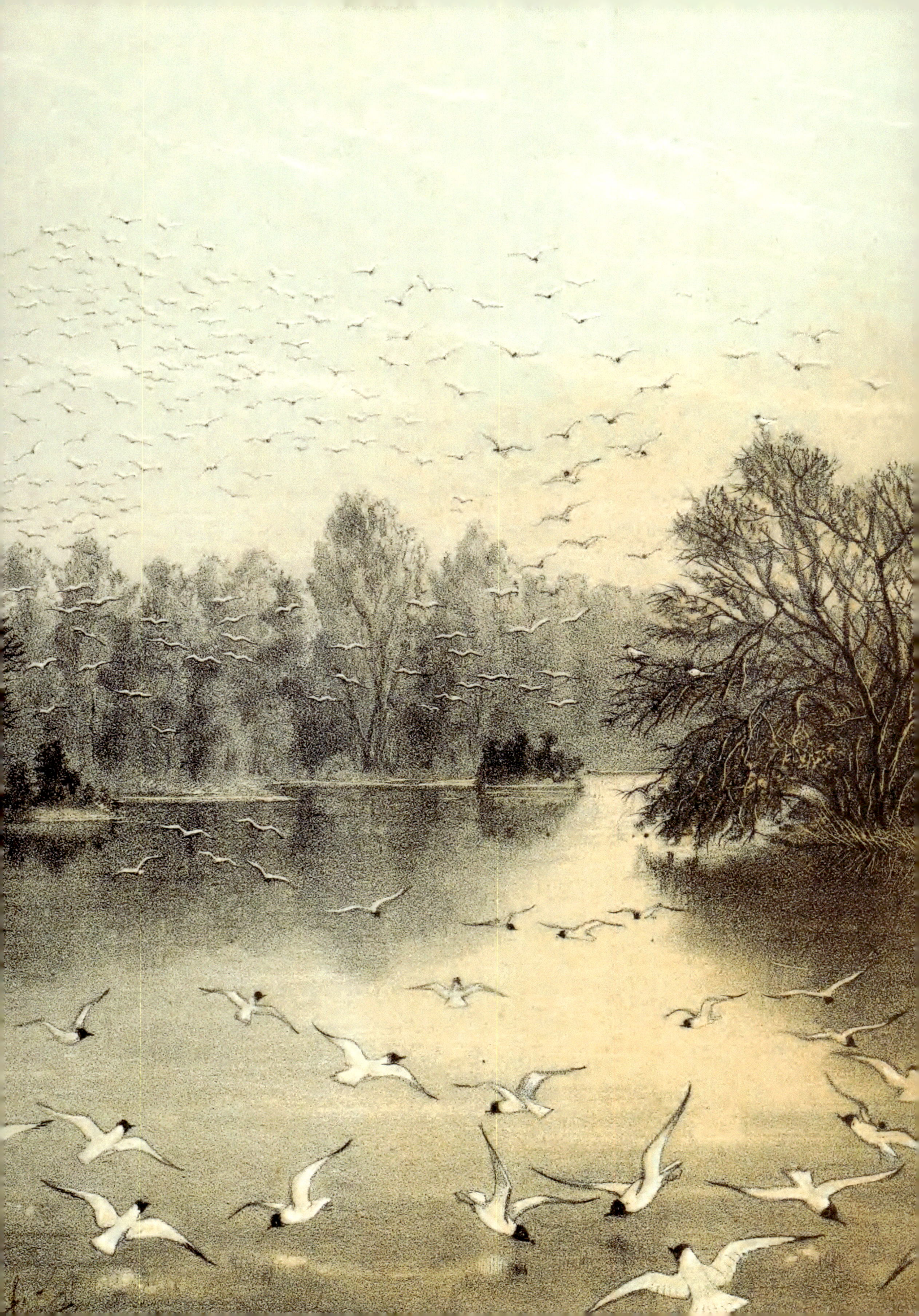

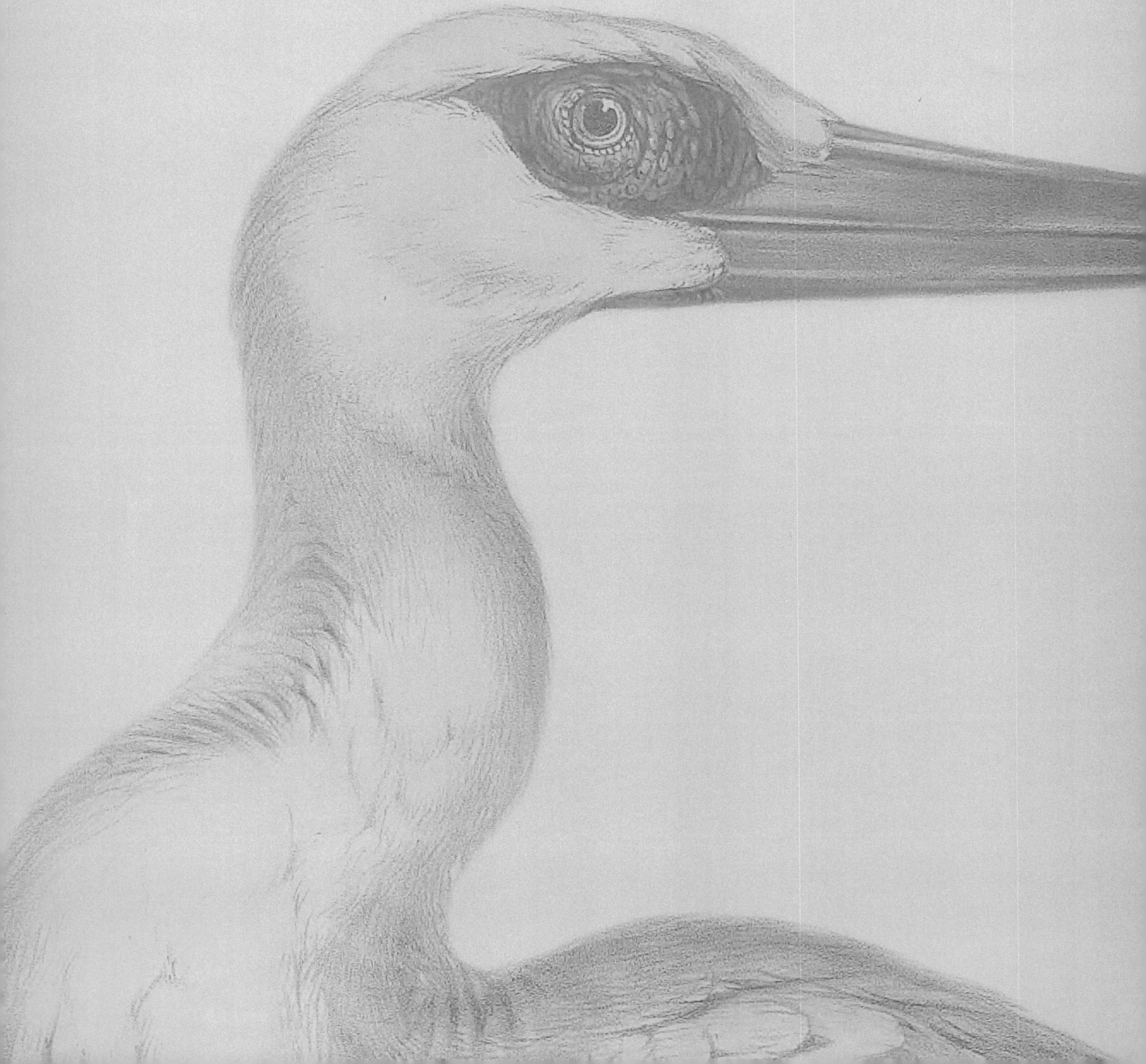

Shorebirds, Wading Birds & Rails

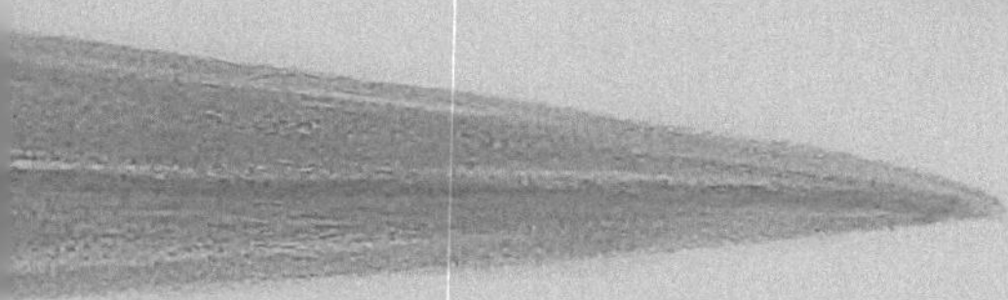

A miscellany of birds can be found foraging at the water's edge. Many have long legs for running or wading, and many also have long bills for probing, grabbing or reaching for prey. Some are tall and elegant, while others are much smaller and scamper over the surface of mud and sand, or in the shallows.

Around the shorelines and tidal flats, in the marshes and in the shallows of the world, there are birds that specialize in feeding at the interface between land and water. There are many of these, including some ducks and wildfowl (see pp. 96–127); the various families of shorebirds (called 'waders' in Britain); herons and egrets and their allies; flamingos, rails, cranes and their relatives. They are almost invariably long-legged, while many also have a long neck, and most have a long bill. Some feed on fish, some on worms, crustaceans and molluscs. The only characteristic that unites them is their amphibious nature.

In this chapter, we will look at these birds. A large number of the small- to medium-sized shoreline birds belong to a diverse order known as *Charadriiformes*. This order includes gulls and auks and also includes a large family known as *Scolopacidae*. This family includes many waders, such as curlews, sandpipers, snipes, woodcocks and godwits. These birds typically feed on the shore and they each have different sized bills that allow them to eat in the same place without having to compete with one another for food. Many of these birds have long bills that are perfect for foraging in the soft ground, and some even have sensitive nerve endings at the end of their bill that can help them to detect small prey effectively and efficiently.

Larger birds such as herons, egrets and cranes also have impressive bills. These birds are skilled hunters. Aided by their long and agile necks, they have strong jaws and dagger-shaped bills that enable them to grab and hold on to fish. Often seen wading in the water, they catch their prey with incredible agility and precision. Instead of constantly foraging for food, they will often stand alert and motionless by the water, waiting for their prey to come to them. During the breeding season, these birds develop elaborate plumes, and their stillness while hunting makes them appear incredibly elegant and stately.

One might assume that the stillness of such birds would make them ideal subjects for ornithologists to draw in the wild – yet, as we know, this was very rarely the case. While some naturalists would have certainly studied birds in their natural habitat, the majority of illustrations were produced using specimens. The stillness that we see in these illustrations does not necessarily reflect the personality of the birds that are depicted. Instead, what we are looking at is an illustration of a dead specimen. Indeed, an illustrator may not have even encountered certain species in real life. These are birds that have been shot, stuffed and mounted. While this might horrify many nature-lovers today, it was the standard practice at the time.

It is important to remember that tools such as the box easel and the portable paint tube were not invented until the mid-nineteenth century. Considered somewhat simple tools today, their invention revolutionized the art of image-making and brought painters outside to paint *en plein air*. Prior to this, art and image-making was a practice that existed almost exclusively in the studio. In the era of ornithological illustration, those who did sketch in the field would have typically done so in journals, creating preliminary drawings using watercolour, pencil or pastel. These sketches were designed to inform a final drawing, typically while the illustrator worked from a specimen. Field observations helped to enrich the quality of the work and could inform illustrators about the personality of the bird and the habitat it lived in.

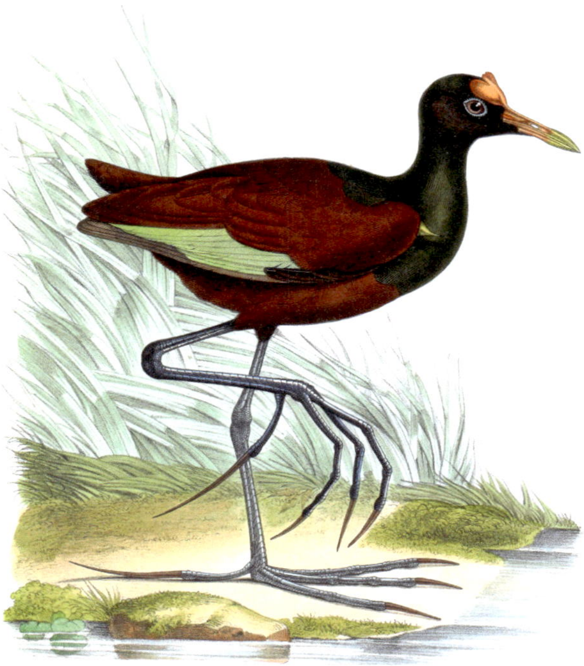

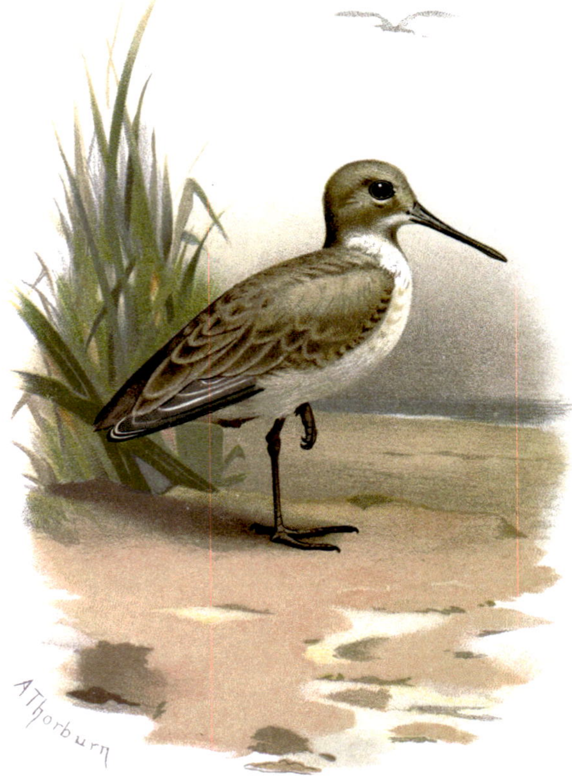

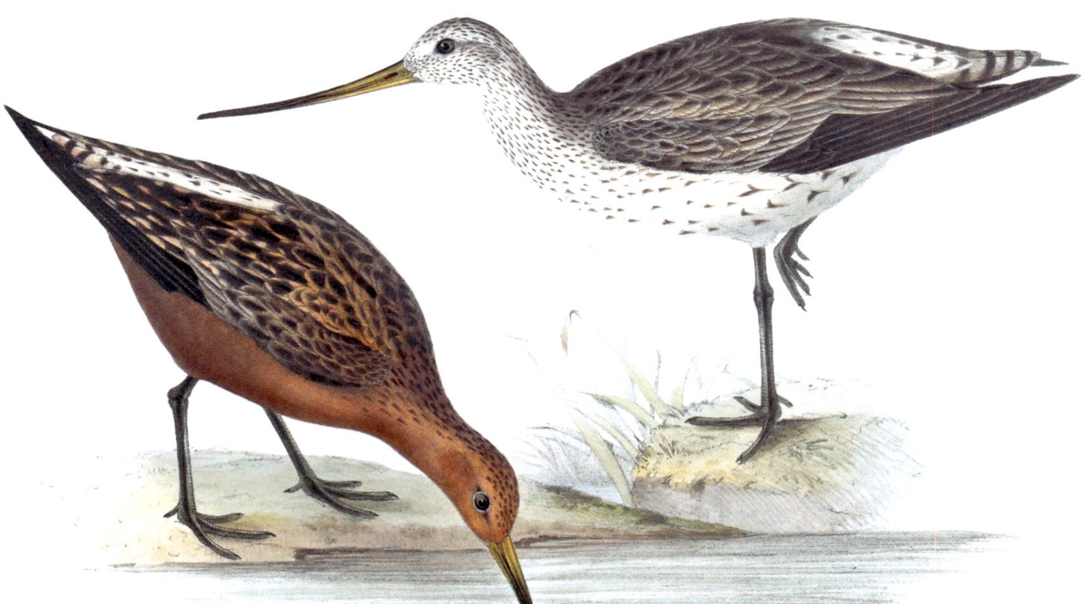

TOP LEFT:

WATTLED JACANA
Jacana jacana

Paul Louis Oudart, *Iconographie ornithologique*, 1845–9

Jacanas are identifiable by their large feet which allow them to walk on floating vegetation such as lilypads. Females have more than one mate and males rear the chicks.

TOP RIGHT:

DUNLIN
Calidris alpina

Archibald Thorburn, *Familiar Wild Birds*, 1883

Dunlins tend to occur in very large flocks and are typically seen in muddy sections of estuaries and close to the tide edge.

BOTTOM:

BAR-TAILED GODWITS
Limosa lapponica

John Gould & Elizabeth Gould, *The Birds of Europe*, 1832–7

Godwits have long beaks that are ideally suited to probing in the sand for worms, molluscs and small crustaceans.

1. VIRGINIA RAIL
Rallus limicola

2. CLAPPER RAIL
Rallus crepitans

3. REDDISH EGRET
Egretta rufescens

4. SNOWY EGRET
Egretta thula

Alexander Wilson,
American Ornithology,
1808–14 (this edition: 1871)

This plate is by Scottish-American ornithologist, illustrator and poet Alexander Wilson. Often called the Father of American Ornithology, his book *American Ornithology* (1808–14) is considered to be the first comprehensive work on American natural history. Wilson travelled extensively through America to create his work, producing paintings and collecting specimens. The final publication spanned nine volumes and contained seventy-six full-page illustrations. Wilson typically included more than one bird per illustration and his work documents 268 unique species.

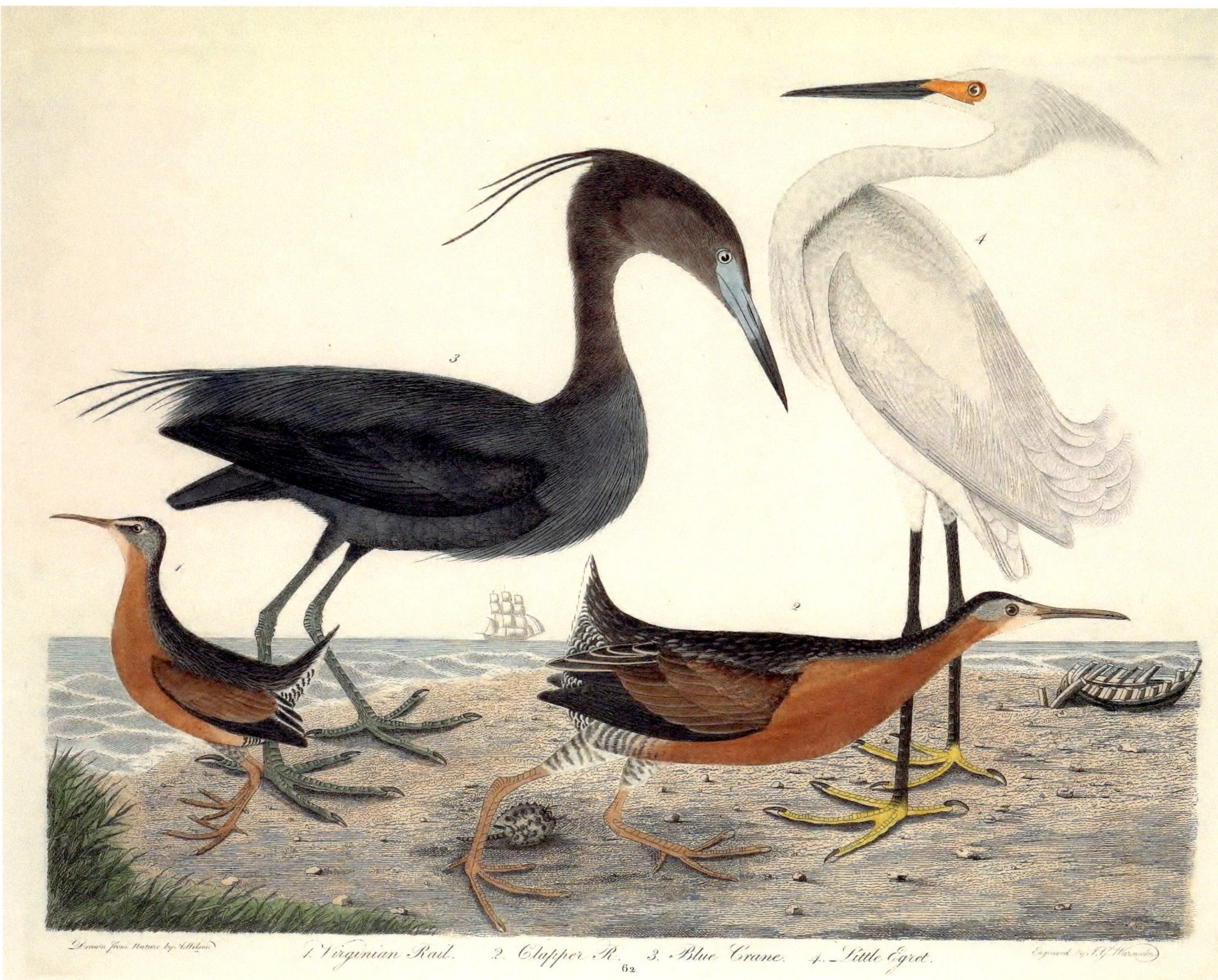

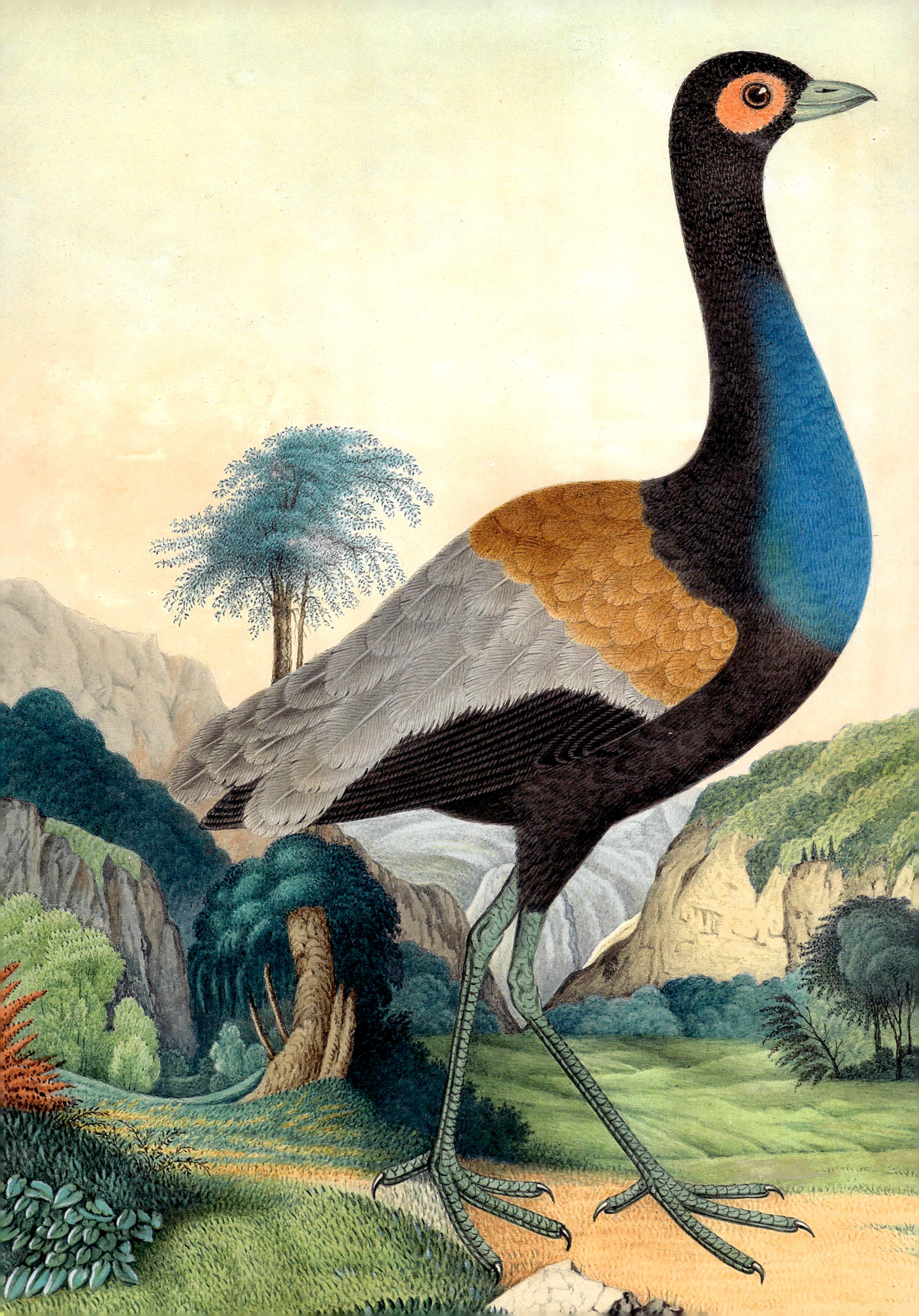

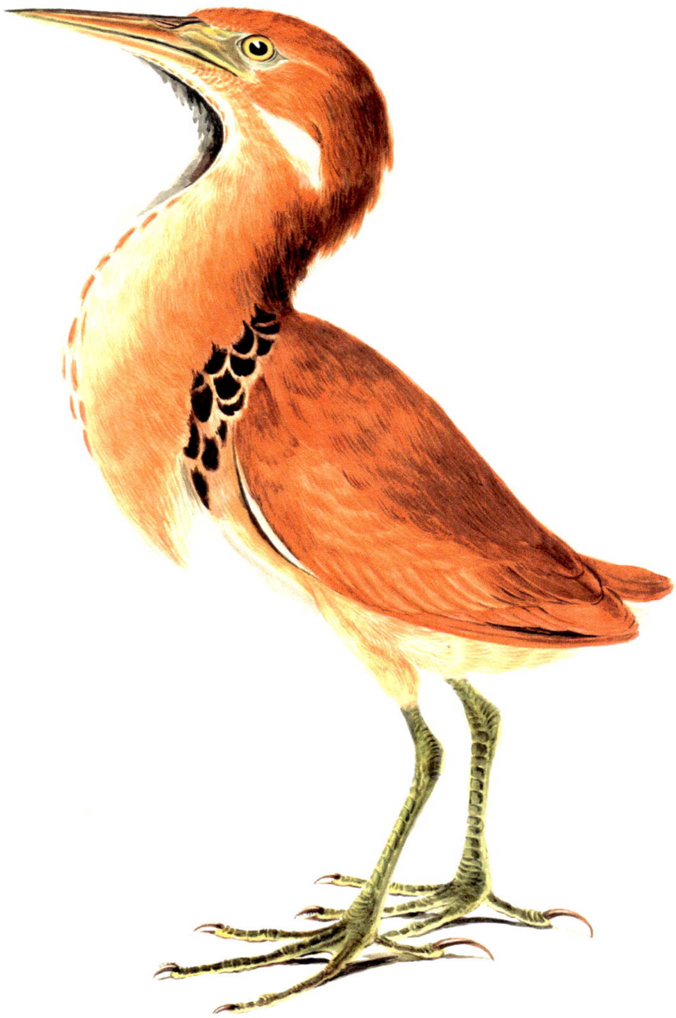

OPPOSITE:

GREY-WINGED TRUMPETER
Psophia crepitans

Aloys Zötl, watercolour, 1877

This plate is by Austrian artist Aloys Zötl. Zötl was relatively unknown in his lifetime but was later re-discovered by the surrealist André Breton, who saw a distinctive surrealist sensibility in much of his work.

CINNAMON BITTERN
Ixobrychus cinnamomeus

Elizabeth Gwillim, watercolour, 1801

Gwillim's watercolour of this cinnamon bittern captures the distinctive rich cinnamon colour of this very small heron. Its habitat is wet and grassy areas such as meadows and paddy fields; it is commonly found in many parts of tropical and subtropical Asia.

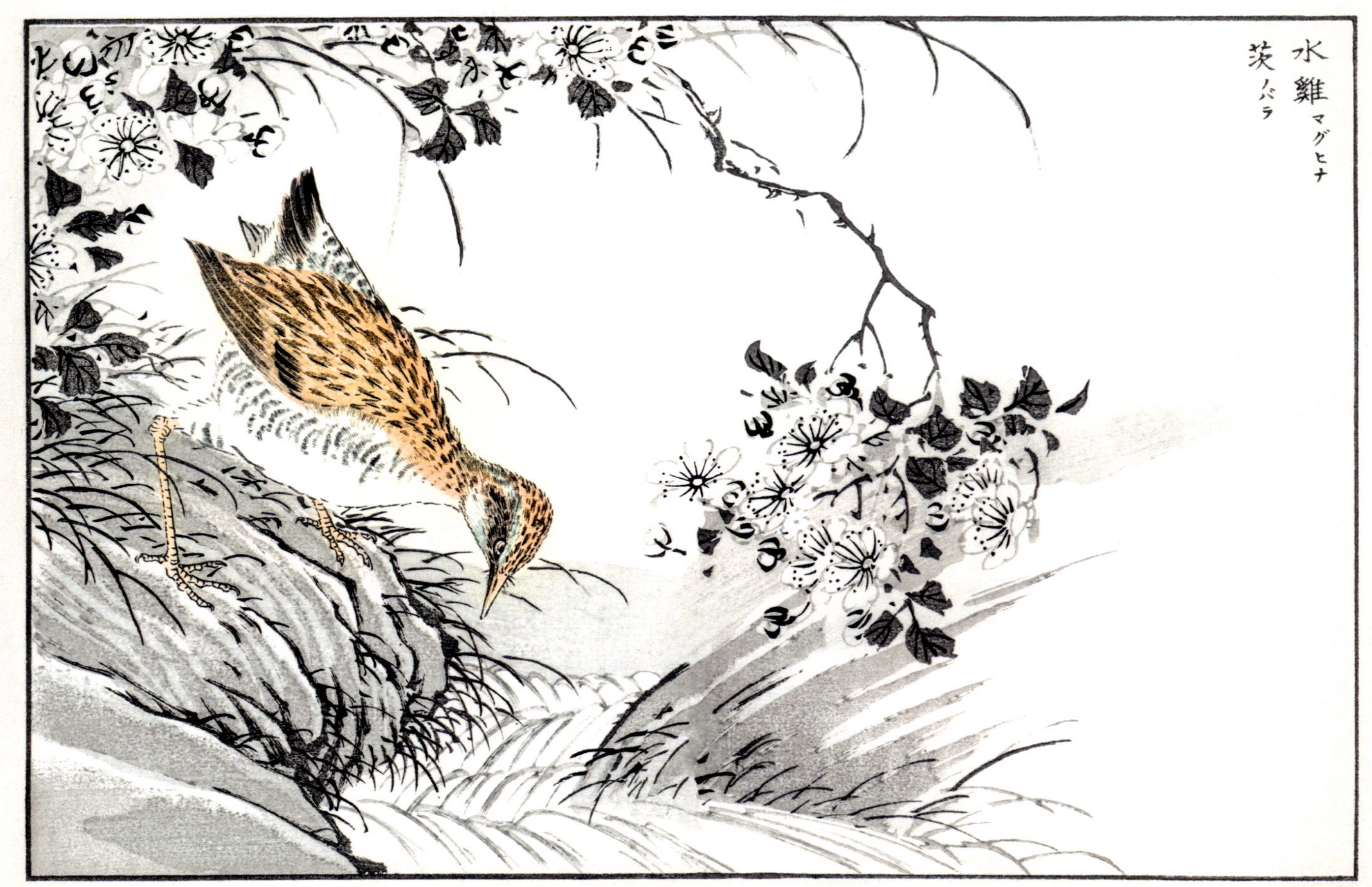

***PERHAPS* BAILLON'S CRAKE**
Zapornia pusilla

Numata Kashū, *Shūchō gafu*, 1885

The industrialization of Japan led to a period of great economic growth in the 1880s, which helped to bring about a renewed enthusiasm for colour woodblock prints. During this time the artist Numata Kashū published a lavish three-volume portfolio that featured depictions of birds and flowers. Soon after its publication, his publisher went out of business and so Kashū's original work became incredibly rare. Dealers responded by cutting up his original work, as selling prints individually brought them higher returns. For this reason, most complete sets have now been destroyed, making copies of his original portfolio incredibly valuable.

UPLAND SANDPIPERS
Bartramia longicauda

Alexander Pope Jr., *Upland Game Birds and Water Fowl of the United States*, 1877–8

Alexander Pope Jr. was an American sporting artist who had an avid passion for painting game birds. This illustration is taken from his book *Upland Game Birds and Water Fowl of the United States*, which featured twenty chromolithograph versions of his watercolour paintings. He was also a successful artist who helped to popularize still-life painting in America during the late nineteenth century. His paintings typically showed hunting scenes with compositions featuring dead birds and other animals that he rendered in the *trompe-l'œil* style of painting.

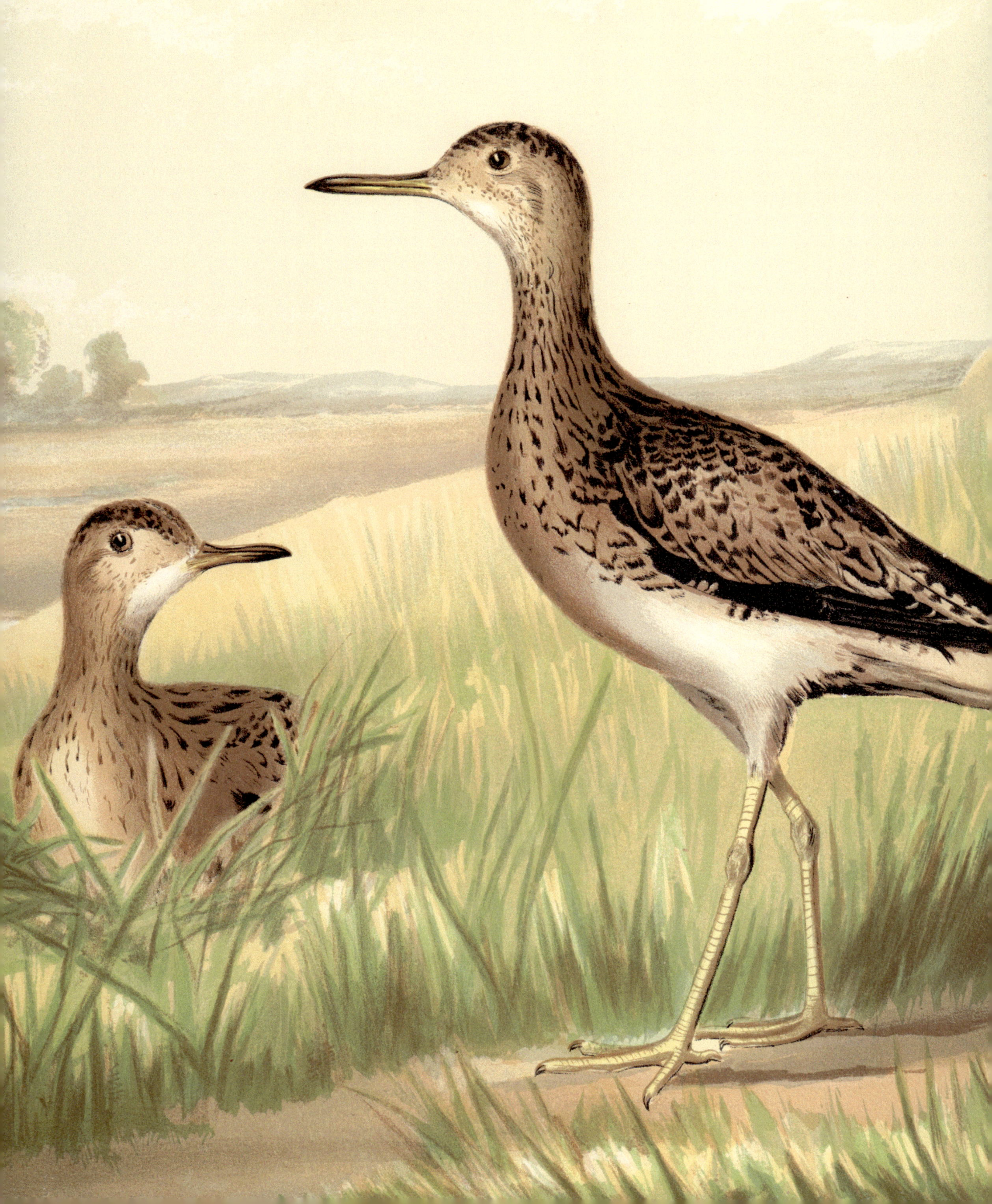

HORNED SCREAMERS
Anhima cornuta

Uncredited, *Bilder-atlas*, 1860–4

Found in freshwater marshes, rivers and wetlands in the lowlands of the Amazon, the horned screamer is named both for its distinctive horn and its impressive loud call. Its horn is composed of cartilage and has led many to call it the unicorn of the avian world. So unusual is this bird that ornithologists have often struggled classifying it. Related to ducks, geese and swans, screamers have a different bill more closely resembling that of chickens or other gamebirds.

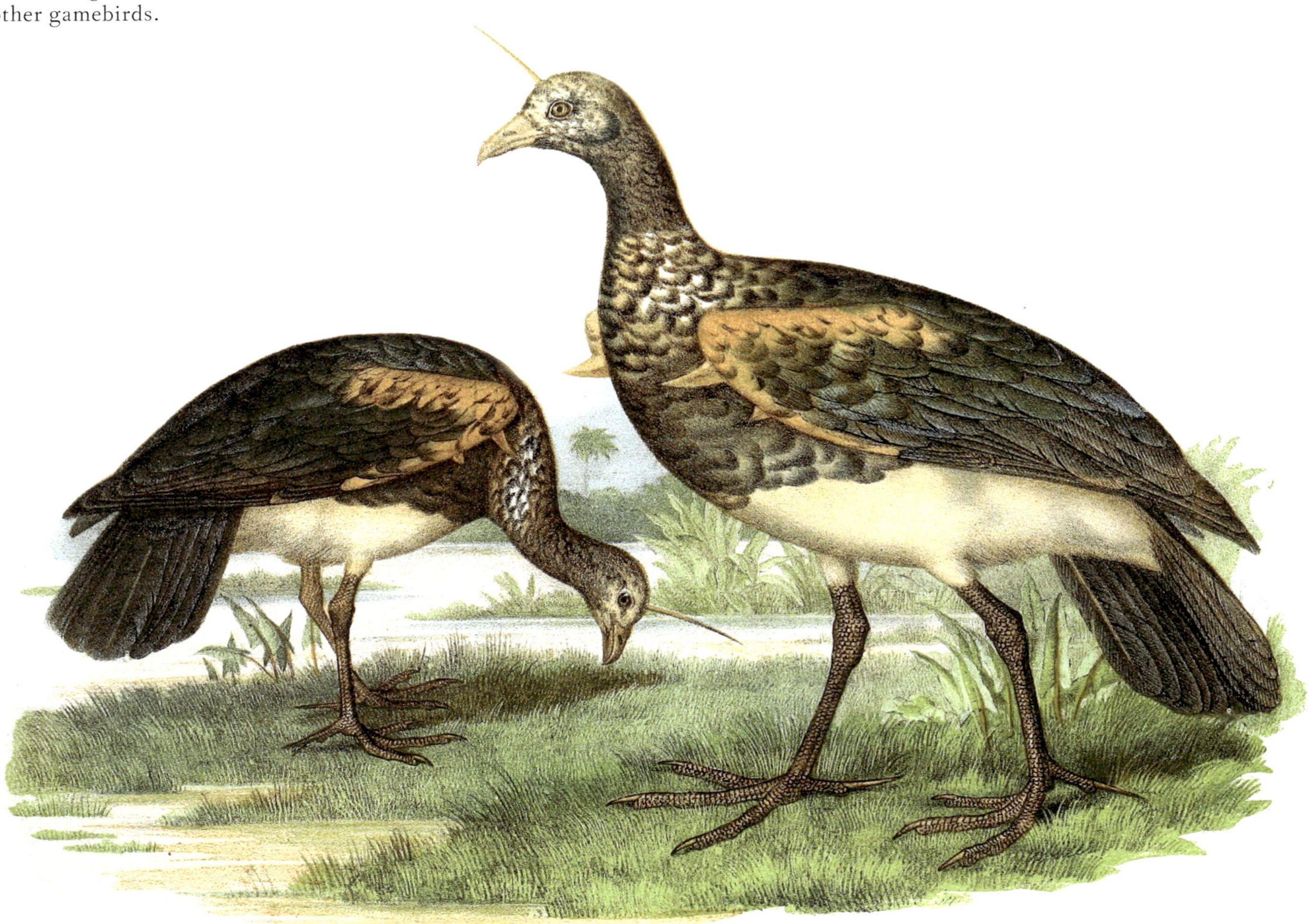

DEMOISELLE CRANE
Anthropoides virgo

William Daniell, *Zoography*, 1807

William Daniell's depiction of a demoiselle crane comes from William Wood's three-volume work *Zoography*. Better known as a landscape and marine painter, Daniell travelled extensively during his life and achieved success for his collections of aquatints. These depicted scenes from both his native Britain and his travels in India.

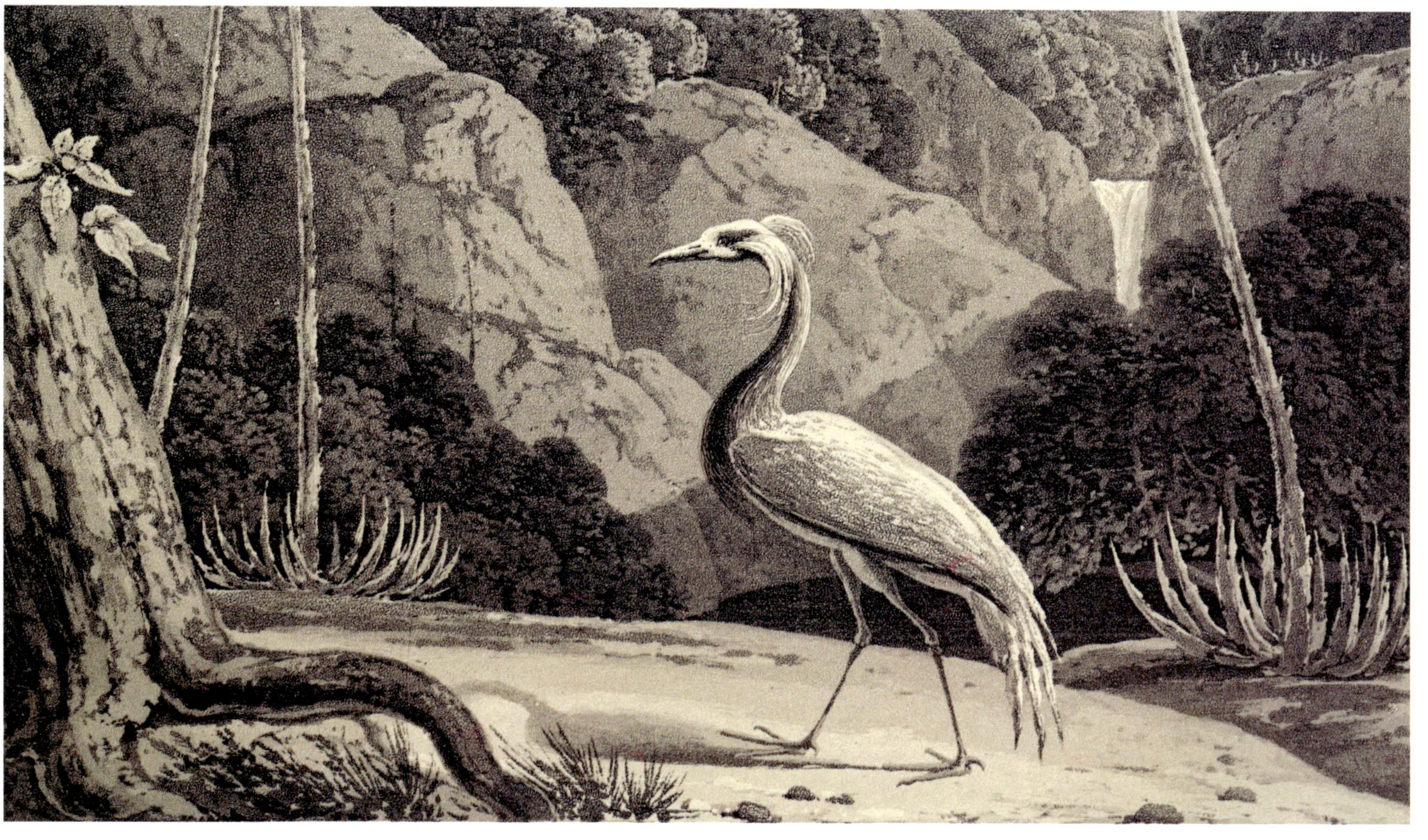

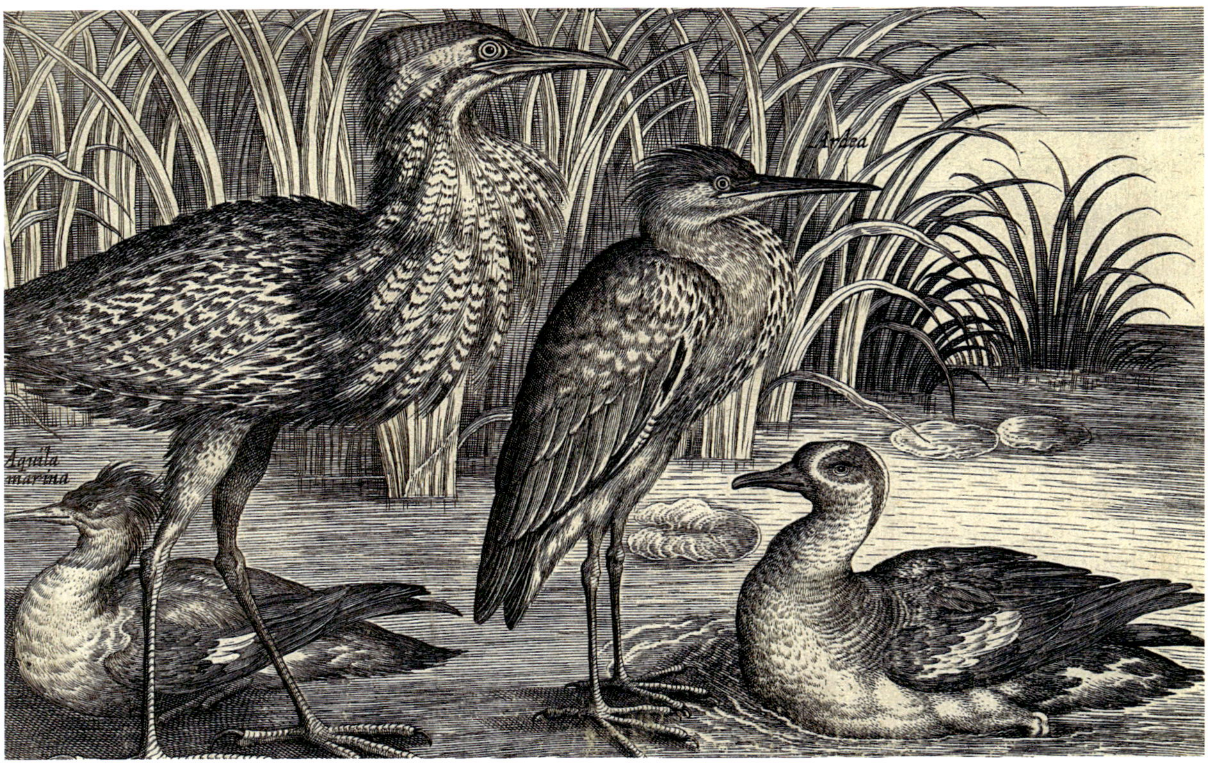

LEFT TO RIGHT

MERGUS, BITTERN, HERON AND TEAL

Mergus, *Botaurinae*, *Ardeidae*, *Teal*

Adriaen Collaert, *Vier waadvogels langs een oever*, engraving, *ca*. 1598–1602

This etching of four waders along a shore is by Flemish artist Adriaen Collaert. Born in Antwerp around 1560, he published prints, illustrated books and was a reproductive engraver.

AFRICAN SACRED IBIS
Threskiornis aethiopicus

John Gerrard Keulemans, *A History of the Birds of Europe*, 1871–81

Throughout history, birds have been central figures in many myths and theologies. The ibis held a particular significance in the belief system of the ancient Egyptians, who often linked it to Thoth – the god of wisdom, knowledge and writing. Today, the bird is extinct in Egypt but is commonly found in Sub-Saharan Africa as well as in southeastern parts of Iraq.

OVERLEAF:

HERONRY IN GREAT SOWDEN WOOD, SUSSEX

John Gerrard Keulemans, *Ornithological Miscellany*, 1876–8

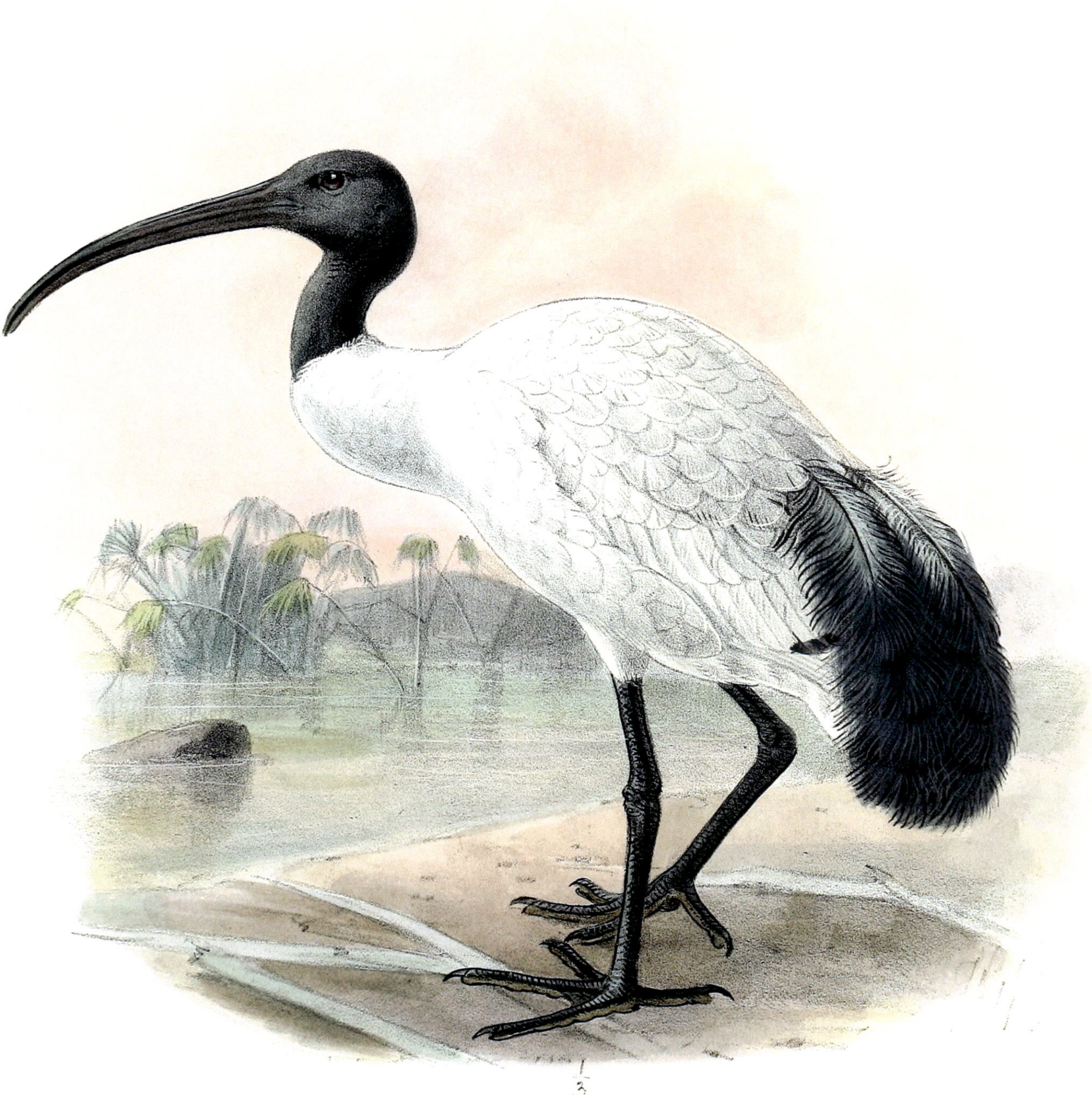

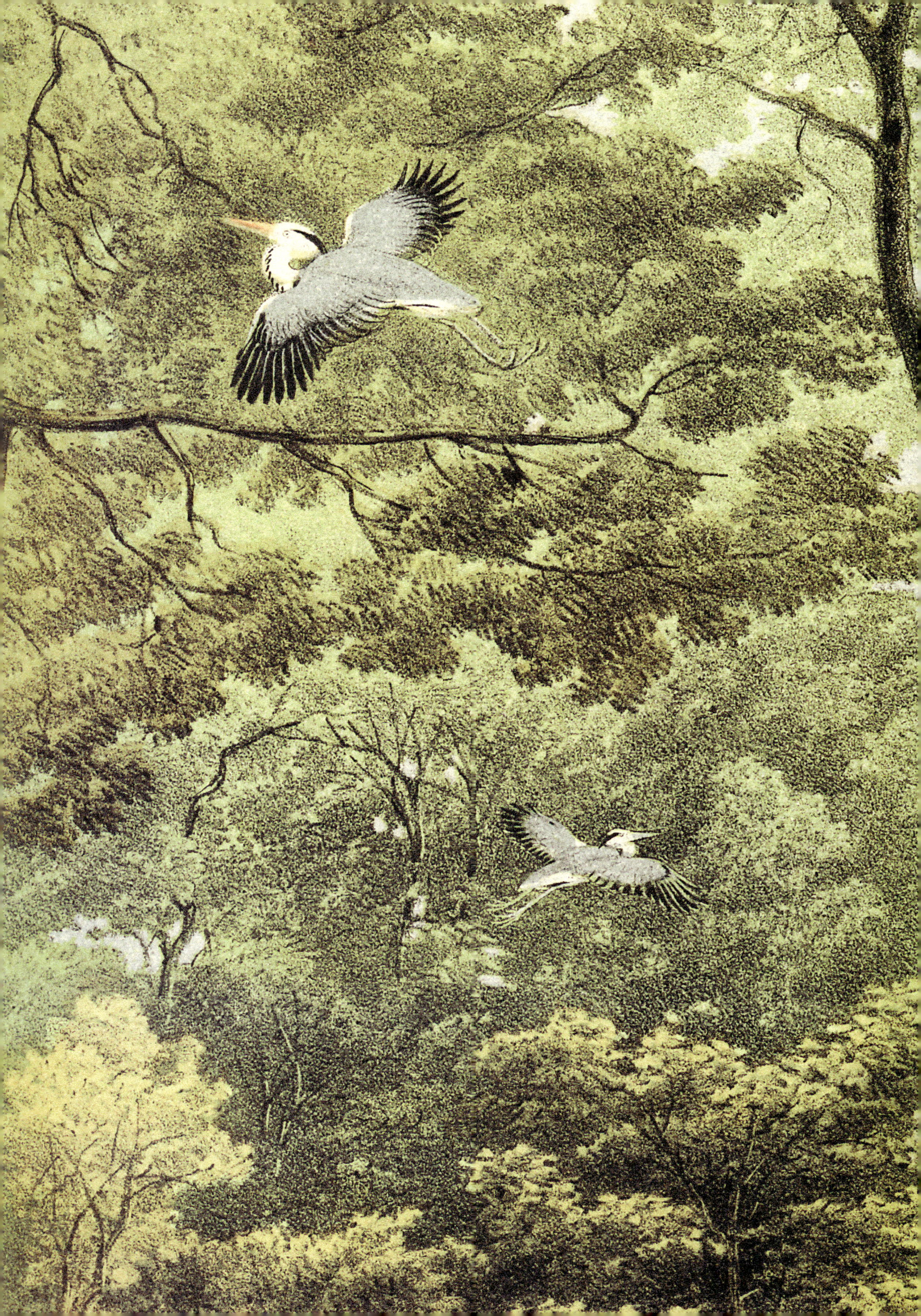

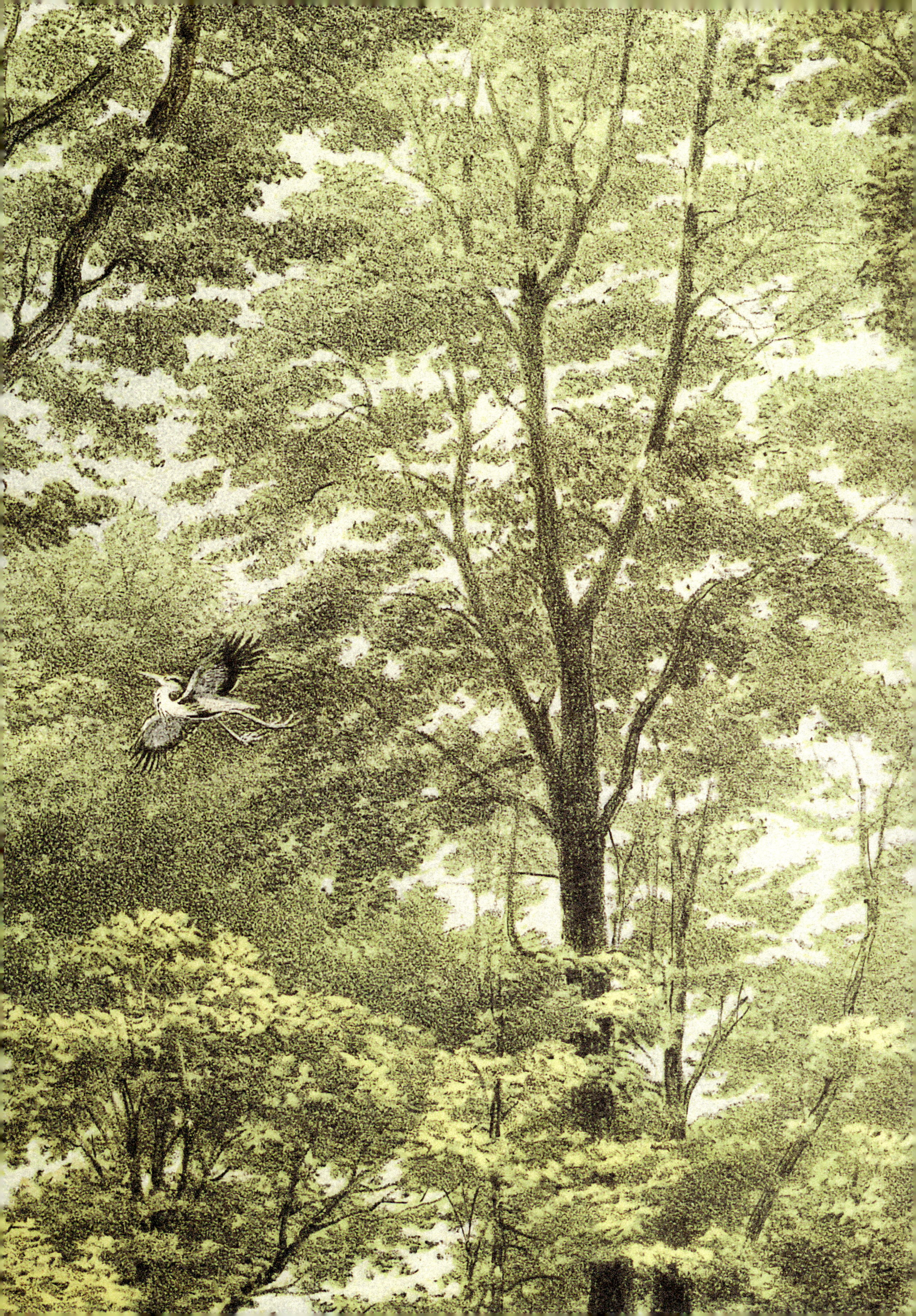

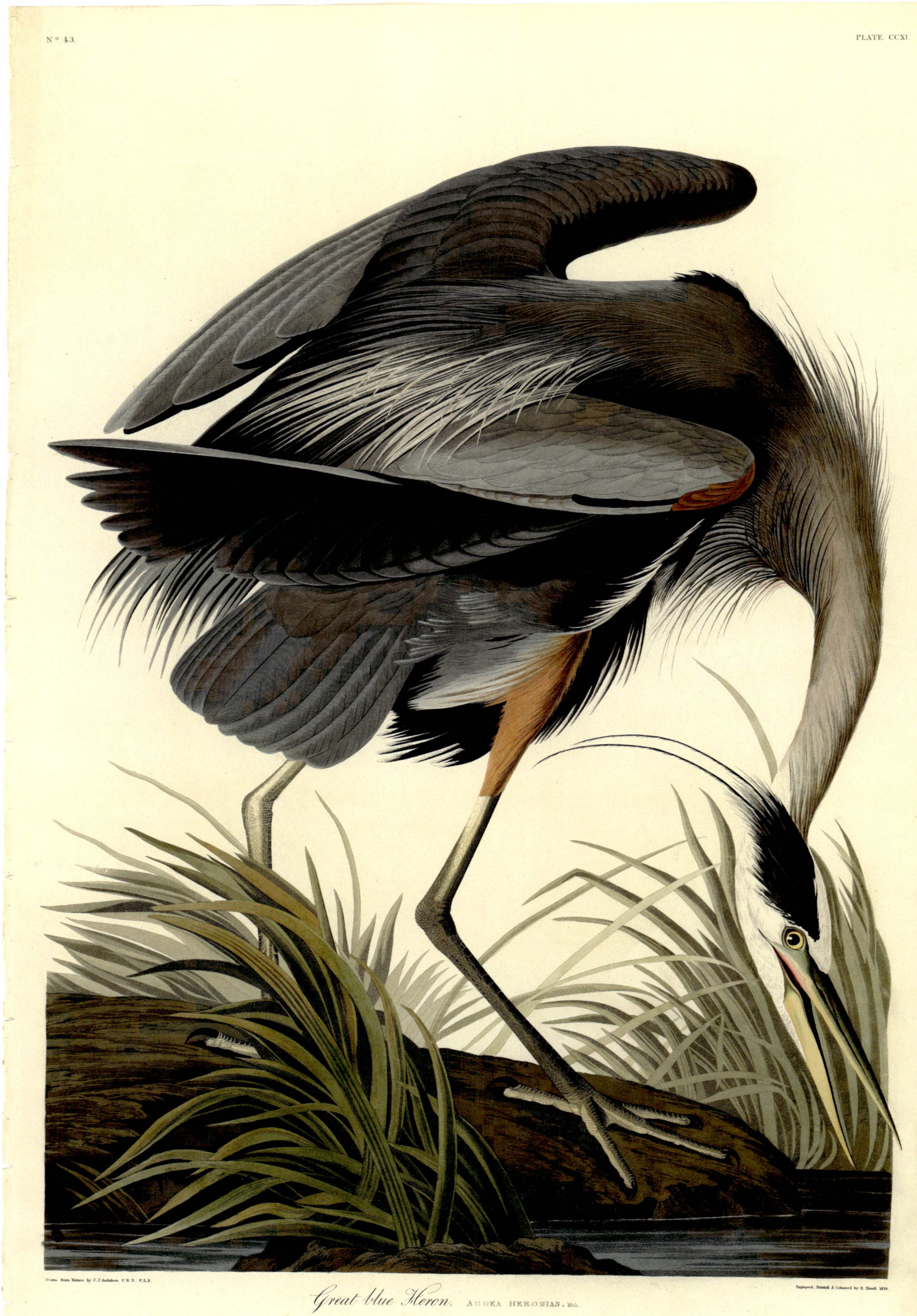
N° 43.
PLATE CCXI.
Drawn from Nature by J. J. Audubon, F.R.S. F.L.S.
Great blue Heron. ARDEA HERODIAS. Male.
Engraved, Printed & Coloured by R. Havell. 1834.

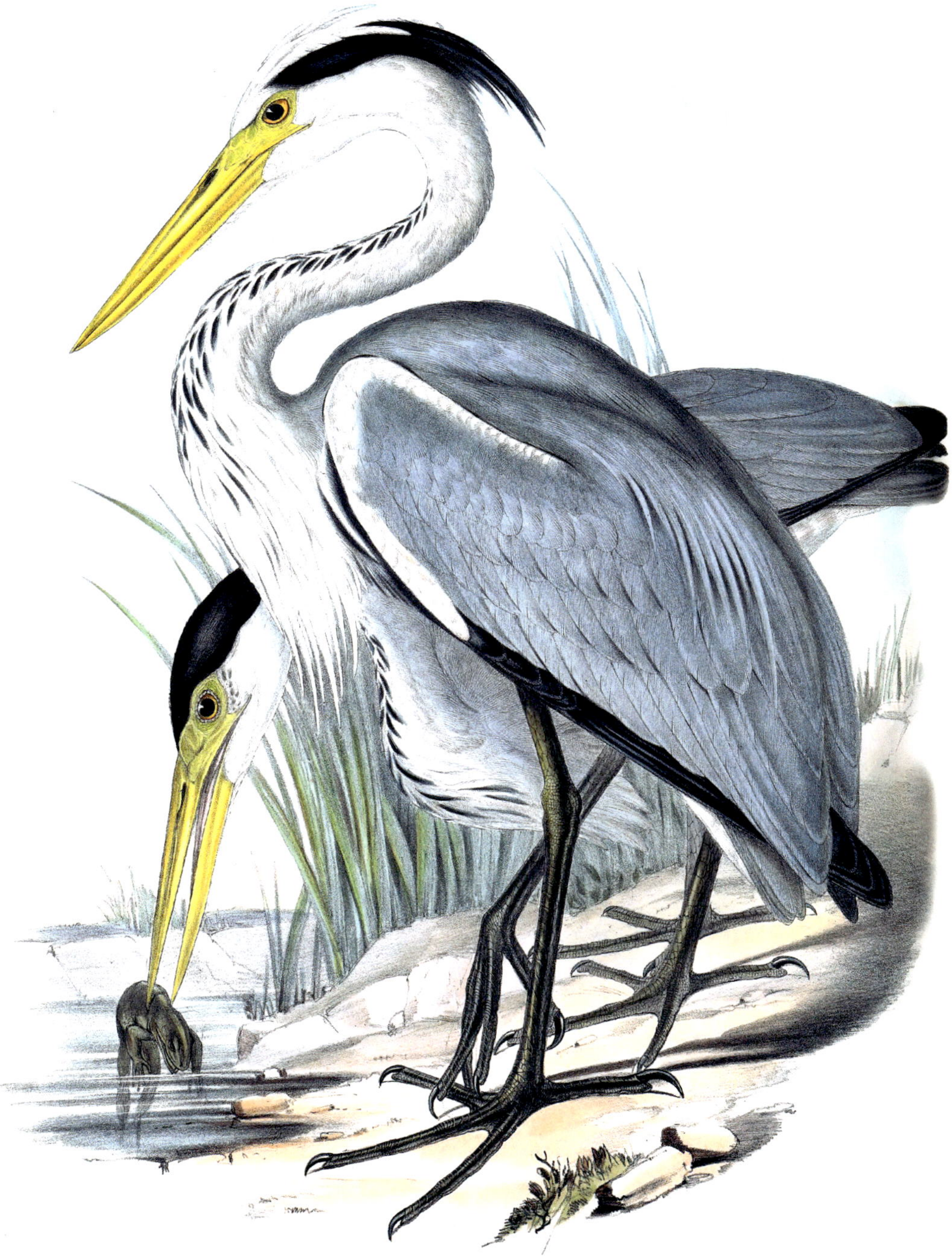

OPPOSITE:

GREAT BLUE HERON
Ardea herodias

John James Audubon,
The Birds of America, 1827–38

The great blue heron is the largest and most common heron in North America. Audubon noted that he had found it in every state in which he had travelled, but this particular bird was most likely seen near Henderson, Kentucky, just south of the Ohio River. Audubon lived in this small frontier village for several years during the 1810s and today the area is home to the John James Audubon State Park.

GREY HERONS
Ardea cinerea

John Gould & Henry Constantine Richter, *The Birds of Australia*, 1840–8

This plate from Gould's *Birds of Australia* mistakenly identifies these birds as a new species that he calls the Australian heron (*Ardea Leucophæa*). In fact, they are simply grey herons, the common heron of Europe. In Western Australia, the bird is a vagrant – the result of migrating birds that have lost their way and end up appearing outside of their normal range.

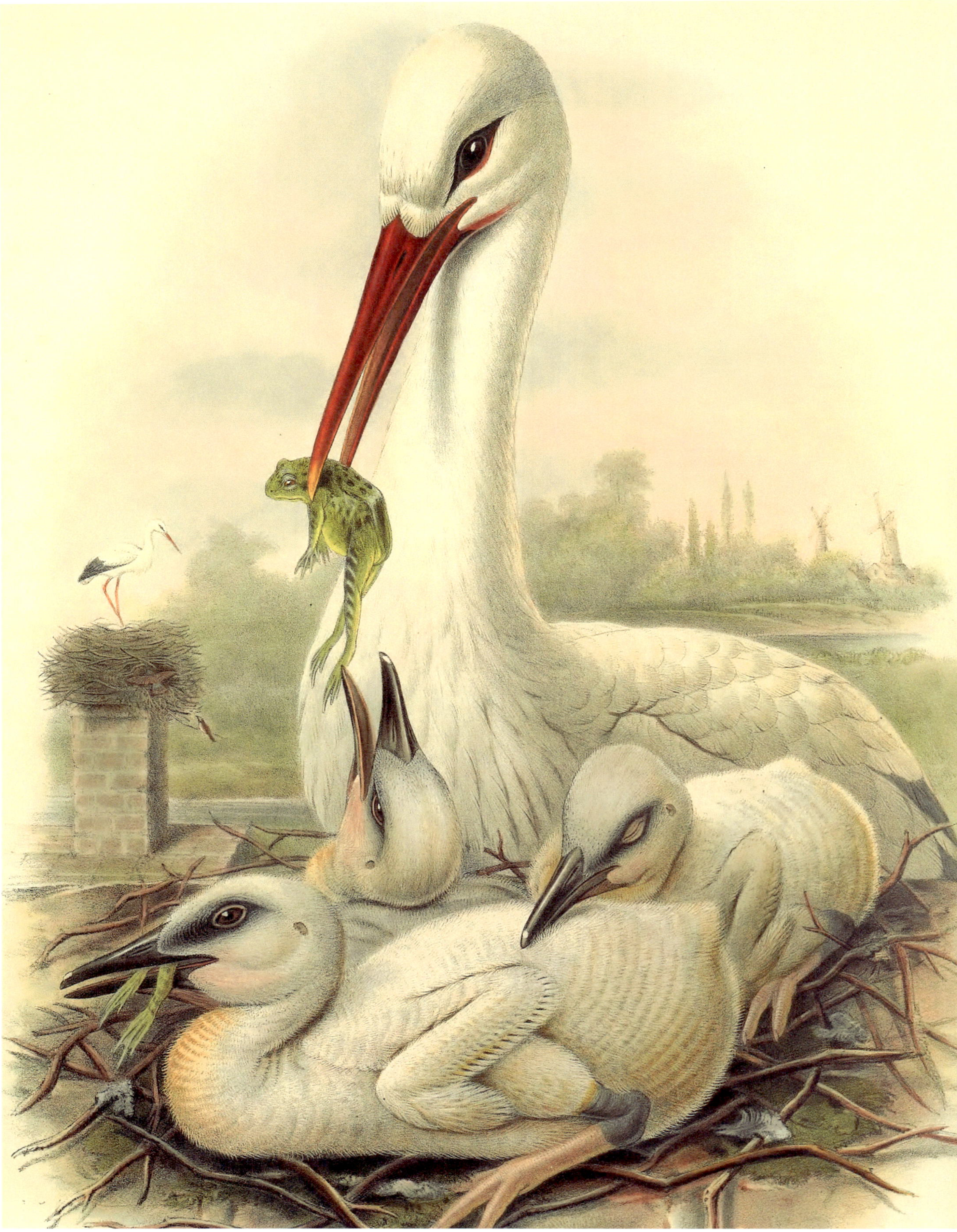

WHITE STORK
Ciconia ciconia

John Gould & Henry Constantine Richter, *The Birds of Great Britain*, 1862–73

For centuries, many cultures have associated the stork with birth. In European folklore, the stork is thought to deliver babies to their parents. This belief is said to have originated in Germany, where migratory storks would return to nest in Europe during springtime. Their return came nine months after the summer solstice. In Germany, this midsummer festival celebrated marriage and fertility, so babies born during the stork's return were likely conceived in June of the previous year.

WHITE STORK
Ciconia ciconia

Oluf August Hermansen,
Danske billeder, 1892

Storks are known for their parenting skills. Although they do not mate for life, they are monogamous breeders and both partners will work together to build large nests in trees, chimneys or on rooftops. Storks often nest close to human habitation, but industrialization, pest control and changes in farming have led to a decline in their species. This is an educational wall chart by Danish illustrator O. A. Hermansen. Published in 1892, the image comes from his series *Danske Billeder*. Today, the stork has completely died out in Denmark.

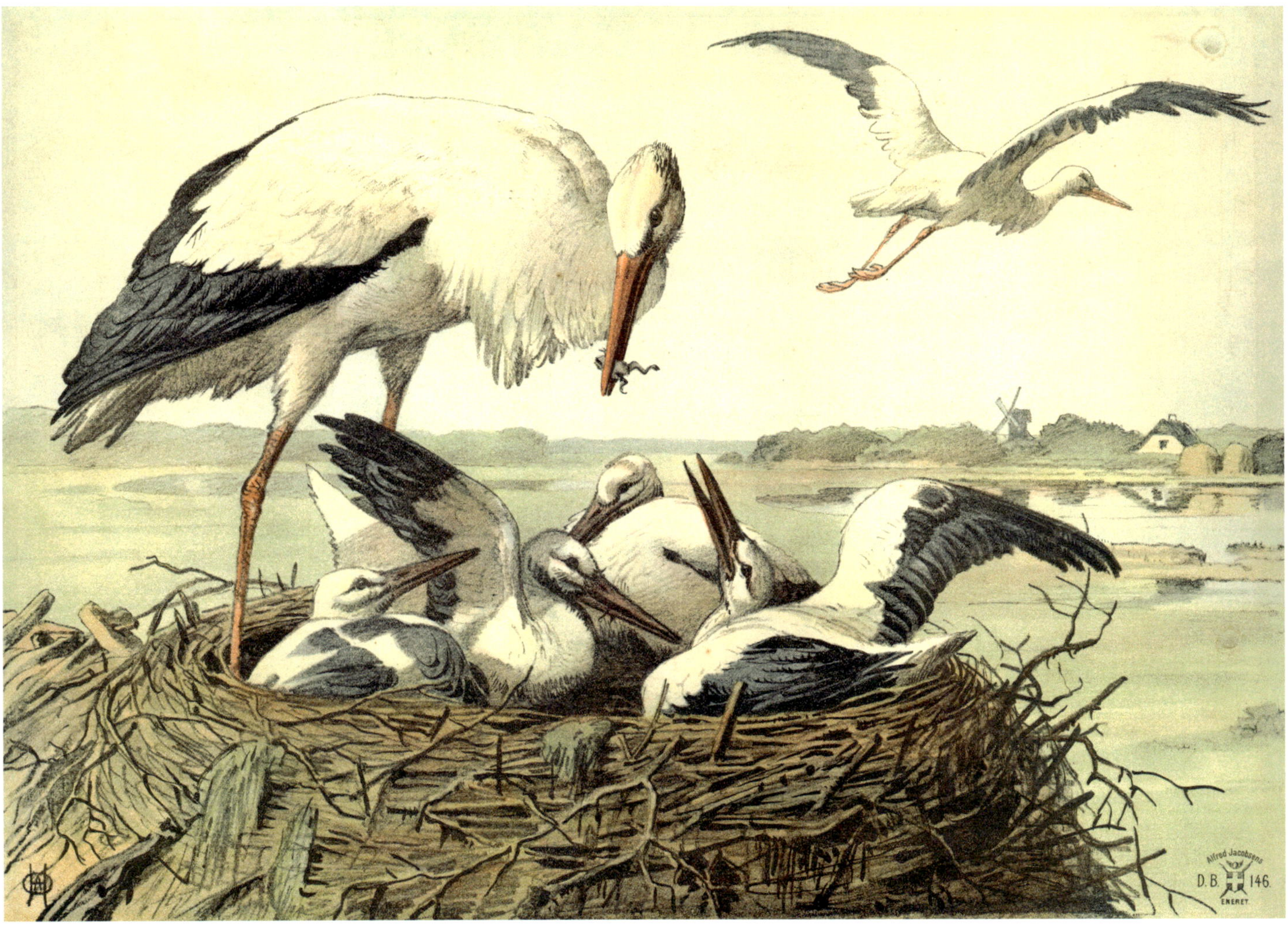

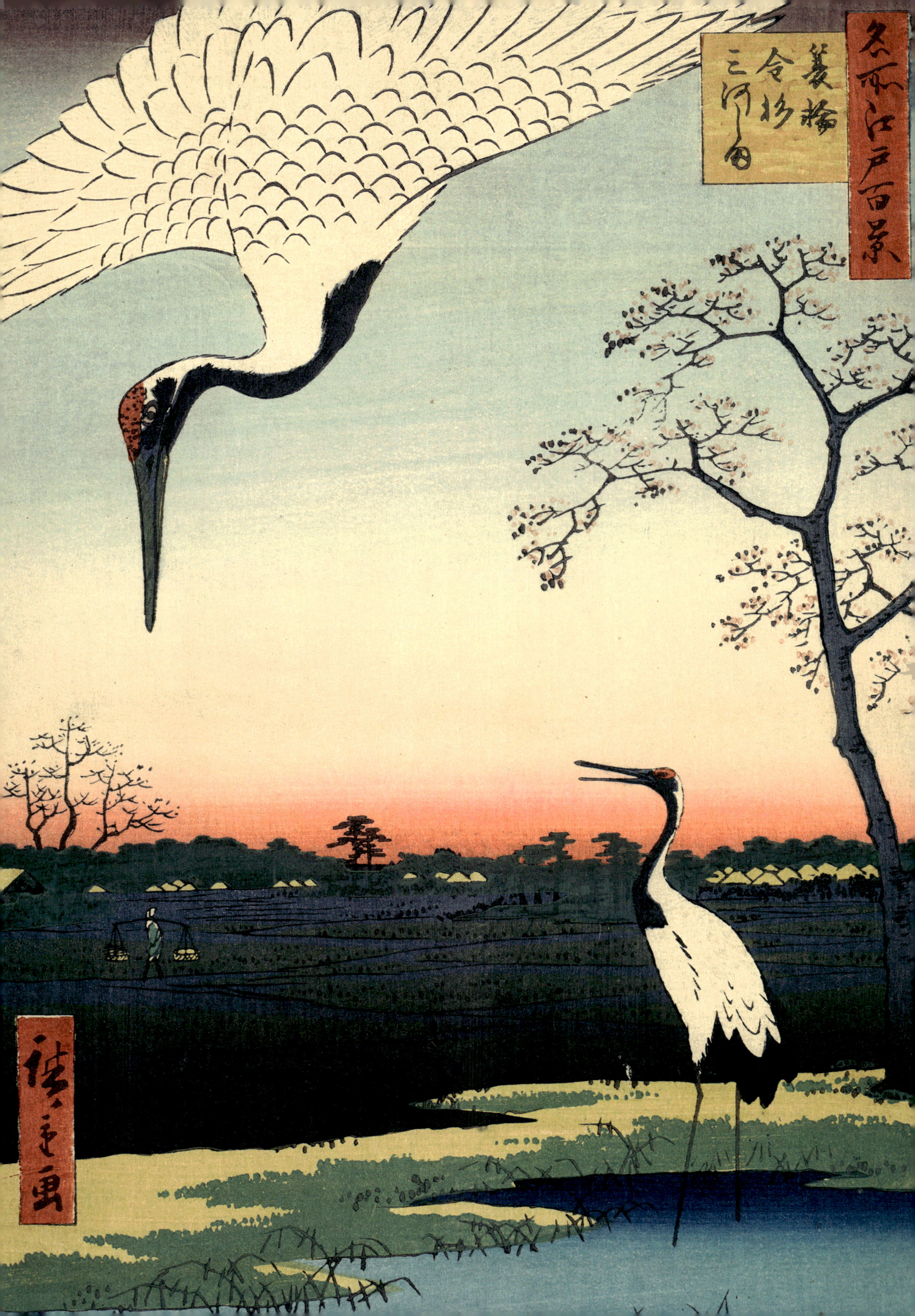
名所江戸百景
蓑輪金杉三河しま
廣重画

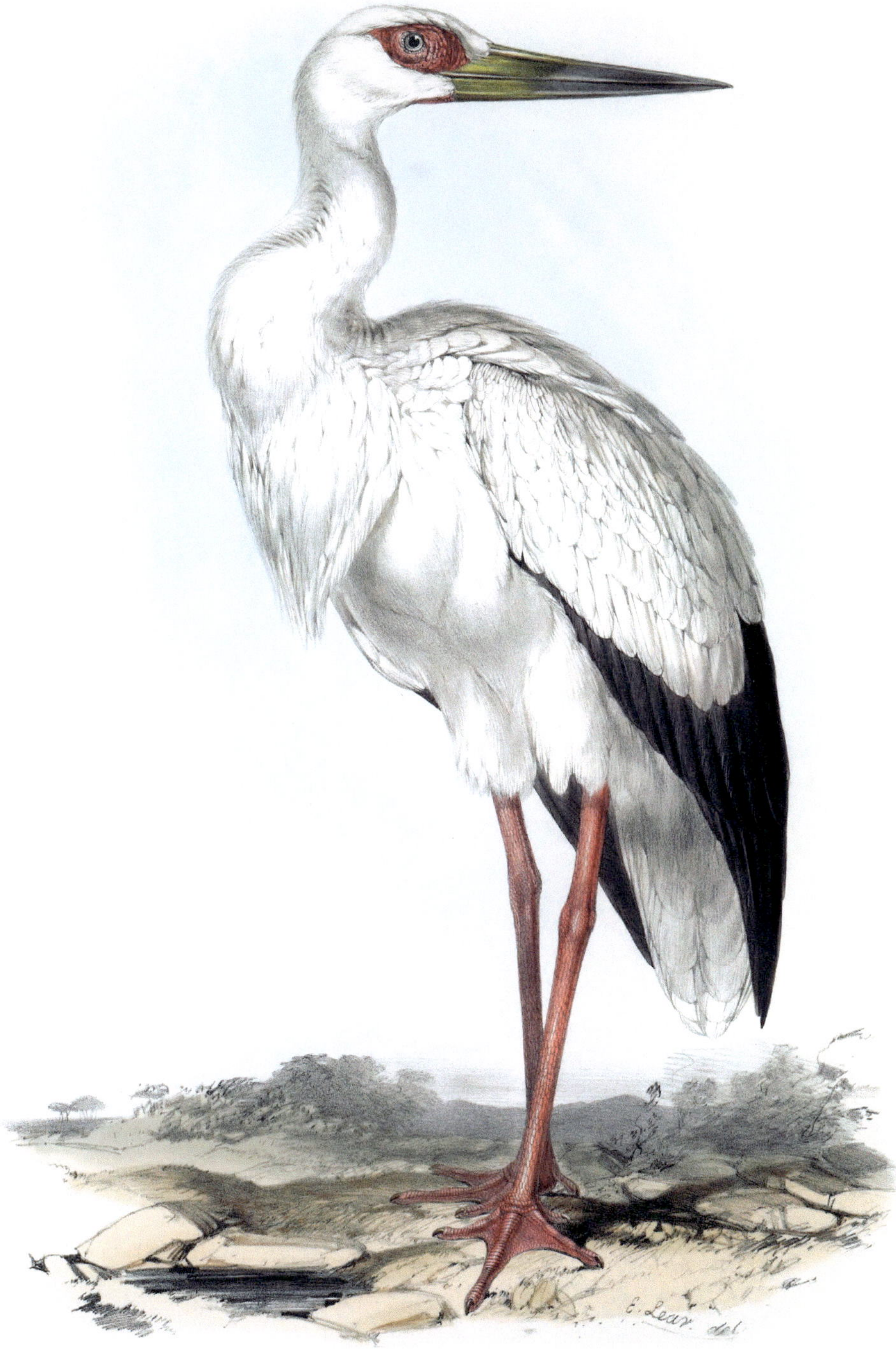

OPPOSITE:

RED-CROWNED CRANES
Grus japonensis

Utagawa Hiroshige, *Minowa, Kanasugi, Mikawashima*, No. 102 from *One Hundred Famous Views of Edo*, 1857

This print depicts two red-crowned cranes and comes from the series *One Hundred Famous Views of Edo* by Hiroshige. It depicts Mikawashima, an area northwest of Yoshiwara, where the shogun would hunt cranes with eagles and present these birds to the imperial court for a ceremonial meal.

MAGUARI STORK
Ciconia maguari

Edward Lear, *The Birds of Europe*, 1832–7

This maguari stork was illustrated by Edward Lear for Gould's *The Birds of Europe*. It is an interesting inclusion since it is a South American bird. It is similar in appearance to the white stork but is slightly larger, and its lores – the area between its bill and eyes – are distinctly red in colour.

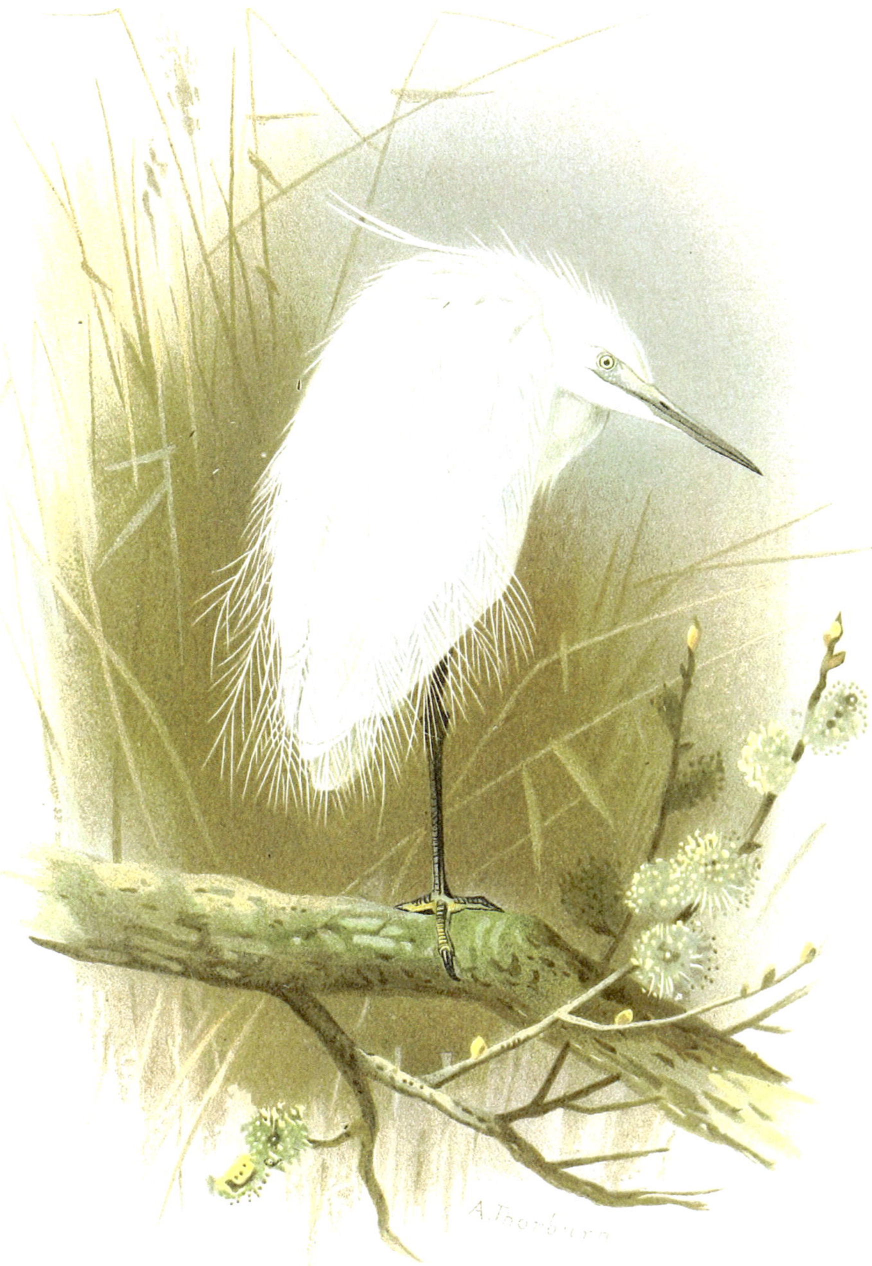

LITTLE EGRET
Egretta garzetta

Archibald Thorburn, *Coloured Figures of the Birds of the British Islands*, 1885–97

The little egret is a species of small heron. It appears here in an illustration from *Coloured Figures of the Birds of the British Islands* (1885–97). This book was written by British ornithologist and aristocrat Thomas Powys, 4th Baron Lilford. In 1858, Lilford was one of the eight founders of the British Ornithologists' Union.

OPPOSITE:
BLACK STORK
Ciconia nigra

Elizabeth Gwillim, watercolour, 1801

This black stork would have been observed by Elizabeth Gwillim during her time in India. The black stork has one of the widest geographic distributions of any bird in the stork family (*Ciconiidae*). During nesting season it can be found in East Asia and as far west as some of Europe's warmer parts. In autumn, it migrates south to Sub-Saharan Africa as well as in and around India and China.

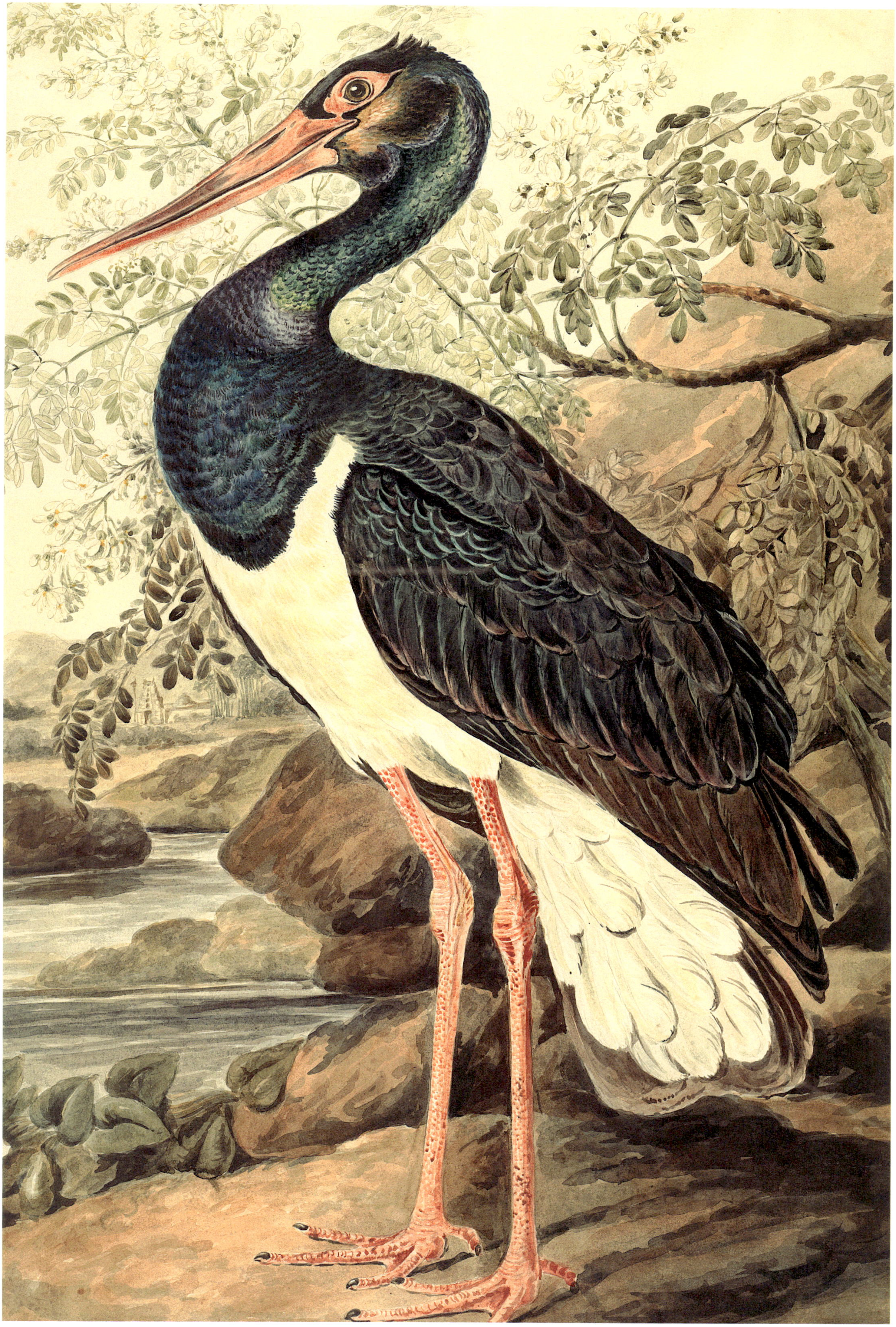

1. WOOD STORK
Mycteria americana

2. SCARLET IBIS
Eudocimus ruber

3. AMERICAN WHITE IBIS
Eudocimus albus

4. AMERICAN FLAMINGO
Phoenicopterus ruber

Alexander Wilson, *American Ornithology*, 1808–14

Alexander Wilson's *American Ornithology* catalogued and illustrated the birds of America almost two decades before Audubon's more celebrated *Birds of America*. In March of 1810, the pair met at Audubon's general store while Wilson was attempting to sell subscriptions for his book. It was there that he saw some of Audubon's work and encouraged him to have it published. Fourteen years later, Audubon travelled to Philadelphia to seek a publisher but was met with rejection and hostility. The city's Academy of Natural Sciences had had a strong connection with Wilson and their president, George Ord, even completed the final volumes of *American Ornithology* after Wilson's death in 1813. Ord felt that Audubon was actively trying to usurp Wilson and that his bird illustrations were too romantic and over-dramatic. Despite this rejection, it was in Philadelphia that Audubon met French ornithologist Charles Lucien Bonaparte, who suggested that he travel to Europe and have his illustrations engraved there.

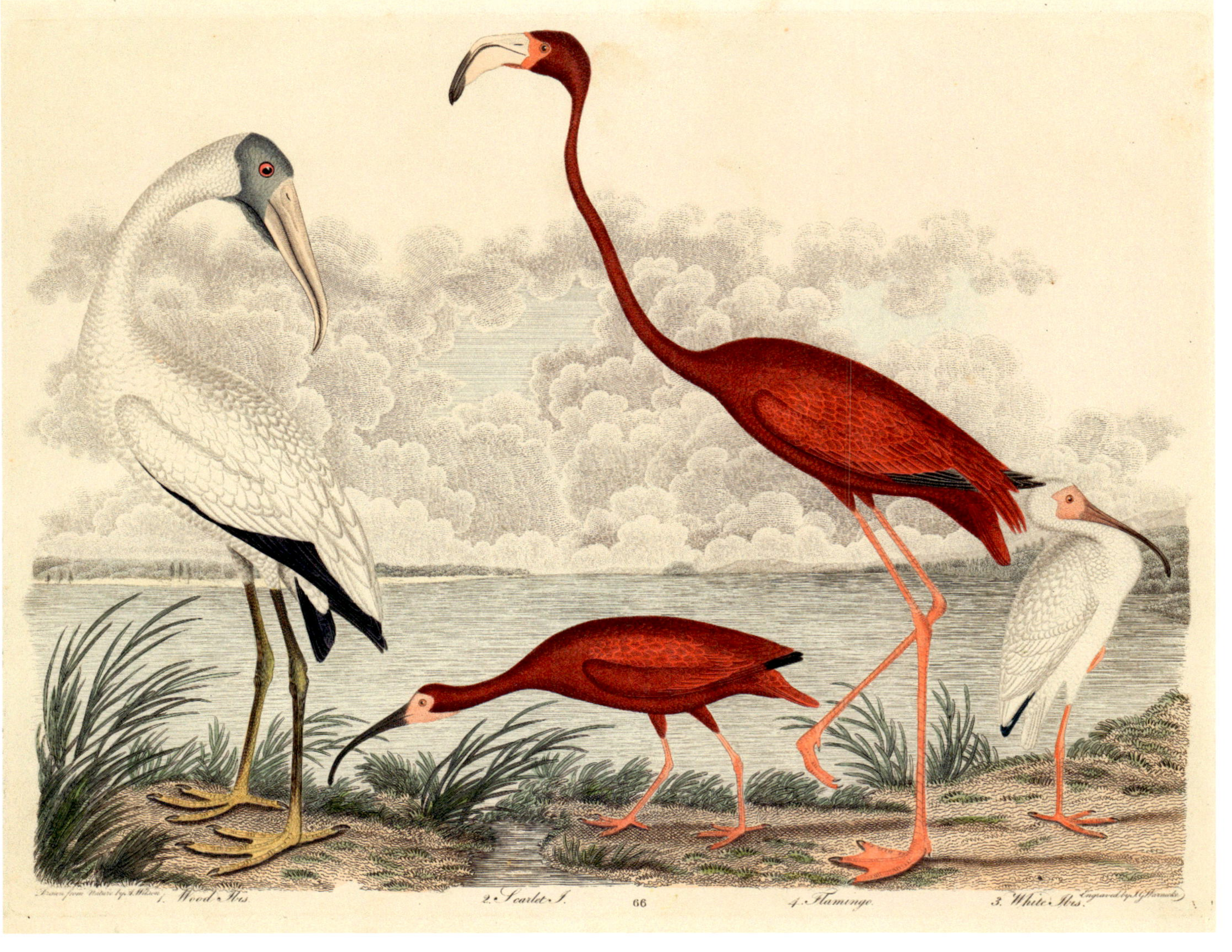

OPPOSITE:
AMERICAN FLAMINGO
Phoenicopterus ruber

John James Audubon, *The Birds of America*, 1827–38

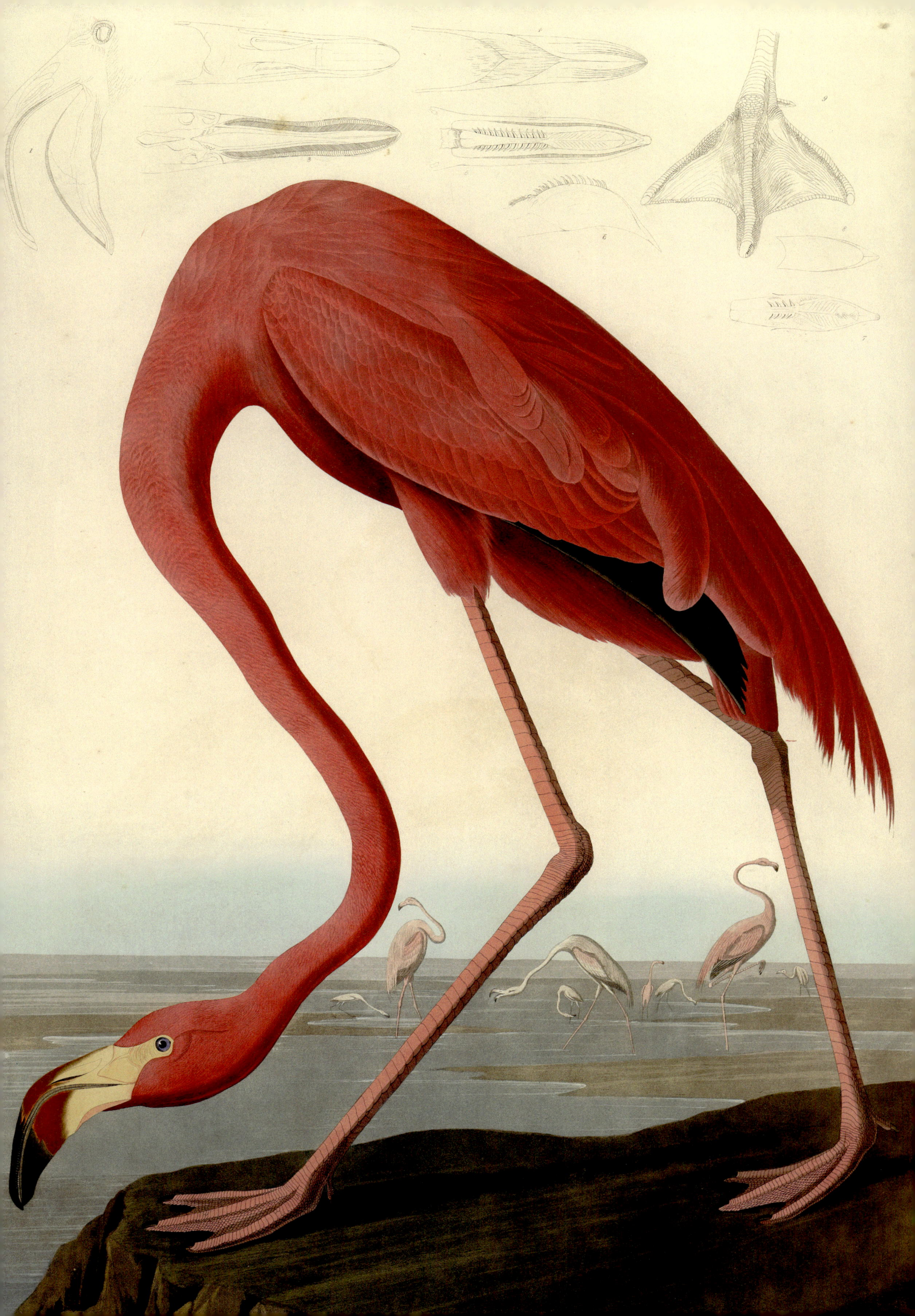

GREATER FLAMINGO
Phoenicopterus roseus

Marinus Adrianus Koekkoek II, *Ornithologia neerlandica*, 1922–35

The greater flamingo is the largest and most widespread species of the flamingo family (*Phoenicopteridae*). Their pink colouring comes from the carotenoid-rich algae and organisms that they eat. Levels of carotene can differ depending on where a bird feeds, so one can see a noticeable difference in colouration between birds. Flamingos of South America, for example, tend to be redder than those from North America.

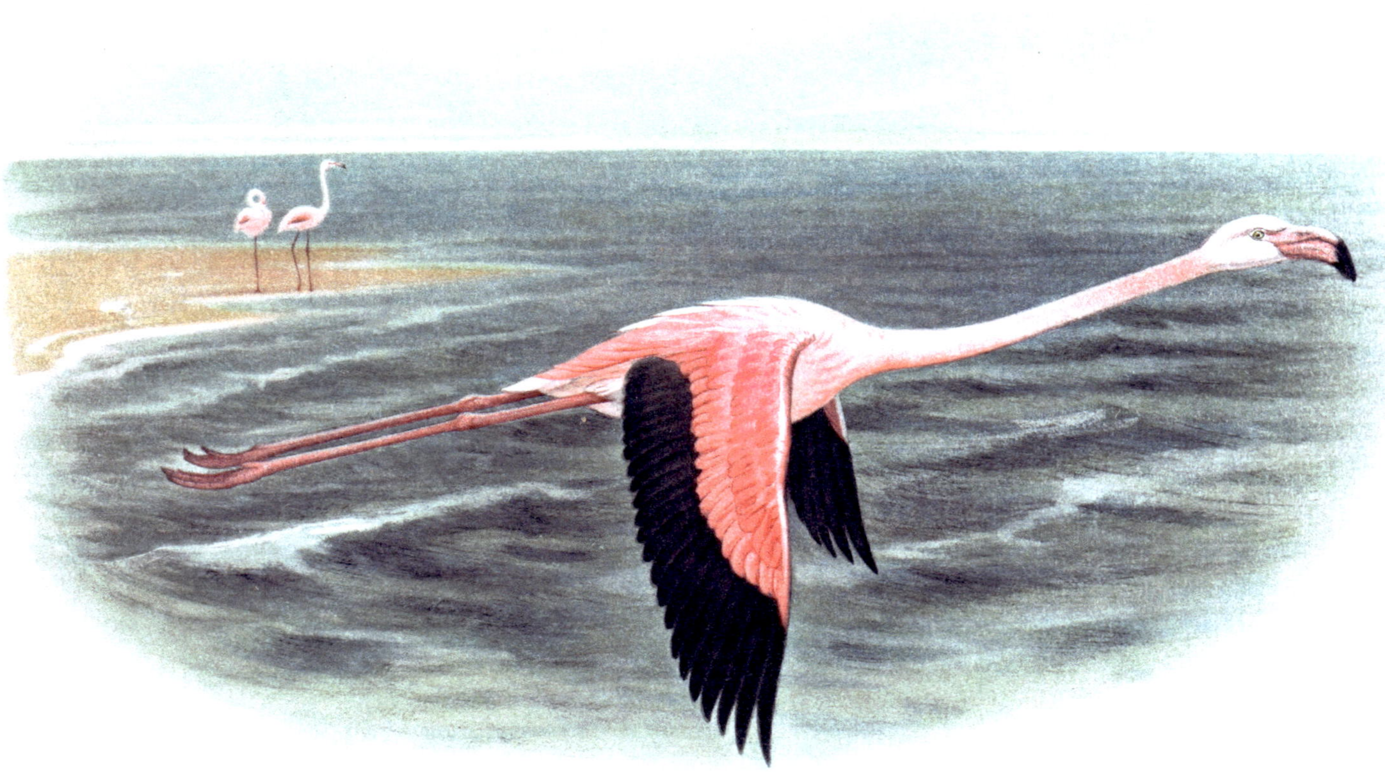

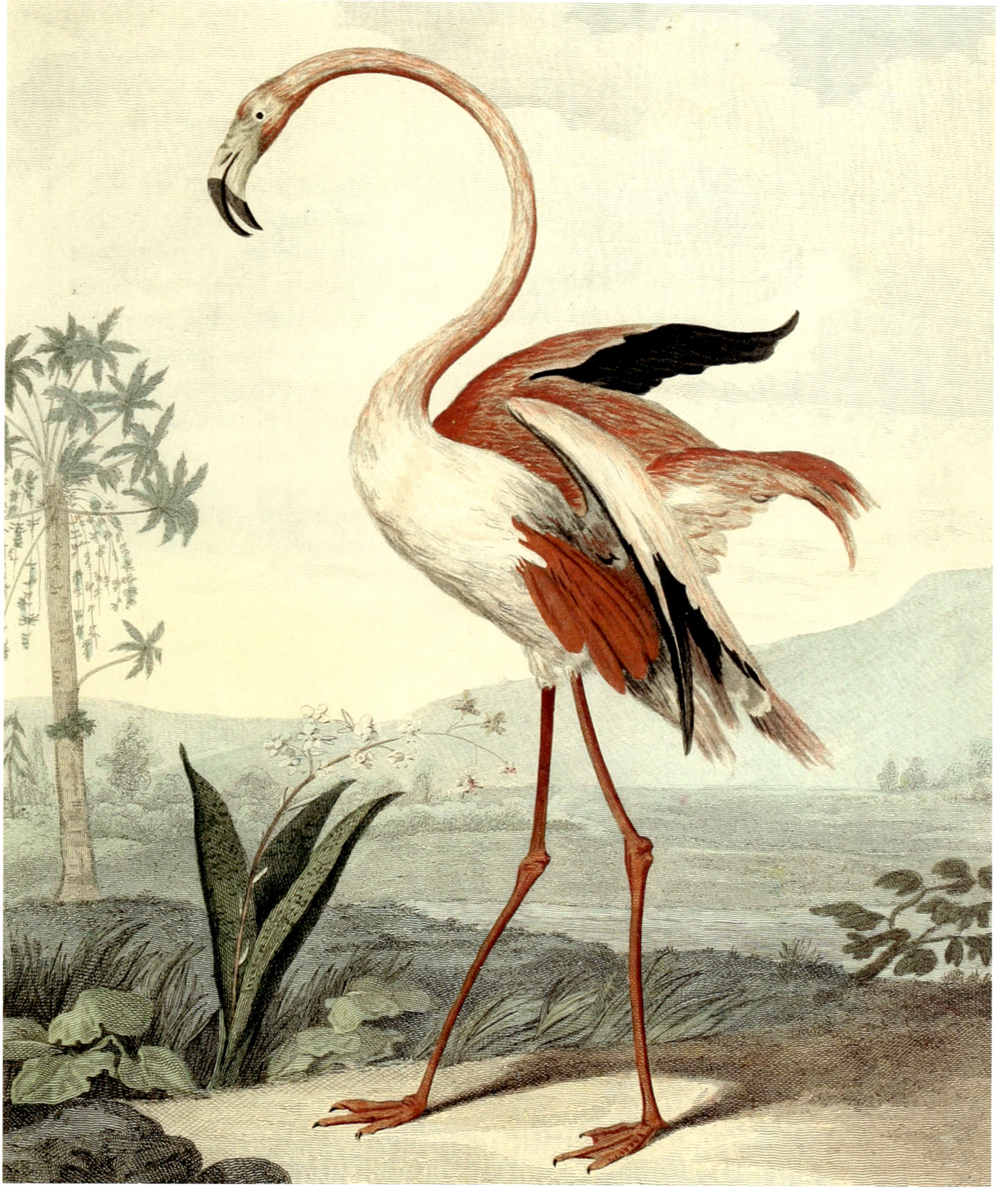

AMERICAN FLAMINGO
Phoenicopterus ruber

Charles Reuben Ryley,
Museum Leverianum, 1792–6

This flamingo comes from George Shaw's book *Museum Leverianum*. Published between 1792 and 1796, it contains several specimens from the museum of Sir Ashton Lever. An obsessive collector of artefacts, animals and specimens, Lever's was one of the largest private collections in the world. In 1771, he opened his first museum on the outskirts of Manchester. At the time, this was a radical idea. Over the years his collection continued to grow and later included many objects brought back from Captain Cook's voyages. His obsession with collecting led to his financial ruin and in 1806 he was forced to sell his entire cabinet of curiosities in a public auction that lasted sixty-five days.

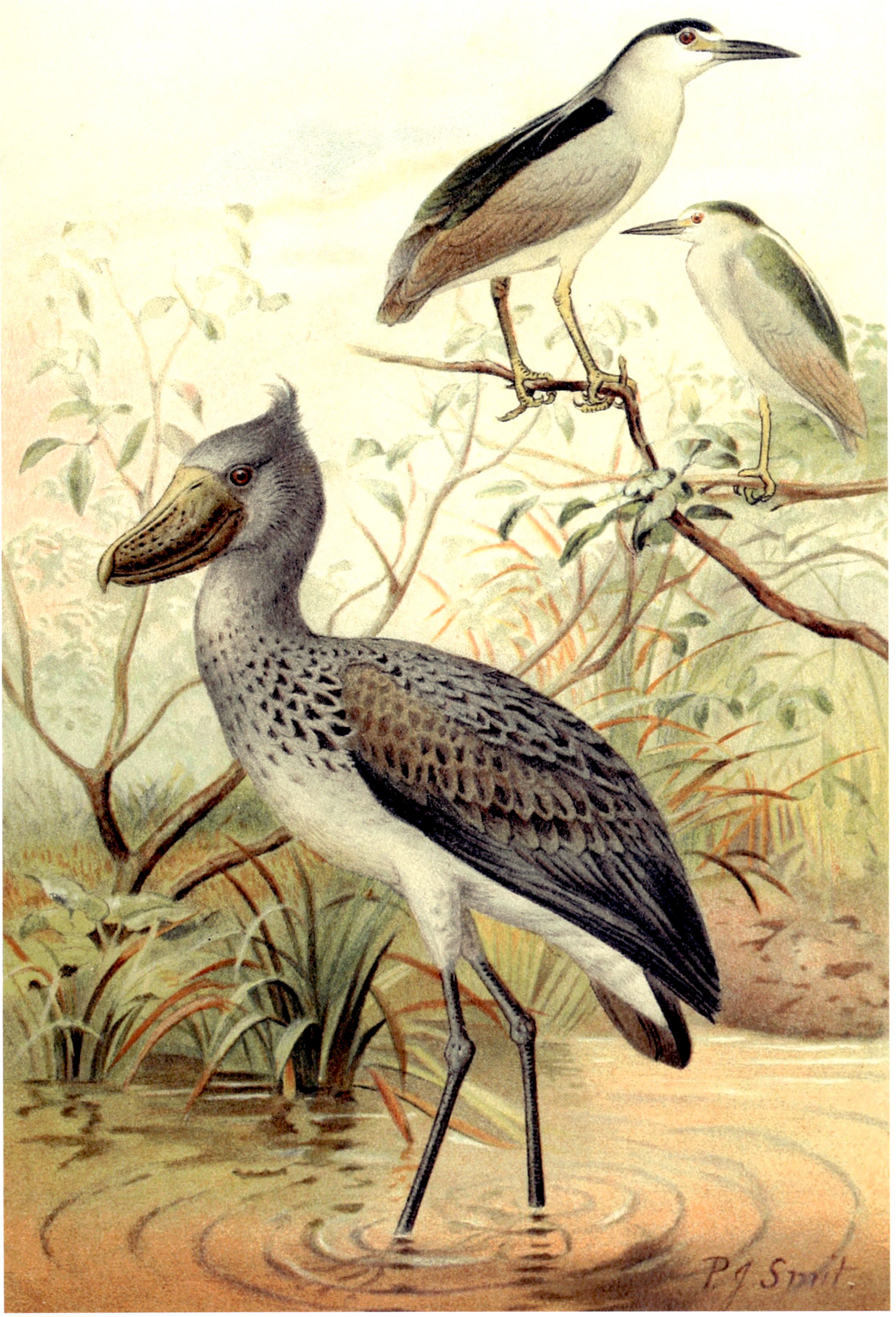
P. J. Smit

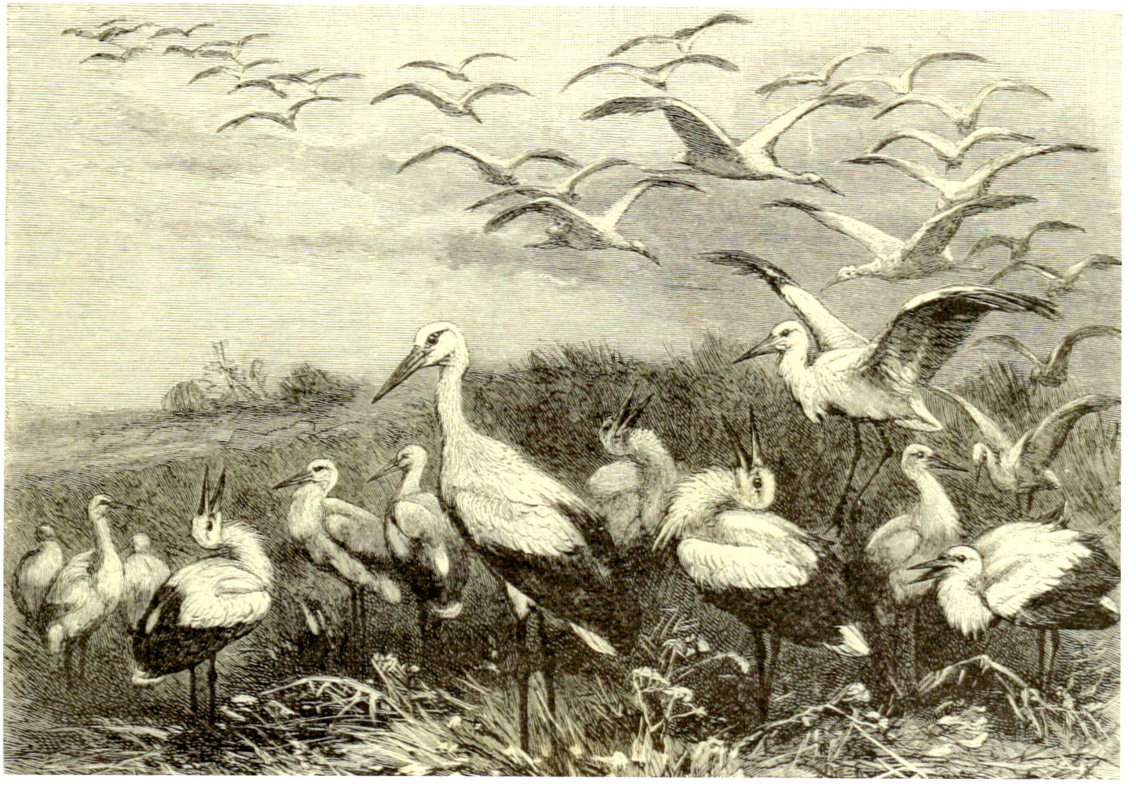

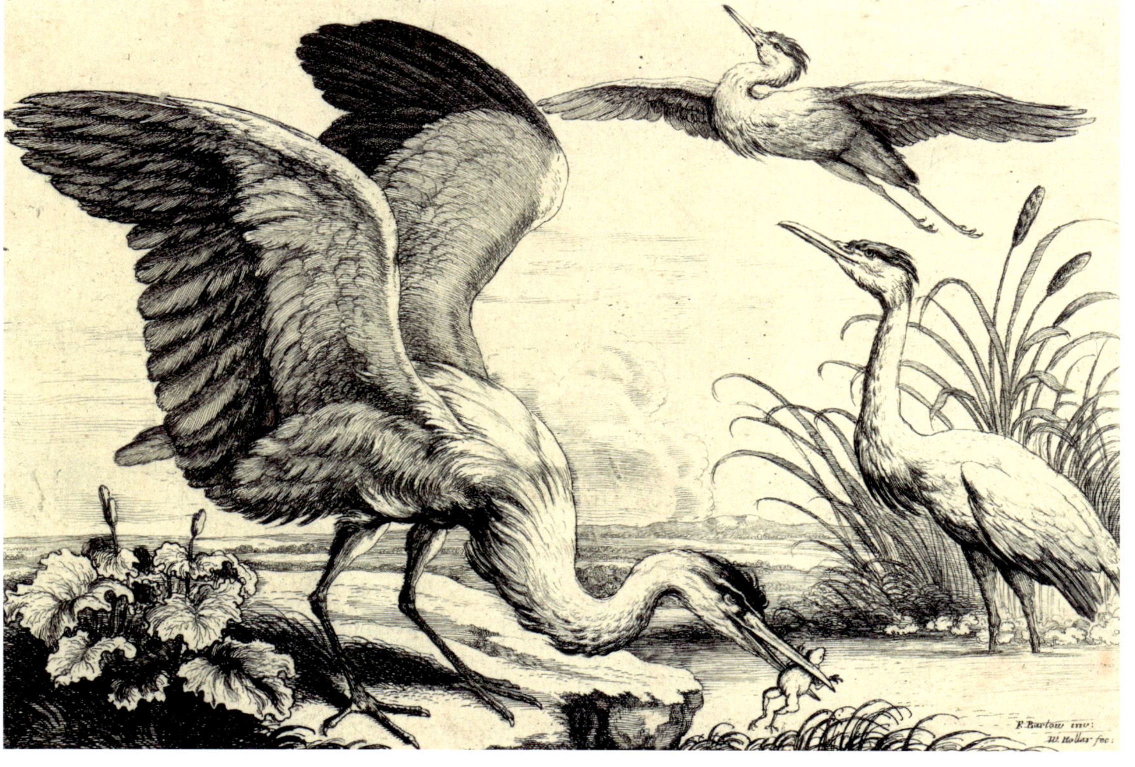

OPPOSITE:

BLACK-CROWNED NIGHT HERONS AND SHOEBILL

Nycticorax nycticorax and Balaeniceps rex

Pierre Jacques Smit, *The Royal Natural History*, 1893–6

The Royal Natural History consists of six volumes and covers mammals, birds, reptiles, fish and invertebrates. The larger bird seen here is a shoebill, an impressive stork-like bird that can only be found in eastern parts of Africa.

TOP:

STORKS ASSEMBLING FOR MIGRATION

The Royal Natural History, 1893–6

Storks assemble for migration in this unsigned etching from *The Royal Natural History*. In 1822, a stork was found near the town of Klütz in north eastern Germany that had been carrying an arrow in its neck all the way from central Africa. This specimen provided the first concrete evidence for long-distance bird migration.

BOTTOM:

THREE HERONS

Francis Barlow, etching, *ca.* 1654–62

Francis Barlow was one of the first professional illustrators and etchers in Britain and is often called the father of British sporting painting. This etching of herons was made after Barlow by the Bohemian etcher Wenceslaus Hollar, who was an accomplished artist in his own right.

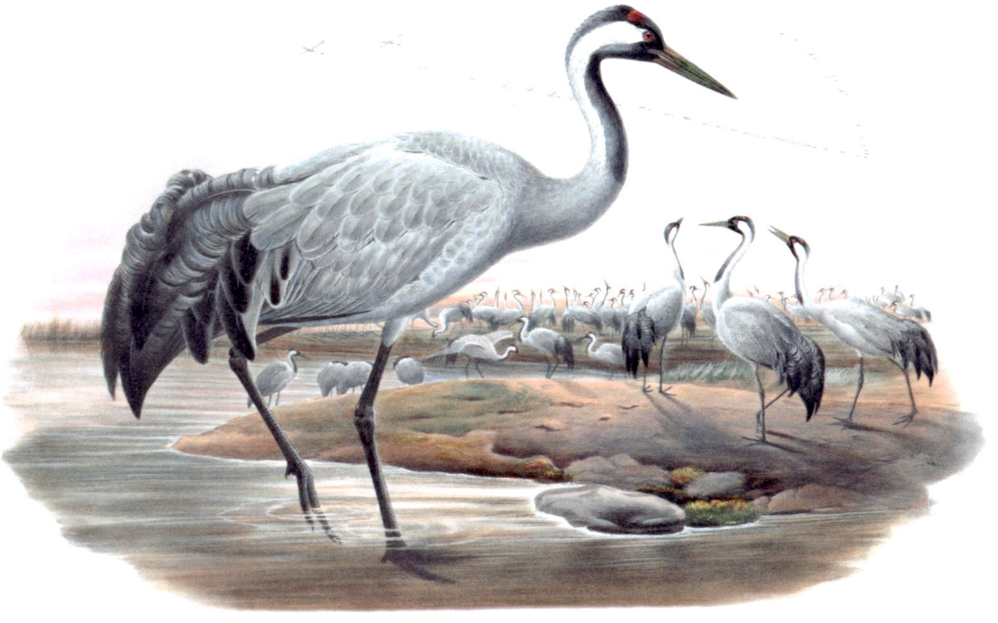

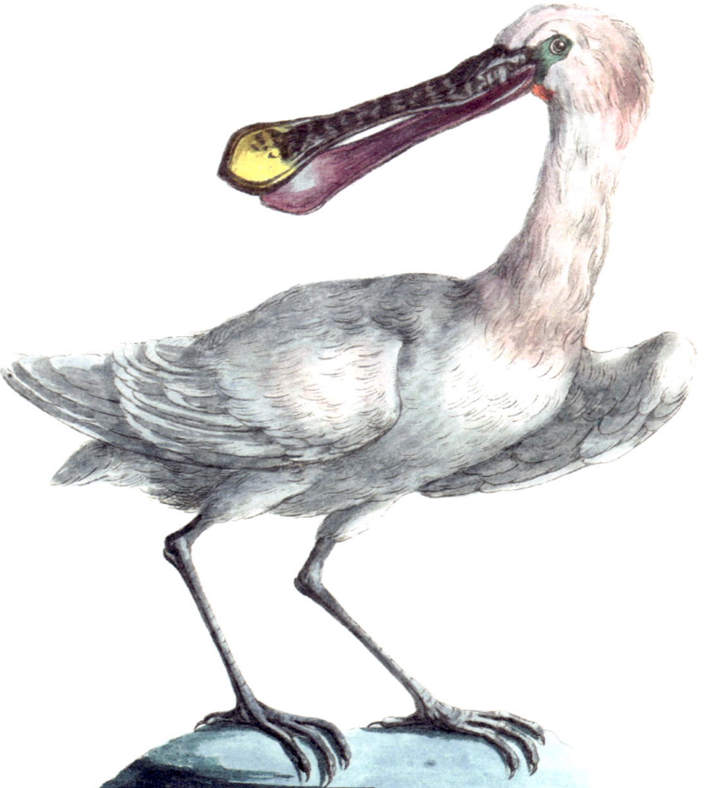

TOP:

COMMON CRANE
Grus grus

Joseph Wolf & William Matthew Hart, *The Birds of Great Britain*, 1862–73

Although this plate appears in John Gould's *The Birds of Great Britain*, the common crane is not a common sight on the island. Indeed, Gould himself is quick to acknowledge this in his writing, highlighting that the crane had once nested in Britain but now appears only as a rare sighting. Crane populations were once widespread, but in the 1600s they became extinct through over-hunting and the destruction of wetlands. In 1979 a small group of migrant cranes established themselves in the Norfolk Broads, leading to a conservation effort that has now ensured that the crane remains a bird of Great Britain.

ROSEATE SPOONBILL
Platalea ajaja

Lorenzo Lorenzi & Violante Vanni, *Ornithologia methodice digesta*, 1767–6

Saverio Manetti's text identifies this bird as a pink-backed pelican (*Pelecanus rufescens*), but its distinctive flat bill suggests that it may, in fact, be a roseate spoonbill (*Platalea ajaja*). The 600 illustrations in Manetti's *Ornithologia methodice digesta* were all based on the birds in the collection of the Marquis Giovanni Gerini of Florence, who sponsored the publication of the book.

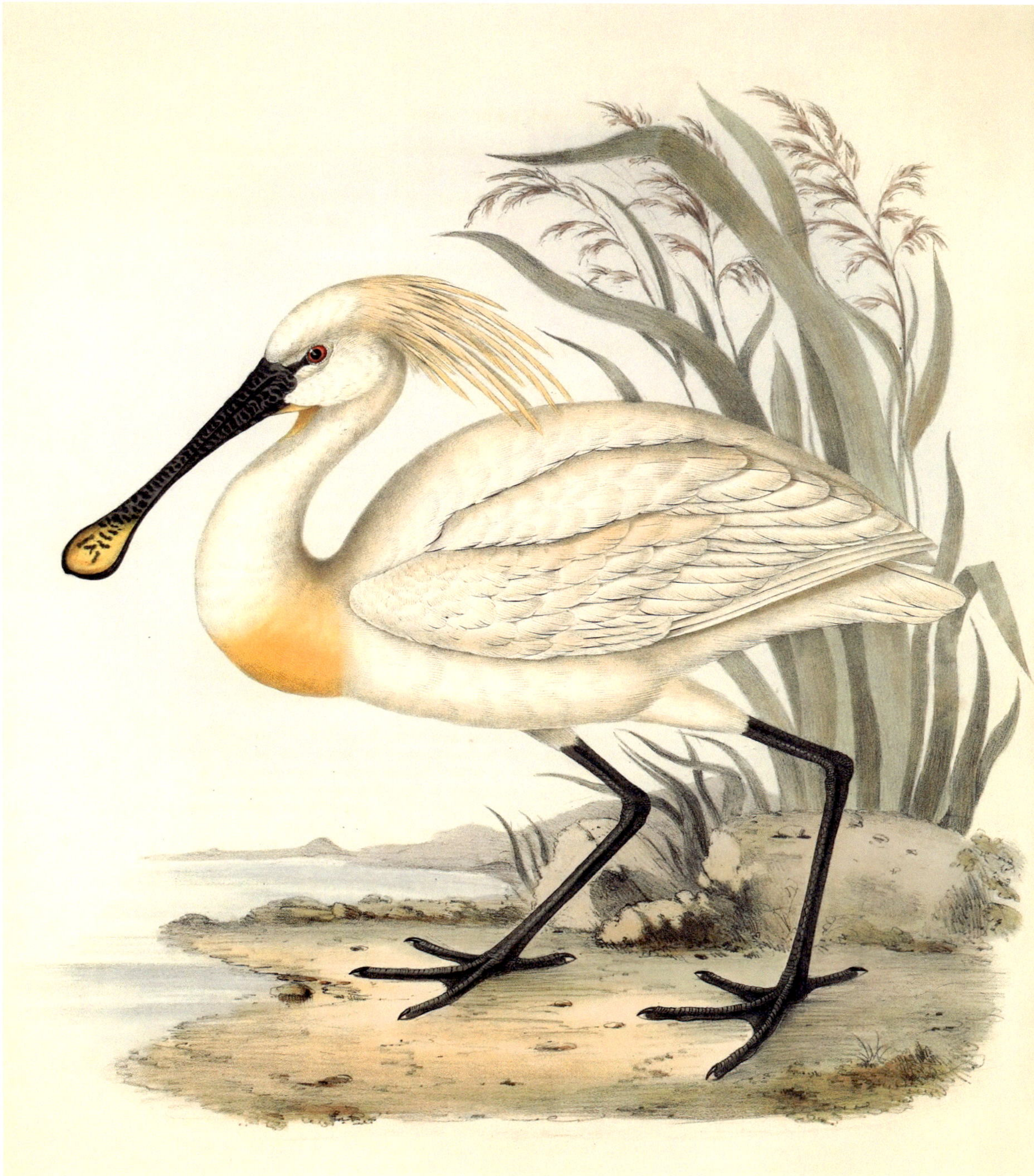

EURASIAN SPOONBILL
Platalea leucorodia

John Gould & Elizabeth Gould, *The Birds of Europe*, 1832–7

This fetching spoonbill appears in John Gould's *Birds of Europe*. Spanning five volumes, the work consists of 448 hand-coloured lithographs – the majority of which were drawn and lithographed by Elizabeth Gould based on sketches by her husband. Many of the foregrounds in the book were produced by Edward Lear, who also contributed 67 plates to the publication.

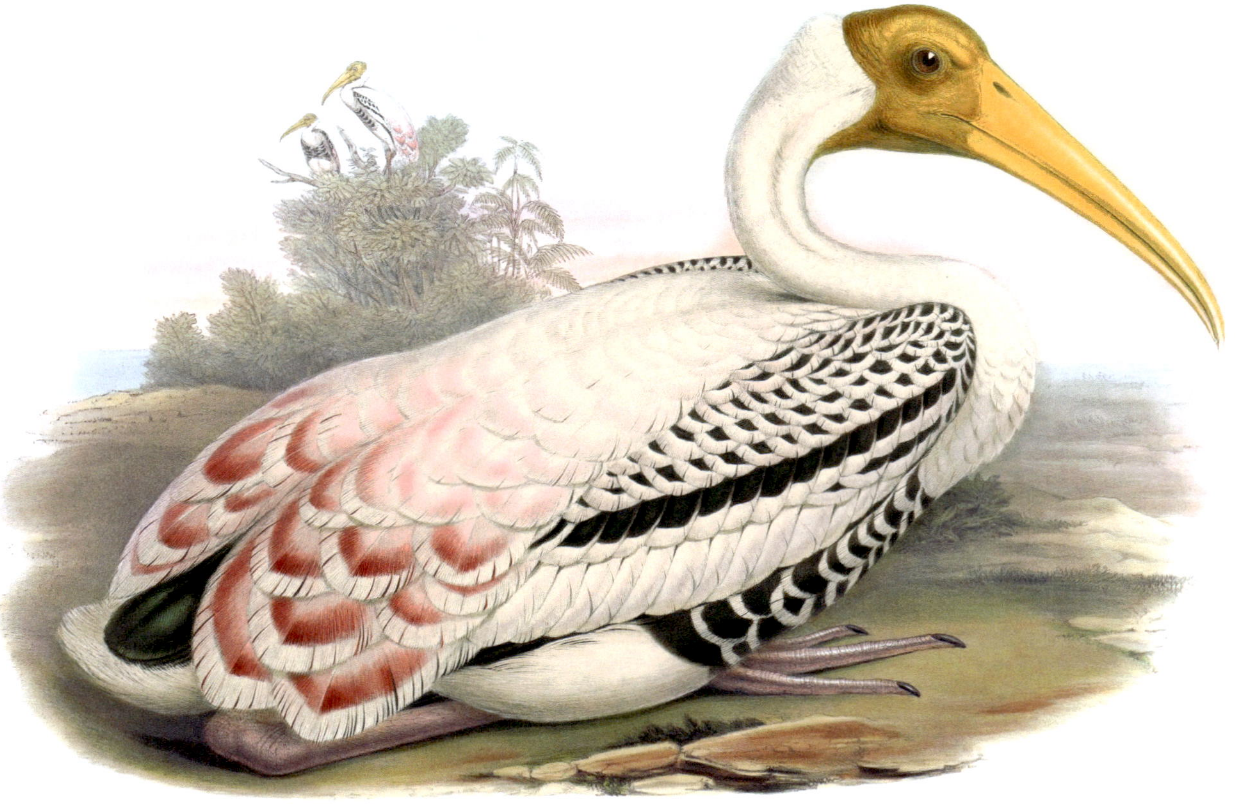

PAINTED STORK
Mycteria leucocephala

John Gould & Henry Constantine Richter, *The Birds of Asia*, 1850–83

The pink tertial feathers of this large stork look almost as though they have been added with a brush, earning the bird its name – the painted stork. Found predominantly in India, it can be seen in the wetlands south of the Himalayas as well as in many parts of Southeast Asia.

OPPOSITE:
LORD HOWE SWAMPHEN
Porphyrio albus

Sarah Stone, *A Journal of a Voyage to New South Wales*, 1790

Sarah Stone was a natural history illustrator commissioned by Sir Ashton Lever to document the objects in his collection at the Leverian Museum. She later created a series of watercolours based on the specimens sent from Australia by the Irish surgeon and botanical collector John White. These became copper plate engravings when White published his account as *A Journal of a Voyage to New South Wales* in 1790. The Lord Howe swamphen is now extinct.

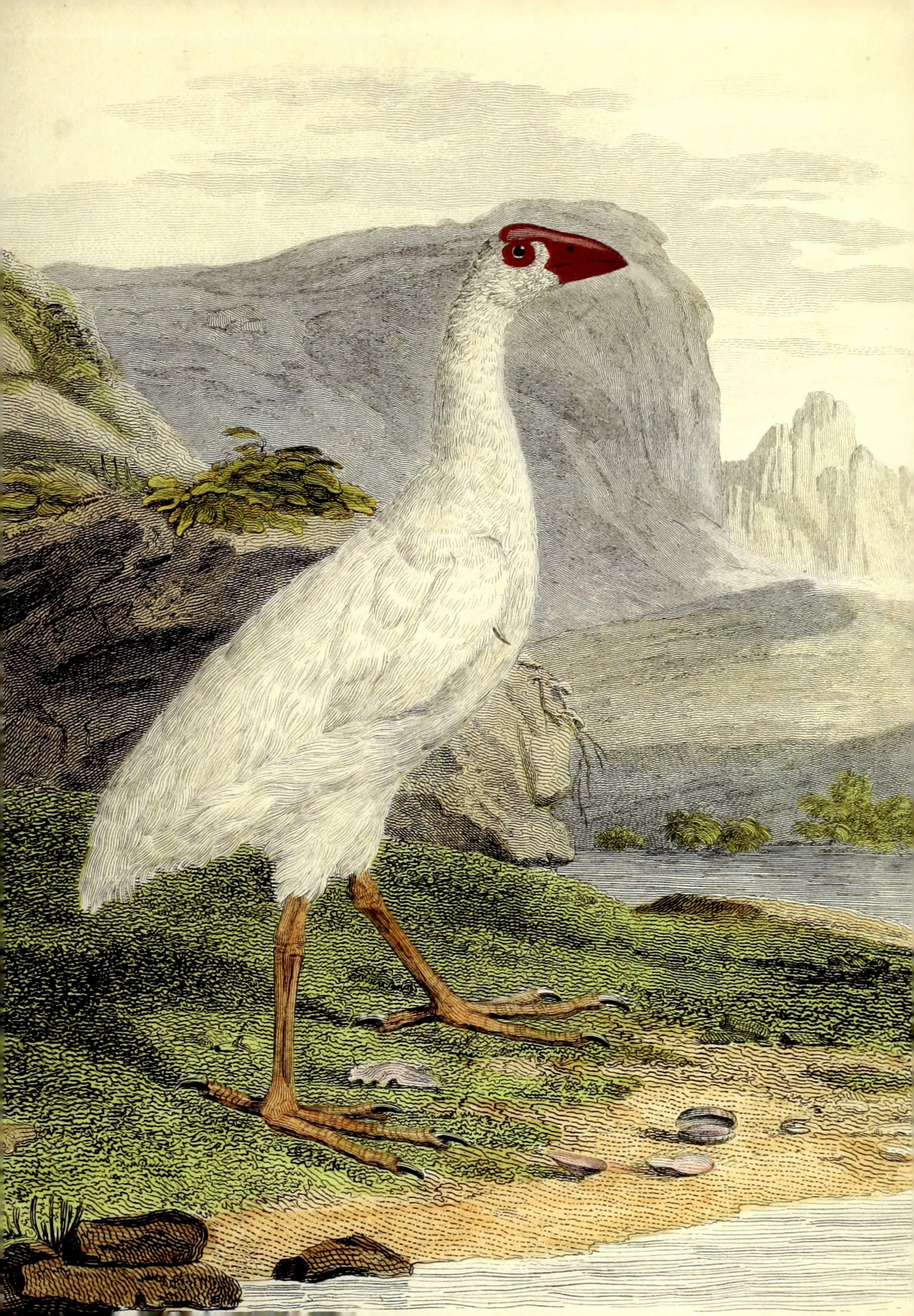

AUSTRALASIAN SWAMPHENS
Porphyrio melanotus

John Gould & Henry Constantine Richter, *The Birds of Australia*, 1840–8

These swamphens have red frontal shields on their heads, a feature that is common in several bird species. These shields can be used when birds get territorial, with males occasionally biting and kicking one another, particularly during the breeding season.

Australasian swamphens also have impressive plumage, and their heads, breast and throat typically are a deep blue. This is uncommon in wetland species which tend to have dull plumage to aid in camouflaging against wetland vegetation.

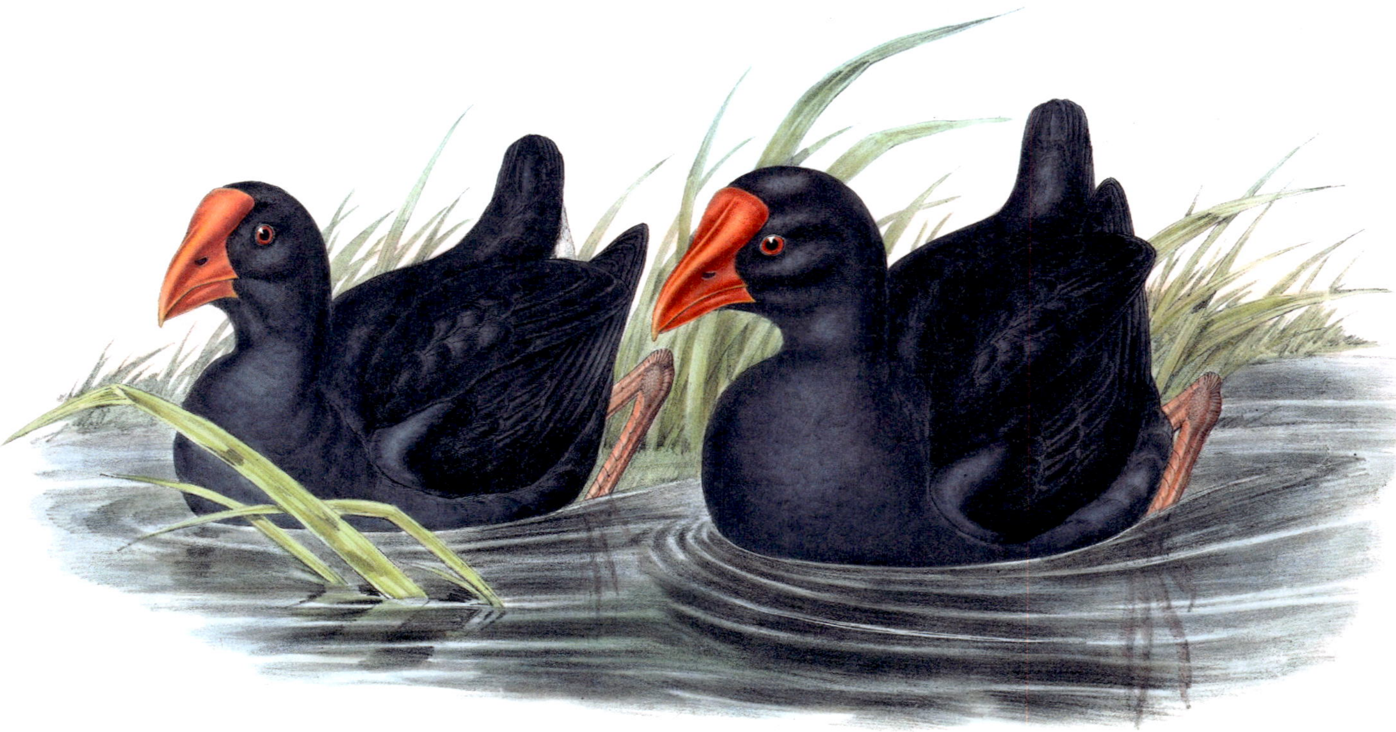

OPPOSITE:
WESTERN SWAMPHEN
Porphyrio porphyrio

Nicolas Huet II, *Nouveau recueil de planches coloriées d'oiseaux*, 1820–39

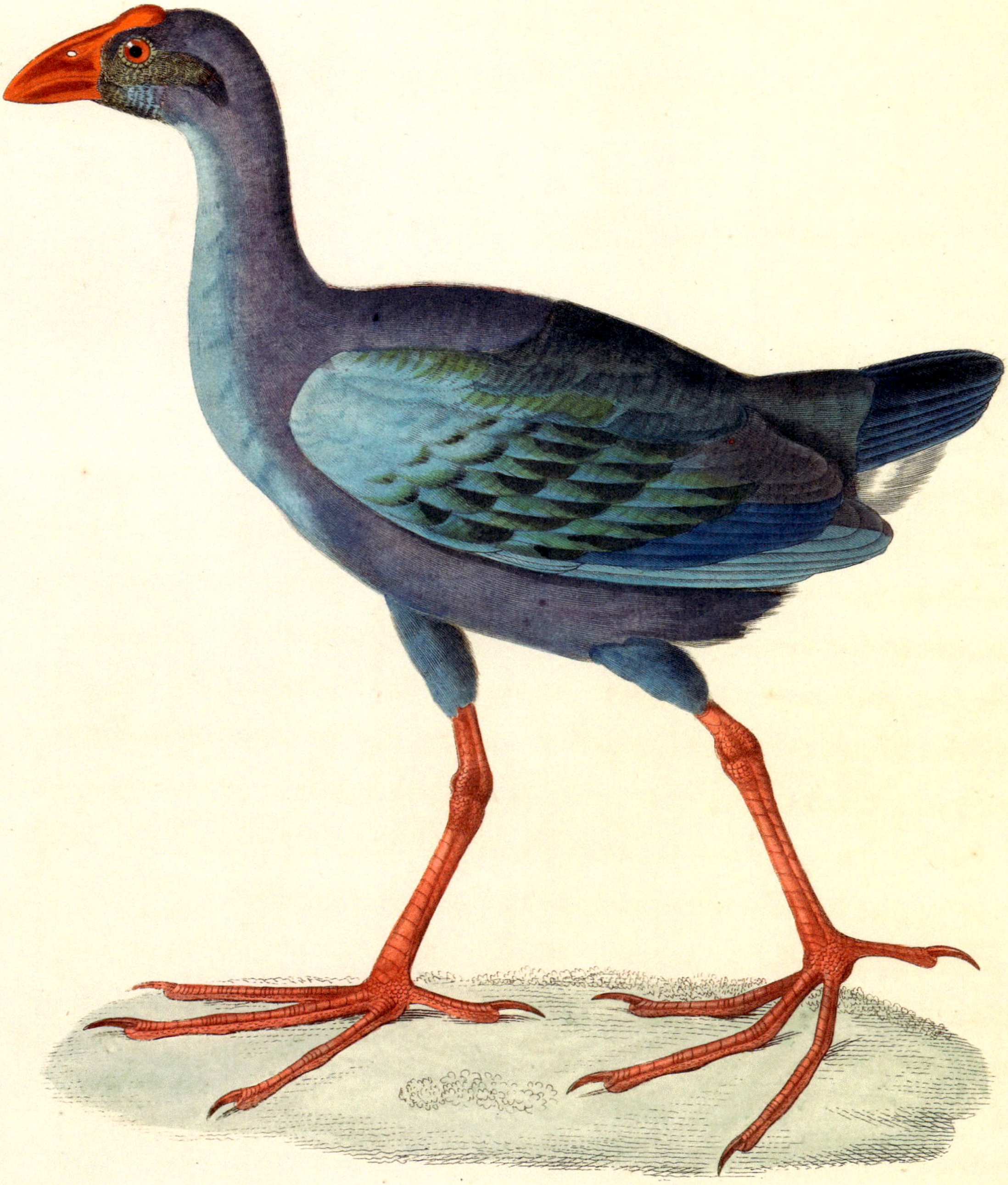

Talève *émeraudin*, *adulte*.

Eleazar & Elizabeth Albin

fl. 1708–*ca.* 1742; 1708–1741

Authors of A *Natural History of Birds* – one of the earliest illustrated bird books – this father and daughter team helped pioneer the use of colour illustrations and the publication of zoological books for the general reader.

Eleazar Albin was an artist and art teacher who produced a number of books during the eighteenth century. Little is known of his early life, although it is believed that he was born in Germany to a family named Weis. By 1708 he had moved to London and had changed his name to Albin. There, he lived in Piccadilly, got married, had a family and taught watercolour painting. He earned much of his living by producing watercolour drawings of natural history objects for collectors such as Sir Hans Sloane, the wealthy physician whose collection formed the basis of the Natural History Museum, the British Museum, and the British Library.

Albin was a friend of Mark Catesby and famously declined the invitation to travel to America that Catesby eventually accepted. Both men's outputs are similar in that they cover more than just ornithology. As a writer and illustrator, Albin's works include *A Natural History of English Insects* (1720), *A Natural History of Spiders, and Other Curious Insects* (1736), *A Natural History of English Songbirds* (1737), and *The History of Esculent Fish* (1794).

His most celebrated work, though, is the highly prized *A Natural History of Birds*. Published from 1731 and 1738, it spanned three volumes and was released at approximately the same time as Catesby's account of the New World (1731–43). Noted for being one of the earliest illustrated bird books, the work consists of over 300 copper plates which are all individually coloured by hand using watercolour. In the introduction to the book, Albin notes that the illustrations were produced by his own hand as well as that of his daughter, Elizabeth Albin, whom, he notes, he taught to draw and paint. Elizabeth's contribution to the work is quite sizable, and while she clearly was not the first women artist to have drawn birds, she does hold the distinction for being the first known recorded woman to have illustrated a bird book.

While the illustrations in *A Natural History of Birds* hold some charm, it would be remiss not to note that the birds' stiff poses feel somewhat lifeless. This is partly down to the challenges of engraving into copper plates, a somewhat rigid process unlike painting or drawing. The resulting work has a naivety to it that perhaps feels even more pronounced today because of our knowledge of the work that followed. Yet the Albins deserve credit for their use of colour. Not only does it enhance the engravings aesthetically, but it also provides vital information for those eager to identify a species. The importance of colour may seem glaringly self-evident today, yet in the early eighteenth century this addition to the work was a breakthrough. Readers were given the option to purchase books either with the original black-and-white plates or pay an additional fee to have these plates hand coloured.

Like many works of the era, *A Natural History of Birds* was sold by subscription and the list of subscribers at the start of each volume demonstrates that this was a popular work. With almost 100 names listed, these entries provide a fascinating overview of the readership for such a title. Even a cursory look presents us with names of prominent lords, ladies, dukes and countesses. There are also mentions of chemists, booksellers, engravers and slave traders. Albin lists one John Brindley – 'bookbinder to her Majesty' and includes a note of 'six books' beside his name. 'Amsterdam' is written beside the name of famed Dutch zoologist and collector Albertus Seba (1665–1736), while Albin's former employer, Sir Hans Sloane, also appears as a subscriber. In addition to subscribers, such projects would also typically have a patron. In the case of this work, the first two volumes were printed under the patronage of Tory politician Sir Robert Abdy (1688–1748), while the third was supported by Richard Mead (1673–1754), physician to the king.

Subscribers and patrons were important for funding such ambitious work, yet creators would also need connections to people with large collections of rare and unusual birds. Many of the birds featured in *A Natural History of Birds* come from the collections of landowner Sir Thomas Lowther and silk-pattern designer Joseph Dandridge. Albin also sought out specimens himself, visiting markets, merchants, aviaries and apothecaries. In the first volume of his book, he even makes a request for birds. In the introduction he writes: 'I shall be very thankful to any Gentleman that will be pleased to send any curious birds to Eleazar Albin near the Dog and Duck in Tottenham-Court Road'.

Text was also sourced for this work, with notes and observations provided by the clergyman and naturalist William Derham (1657–1735). For the descriptions, he notes 'I have done those with all the accuracy I could from the very birds themselves, which I had always by me at that time; and particularly I took care to describe their characteristic notes and specific differences; and to impose on them the most received and proper names both English and Latin in all which I had the assistance of the best ornithologists, particularly of the justly famous Mr. Willougby's *Ornithology* and Mr. Ray's Improvements.'

The work of Albin helped to establish a tradition for illustrated zoological books in England, and specifically books that targeted the general reader. His work helped to create a growing interest in ornithology, and this led to a growing demand for reprints of his work as well as new books from other authors.

GREATER PAINTED-SNIPE
Rostratula benghalensis

Eleazar Albin & Elizabeth Albin, *A Natural History of Birds*, 1731–8

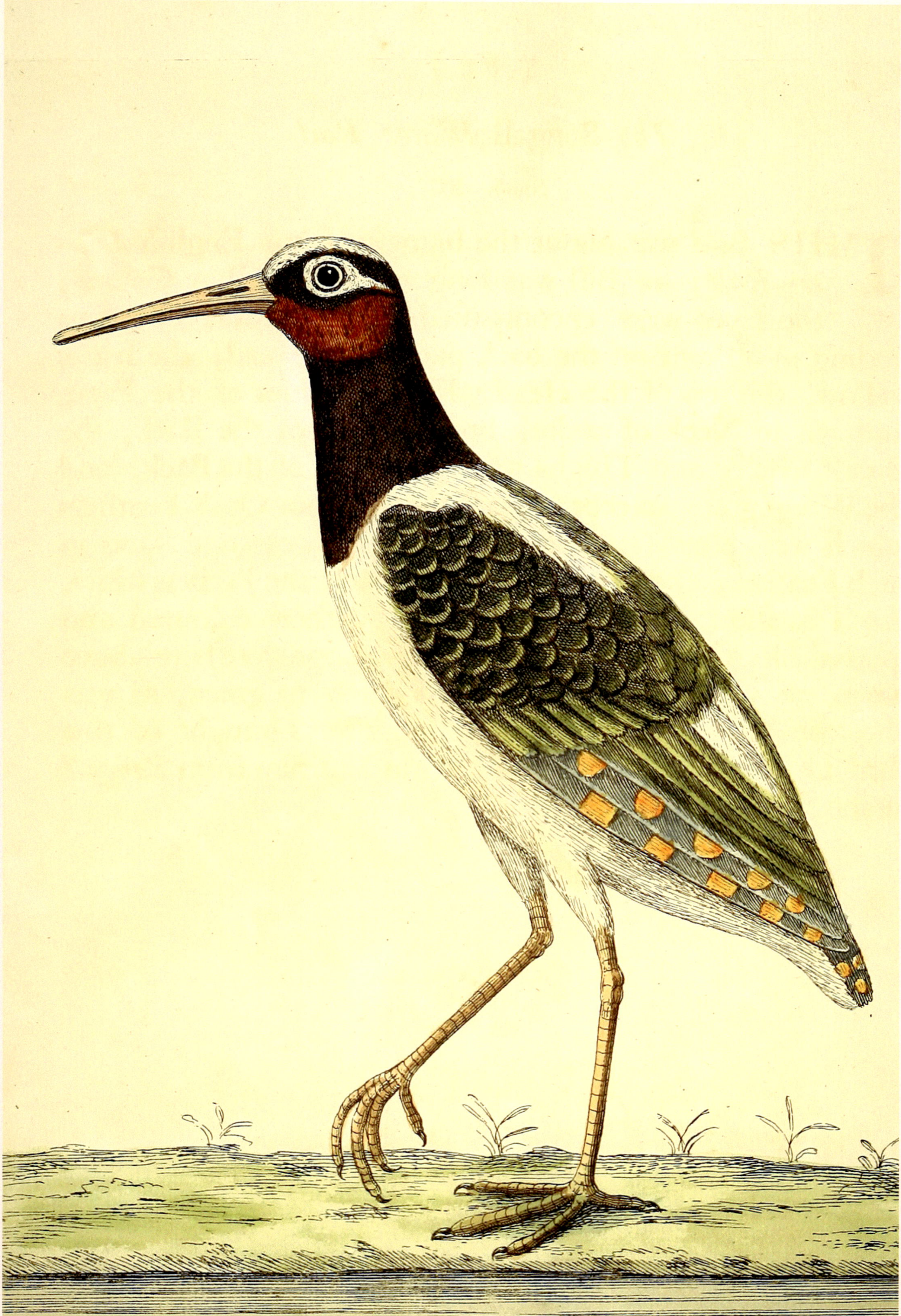

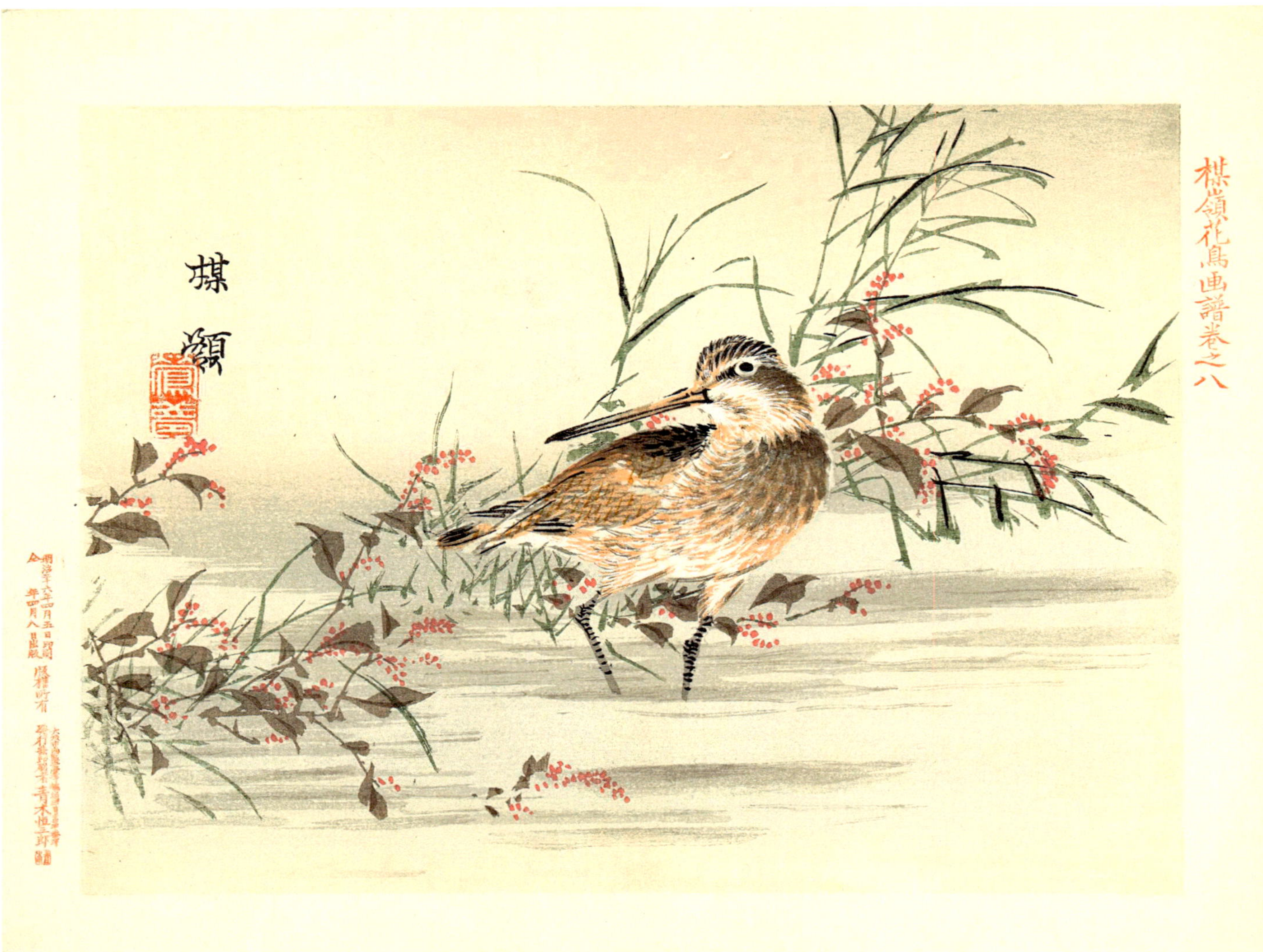

SHOREBIRD, *POSSIBLY* SNIPE
Gallinago sp.

Kōno Bairei, *Bairei kacho gafu*, 1893

Ukiyo-e was a school of Japanese art that flourished during the seventeenth, eighteenth and nineteenth centuries. Artists produced work that depicted everyday life and explored the interests of ordinary Japanese people. Prints that depicted birds and flowers were called *kachō-e*, and this work by Kōno Bairei is one such example. Produced during Japan's Meiji period, it utilizes some aspects of Western art while still maintaining its Japanese spirit. This work comes from Bairei's most well-known work *Bairei kacho gafu*. This album covers an entire year with representations of birds and flowers throughout the four seasons.

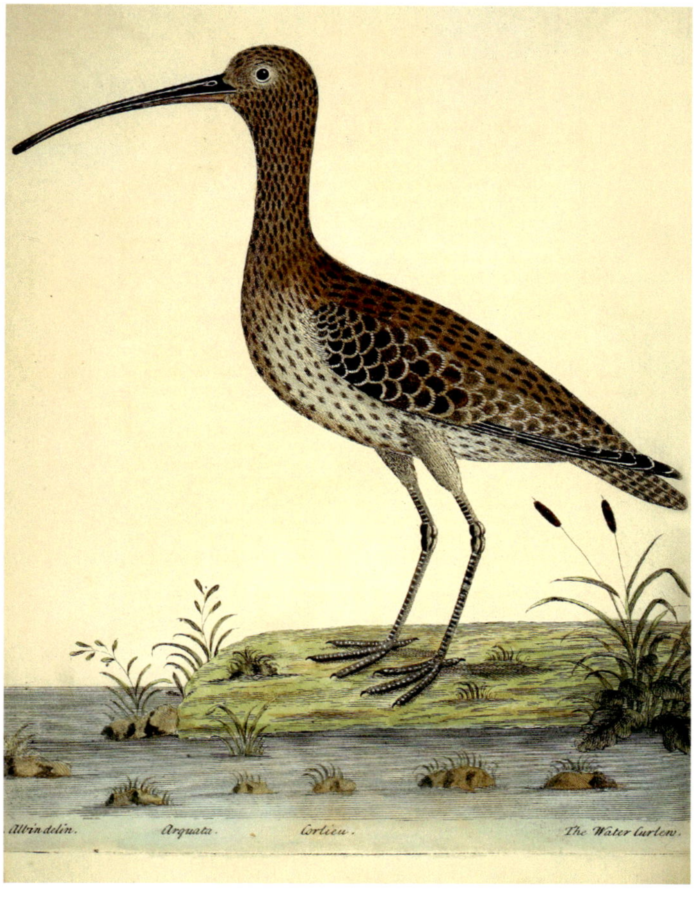

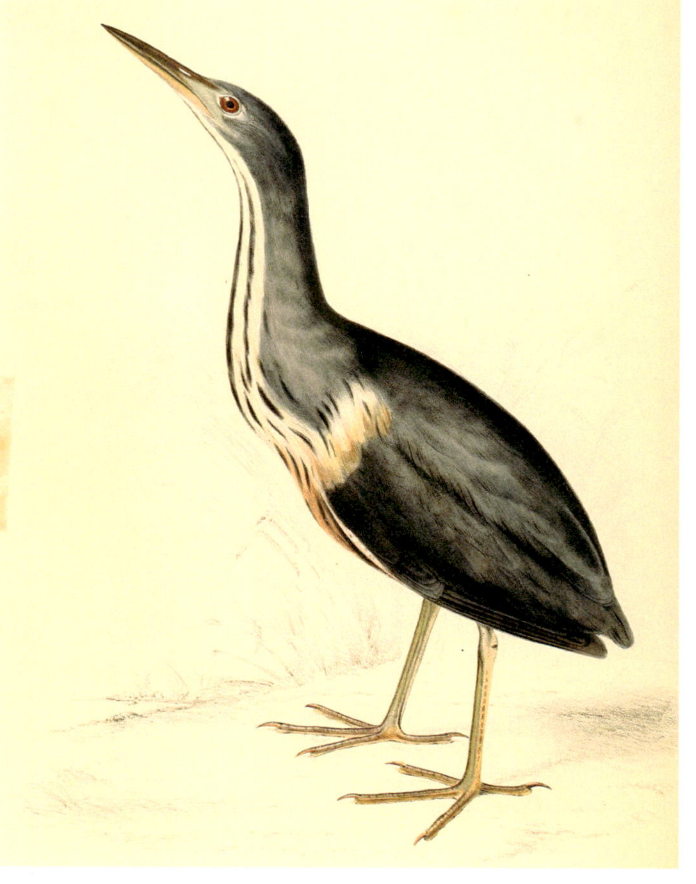

LEFT:

LONG-BILLED CURLEW
Numenius americanus

Eleazar Albin & Elizabeth Albin,
A Natural History of Birds, 1731–8

RIGHT:

DWARF BITTERN
Ixobrychus sturmii

Andrew Smith, *Illustrations of the Zoology of South Africa*, 1838–49

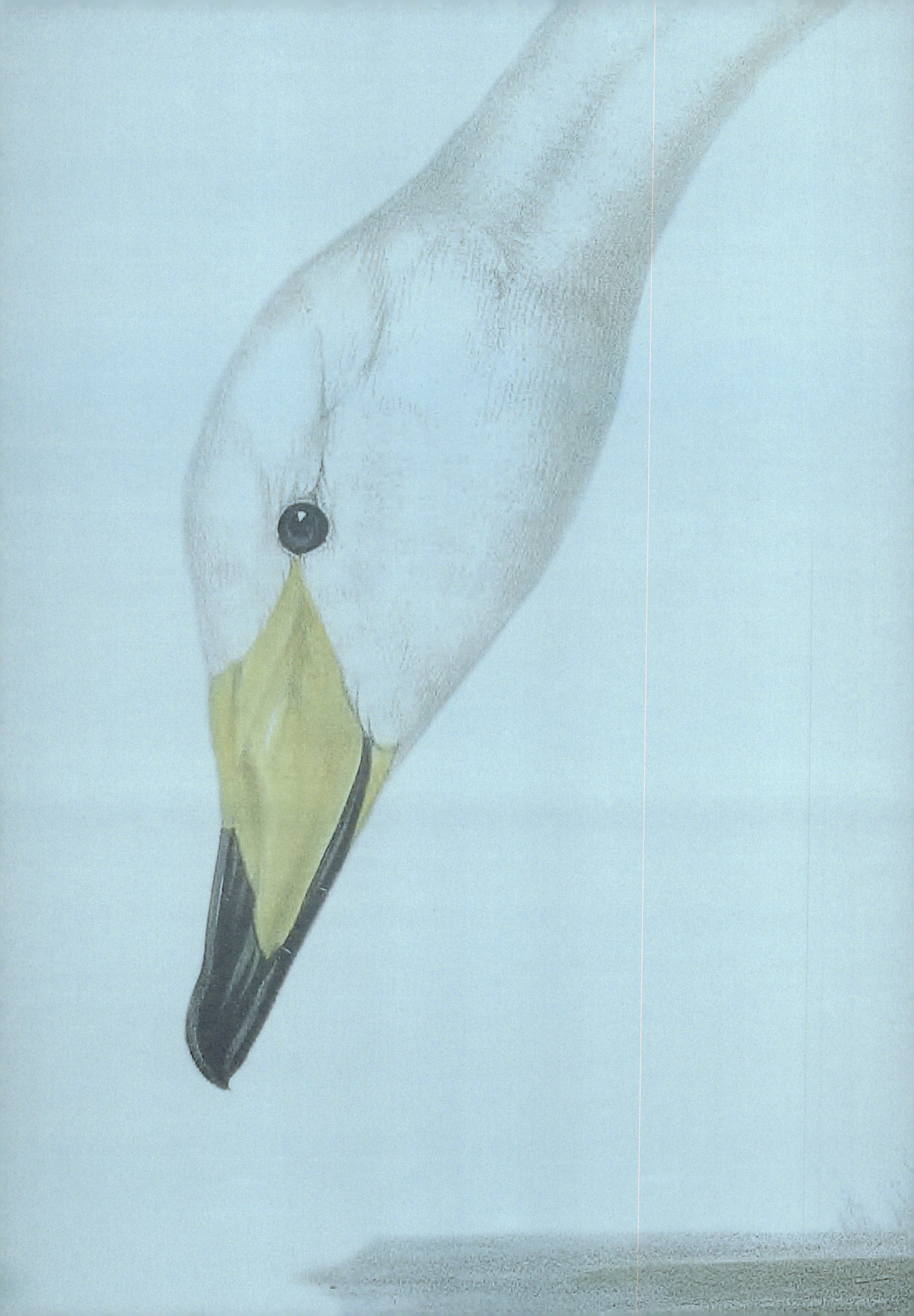

Waterfowl

These birds rank among some of the most recognizable birds in the world. Found all around the globe, they often live close to human habitats, commonly nesting near freshwater lakes and ponds. Some species, such as geese, have been domesticated by humans and are kept for their meat, eggs and feathers. Ducks, geese and waterfowl are medium to large in size and are characterised by their long necks, webbed feet and broad bills.

Ducks, geese and swans are all water birds that belong to a family known as *Anatidae*. All of the species within this family have webbed feet. This allows them to spend a lot of their life either in or near water. Many ducks are also characterized by their plump bodies, long necks and waterproof feathers. Their webbed feet are uniquely designed to help them move through the water, with the majority of species spending most of their time living on ponds or lakes.

Some waterfowl are vegetarian and eat grass or water weeds. Others hunt for fish, insects or small molluscs. When it comes to feeding, there are three key methods that these birds will use. Many species, including seaducks such as eiders, scoters, pochards and mergansers, primarily dive for underwater food. Surface-feeding ducks specialize in foraging in the shallows, either dabbling or upending. Many others, including ducks, geese and swans, feed on land, often literally grazing, but also rooting up tubers and other plant material.

Most species of waterfowl are migratory and will travel great distances each year in search of food and habitat. Roughly one-third of all birds migrate, with many ducks, geese and swans leaving their summer breeding grounds in the autumn to head to warmer climates. While some birds will fly together in loose flocks, many waterfowl will form tight V-shaped formations.

Geese are particularly adept at this type of flying, and some travel distances of up 5,000 kilometres (roughly 3,000 miles) every year. In this formation, each bird flies slightly ahead of the other to provide a slight updraft for the bird behind. This formation may also help the flock to coordinate and communicate, making it easier for them to keep track of each other as they make their long journey. Formations like this will also have a leader, who will rotate back into the formation when it tires; another goose will then take the point position.

Many ducks also demonstrate sexual dimorphism. This is a condition where two sexes of the same species will exhibit distinctive differences in their physical appearance. Sexual dimorphism is something that occurs in many birds, as well as in animals and even some plants. When we think of this condition, one of the most common examples of this in the northern hemisphere is the mallard duck (*Anas platyrhynchos*). Here, we can see a dramatic difference between the sexes: the male mallard has a glossy green head, a brown breast and a white ring around its neck, while the female is far less ornamented, simply having a mottled brown plumage.

This sexual dimorphism is something we can see in many of the illustrations collected in this book. Some illustrations are better than others at showing the physical differences between the sexes, but it is important to remember that many illustrators were simply working with the specimens that they had available to them. John Gould (see page 292) was particularly interested in having illustrations that depicted the sexes of a species and this can be clearly seen in the books that he produced. His works tend to depict more than one bird in each scene and in instances where both genders appear identical, he will show them in two distinctive poses, allowing the viewer to get a clearer idea of their shape, size and even personality of these birds.

Ducks, geese and waterfowl can be found in wetlands, coastal marshes and lowland lakes throughout the world. Indeed, with the exception of Antarctica, they can be found on every continent on the planet. This ubiquity is demonstrated well by the variety of work featured throughout this chapter. Illustrations come from such works as Audubon's *Birds of America* (1827–38), Gould's *Birds of Europe* (1832–7) and Buller's *History of the Birds of New Zealand* (1872–3). These illustrations demonstrate the sheer variety and splendour that can be found within this large and diverse biological family.

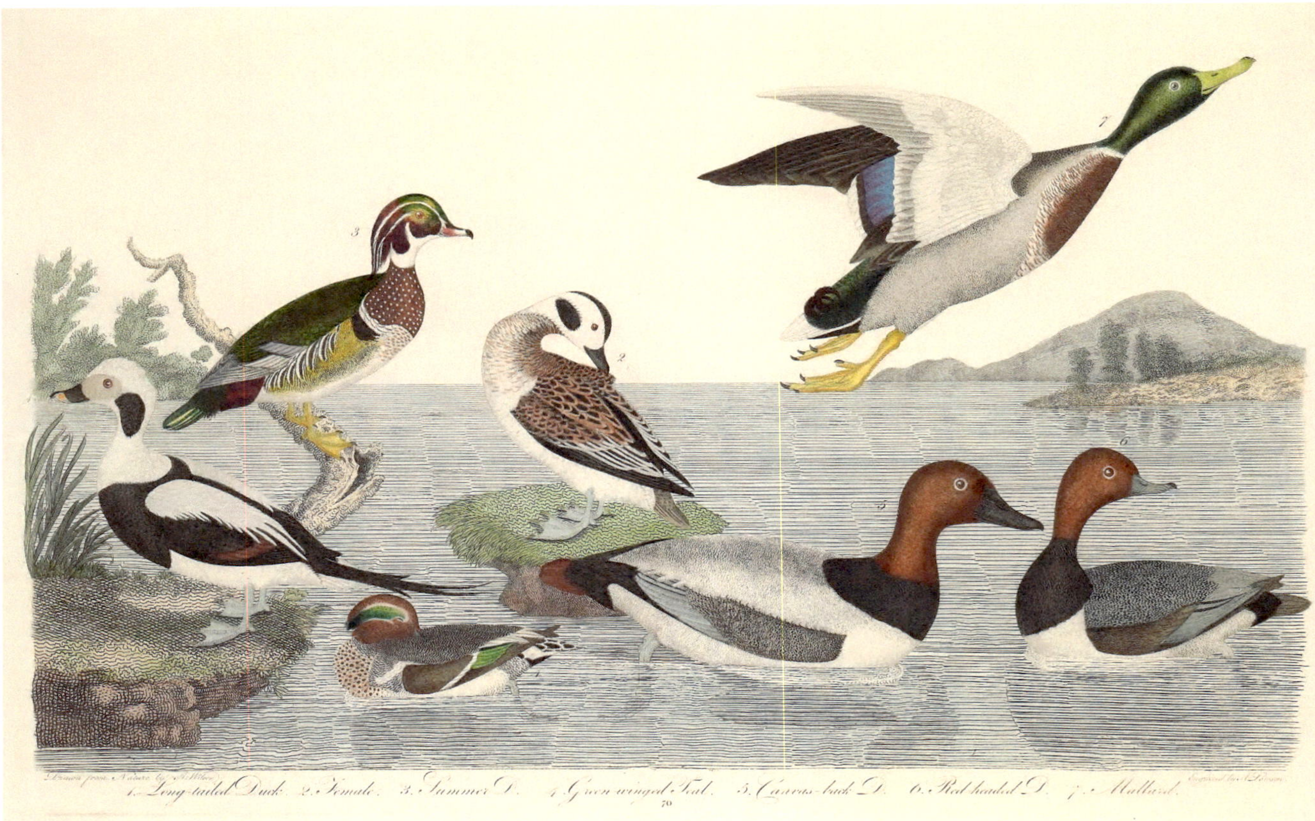

1. LONG-TAILED DUCK (MALE)

2. LONG-TAILED DUCK (FEMALE)
Clangula hyemalis

3. WOOD DUCK
Aix sponsa

4. GREEN-WINGED TEAL
Anas carolinensis

5. CANVASBACK
Aythya valisineria

6. REDHEAD
Aythya americana

7. MALLARD
Anas platyrhynchos

Alexander Wilson, *American Ornithology*, 1808–14 (this edition: 1871)

This plate shows ducks standing, swimming and in flight. It is a lively composition – particularly by Wilson's standards – and there is a real sense that he is eager to show the feet and legs of these ducks. Birds sit on the water with their legs raised slightly higher than normal, while the feet of the mallard in flight look especially pronounced. For illustrators like Wilson, the primary purpose of this work was to convey information and to aid in the identification of species.

OPPOSITE:
PARADISE SHELDUCKS
Tadorna variegata

John Gerrard Keulemans, *A History of the Birds of New Zealand*, 1872–3 (this edition: 1887–8)

Keulemans' depiction of the paradise shelduck shows both a female (left) and male (right). This large goose-like bird is endemic to New Zealand and demonstrates how distinctive sexual dimorphism can be between the two sexes. Unusually for ducks, it is the female that is more distinctive than the male; females have a pure white head and a chestnut-brown body, while males have a black head and a dark grey body.

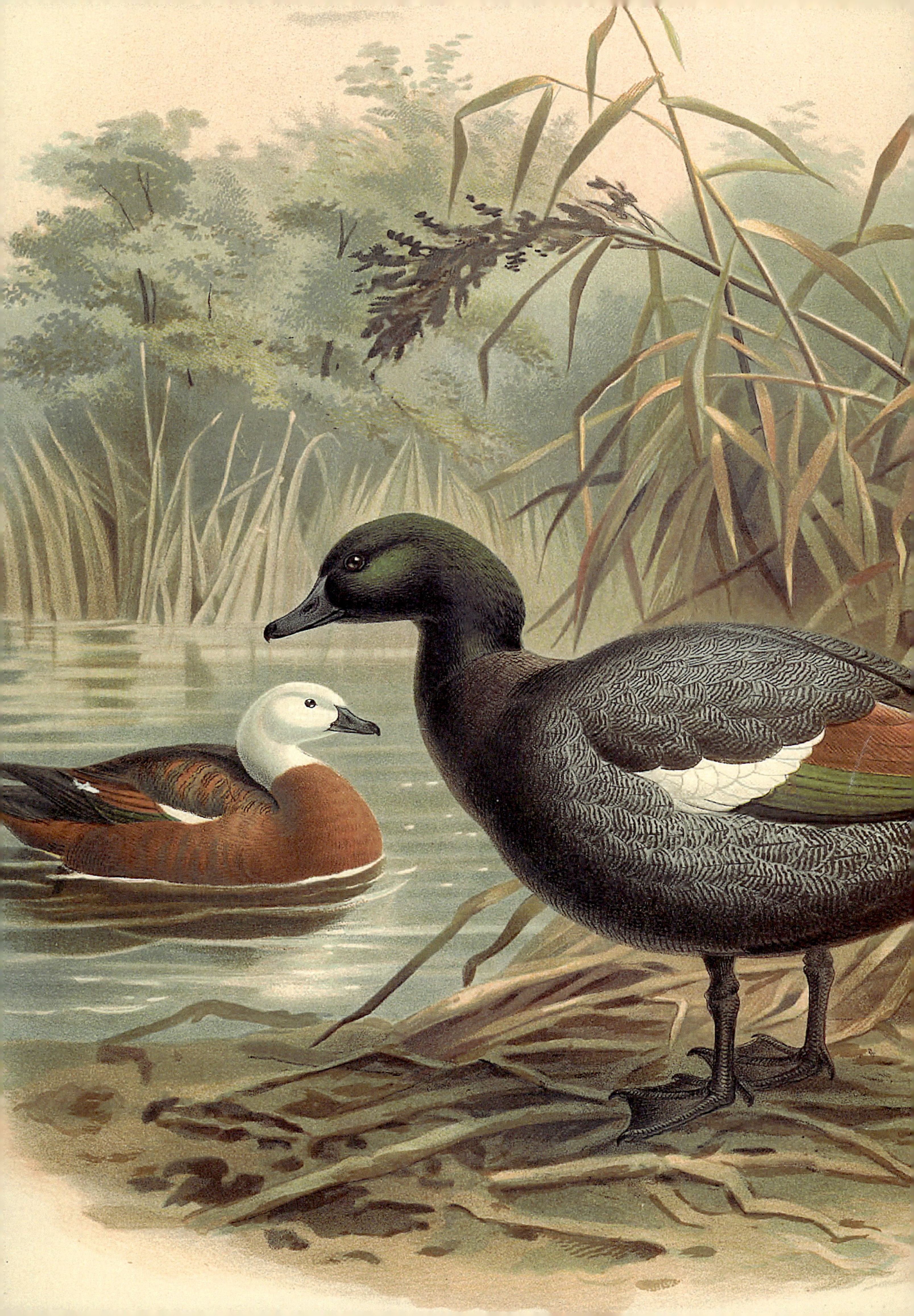

Alexander Wilson

1766–1813

Poet, ornithologist and naturalist, Wilson is often dubbed the Father of American Ornithology. His work depicted the birds of the United States of America in the hopes of fostering a shared sense of cultural identity for the then-new nation.

Alexander Wilson was a Scottish-American poet, ornithologist and naturalist. Today, his work is somewhat overshadowed by Audubon's *Birds of America*, yet his work predates Audubon by some years, and his masterwork, *American Ornithology* (1808–14), earned him the title the Father of American Ornithology. On publication, this work was the most comprehensive work on the birds of the continent and featured the first coloured ornithological illustrations to be published in America.

Born in the Scottish town of Paisley in 1766, Wilson initially trained and worked as a weaver. It was a time of social and political upheaval in both Europe and America, and within these changing times, Wilson probably thought of himself as part of both the enlightened and labouring classes. He was interested in writing poetry, and he often wrote about the unfair treatment of the weavers. This work was often deemed inflammatory and he even found himself in legal trouble when a libel case was taken against him. In 1794, he emigrated to the United States in the hopes of finding a better life. There he settled in Pennsylvania where he supported himself for several years working as a teacher. However, nature was his greatest love, and he was determined to publish a collection of illustrations of the birds of North America.

Much like Mark Catesby, he was drawn to observing birds in nature and spent much of his time travelling extensively throughout the United States. For the most part, he travelled on foot, taking time to paint, write notes and collect specimens. Unlike Catesby, he had the resources to hire a professional engraver and so his field drawings were transferred onto copper plates for printing by a Scottish engraver by the name of Alexander Lawson (1773–1846).

While Wilson's illustrations demonstrate keen attention to detail, his work is of varying quality overall. Often, he would depict several birds on one page as a means to save money and this led to many of his compositions looking stiff, awkward and even clumsy. Yet despite his somewhat limited artistic ability, his true skills lay in writing. Wilson had an interest in poetry from a young age and was a contemporary of the celebrated poet Robert Burns (1759–96). His writing in *American Ornithology* reflects his love of the written word, but also shows his incredible knowledge of birds, their habitats and behaviours. Wilson was also the first ornithologist in America to describe and classify birds using the Linnaean system. Of the 268 species that are contained in this work, twenty-six had never been described before.

Wilson saw himself, not simply as an American ornithologist, but as the first American ornithologist, and for him this project was underlined with a sense of patriotic duty. His aim was to use the nation's birds as a means to foster a shared sense of cultural identity. When he arrived in the United States, the country had only been independent for eighteen years. The American War of Independence, fought primarily between Britain and the Thirteen Colonies in America, had led to British rule being overthrown and the establishment of a new country. The principles of nationalism are nearly always linked with the ideas of war, and Wilson's hope for his work, created against the backdrop of a this newly evolving nation, was that this new country's natural history could help to unify its people. Not only was it important for this work to be a 'national collection' of birds, but the project itself was an all-American undertaking: framed as a testament to American ingenuity and manufacturing, its aim was to show that America could compete with the Old World in producing books.

As a project, it was also deeply intertwined with the emerging cultural belief in 'manifest destiny'. While this term may not have been coined until 1845, the philosophy behind it was very much starting to take hold. When Wilson produced this book, the United States was beginning to expand vigorously across North America, acquiring new territories as well as displacing and murdering Native Americans. Many justified this expansion through a belief that they had been destined by God to expand their dominion over this continent. To produce a work like this was to embrace westward expansion and to prove dominion over nature. It was a tireless effort for the naturalist, and it saw him travel within all the states and territories that had been established at this time. He drew more than 268 species of birds, many of which he observed first-hand in nature. He also relied on specimens that had come from other explorers. Many of these came from the collection at the Peale Museum in Philadelphia, a collection built on specimens collected by explorers Lewis and Clark on their famed 1804–6 expedition across the continent.

Like many authors at that time, Wilson relied on a subscription model to fund the publication of this work and, when the first volume was published, he hand-delivered a copy to one of its early subscribers: Thomas Jefferson (1743–1826), then president of the United States. In total, *American Ornithology* spanned nine volumes. The first seven of these were produced in just six years – a remarkable achievement for such a mammoth project. Yet the workload was arguably too much for its creator, who overworked himself while living in poverty. Wilson died during the preparation of the final volume and the work was eventually completed by his friend, naturalist George Ord (1781–1866). Despite the tragic end to his story, he achieved many of his ambitions in life and today a number of birds are named in his memory including Wilson's storm-petrel, Wilson's plover and Wilson's snipe.

DETAIL, SHOWING:

1. GADWALL
   Mareca strepera
2. COMMON EIDER (MALE)
3. COMMON EIDER (FEMALE)
   Somateria mollissima
4. SMEW
   Mergellus albellus

Alexander Wilson,
American Ornithology, 1808–14
(this edition: 1871)

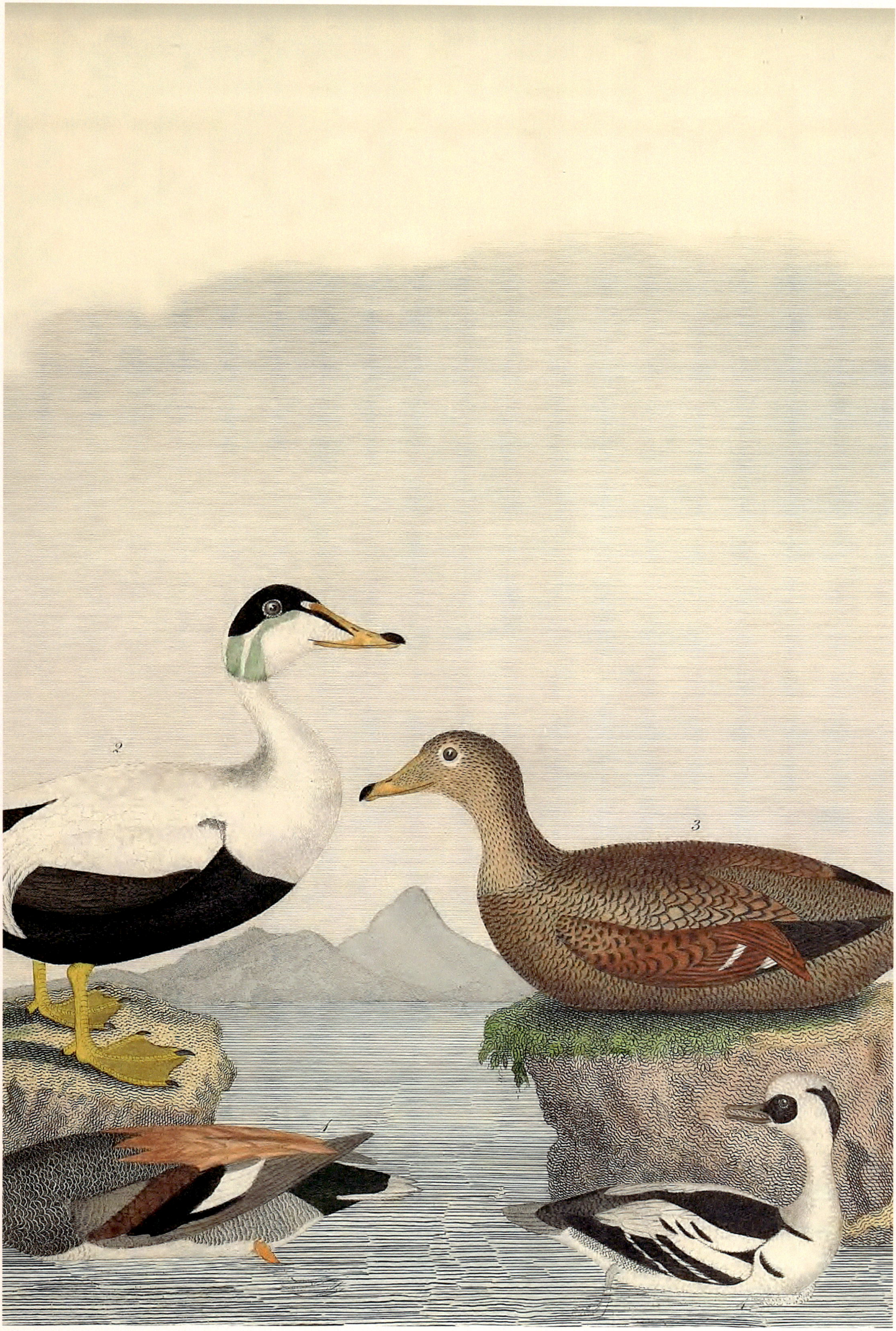

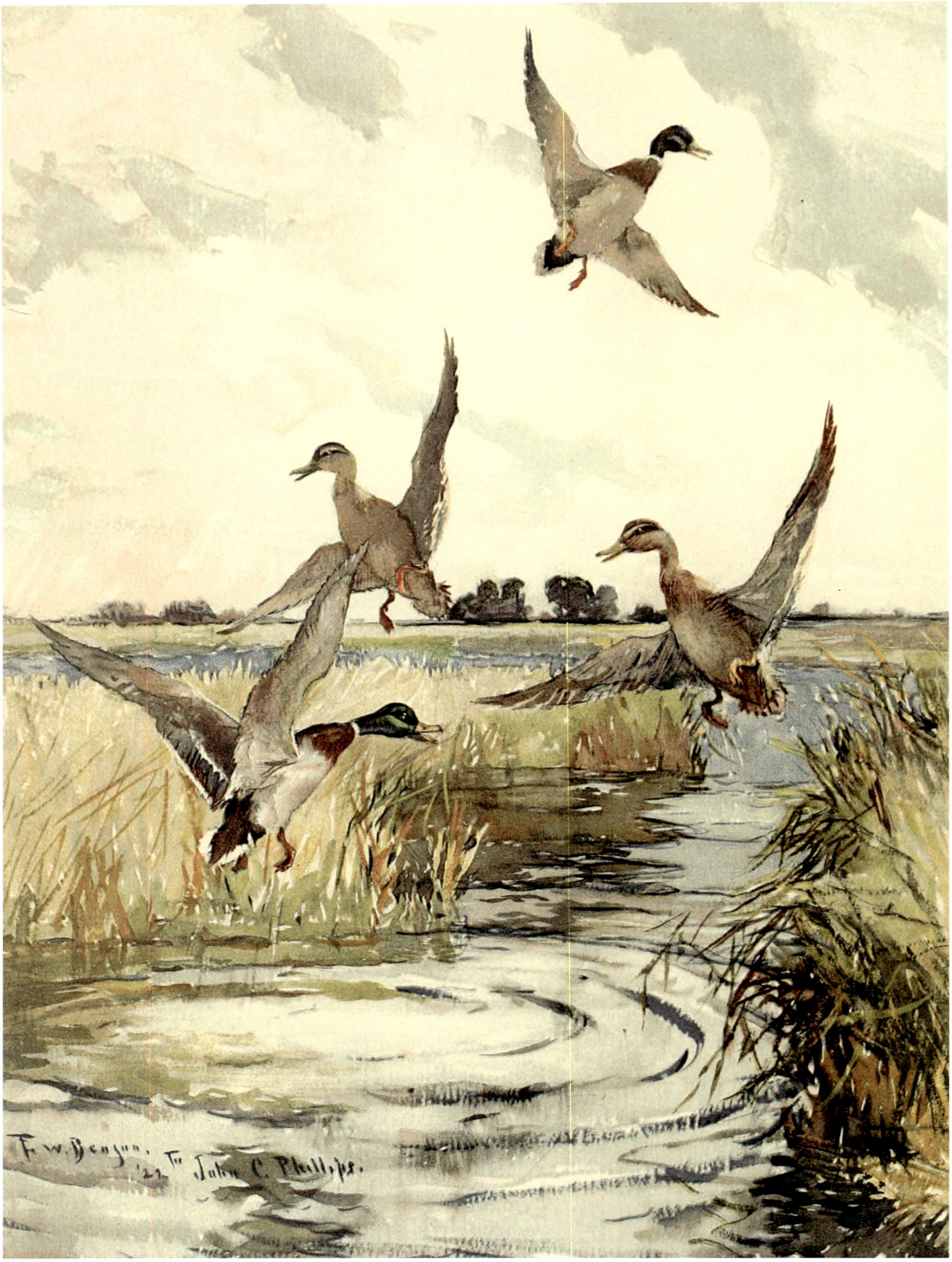

MALLARDS
Anas platyrhynchos

Frank W. Benson,
A Natural History of the Ducks, 1922–6

The illustrations across both these pages come from *A Natural History of the Ducks* by John Charles Phillips. A prolific writer, Phillips published more than 200 books and articles that spanned a wide range of subjects including experimental animal breeding, wildlife conservation, genetics, hunting and ornithology. His *Natural History of the Ducks* is arguably his most important work, and many believe it to be the best monograph of its kind on the subject.

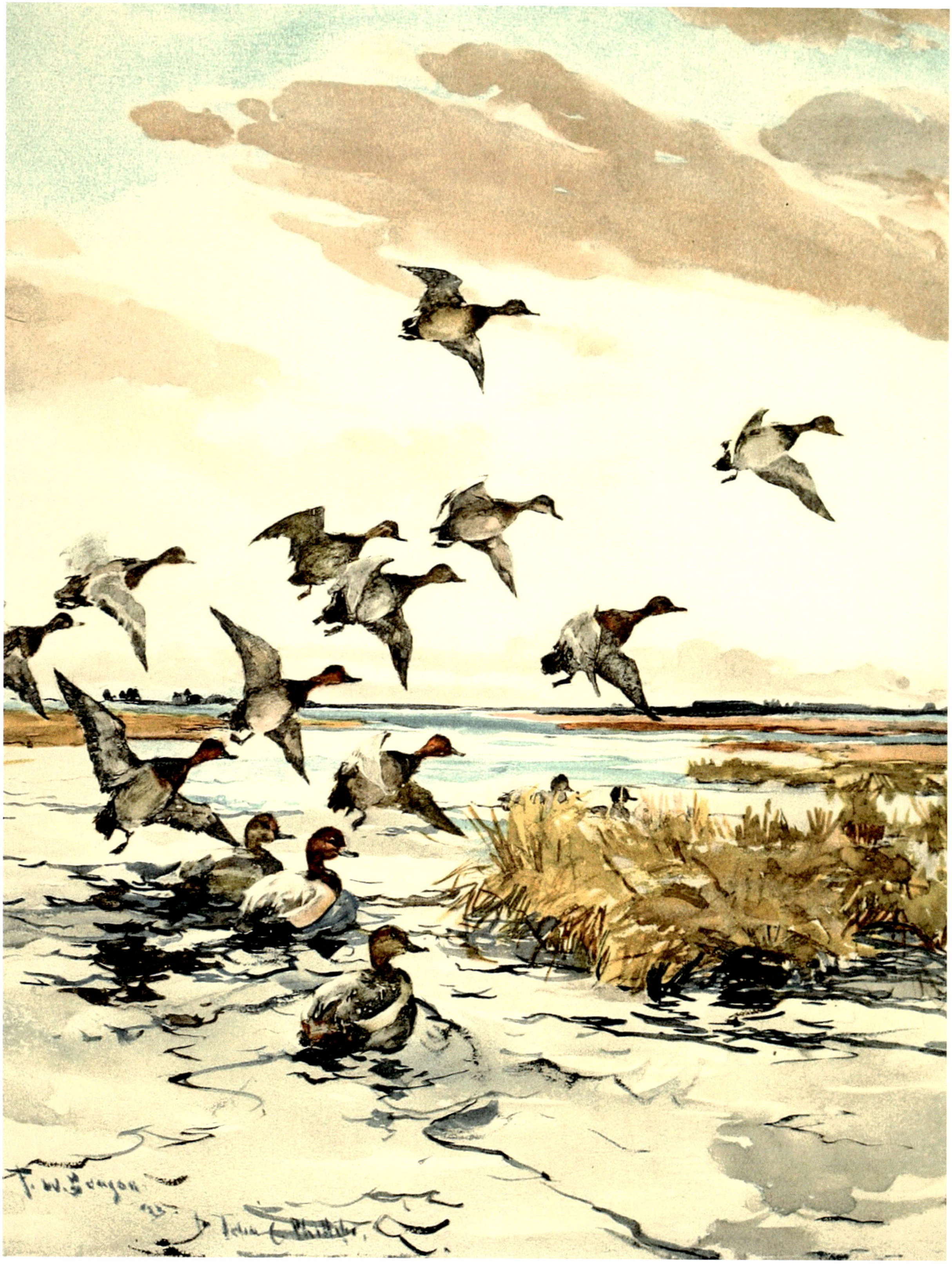

REDHEADS
Aythya americana

Frank W. Benson, *A Natural History of the Ducks*, 1922–6

One of the illustrators who contributed to *A Natural History of the Ducks* was American artist Frank W. Benson. He provided lively watercolour illustrations of ducks in nature. An American impressionist, Benson was a well known and financially successful painter, print-maker and teacher from Salem, Massachusetts. Despite his success as an artist, his boyhood dream had been to become an ornithological illustrator. He was an avid hunter, fisherman and nature lover and he garnered acclaim in his early years through his *plein air* paintings of women and children. In mid life he returned to his original passion and produced a number of impressive paintings, drawings and etchings of wildlife – particularly of waterfowl.

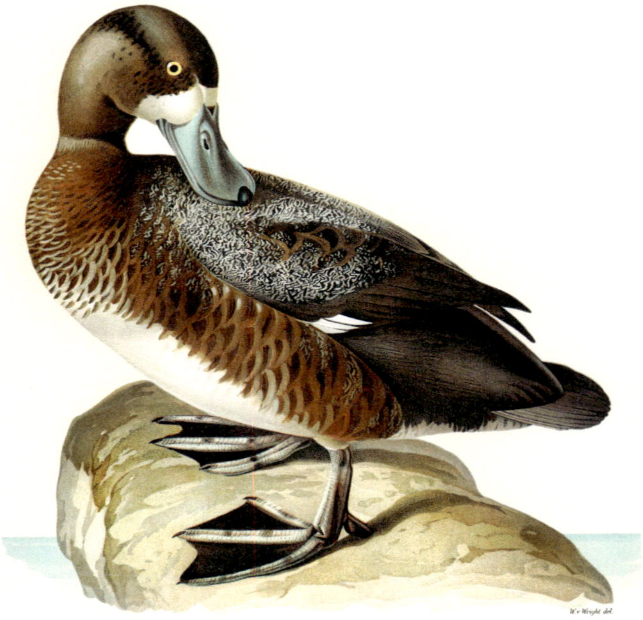

NYROCA MARILA LIN. 7/10
Bergand, ung ♂ i nov.

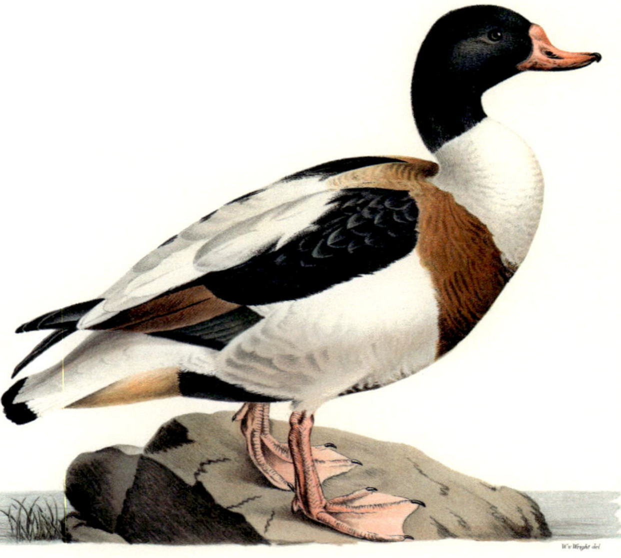

TADORNA TADORNA LIN. 5/10
Grafand.

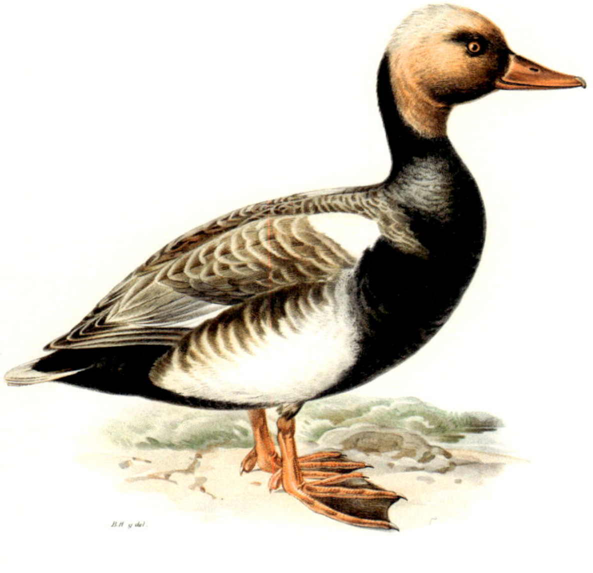

NETTA RUFINA KAUP.
Rödhufvad Dykand, ♂ 5/10

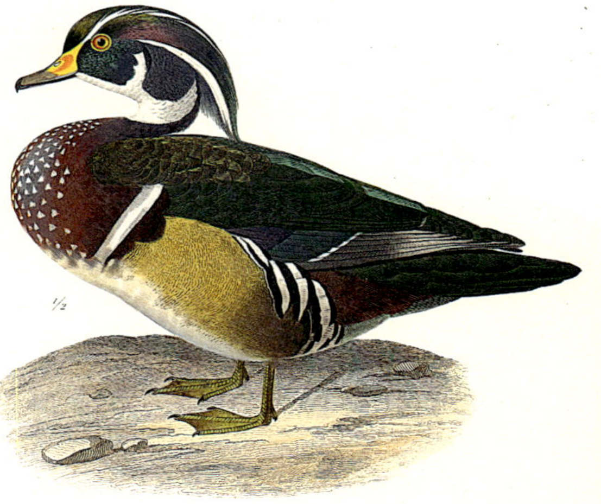

ANAS Sponsa, _Linn.
Caly. HUYUYO.

OPPOSITE TOP LEFT:
GREATER SCAUP (FEMALE)
Aythya marila

Wilhelm von Wright, *Svenska Fåglar*, 1828–38 (this edition: 1929)

OPPOSITE BOTTOM LEFT:
RED-CRESTED POCHARD (MALE)
Netta rufina

Wilhelm von Wright, *Svenska Fåglar*, 1828–38 (this edition: 1929)

OPPOSITE TOP RIGHT:
COMMON SHELDUCK (MALE)
Tadorna tadorna

Wilhelm von Wright, *Svenska Fåglar*, 1828–38 (this edition: 1929)

OPPOSITE BOTTOM RIGHT:
WOOD DUCK (MALE)
Aix sponsa

Édouard Traviès, *Historia física política y natural de la Isla de Cuba, Botánica*, 1838–61

TORRENT DUCK (TOP: MALE, BOTTOM: FEMALE)
Peruvian torrent duck, *turneri* morph
Merganetta armata

Joseph Smit, *Exotic Ornithology*, 1869

A male torrent duck stands on a rock while its female counterpart swims in the water. This plate was produced by Dutch zoological illustrator Joseph Smit for Philip Lutley Sclater's *Exotic Ornithology*. Smit was born in the Dutch town of Lisse and emigrated to England in 1866 when Philip Lutley Sclater – then secretary of the Zoological Society of London – invited him to produce drawings and hand-coloured lithographs for the book.

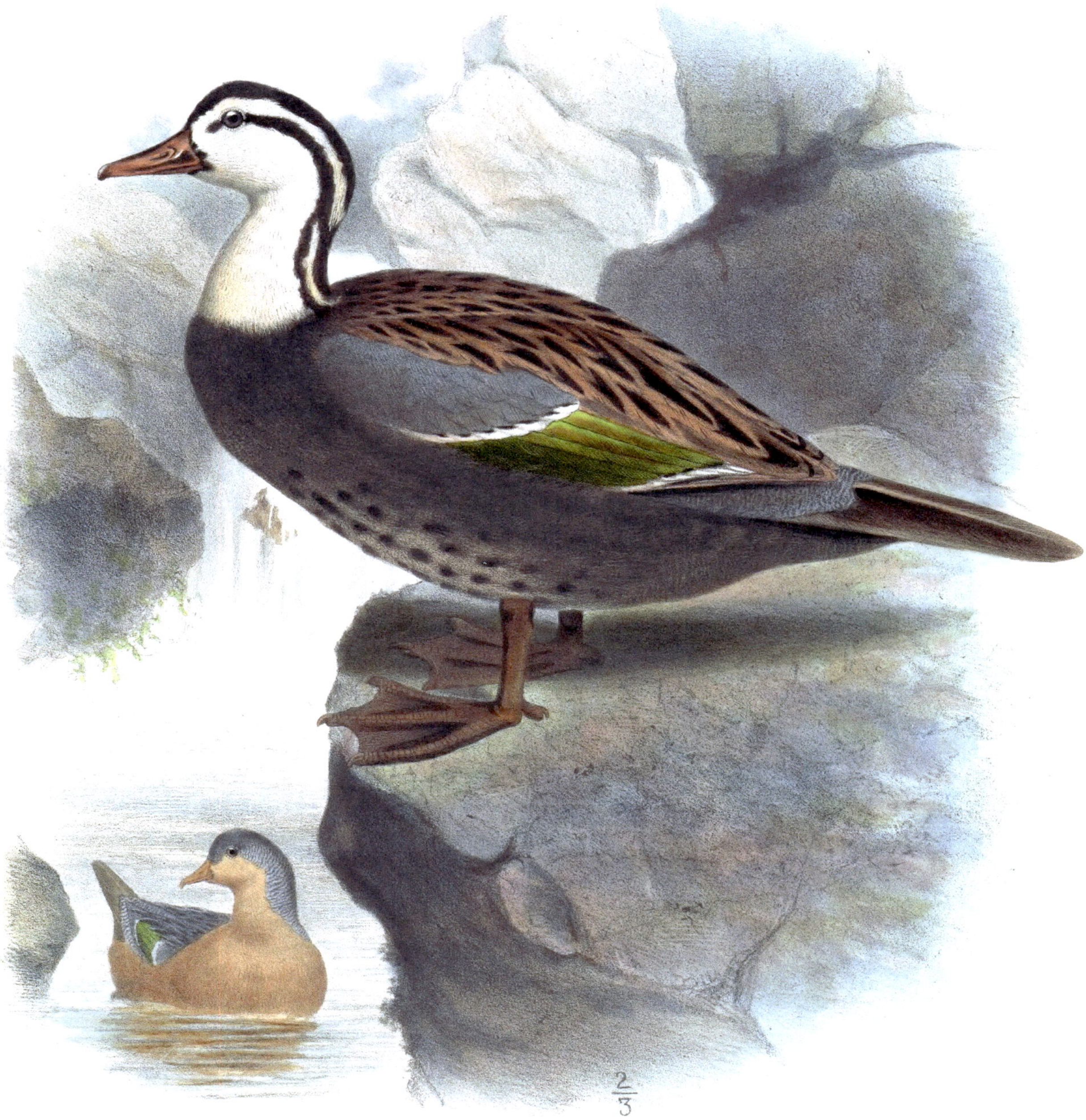

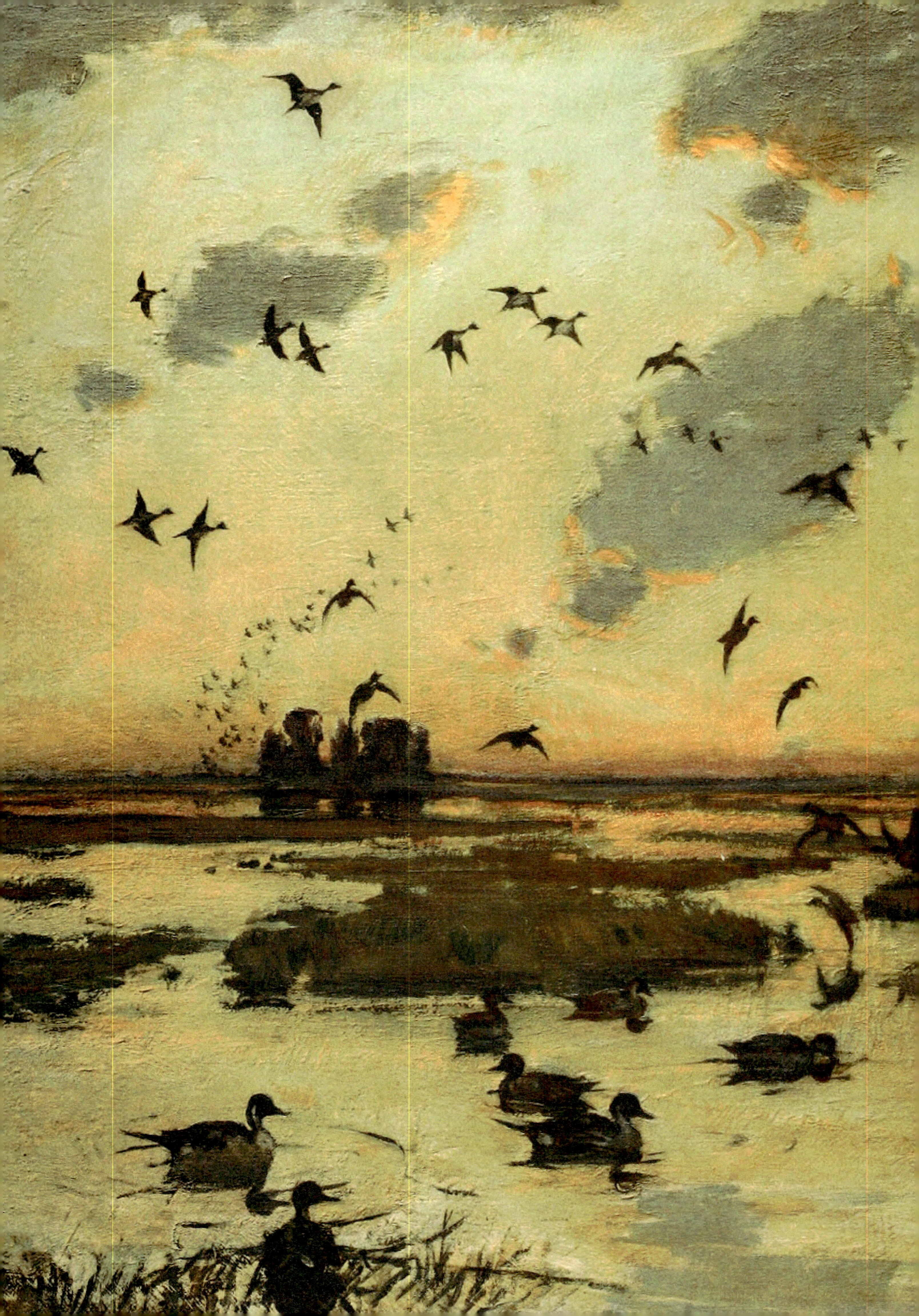

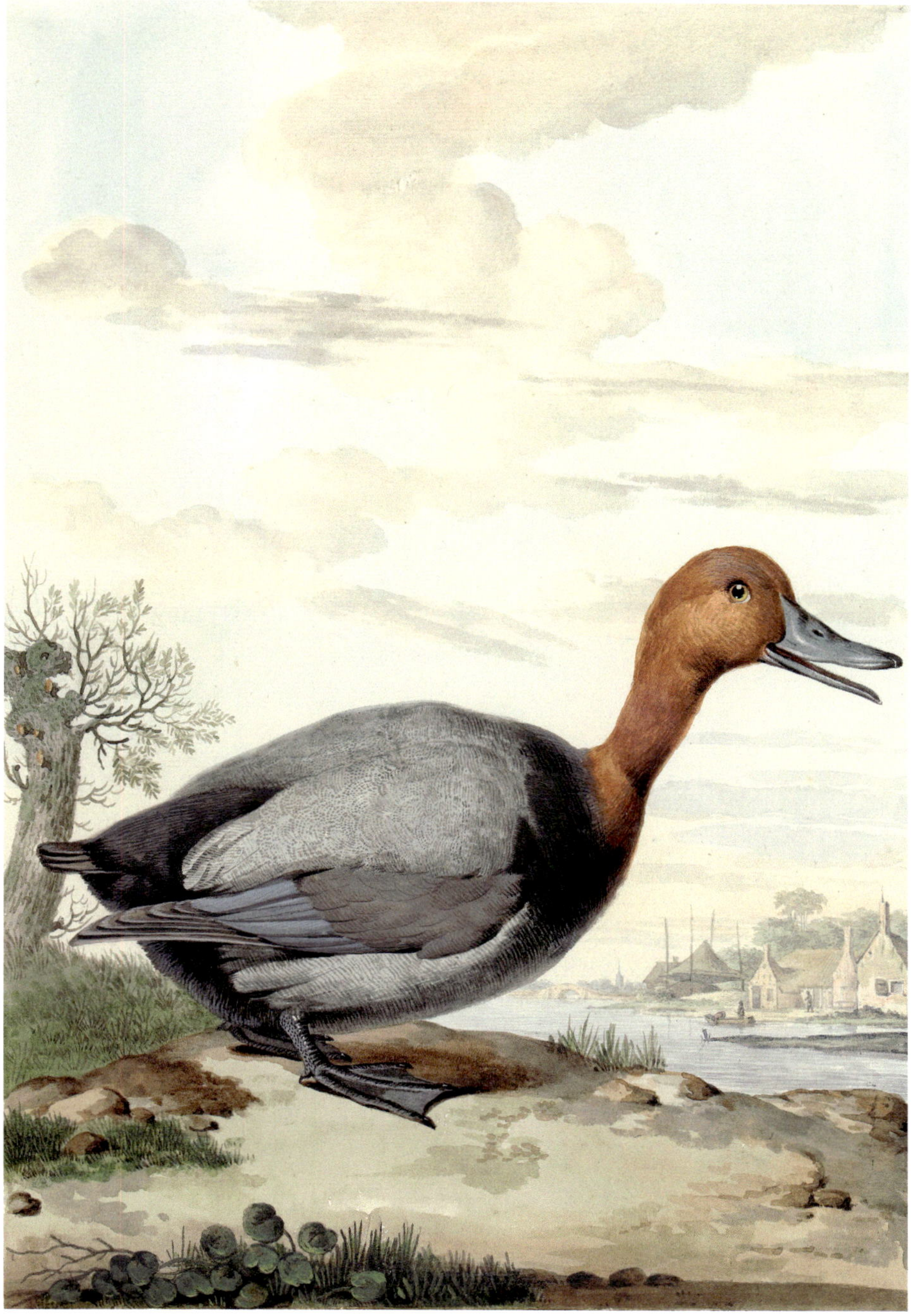

OPPOSITE:
THE DUCK MARSH
Frank W. Benson,
A Natural History of the Ducks, 1922–6

COMMON POCHARD
Aythya ferina

Jabes Heenck, watercolour, 1777

This painting is the work of Dutch painter Jabes Heenck. While little is known about this artist, we do know that he was a student of the naturalist painter Aert Schouman and it is clear to see this influence. Much like Schouman, Heenck's paintings often depict birds as their central subject. In addition to his work as a painter, he also worked as an etcher of genre subjects and – for a time – worked as an art dealer. His painting demonstrates the distinctive colouring of the male common pochard, showing its bright reddish-brown head, black breast and tail and pale grey body.

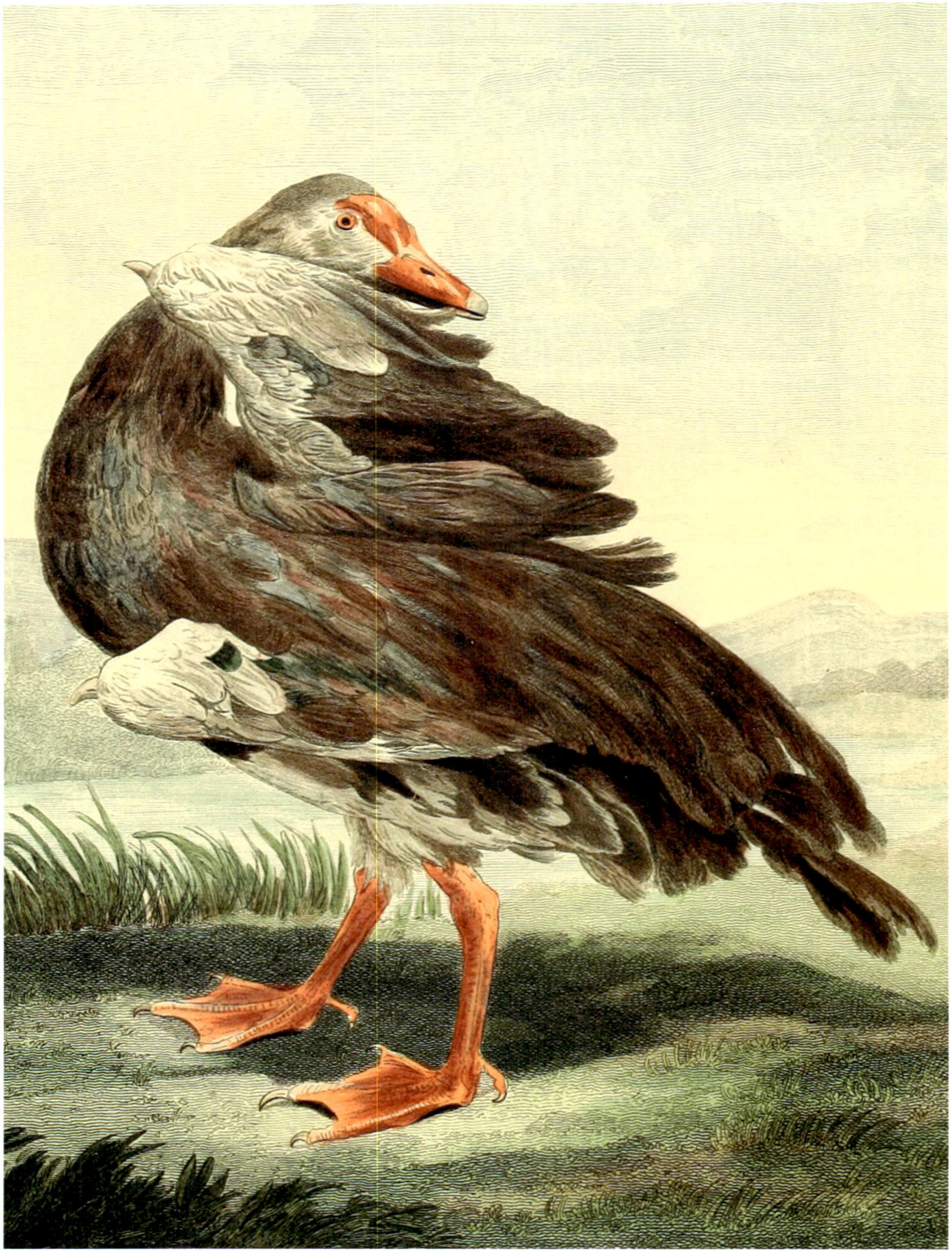

SPUR-WINGED GOOSE
Plectropterus gambensis

Charles Reuben Ryley,
Museum Leverianum, 1792–6

Several artists contributed to George Shaw's *Museum Leverianum*, but the majority of the plates were produced after drawings by Charles Reuben Ryley. Born in London in around 1752, Ryley studied engraving at the Royal Academy Schools, and it was there that he was awarded the silver medal. From 1772 until 1775 he decorated Goodwood House for the Duke of Richmond, then travelled to Ireland where he decorated the walls in the Long Gallery at Castletown House for Lady Louisa Conolly. Upon returning to London in 1777, he began studying once more, this time painting at the Royal Academy under John Hamilton Mortimer. In 1778, he was awarded a gold medal from the Royal Academy for an oil painting depicting the sacrifice of Iphigenia and this work was exhibited the following year. He continued to be a regular exhibitor at the Royal Academy but being stricken with poor health from birth limited his output, so he often relied on teaching and book illustration to support himself. He passed away in 1798 – two years after the completion of Shaw's impressive two-volume work.

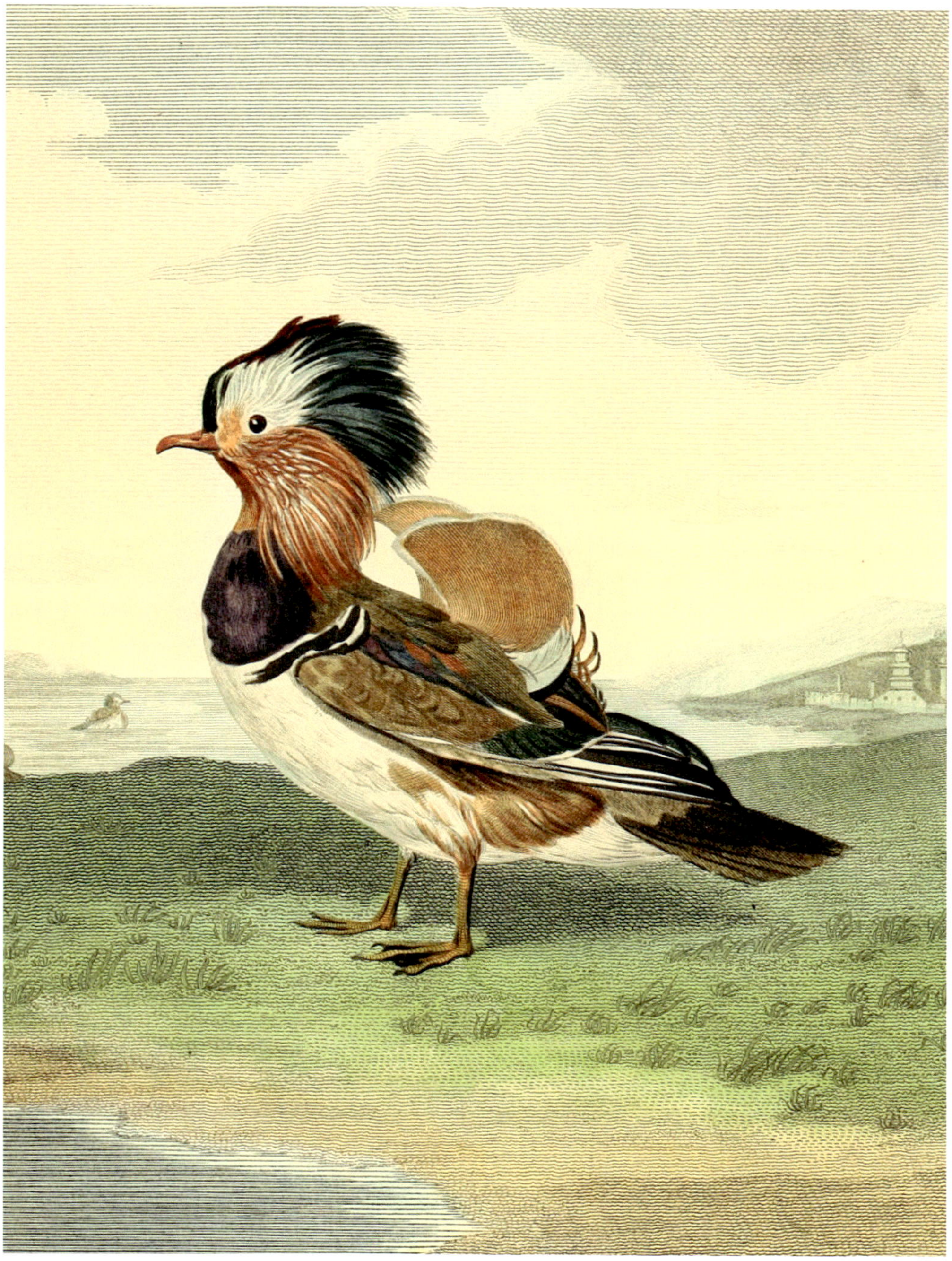

MANDARIN DUCK
Aix galericulata

Charles Reuben Ryley,
Museum Leverianum, 1792–6

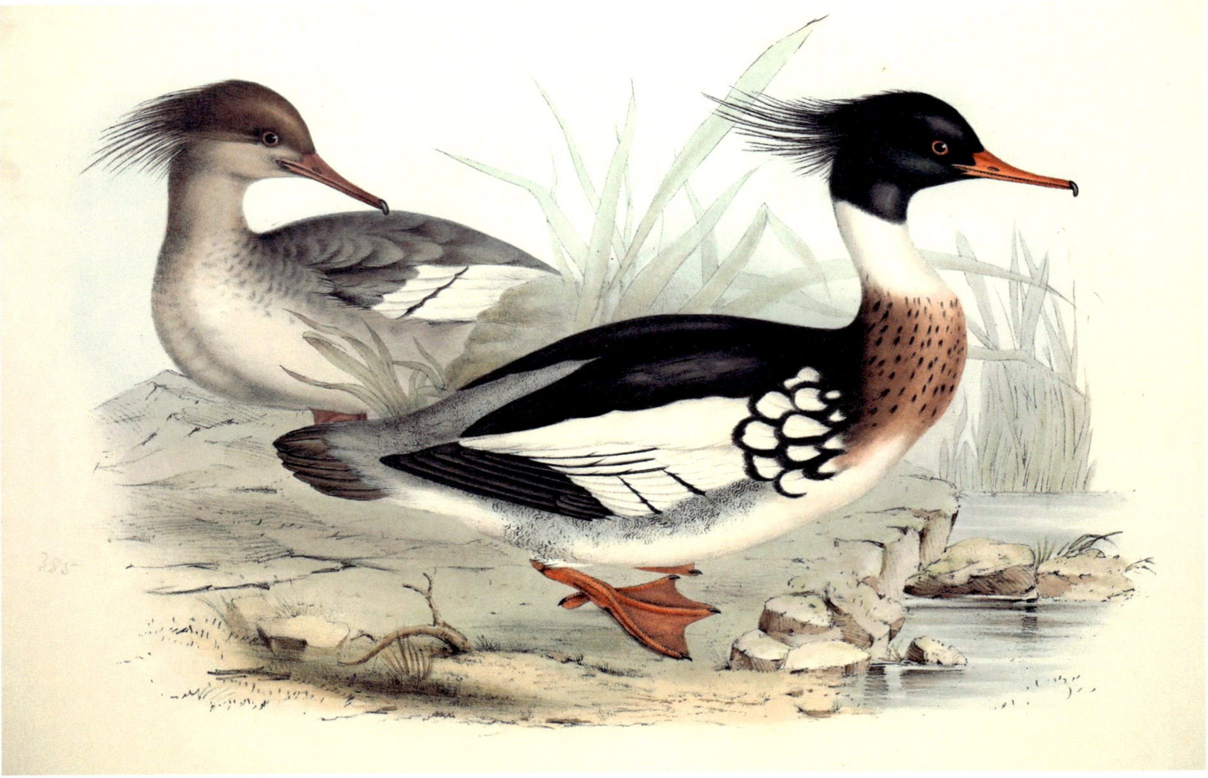

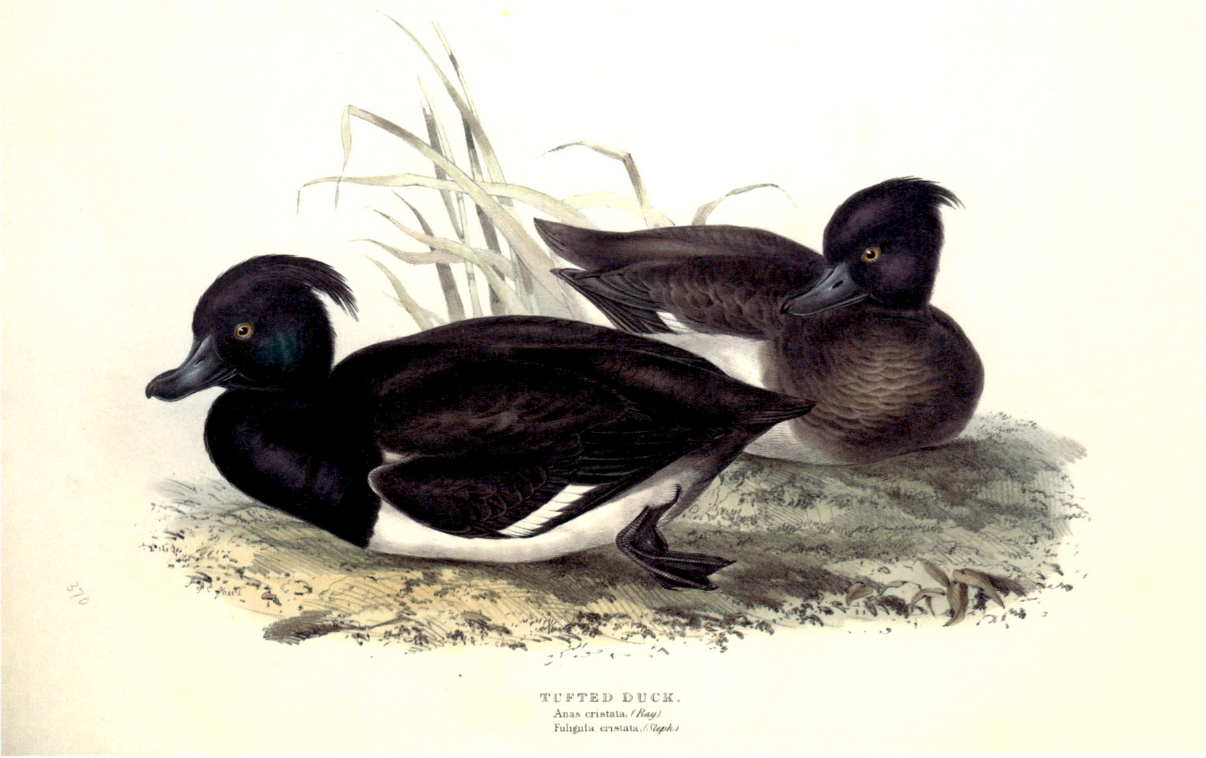

TOP:

RED-BREASTED MERGANSERS
Mergus serrator

John Gould & Elizabeth Gould,
The Birds of Europe, 1832–7

BOTTOM:

TUFTED DUCKS
Aythya fuligula

John Gould & Elizabeth Gould,
The Birds of Europe, 1832–7

Both of these illustrations appear in Gould's *Birds of Europe*. Published at a time when foreign birds dominated ornithology, Gould's ambition for this work was to redress the imbalance and reignite a popular interest in the native birds of Europe.

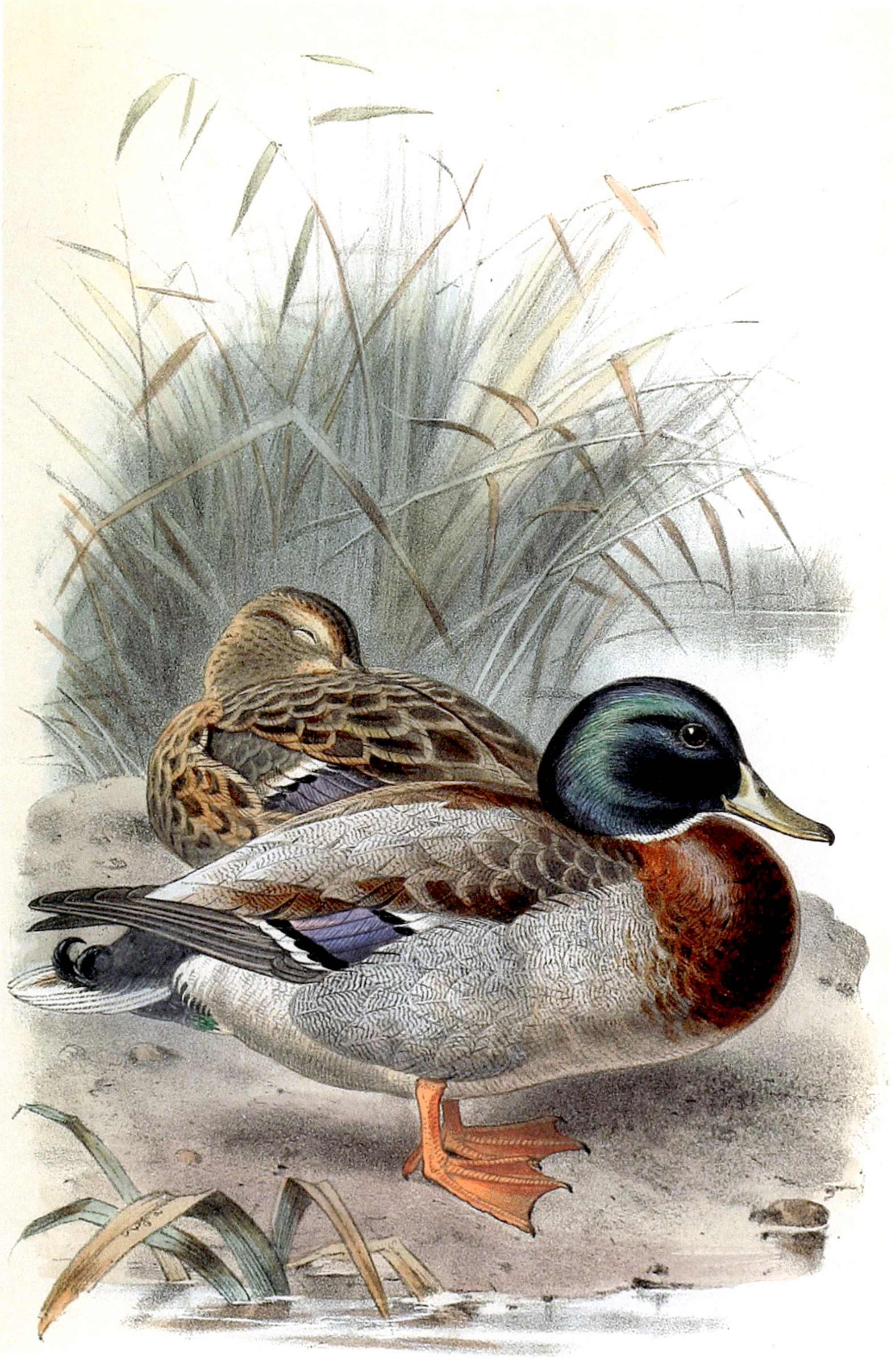

MALLARDS
Anas platyrhynchos

John Gerrard Keulemans, *Coloured Figures of the Birds of the British Islands*, 1885–97

The mallard is a ubiquitous and easily identifiable bird. So common is it, in fact, that the original text that accompanies this plate notes that 'it would be difficult to write anything that is not already perfectly well known'. They are the most abundant and most widespread of duck species, and are considered to be invasive in some areas. They are an incredibly adaptable bird and can commonly be found in urban areas, where they live on ponds, lakes and wetlands.

CANADA GOOSE
Branta canadensis

Mark Catesby, *The Natural History of Carolina, Florida and the Bahama Islands*, 1729–47

This portrait of a Canada goose emphasizes the bird's prominent white chinstrap and cheeks. These markings distinguish it from other species of geese. The plant in the background is identified as a rong bush (*Wedelia bahamensis*), a plant native to the Bahamas.

OPPOSITE:
SURF SCOTER
Melanitta perspicillata

George Edwards, *A Natural History of Uncommon Birds*, 1743–51

In 1745 George Edwards came into the possession of a number of stuffed specimens through the naturalist James Isham. Born in England, Isham worked as a trader for the Hudson's Bay Company in Canada. This surf scoter was one of several newly discovered specimens that Edwards included in his book, a work that later became one of the main sources of information for Linnaeus when he compiled his list of definitive species in 1758.

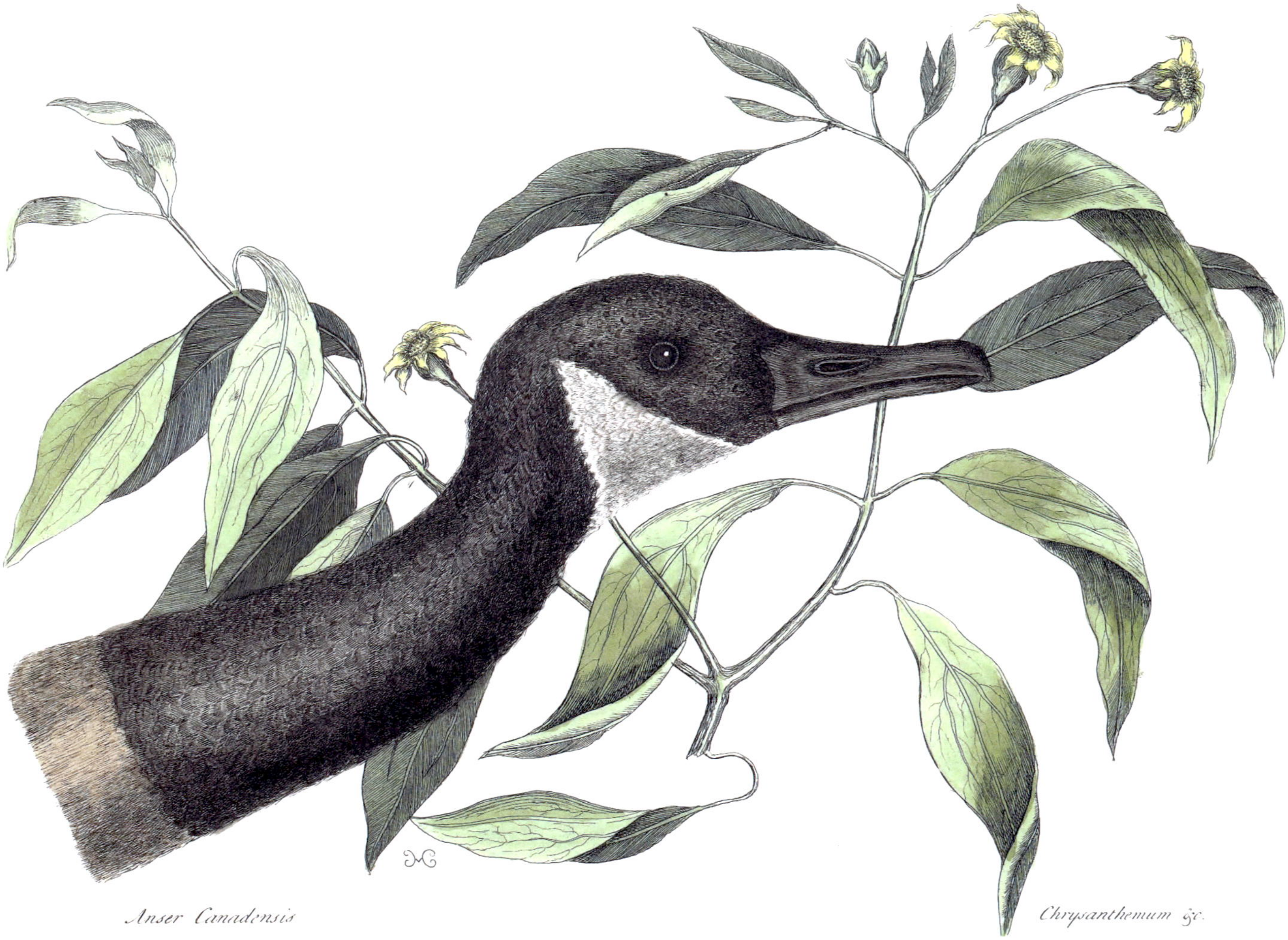

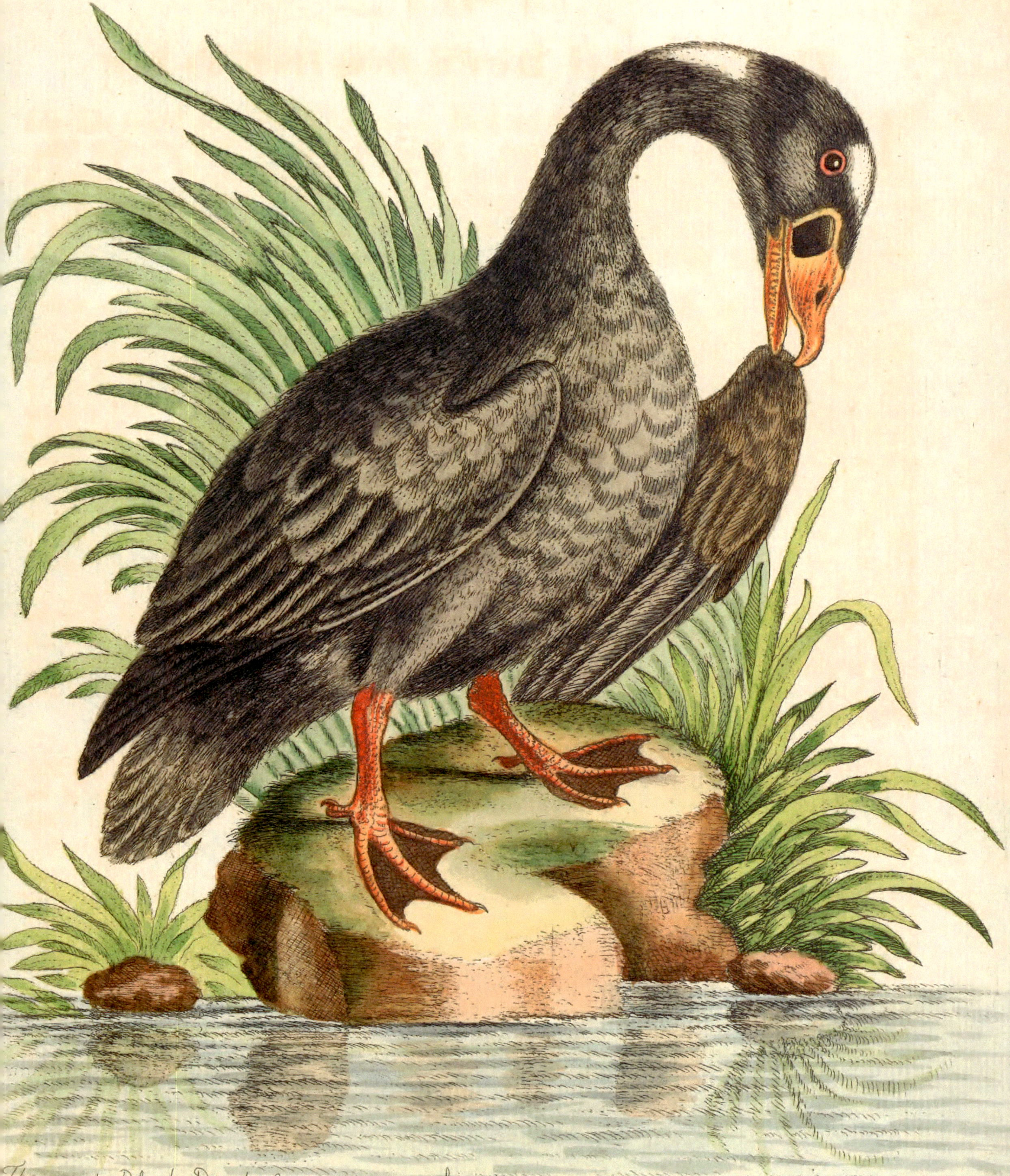

The great Black Duck from Hudsons bay 1749 G Edwards

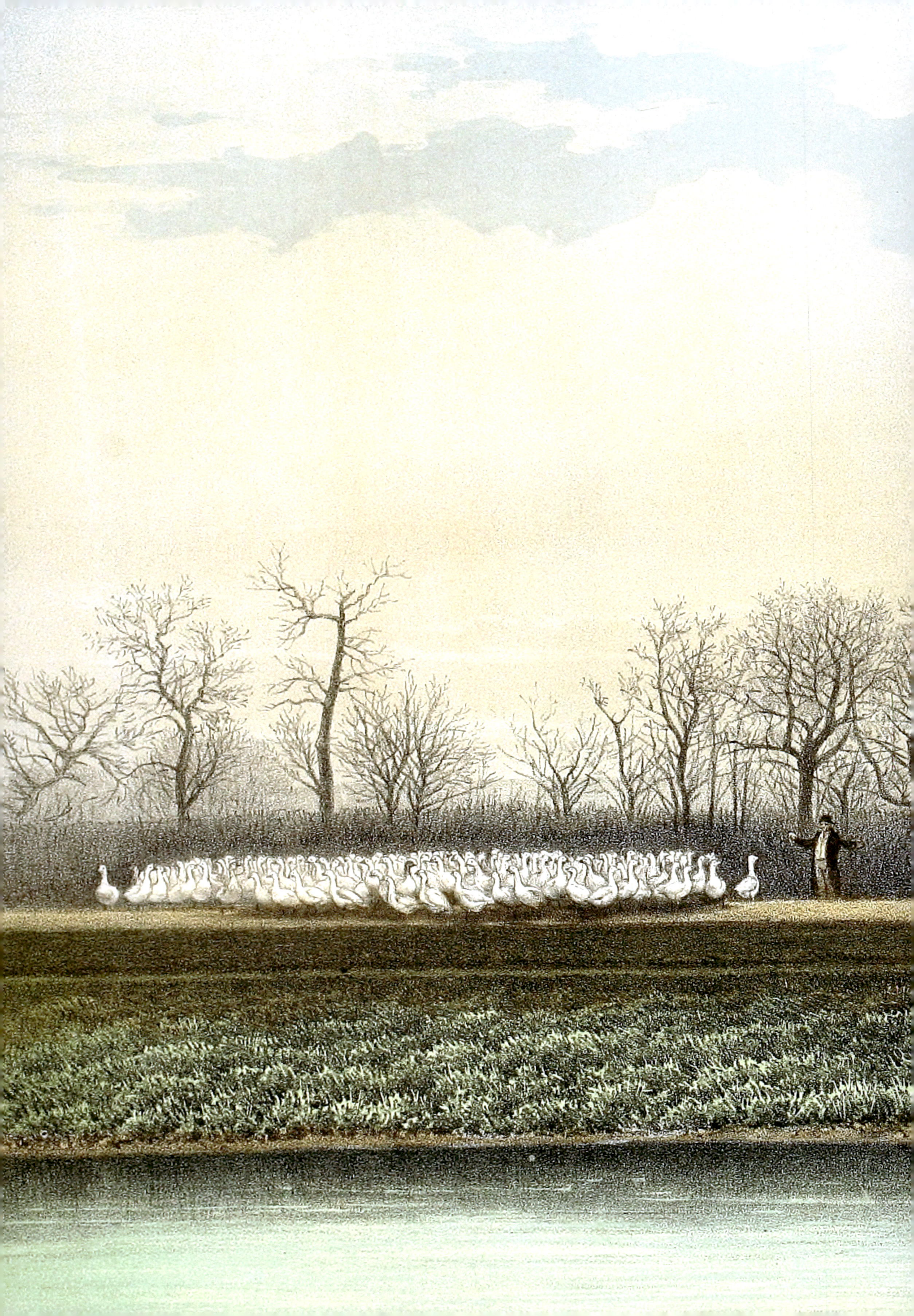

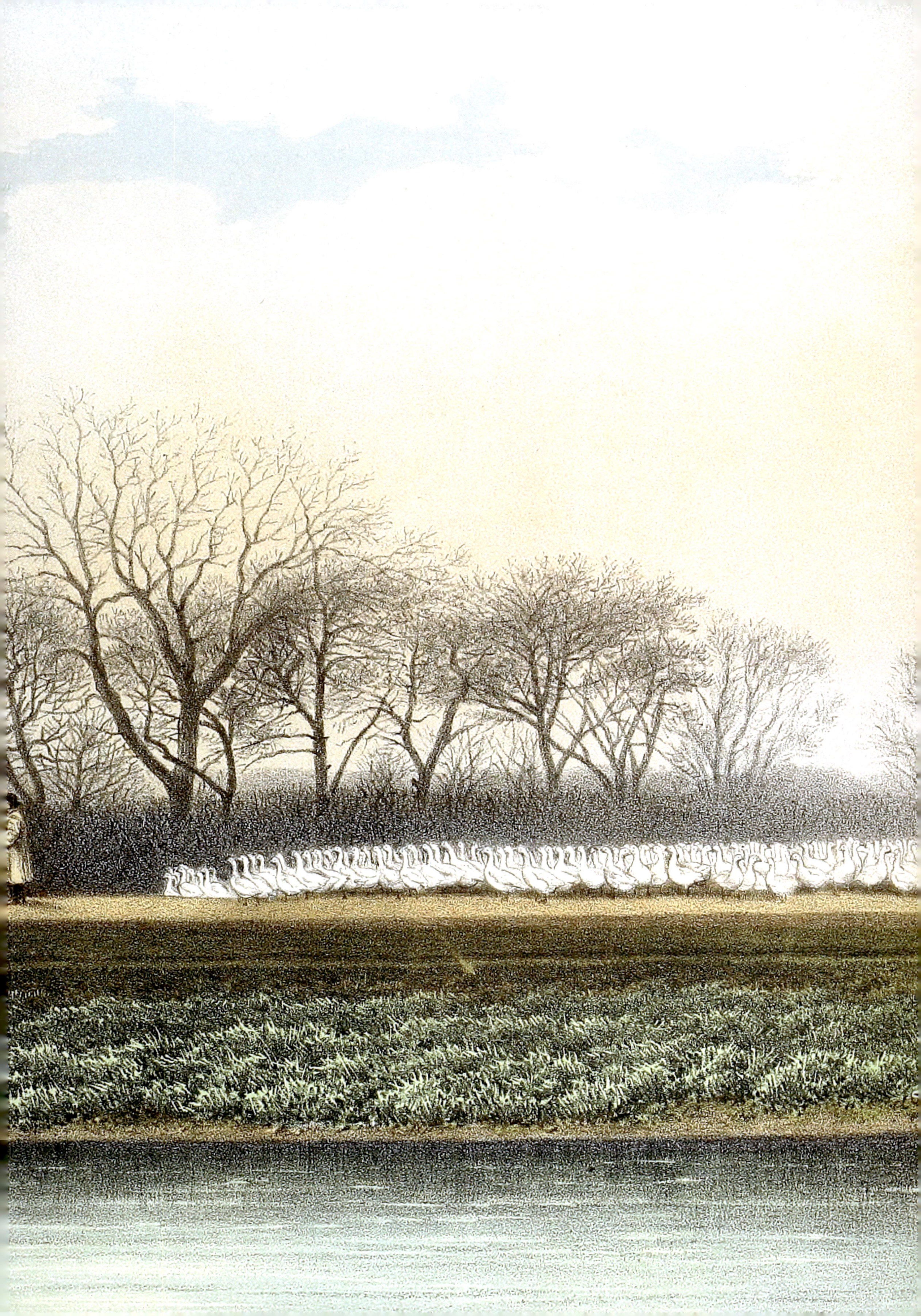

PREVIOUS PAGES:

MR. JOHN TAYLOR'S FLOCK OF 800 GEESE ON THE HORNCASTLE ROAD BY BARGATE DRAIN, BOSTON

Joseph Smit, *Ornithological Miscellany*, 1876–8

BLACK SWAN

Cygnus atratus

John Gould & Henry Constantine Richter, *The Birds of Australia*, 1840–8

Black swans are widespread throughout much of Australia. In 1697, the Dutch navigator Willem de Vlamingh was on a rescue mission on the Australian west coast, close to present-day Perth. While venturing up a river, he became the first European to encounter this stately black bird. So taken was he by the discovery that he named the river *Swarte Swaene-Revier*, or Swan River, and even attempted to bring some swans back to Amsterdam, but they died on the voyage.

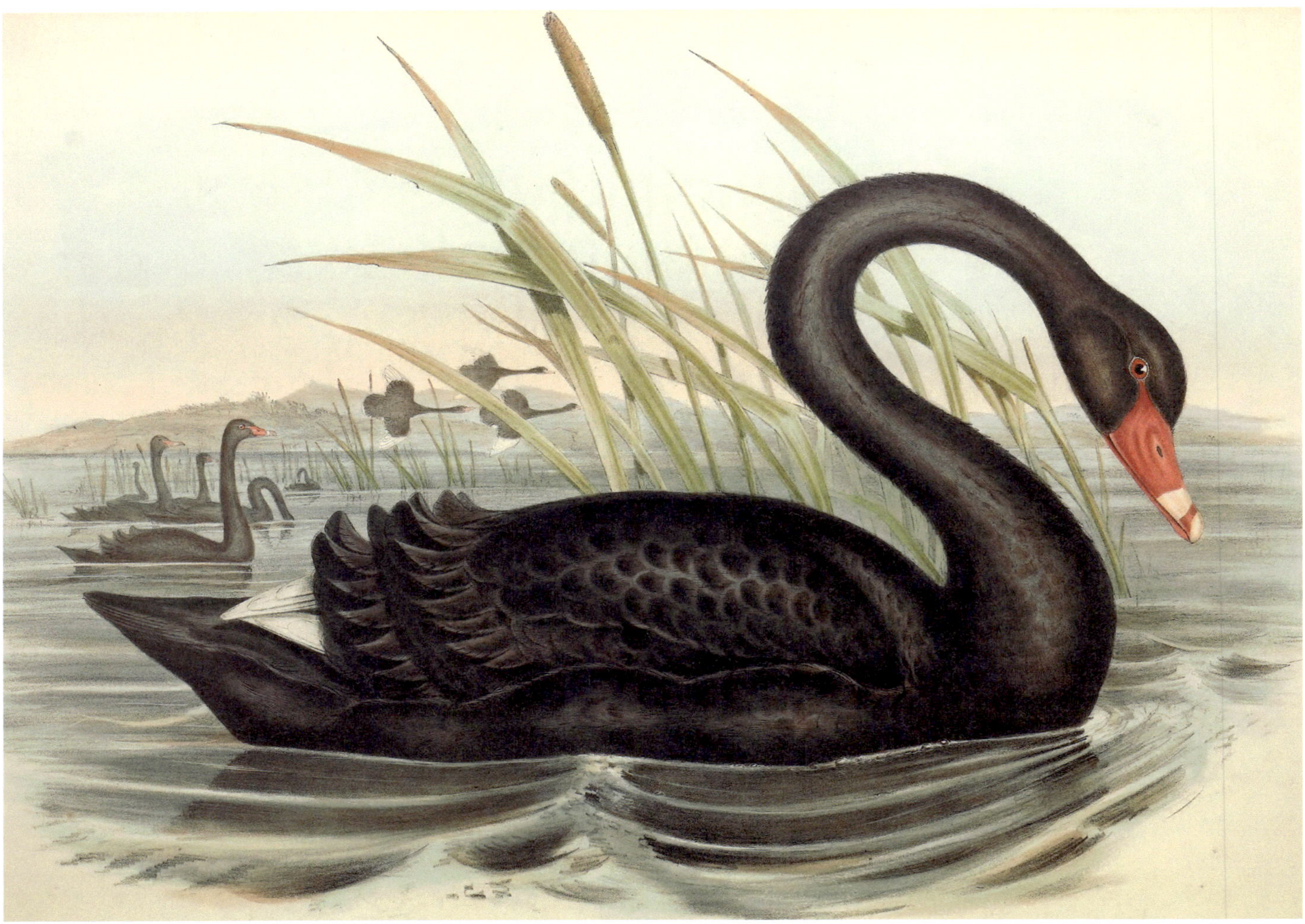

WHOOPER SWAN

Cygnus cygnus

Edward Lear, *The Birds of Europe*, 1832–7

While Edward Lear only contributed 62 of the 448 plates in Gould's *Birds of Europe*, his overall contribution was revelatory. Lear was a pioneer in how he used lithography for his illustrations and even learned to draw on the stones himself. His skill inspired Gould to enlist him in teaching Elizabeth Gould his lithographic methods. Through this partnership, Elizabeth's work transformed into impressive, expressive images that captured the character of each bird.

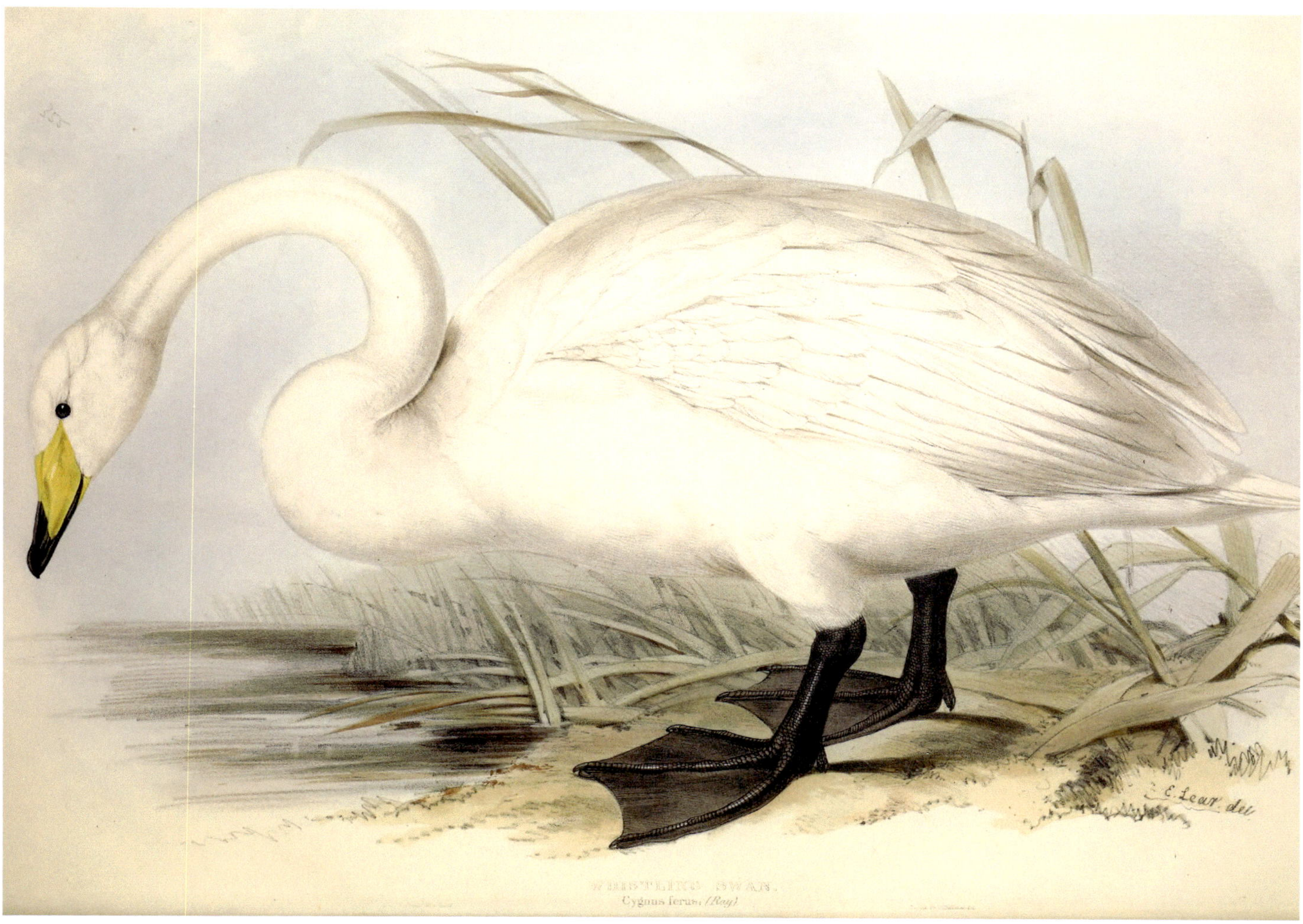

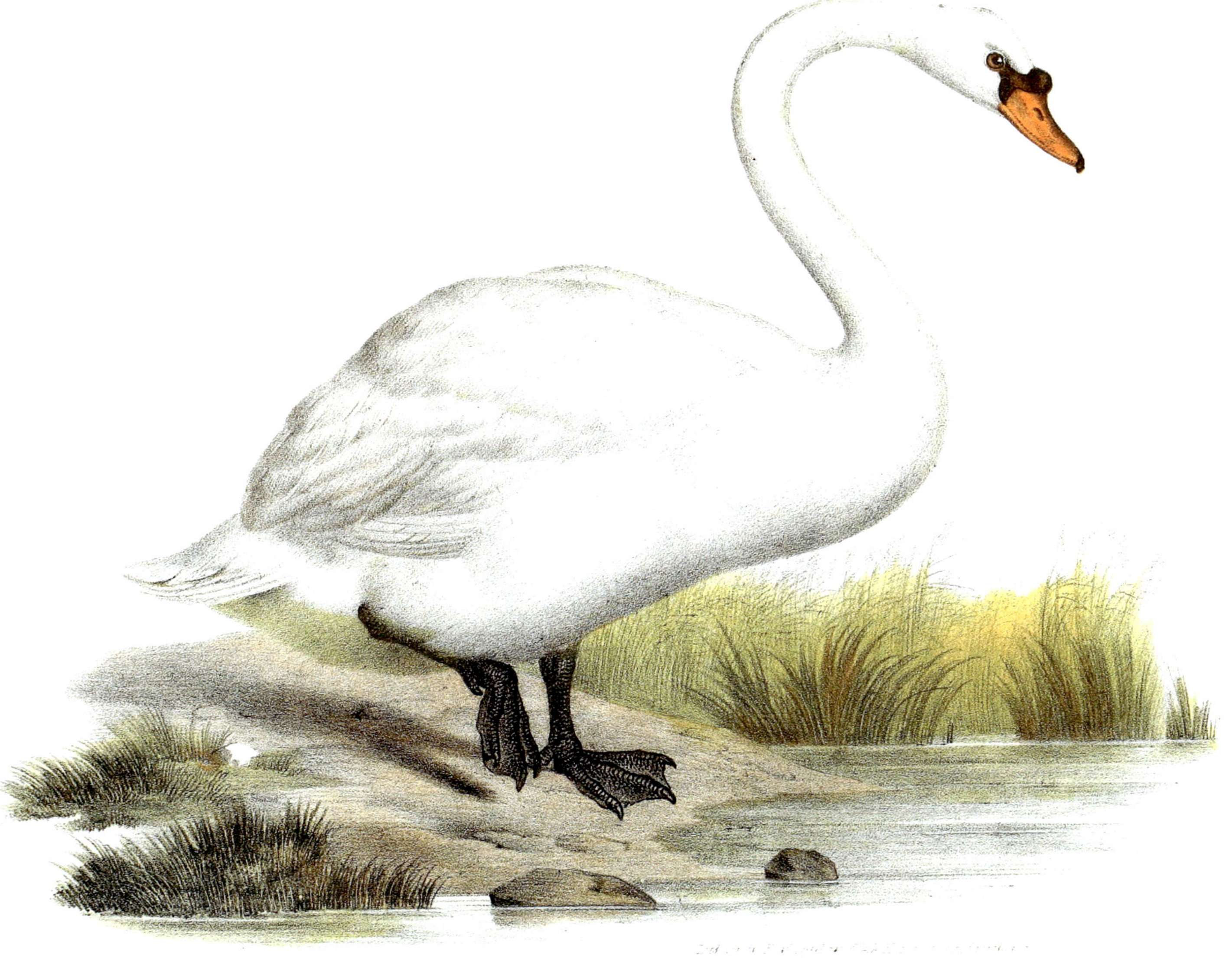

MUTE SWAN
Cygnus olor
Uncredited, *Bilder-atlas*, 1860–4

The phrase 'swan song' has its origins in a belief that Aristotle mentions in his *Historia animalium*. He writes that 'swans are musical, and sing chiefly at the approach of death'. This is obviously incorrect, as all swans make vocal sounds, but their calls hardly resemble songs. Even the mute swan – pictured here – produces a variety of sounds, and although they may be less vocal than other swan species, they will make a range of noises when disturbed or threatened.

GREAT CRESTED GREBES
Podiceps cristatus

Prideaux John Selby, *Illustrations of British Ornithology*, 1819–34

Selby's illustration of the great crested grebe shows an adult in both breeding plumage (left) and non-breeding plumage (right). During the breeding season, these birds develop an impressive plume on their head and an orange ruff around their necks. These water birds are not strictly waterfowl. They are noted for their elaborate mating displays, which tend to resemble courtship dances. Birds will pair off in springtime and engage in a bout of graceful head-shaking and feather-ruffling. Often, this leads to both birds diving deep underwater and resurfacing with waterweed in their bill. They then swim rapidly towards each other and, when they meet breast-to-breast, they rise upwards and paddle vigorously to keep their bodies high above the water.

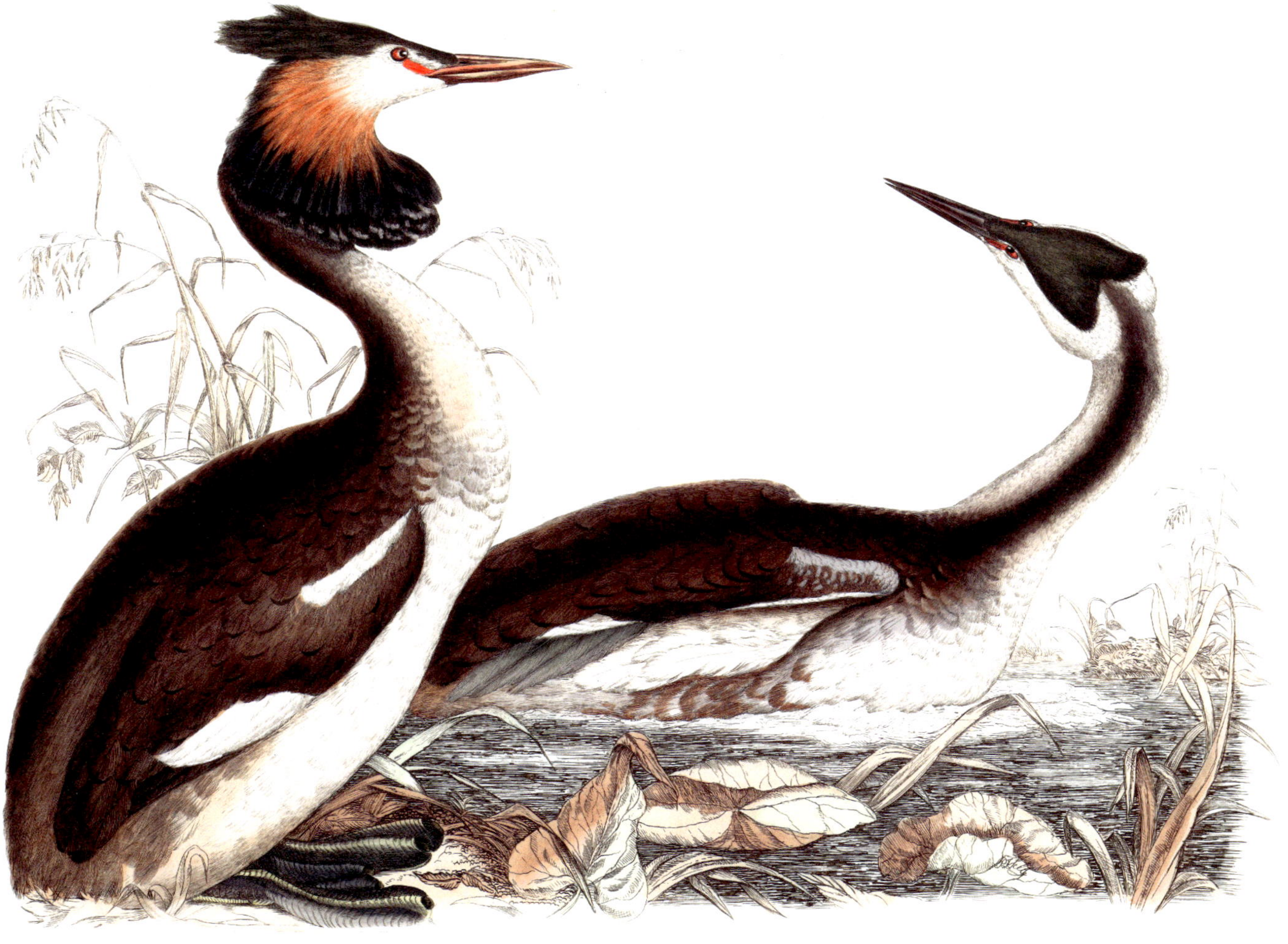

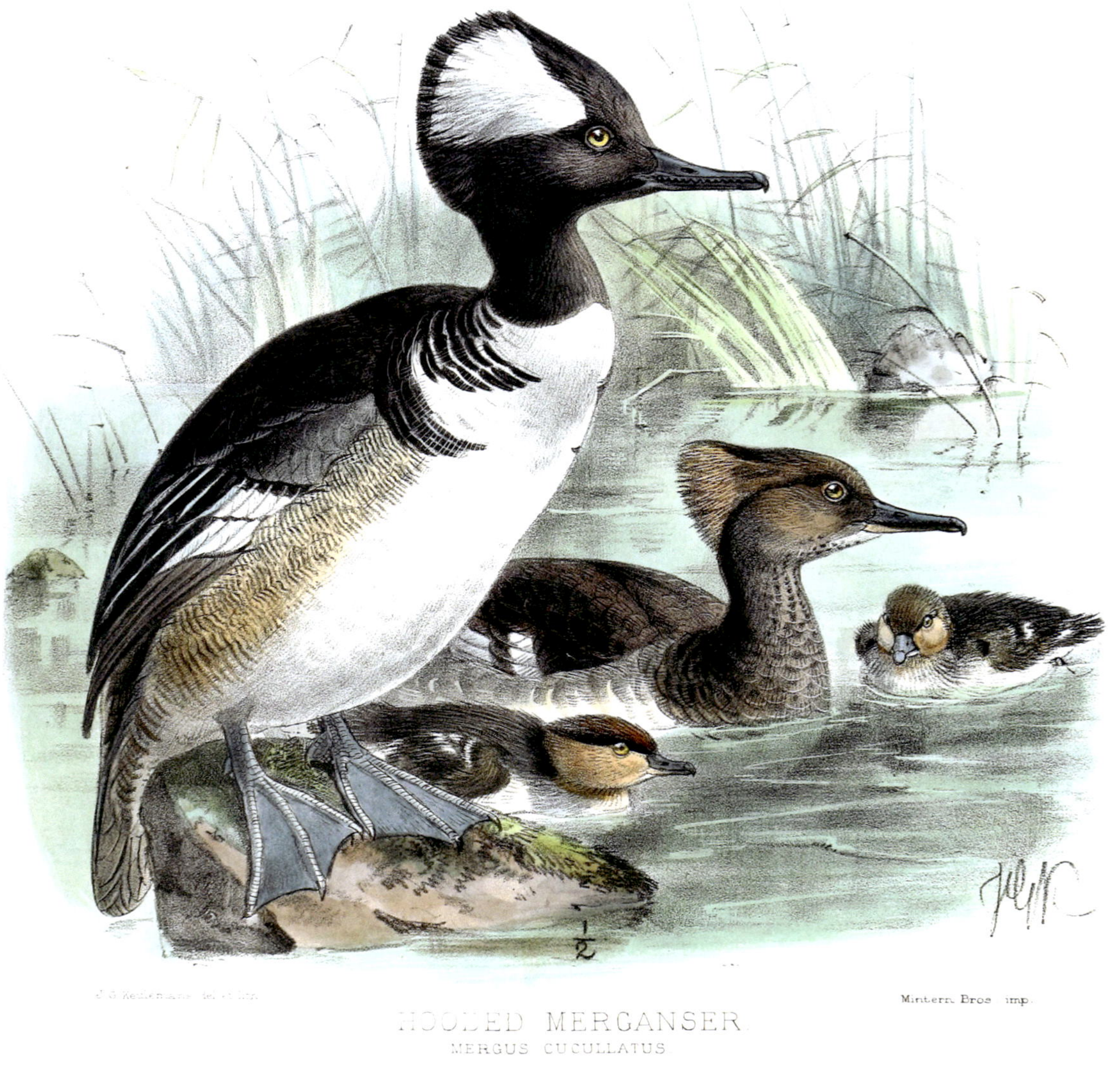

HOODED MERGANSERS
Lophodytes cucullatus

John Gerrard Keulemans, *A History of the Birds of Europe*, 1871–81

Chiefly considered to be a North American bird, the existence of a few rare stragglers in the British Isles has allowed this species to sneak into Henry Eeles Dresser's *A History of the Birds of Europe*. Today, these birds are commonly kept in captivity in Europe, with most records of the bird in the wild believed to be birds that have escaped. While they are short-distance migrants, a small number of European hooded mergansers are accepted as being genuine wild vagrants. Male breeding plumage is particularly distinctive, with an extravagant black-and-white crest that is often even more pronounced than the one featured in this plate.

A WILD DUCK NEAR A SNOW-LADEN SHORE
Utagawa Hiroshige, woodblock print, *ca.* 1843

This woodblock print by Hiroshige shows a mallard duck beside snow-covered reeds. Hiroshige was a prominent Japanese *ukiyo-e* artist who many consider being the last great master of the tradition. He gained popularity producing landscape prints (*fūkei-ga*) and images depicting famous Japanese places in Japan (*meishō*). He also had great success producing *kachō-e* – prints, such as this, that depict birds and flowers. The text above the bird is a *haiku* – a traditional Japanese poem of seventeen syllables that often reflects upon nature. It reads:

Ducks cry and wind
makes the ripples on the water

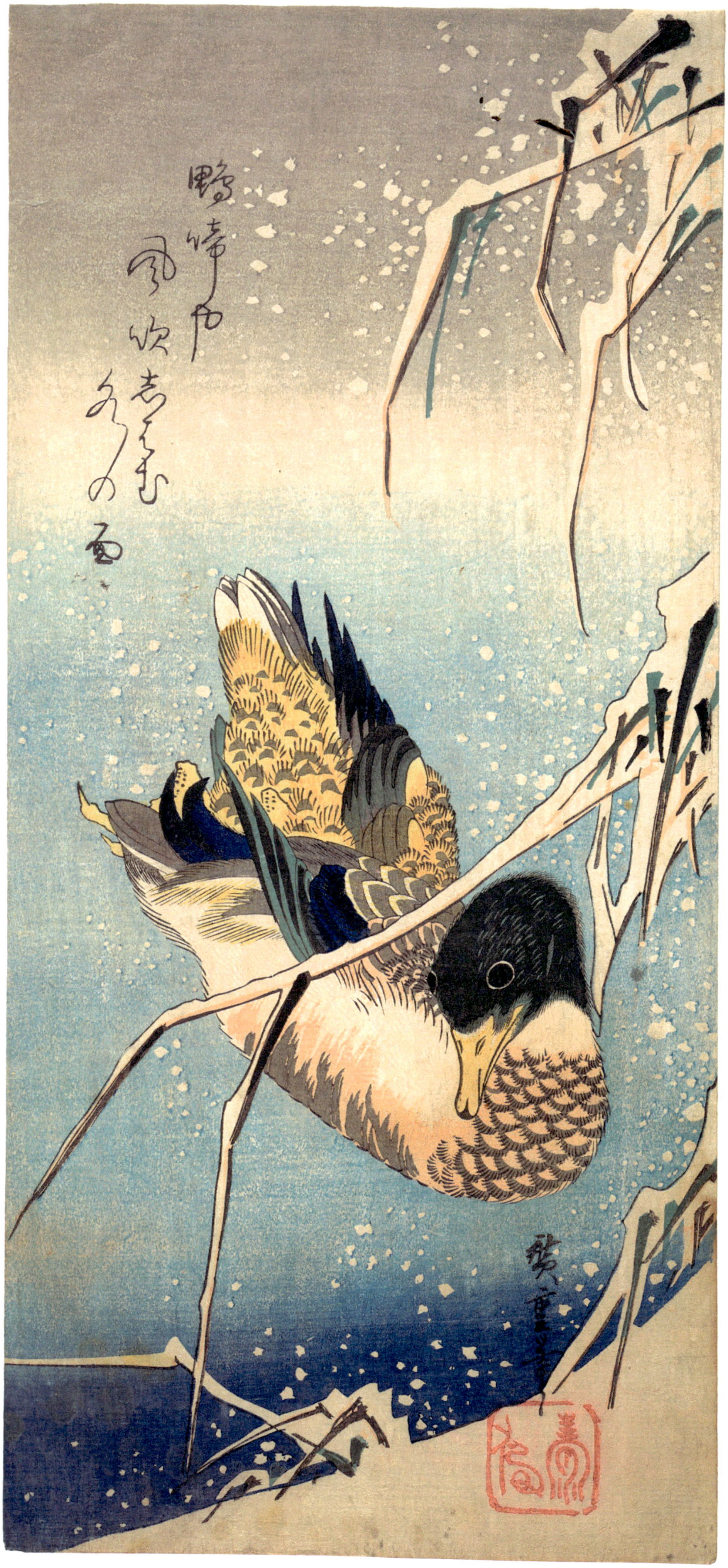

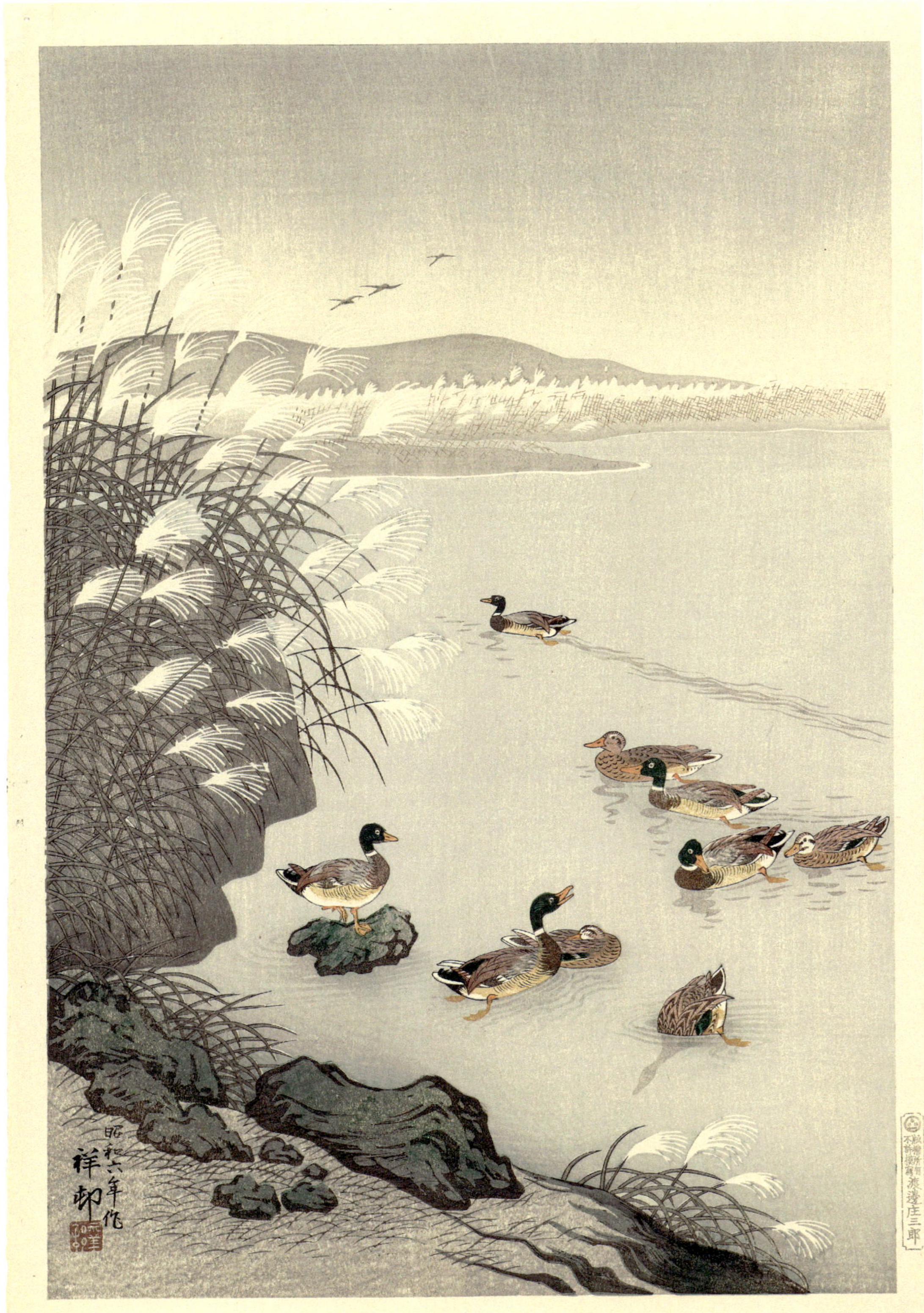

DUCKS IN THE WATER

Ohara Koson, woodcut print, 1931

Influenced by Hiroshige, Ohara Koson was a Japanese printmaker who belonged to the early twentieth century Japanese *shin-hanga* movement. Translated as 'new prints', this movement aimed at reviving the traditional *ukiyo-e* style of printmaking, and Koson was the movement's best-known *kacho-e* artist. During his lifetime he earned little recognition in his native Japan, but his prints were popular abroad, where Japanese art had become a popular commodity with collectors in Europe and the United States.

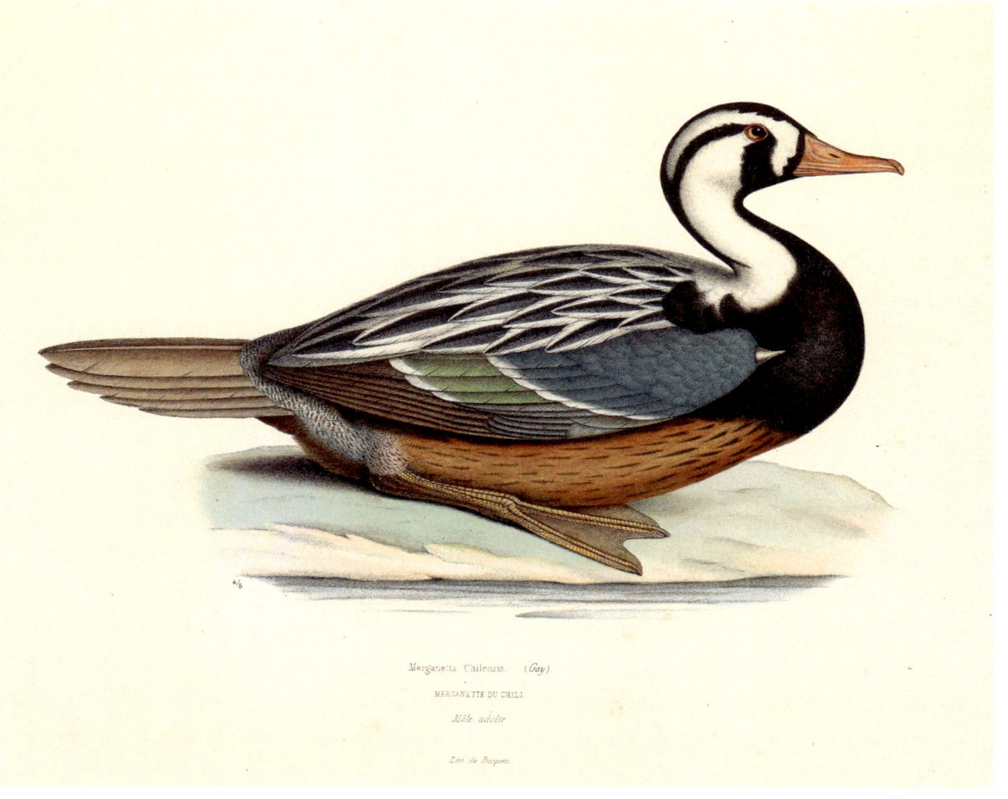

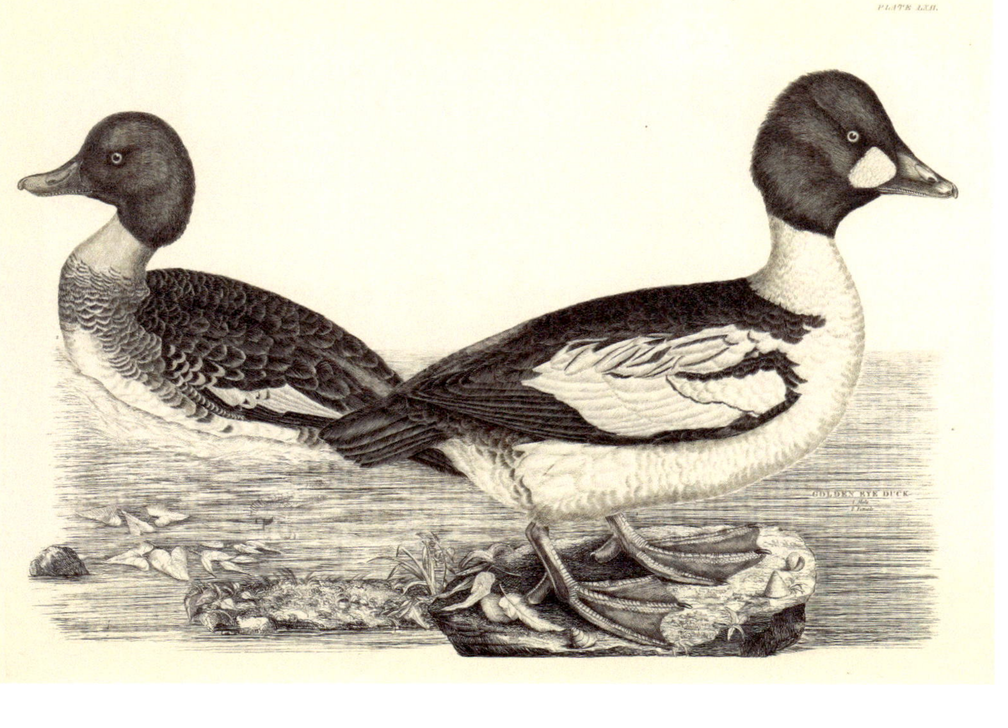

TOP:

TORRENT DUCK
Merganetta armata

Florent Prévost, *Iconographie ornithologique*, 1845–9

A male torrent duck from the French book *Iconographie ornithologique*. Published between 1845 and 1849, it was a work undertaken by the ornithologist Marc Athanase Parfait Œillet Des Murs.

BOTTOM:

COMMON GOLDENEYES
(FEMALE: LEFT, MALE: RIGHT)
Bucephala clangula

Prideaux John Selby, *Illustrations of British Ornithology*, 1819–34

Selby's goldeneye comes from *Illustrations of British Ornithology*. Often the illustrations in this work were hand-coloured, but in this edition the plates remain in black-and-white. While beautiful in their own regard, it means that we lose important visual information, such as this bird's distinctive yellow eyes – from which its name is derived.

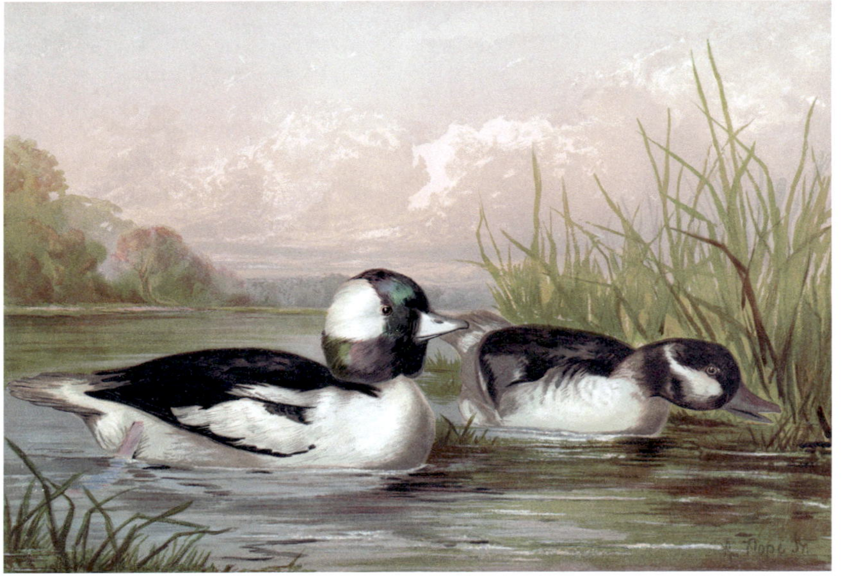

BUFFLEHEADS
(LEFT: MALE, RIGHT: FEMALE)
Bucephala albeola

Alexander Pope Jr., *Upland Game Birds and Water Fowl of the United States*, 1877–8

Buffleheads are small ducks with large bulbous heads and short beaks. Their name is derived from 'buffalo-head', in reference to the large puffy head shape of the male. Both the male and female have a white patch behind their eyes, with the male's patch being large in size. The male's head plumage also has a striking purple-green iridescence, which can be seen here captured effortlessly in this work by Alexander Pope Jr.

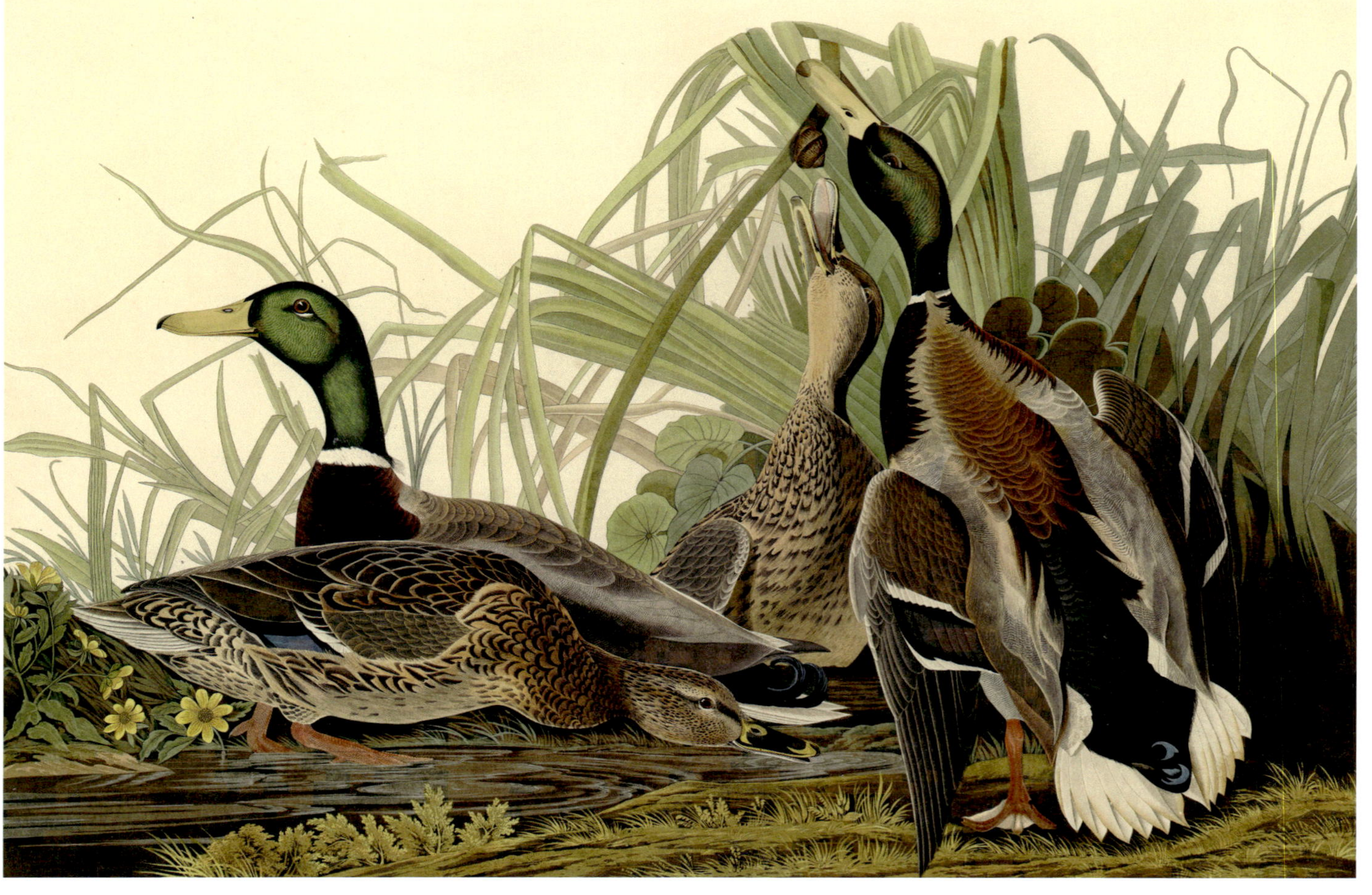

MALLARDS
(1. MALES 2. FEMALES)
Anas platyrhynchos

John James Audubon, *The Birds of America*, 1827–38

Audubon's depiction of this brace of mallards demonstrates his acute ability to capture the liveliness and spirit of his subjects. Unlike other illustrations, Audubon's work never feels stiff or posed. Instead, his illustrations are imbued with life. This naturalistic scene shows the female moments before eating a snail, while the male demonstrates the bird's ability to bend tall reeds. These small details show the artist's skill in observation and highlights the benefits of initially sketching one's subjects in the wild.

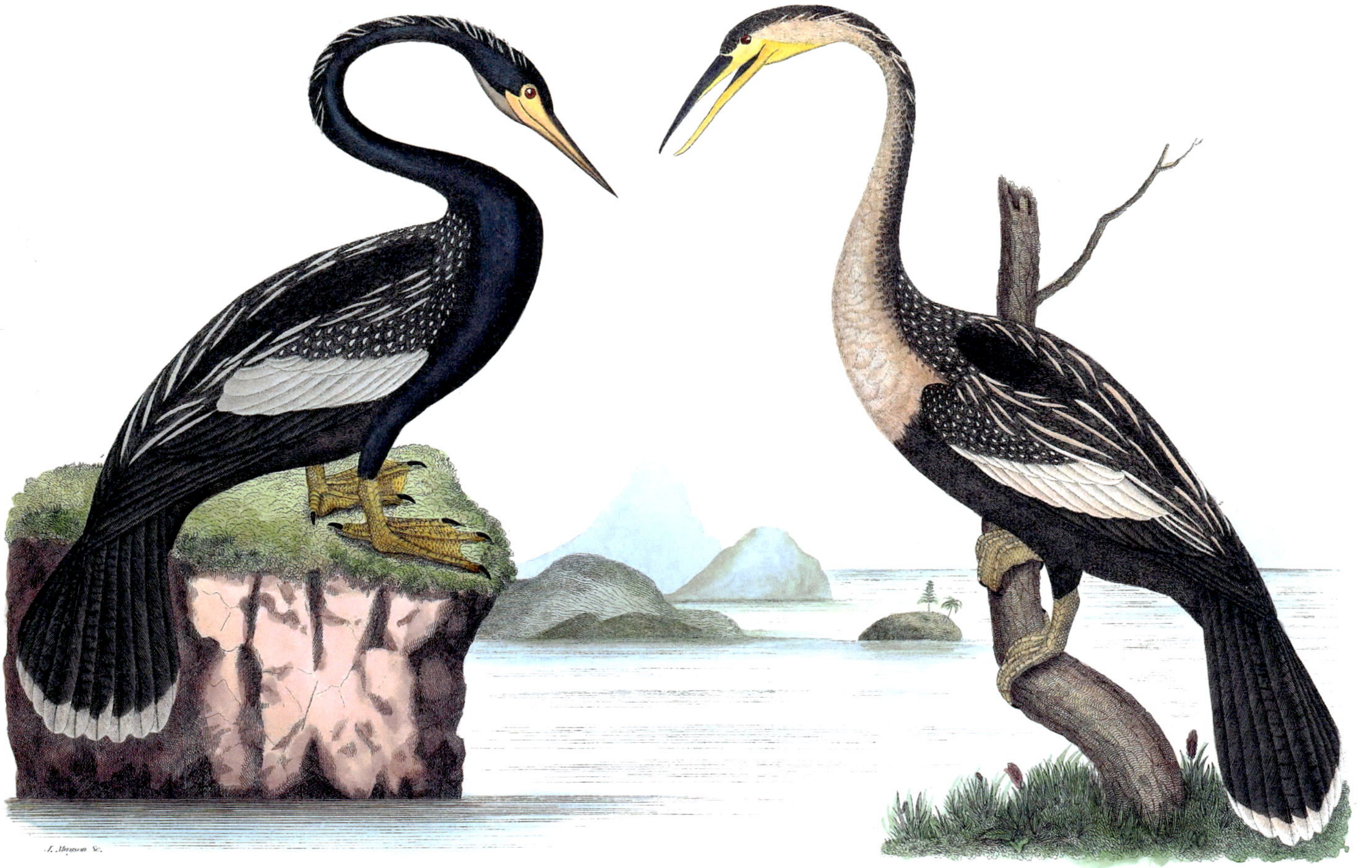

AMERICAN DARTERS
Anhinga anhinga

Alexander Wilson, *Illustrations of the American Ornithology of Alexander Wilson and Charles Lucian Bonaparte*, 1835

This plate by Alexander Wilson shows both a male (left) and female (right) darter. This species is found in warmer parts of the Americas, where the bird is commonly often called the 'snakebird'. This is on account of its long and slender neck, which tends to resemble a snake when the bird is swimming with just its head and neck above the water. Wilson's work described a total of 262 species of bird that he had seen in North America. Earlier naturalists, such as Mark Catesby, had recorded a number of these birds already, but Wilson was the first to describe and classify them using the Linnaean system.

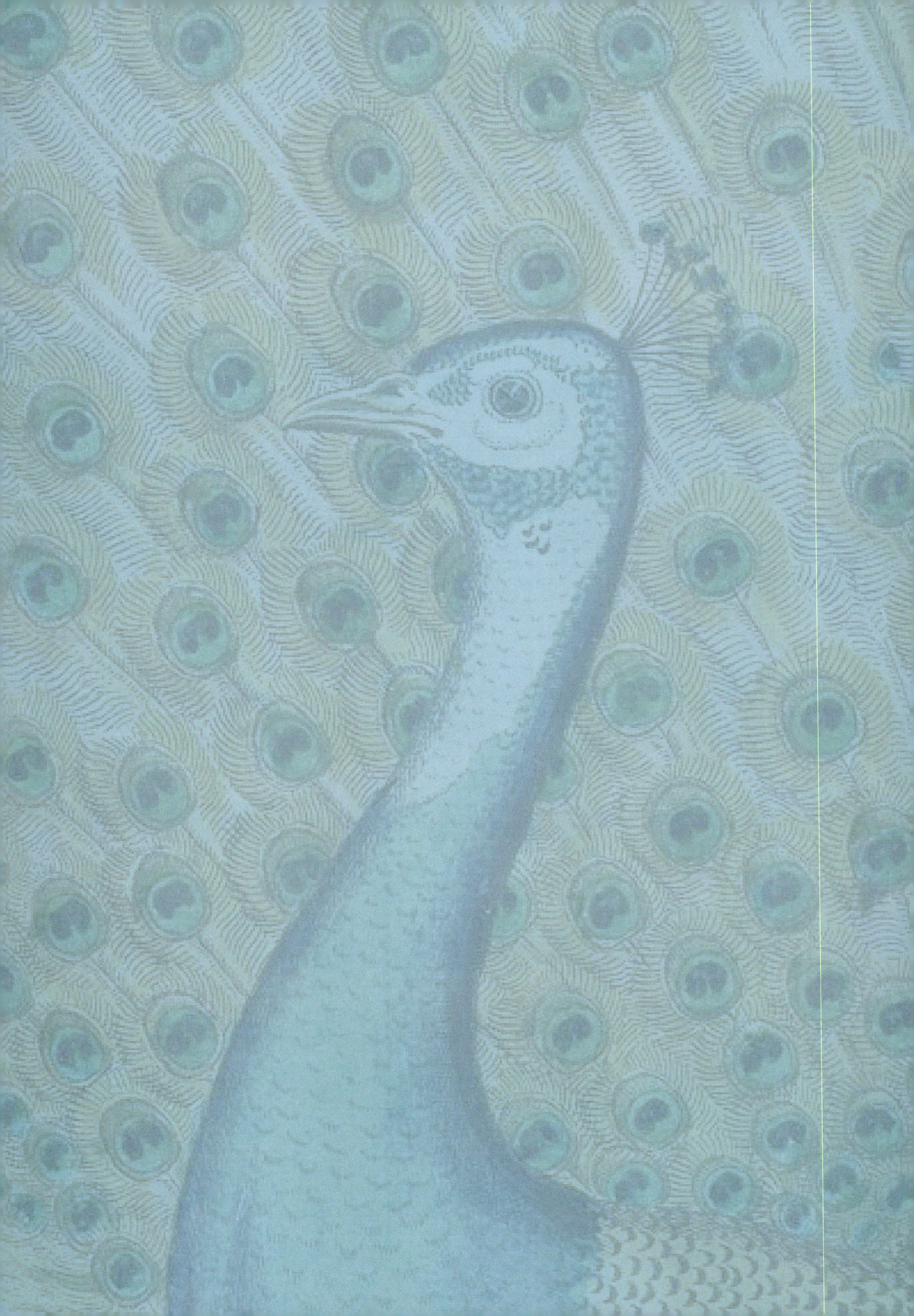

Gamebirds & Domesticated Birds

For millennia, people have hunted birds of all kinds. Owing to their plump bodies and large flight muscles, the most popular targets have often been those we term today as 'gamebirds', a group of related species that includes quails, grouse, pheasants, partridges and turkeys. A number of these birds have also been domesticated worldwide.

As well as being hunted, over the centuries a few choice species of gamebird have also become domesticated for food, eggs or even feathers. Many of these are in the order Galliformes, which includes grouse, guineafowl, pheasants, turkeys and chickens, although several waterfowl (see pp. 96–127) have also been domesticated, along with pigeons and, in places, others such as ostriches. In many cases, the process of domestication can change a bird's appearance through inbreeding. Over the centuries, chickens and pigeons have been bred into all manner of shapes and colours, far removed from what they look like in the wild.

These days, an estimated 70 per cent of all the individual birds on this planet are farmed poultry, well in excess of all wild birds put together. Most of these are chickens.

The illustrations in this chapter come from a time before poultry farming had become the massive industry that it is today. They were made at a time when fowl was either kept in small broods in individual households or was found in the wild. In Europe, households commonly kept birds for their meat and eggs, but it took until the late 1800s for poultry farming to become a large-scale enterprise. Modern industrial poultry farms emerged in the United Kingdom during the 1920s, and it wasn't until after the Second World War that the practice was established in the United States.

Today's domestic chicken (*Gallus gallus domesticus*) evolved from jungle fowl in Southeast Asia, and this species spread to Europe through trade and military campaigns. It was through domestication and intensive factory farming that these birds evolved. Human intervention has found ways to adapt birds purely for commercial gain, with selective breeding being used for enhancing their growth and increasing their egg-laying capacity.

The hunting of birds became a source of entertainment as well as food throughout western Europe during the high Middle

Ages. The hunt became a pastime of the aristocracy and the word 'game' was adopted to define any animal that was hunted chiefly for sport. Pheasants were, and still are, considered to be one of the prime game birds and they are valued greatly for both their meat and their feathers. The quail is another common game bird whose speckled eggs are considered a delicacy throughout much of the world.

When European settlers first arrived in North America, they found that the continent was teeming with wild turkey (*Meleagris gallopavo*). The tradition of American Thanksgiving can be traced back as far as the 1620s, with colonists hunting wildfowl to mark their first harvest in the New World. The sheer size of these birds made them ideal for eating at celebratory meals, and turkeys became a central part of both the traditional American Thanksgiving and the traditional British Christmas meal. The bird also played a role – albeit anecdotally – in American national identity; Benjamin Franklin suggested that the turkey would make a far more respectable national symbol when compared to the bad moral character he saw in the now symbolic 'American' bald eagle (*Haliaeetus leucocephalus*).

When Audubon produced his *Birds of America*, he ensured that the turkey was on the very first pages. Not only did this life-sized illustration demonstrate the need for the book's enormous double-elephant folio, but it also seemed fitting that the book's first plate would be a bird native to North America.

Many of the illustrations in this chapter are taken from works that present an illustrated survey of a specific geographic region, yet a small selection also come from books that focus on specific birds, most notably John Gould's *Partridges of America* (1844–50) and William Beebe's *Monograph of the Pheasants* (1918–22). Such monographs demonstrate the sheer scale of diversity and variation that can be found within this specific order of birds.

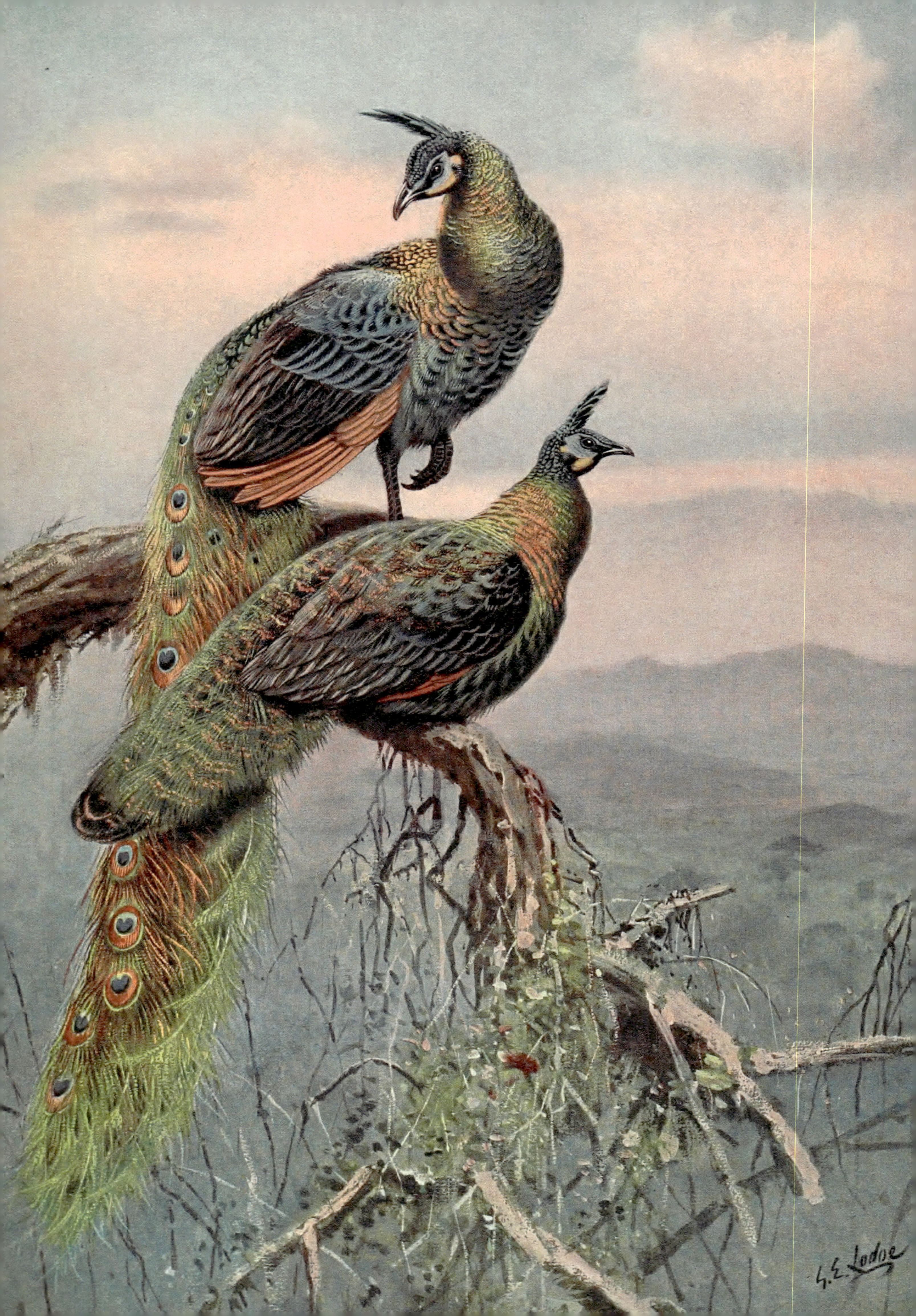
G.E. Lodge

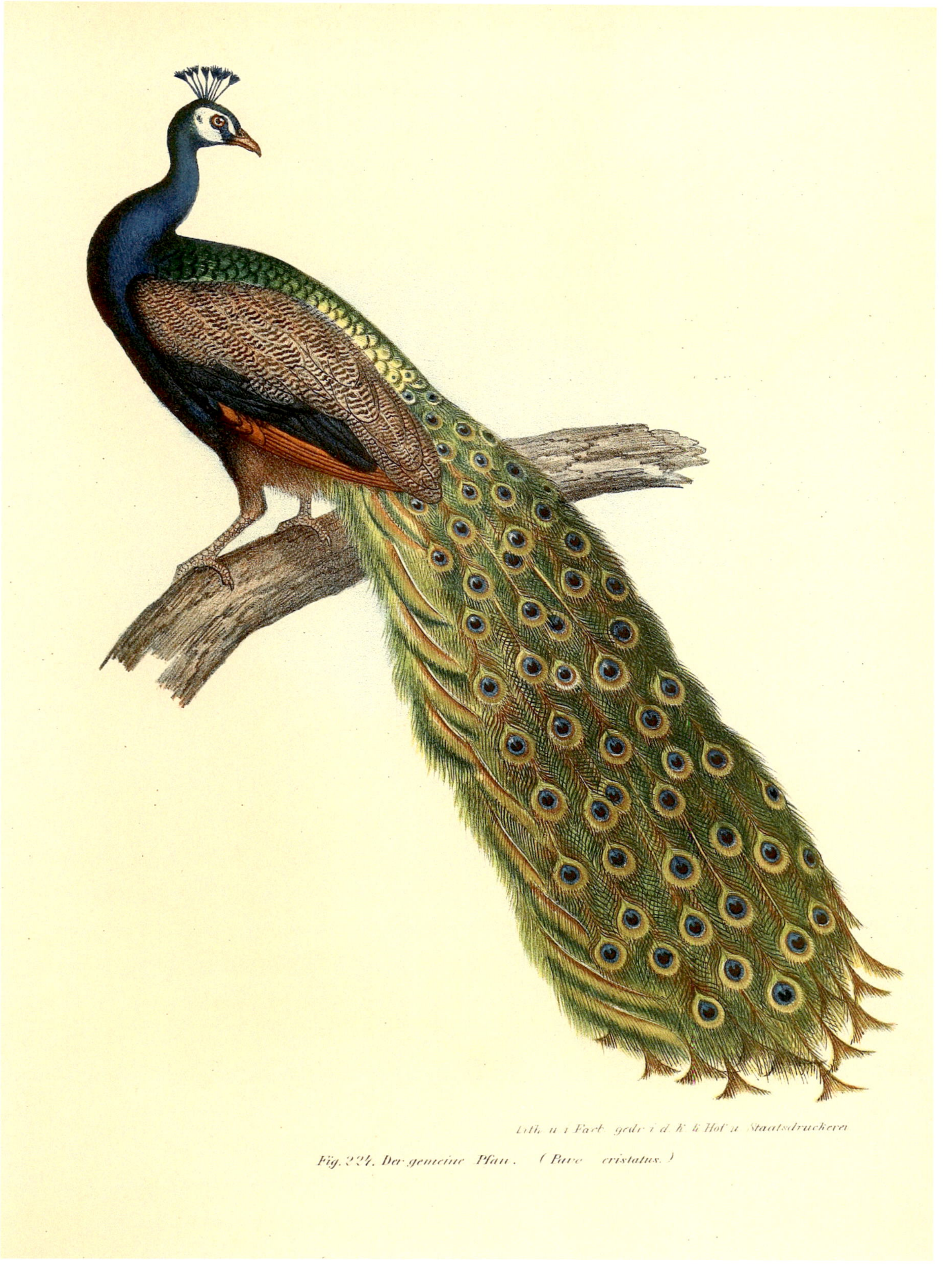

OPPOSITE:

GREEN PEAFOWL (MALE,ABOVE, AND FEMALE, BELOW)
Pavo muticus

George Edward Lodge, *A Monograph of the Pheasants*, 1918–22

INDIAN PEAFOWL (MALE)
Pavo cristatus

Uncredited, *Bilder-atlas*, 1860–4

Peafowl are one of the more spectacular examples of sexual dimorphism. The peacock, the male peafowl, sports a train of long, flamboyant tail feathers which they fan outwards as part of a courtship ritual to attract a mate. The female, or peahen, is more subdued in its appearance, with muted colours and a shorter tail. The peacock's tail is rich with iridescent colour and consists of many multicoloured eyespots that have a purple-black centre surrounded by regions of blue-green and gold-bronze.

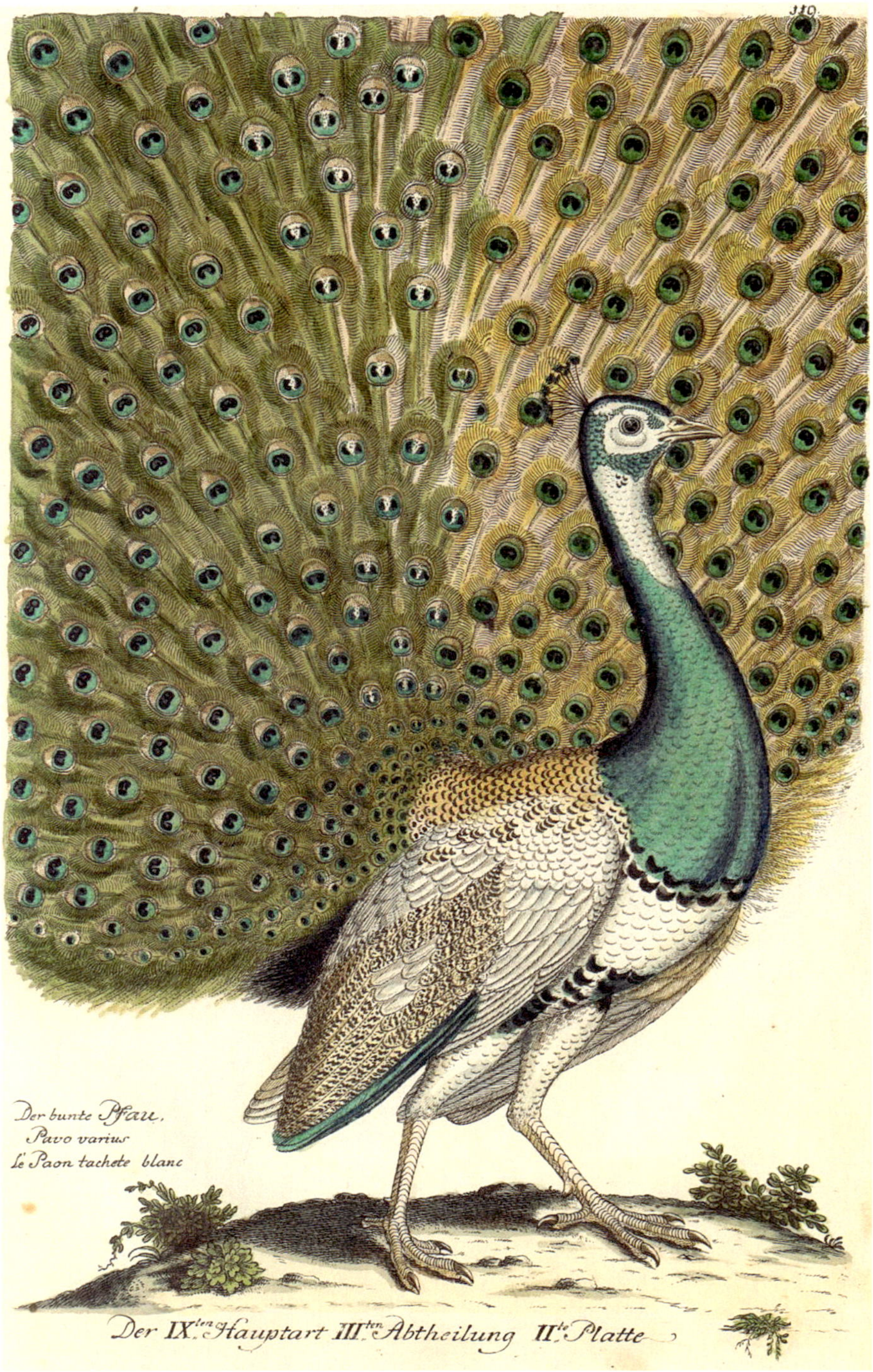

INDIAN PEAFOWL (MALE)
Pavo cristatus

Philipp Jakob Frisch & Ferdinand Helfrich Frisch, *Vorstellung der Vögel Deutschlandes und beyläufig auch einiger Fremden*, 1733–63

One of the first great German bird books, *Vorstellung der Vögel Deutschlandes und beyläufig auch einiger Fremden* was a monumental work that was undertaken by German naturalist and engraver Johann Leonhard Frisch. Published over a period of thirty years, it became a family affair when Frisch died in 1743. From that date forward, his sons, Philipp Jakob Frisch and Ferdinand Helfrich Frisch provided sketches and produced the copper plates, while his youngest son, Leopold Just Frisch, wrote the remaining text.

OPPOSITE:
INDIAN PEAFOWL (FEMALE, LEFT, AND MALE, RIGHT)

Charles Robert Knight, *A Monograph of the Pheasants*, 1918–22

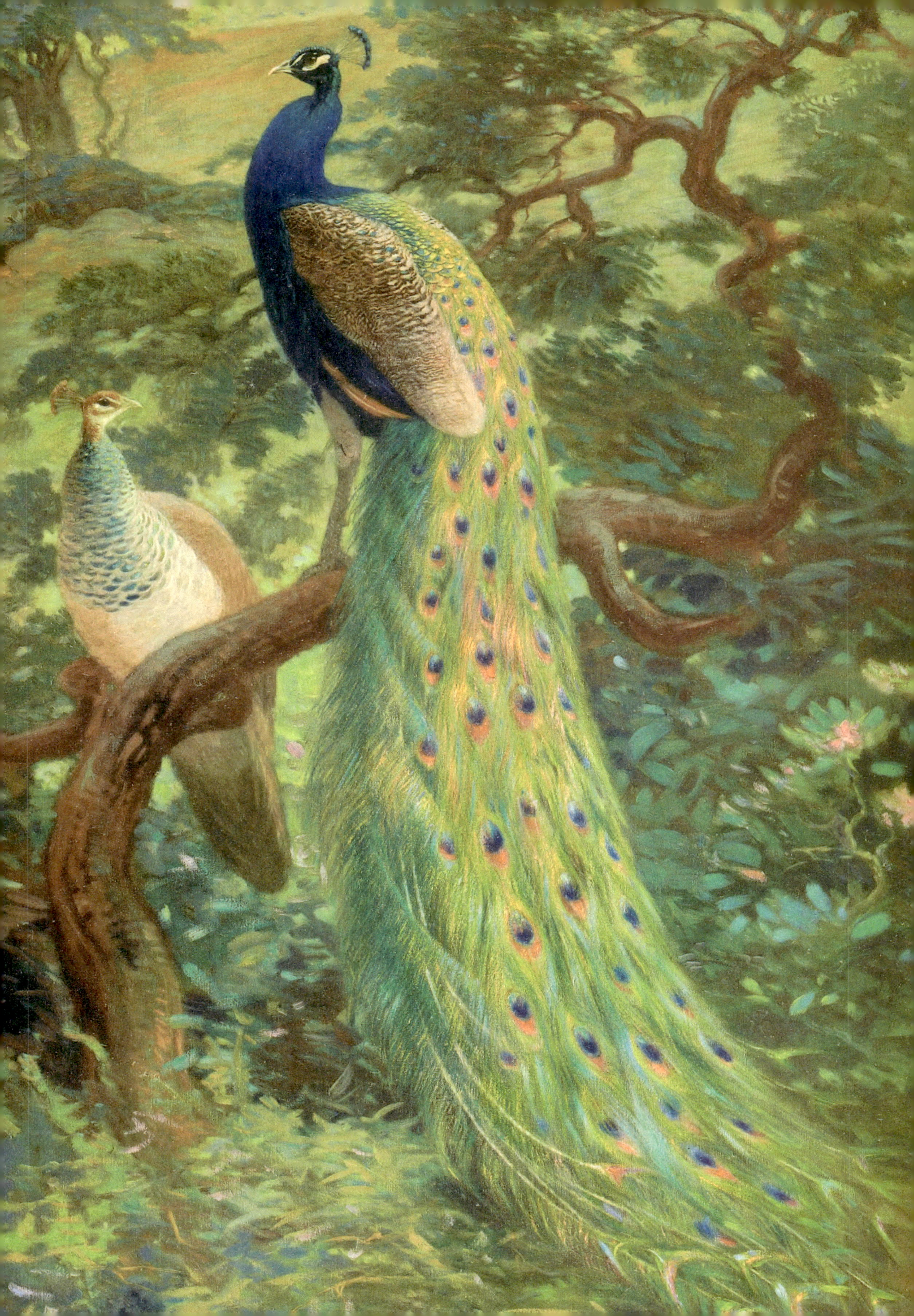

INDIAN PEAFOWL (MALE)
Pavo cristatus

Pieter Pietersz Barbiers, watercolour and pencil on paper, 1759–1842

Pieter Pietersz Barbiers was a nineteenth-century painter and draughtsman from the Netherlands. Born in Amsterdam, he is best known for painting pastoral landscapes, but a number of his drawings and watercolours also survive. While many of these are of nature scenes, a number of them include studies of birds and animals. This work, for example, shows an Indian peafowl. There are two closely related species of peafowl – the Indian peafowl (*Pavo cristatus*), which is native to the Indian subcontinent, and the green peafowl (*Pavo muticus*), which is found in the tropical forests of Southeast Asia.

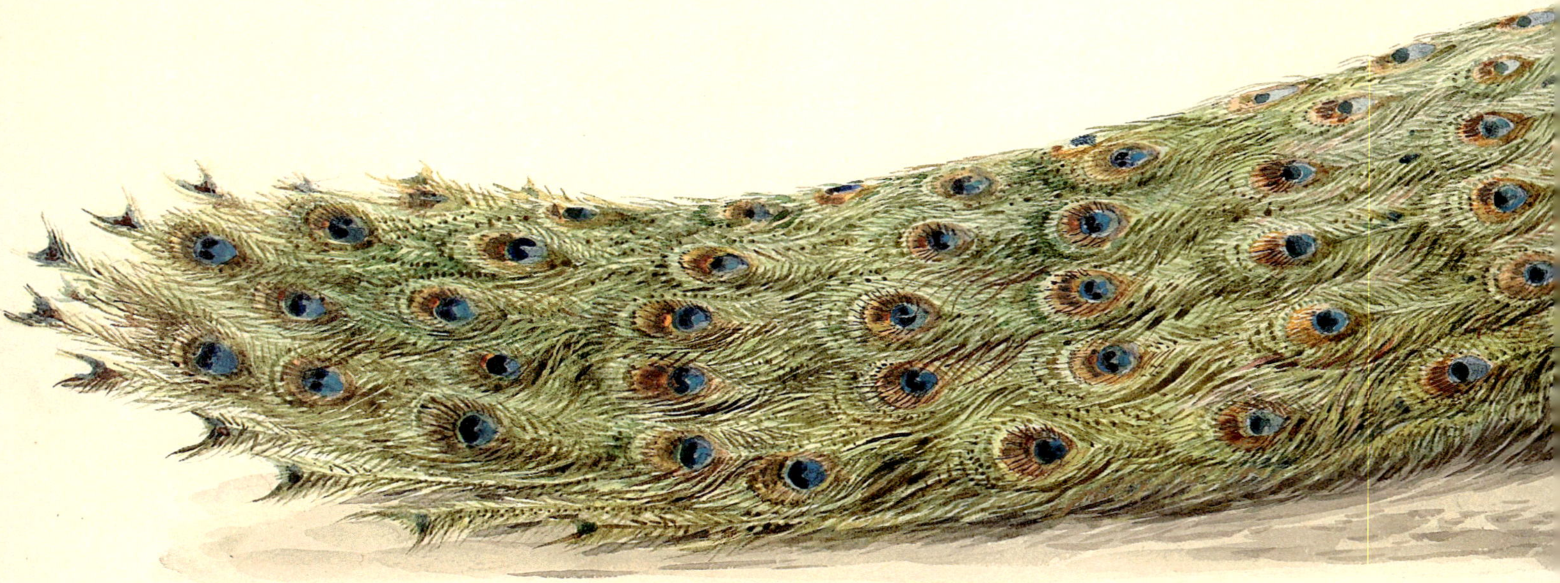

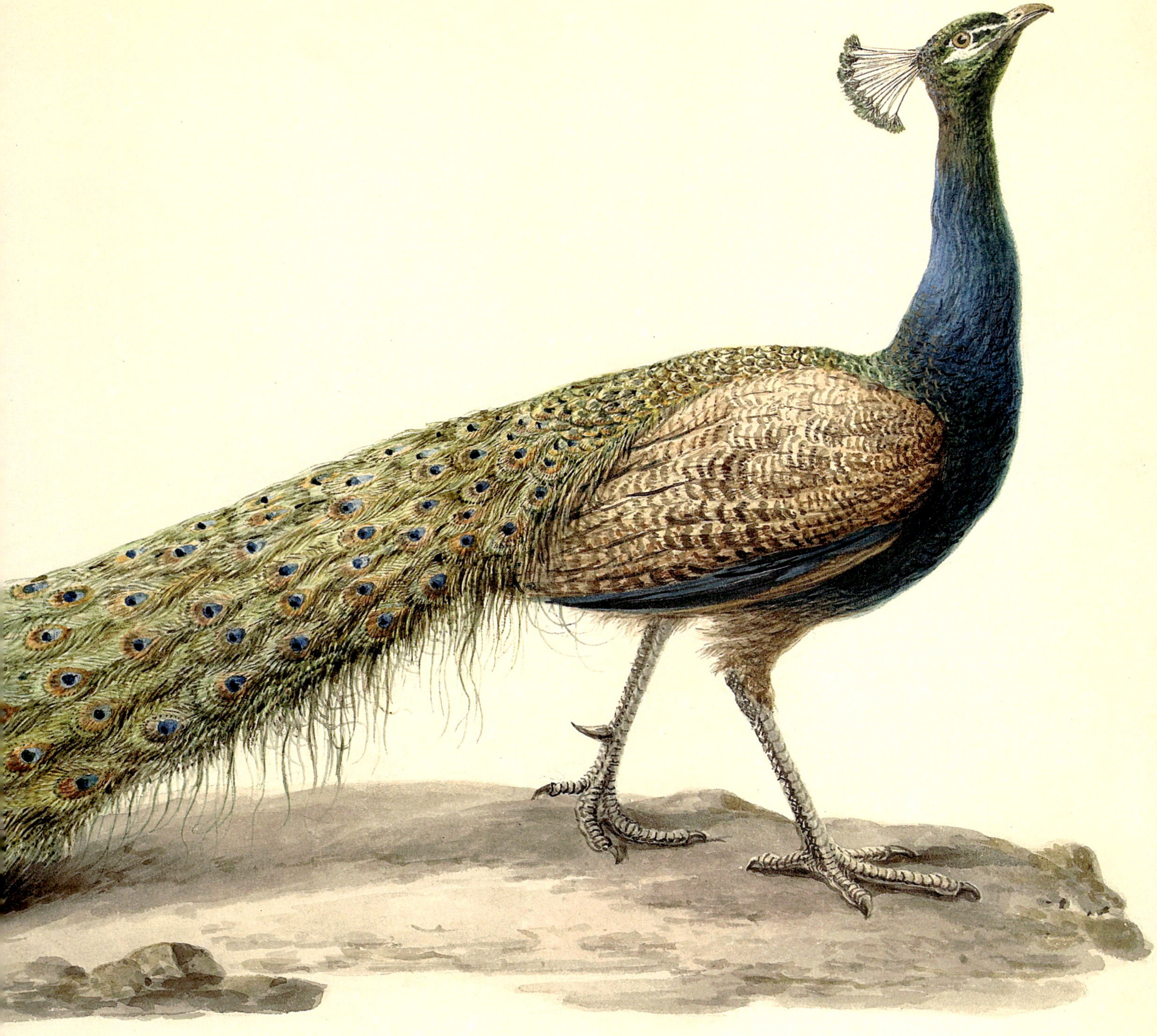

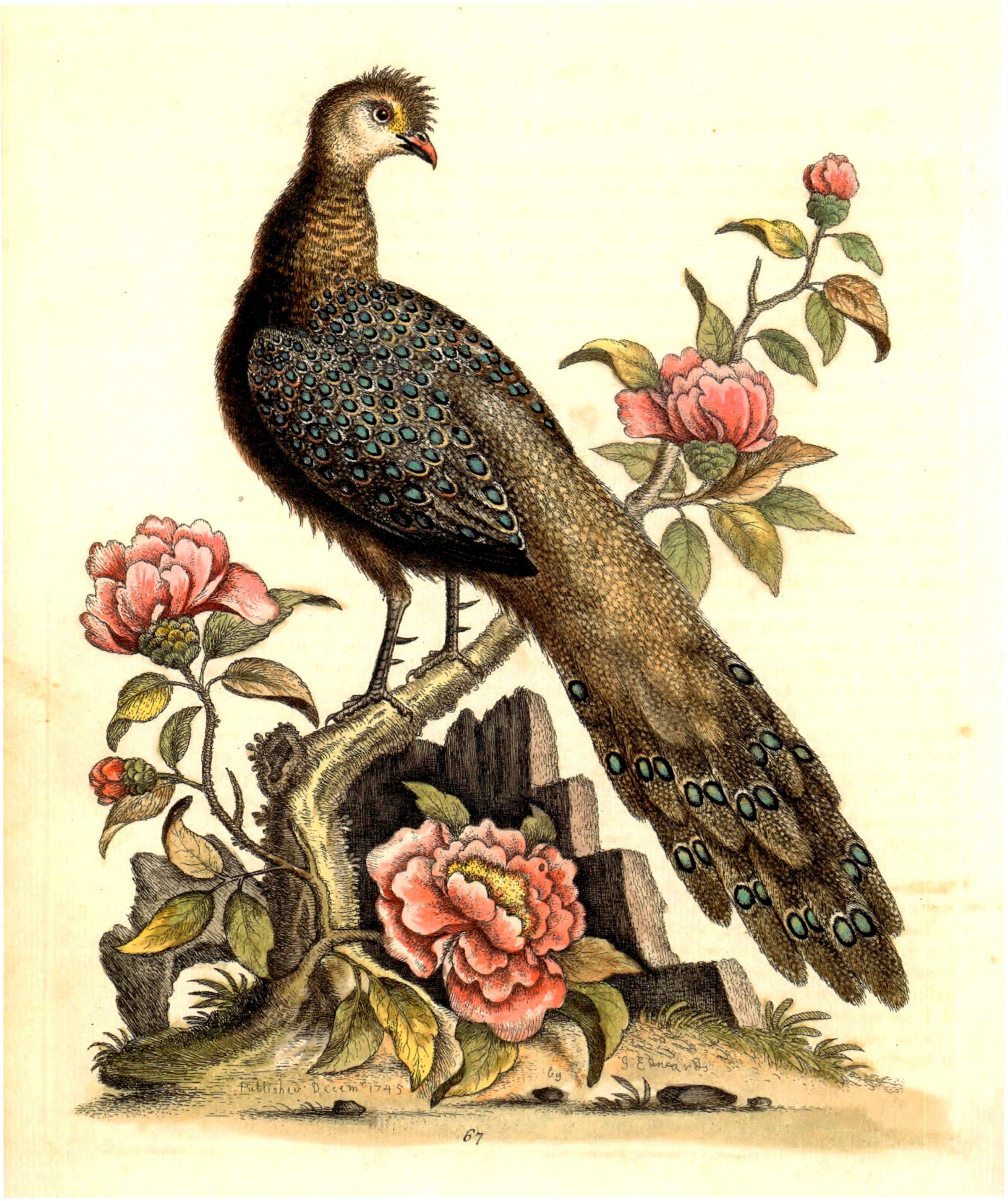

GREY PEACOCK-PHEASANT
Polyplectron bicalcaratum

George Edwards, *A Natural History of Uncommon Birds*, 1743–51

George Edwards was a skilled observer of birds who was able to pursue his passion thanks to the patronage of the collector Sir Hans Sloane. President of Royal College of Physicians, Sloane appointed Edwards as a librarian to the college in 1733. This was an undemanding job and it allowed Sloane the opportunity to employ Edwards to produce drawings for him. Sloane was a renowned doctor among the aristocracy who had built one of the largest private natural history collections in the world. The size of the collection kept Edwards in employment for many years.

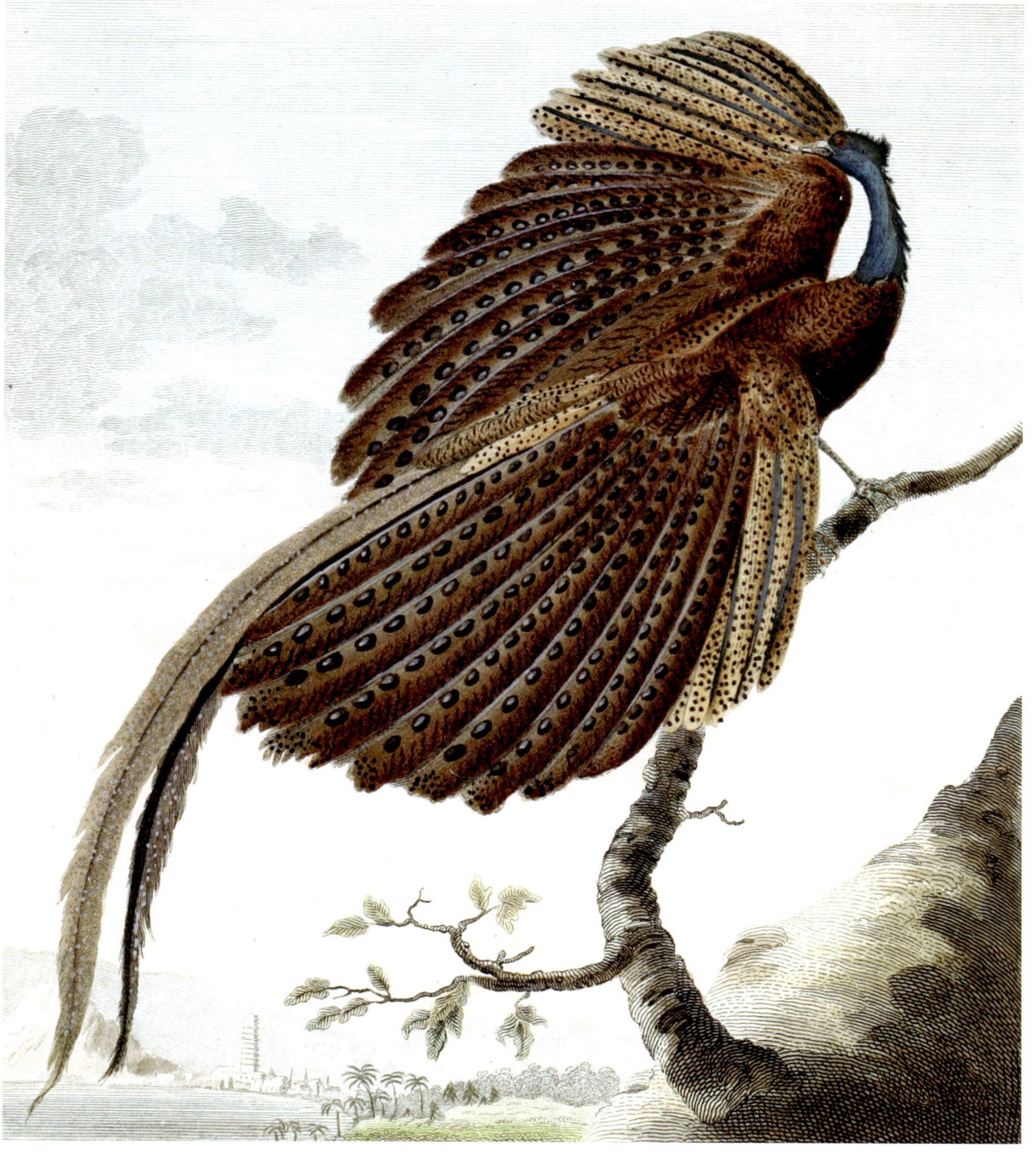

GREAT ARGUS
Argusianus argus

Charles Reuben Ryley,
Museum Leverianum, 1792–6

This species of pheasant boasts an impressive eye-like pattern on its wings. It was Carl Linneaus who chose to call it Argus in reference to Argus Panoptes, a hundred-eyed giant in Greek mythology. Native to Southeast Asia, its habitat is emphasized in this work by Charles Reuben Ryley through the inclusion of a pagoda in the background. Ornithological illustrations that include background details typically tend to emphasize the natural surroundings of the bird that they are depicting and it is rare for an artist to depict any reference to people in their work.

Prideaux John Selby

1788–1867

Life-size illustrations of every bird in the British Isles was the mammoth goal of *Illustrations of British Ornithology* by Prideaux John Selby. A gentleman naturalist, he was also involved with one of the first books that aimed to present ornithology on a global scale.

Often called the British Audubon, Prideaux John Selby was born in the town of Alnwick in the English county of Northumberland just three years after the celebrated American naturalist. While Audubon struggled financially throughout his life, Selby had been born into a family of prominent and prolific landowners. Educated at Durham Grammar school, he later studied briefly at University College, Oxford, before inheriting the family estates after his father's death in 1804.

As landed gentry, Selby did not have to work for a living and instead he could sustain himself entirely by collecting rental incomes from the estates he inherited. This allowed him to live the life of a country gentleman and pursue all the leisure and luxury that such privilege granted him. Yet, despite being afforded such a life of ease, he chose to take an active role in social and political life, serving as a magistrate, High Sheriff and later as Deputy Lieutenant of Northumberland. Despite these roles, his true passion was nature, and this covered all of the gentlemanly pursuits of the day including hunting, fishing, drawing and collecting. In 1819, aged thirty-one, he set himself the task of compiling a set of illustrations that covered every bird found in the British Isles. While similar books about the birds of Britain had been printed before – most notably Thomas Bewick's *A History of British Birds* (1797–1804) – Selby's work was to be the first of its kind to feature life-sized illustrations.

Titled *Illustrations of British Ornithology* (1819–34), this immense project took Selby a total of fifteen years to complete. It contains some 220 plates, most of which were based after Selby's own drawings, and he also etched many of these himself. To help with the etching he enlisted the assistance of his brother-in-law, Admiral Robert Mitford (1781–1870). A distinguished naval officer, Mitford had a passion for ornithology and was also an accomplished draughtsman. In 1816, he had studied under Thomas Bewick and so it was Mitford who taught Selby the art of copper plate engraving and etching. To achieve the life-size drawings that Selby wanted, the etchings had to be printed on elephant-folio size paper – huge sheets that would measure approximately 56 × 71 cm. Despite the immense size of the folio, some species were so large that they needed to be printed at three-fifths of life-size, with the scale change being mentioned in the text.

Selby's illustrations have a sense of clarity and elegance to them. Background details are kept to a minimum and the focus is placed directly on presenting the most life-like depiction of the subject. Indeed, Selby's illustrations are arguably the most accurate depictions of British birds to that date. Many of these were drawn directly from the specimens he held in his own large collection. Unlike some earlier works, a lot of the awkward stiffness has been removed from these drawings, and there is a real sense of the weight of each bird in these illustrations. Their feathers are richly detailed, their stances feel natural and the line work captures delicate shades of tone with confidence and grace.

While *Illustrations of British Ornithology* is certainly Selby's most celebrated work, it was not his only project. He had a long and lasting friendship with prominent Scottish naturalist Sir William Jardine (1800–1874) and the pair collaborated often; most notably in publishing *Illustrations of Ornithology* (1825–43). Together, they were responsible for most of the illustrations in this work. Printed at Royal Octavo size (16 × 25 cm), this smaller work may not perhaps appear quite as impressive as Selby's *British Ornithology*, yet it is admirable nonetheless. Serving as a complementary work to Selby's original publication, it featured birds from around the world. In part, the project was a response to the lack of books that took a global view of ornithology. The scale of the work was unprecedented at the time and involved specimens being sourced from all around the globe. The pair corresponded with foreign naturalists and drew from their own large pool of friends and colleagues. Over the course of the publication, *Illustrations of Ornithology* introduced many species that had never been published before and laid a solid foundation for Jardine and Selby's working relationship.

In 1833, Jardine developed the hugely popular series of illustrated encyclopaedia known as *The Naturalist's Library* (1833–43). This was an inexpensive forty-volume work that made natural history available to middle-class people during the Regency and Victorian eras. Selby contributed a volume on pigeons (1835) and a volume on parrots (1836); the latter included plates by Edward Lear and a memoir on Thomas Bewick. In 1836, Selby, Jardine and naturalist George Johnston (1797–1855) founded the *Magazine of Zoology and Botany*, later renamed the *Annals of Natural History*. He continued to serve as editor of this publication up until his death in 1867.

Throughout his lifetime, Selby earned a reputation for being a well-liked man and he had a large circle of friends within the world of Victorian naturalists. He was an active member of several British natural history societies and would have known the likes of John Gould and Hugh Edwin Strickland, and he had even known Audubon as they briefly shared the same publisher, W.H. Lizars (1788–1859). Through his work, he established many English names of birds, and his *British Ornithology* remains one of the finest illustrated bird books of the early 1800s.

WILLOW PTARMIGAN
Lagopus lagopus

Prideaux John Selby, *Illustrations of British Ornithology*, 1819–34

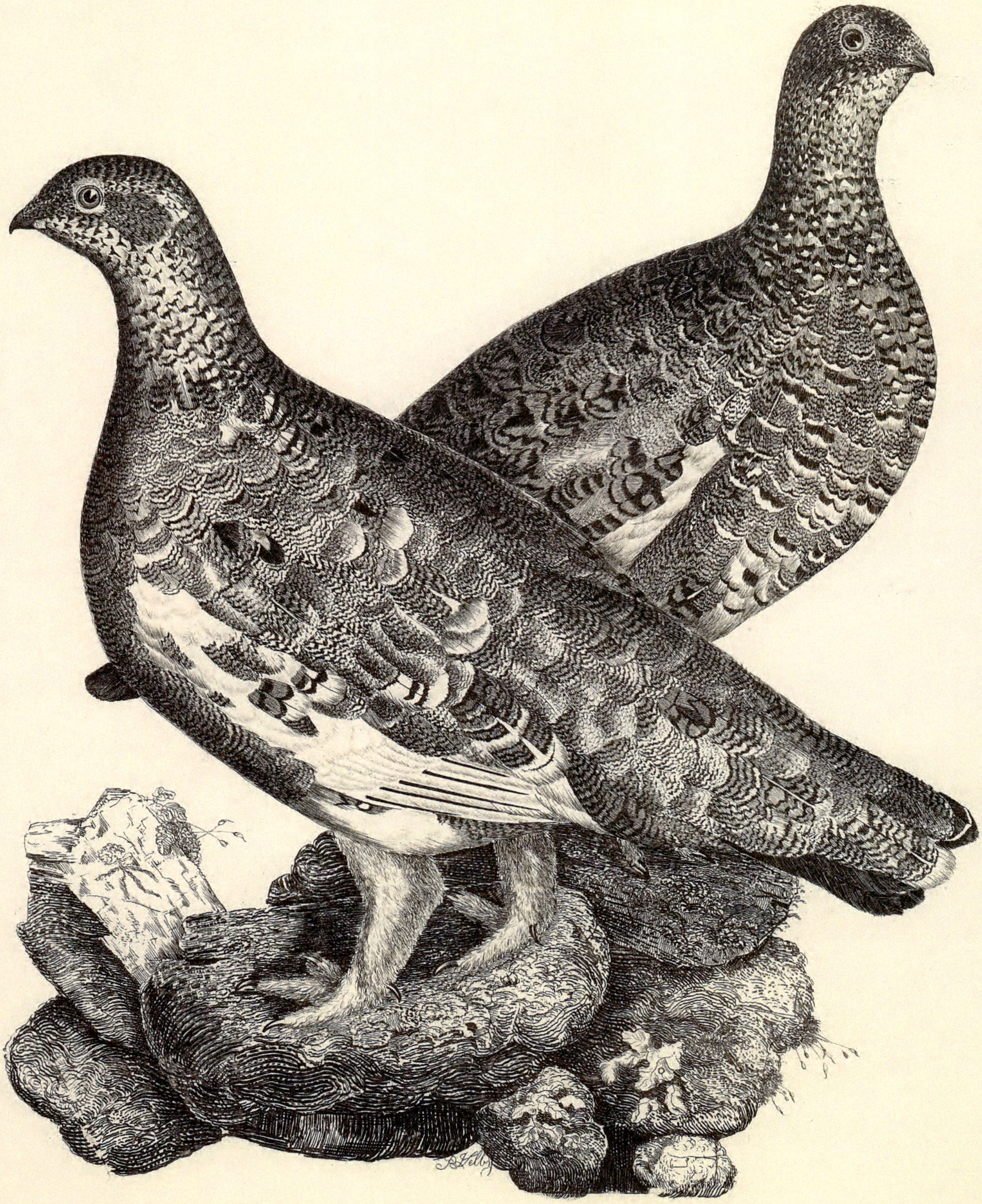

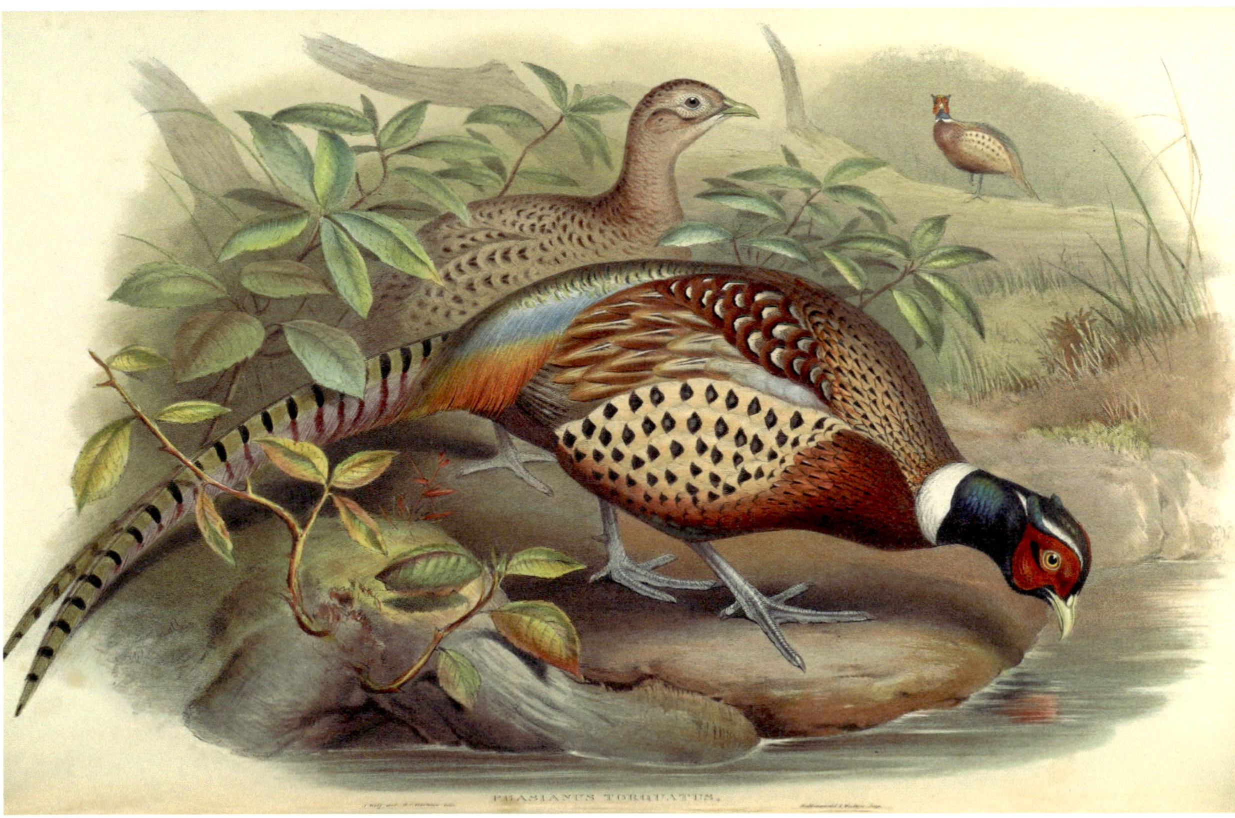

COMMON PHEASANT
Phasianus colchicus

Joseph Wolf, *The Birds of Asia*, 1850–83

The pheasant is one of the most popular game birds and it is valued in the West for both its feathers and its meat. While many of today's naturalists and nature lovers may be horrified by the thought that such birds are hunted as sport, this would not have been a view held during the time in which these illustrations were produced.

Joseph Wolf, who produced this depiction of a male and female pheasant, is just one example of an artist within this book who took an active interest in hunting. His nature illustrations cover scientific and natural histories, as well as literature, educational material and hunting manuals.

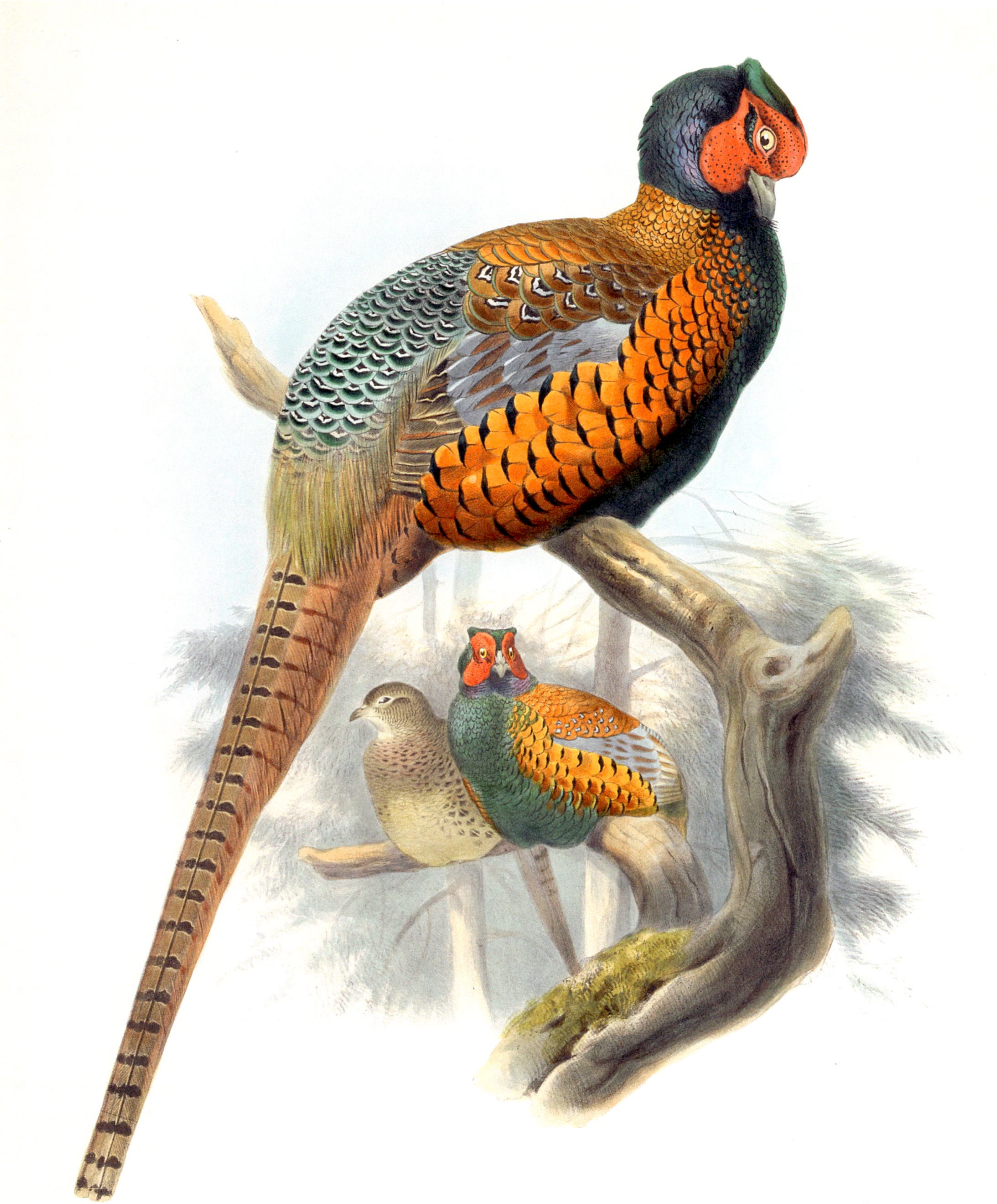

COMMON PHEASANT
Phasianus colchicus

Joseph Wolf, *A Monograph of the Phasianidae*, 1870–2

Wolf was responsible for drawing the majority of the plates in Daniel Giraud Elliot's *A Monograph of the Phasianidae*. First published in New York between 1870 and 1872, it was one of several monographs that the American zoologist produced. Despite being an artist himself, Elliot recognized his own limitations, and so commissioned some of the finest European artists of the time to illustrate his monographs, including Joseph Smit and John Gerrard Keulemans.

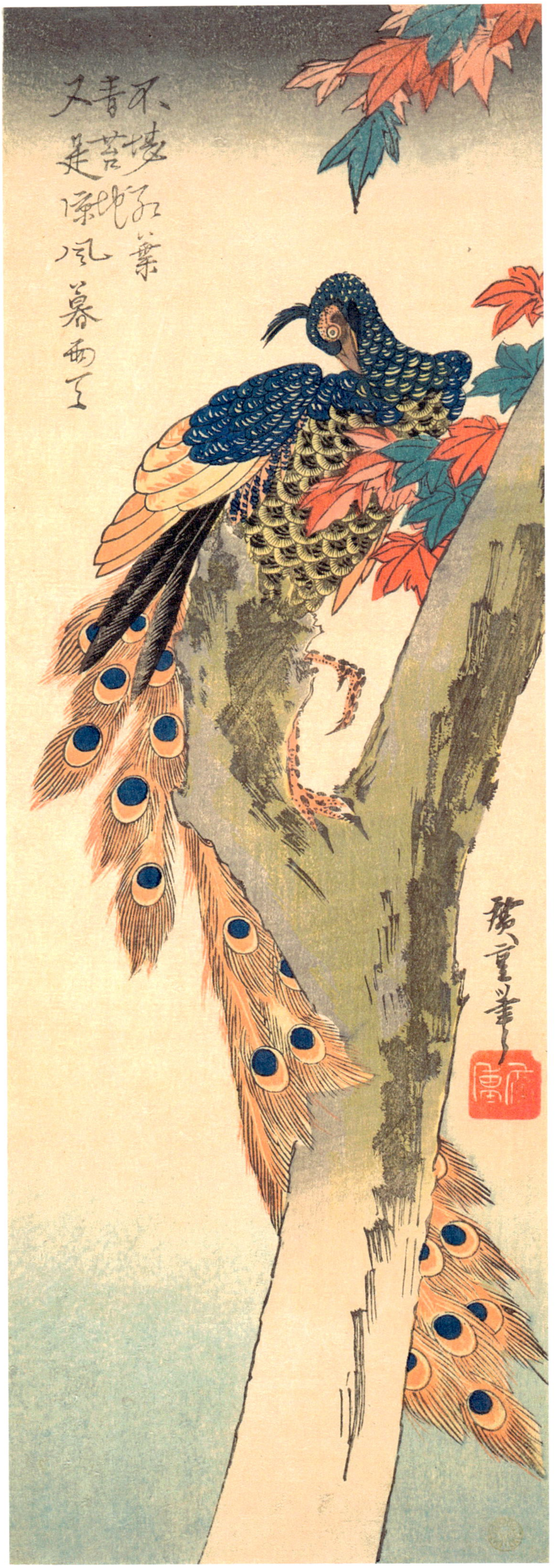

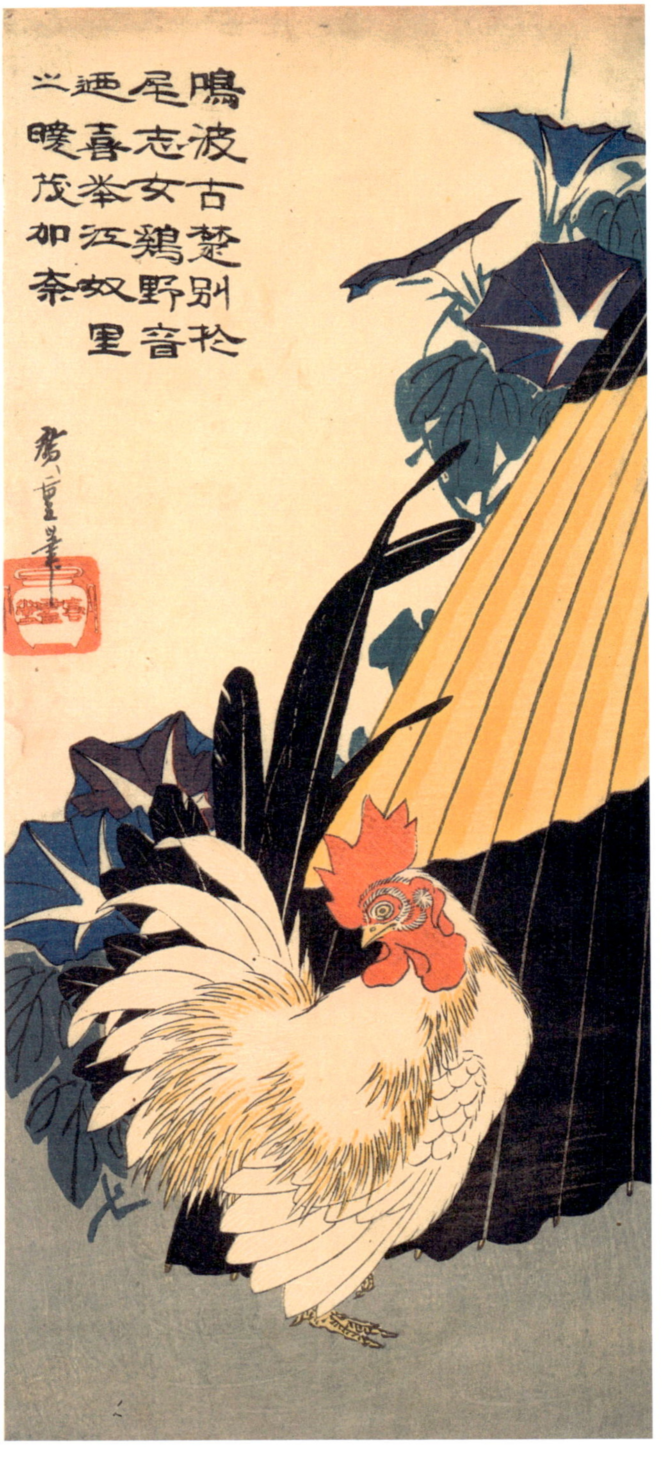

LEFT:

A PEACOCK PERCHED ON A MAPLE TREE

Utagawa Hiroshige, woodblock print, *ca*. 1833

This woodblock print by Hiroshige incorporates a verse by the influential Chinese poet Bai Juyi (772–846 CE).

RIGHT:

ROOSTER, UMBRELLA AND MORNING GLORIES

Utagawa Hiroshige, woodblock print, *ca*. 1830

The poem included on this print is known as *kyōka*, a light-hearted type of poetry that was popular in the Edo period.

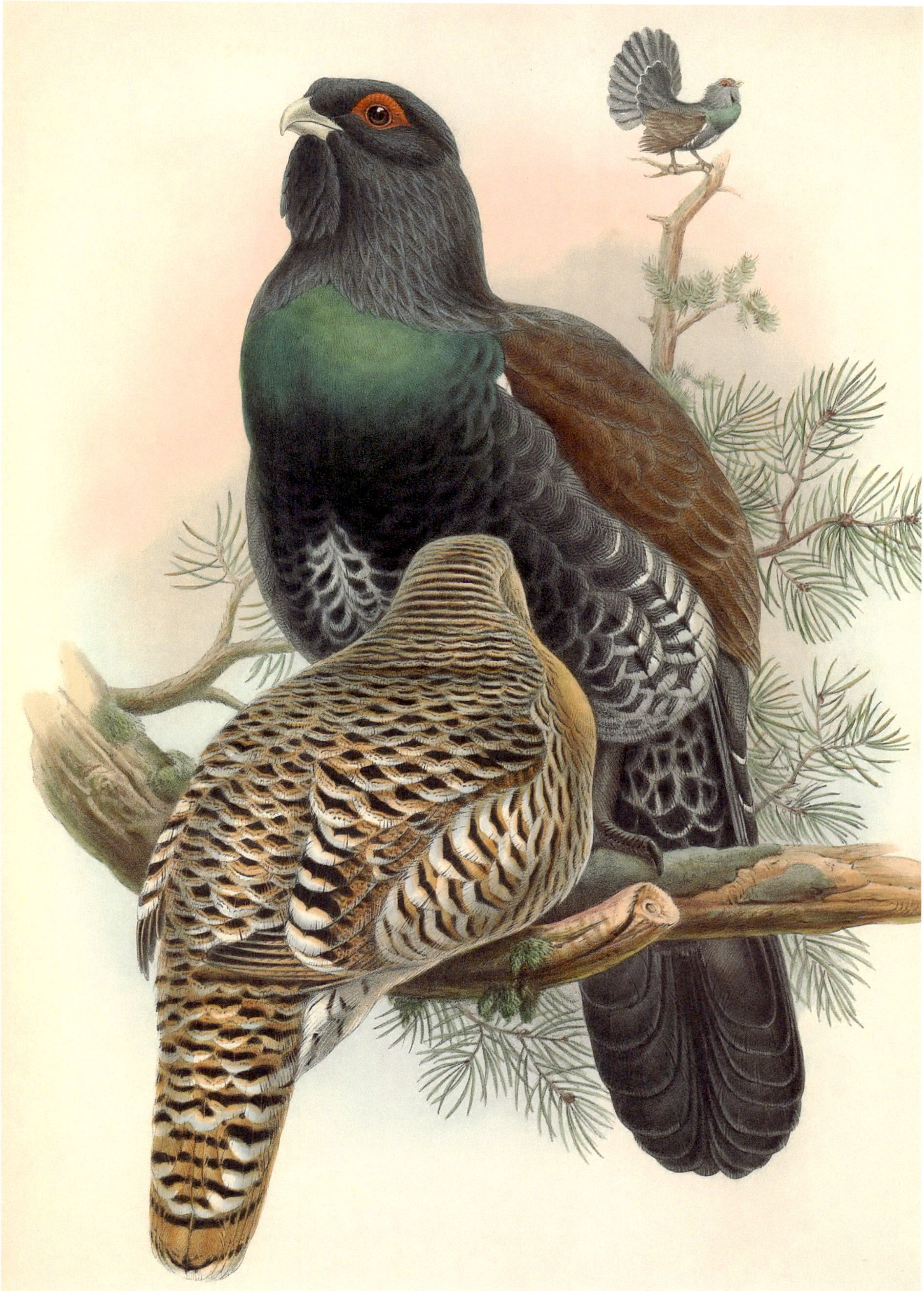

WESTERN CAPERCAILLIE
Tetrao urogallus

Joseph Wolf & Henry Constantine Richter, *The Birds of Great Britain*, 1862–73

The capercaillie is a huge gamebird. It is the largest grouse in the world and can be as large as a turkey. The species shows extreme sexual dimorphism, with the male being twice the size of the female. Its name is derived from Gaelic and translates as 'the horse of the woods'. The origin of this name is debatable, with some attributing it to the size of the bird and others believing that it is in reference to the distinctive cantering sound of the male's mating call.

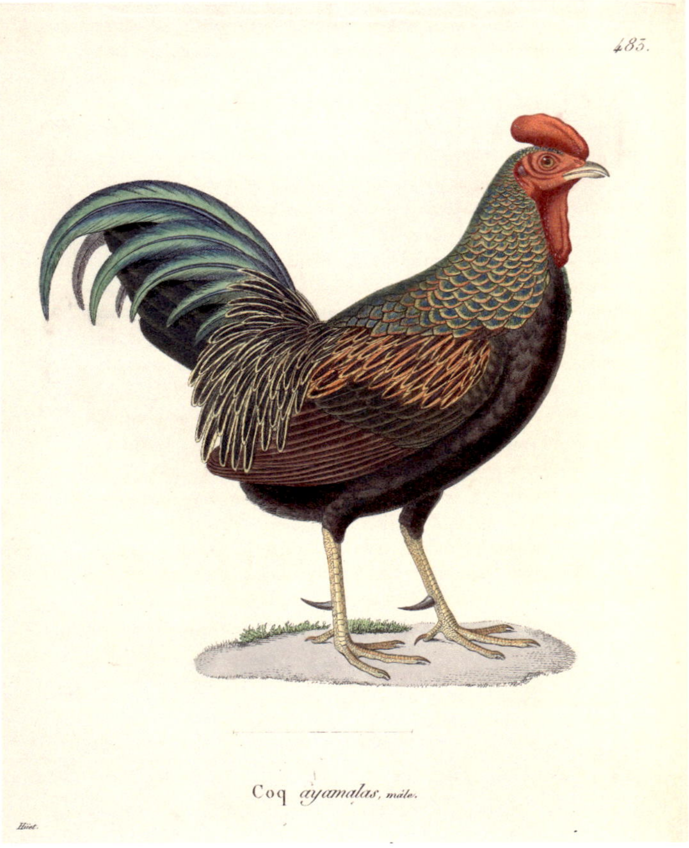

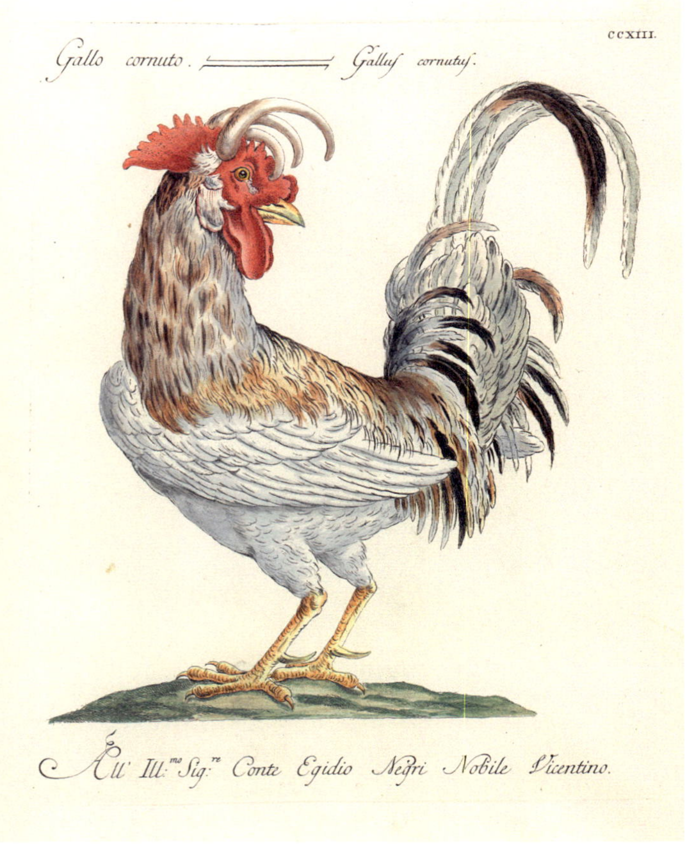

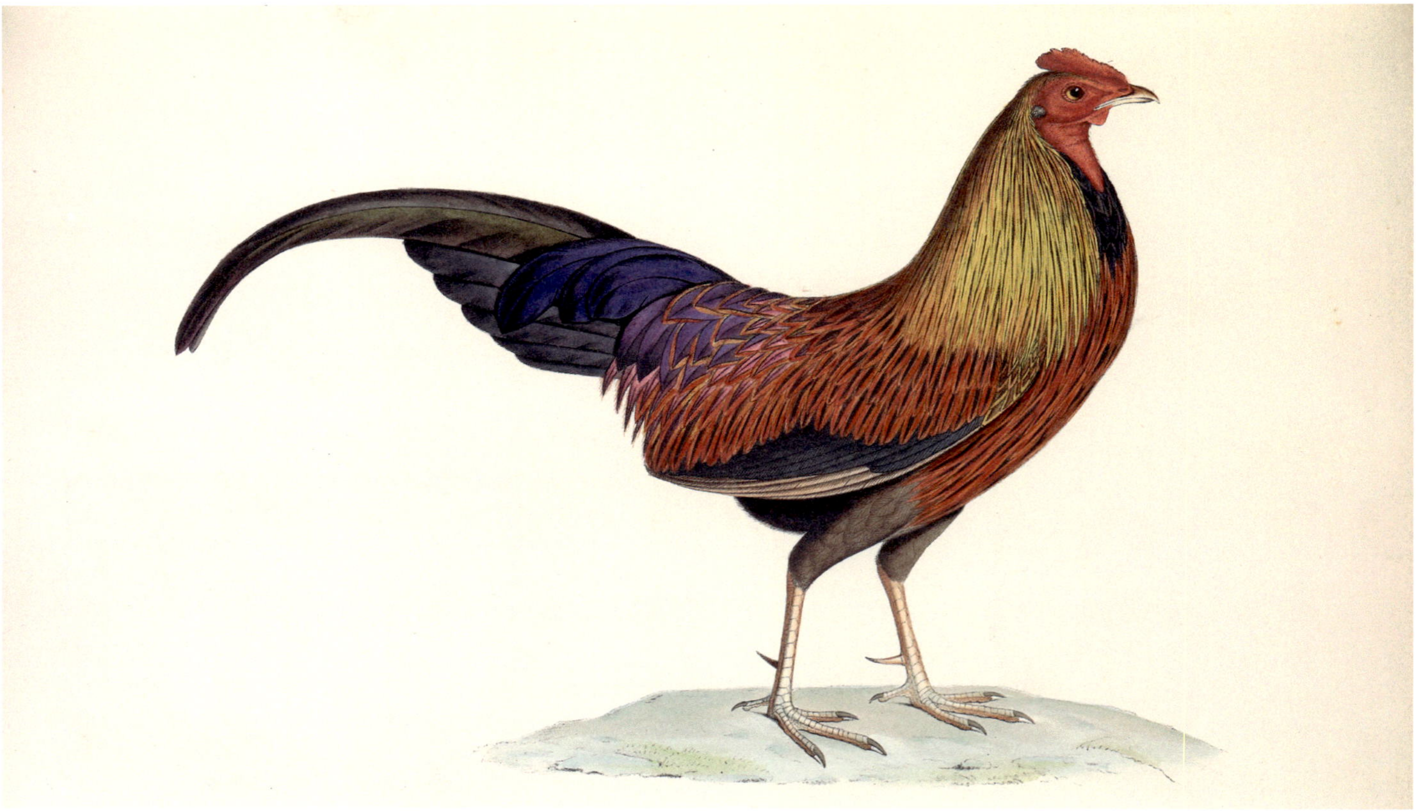

TOP LEFT:
GREEN JUNGLEFOWL
Gallus varius

Nicolas Huet II, *Nouveau recueil de planches coloriées d'oiseaux*, 1820–39

BOTTOM:
SRI LANKAN JUNGLEFOWL
Gallus lafayettii

Florent *Prévost*, *Iconographie ornithologique*, 1845–9

TOP RIGHT:
HORNED COCK
Gallus cornutus

Lorenzo Lorenzi & Violante Vanni, *Ornithologia methodice digesta*, 1767–76

The term 'junglefowl' refers to any of the four Asian birds that are found in the Gallus genus. The domestic chicken (*Gallus gallus domesticus*), is said to have originated from these birds, specifically the red junglefowl (*Gallus gallus*). This bird was first domesticated in India at least 5,000 years ago.

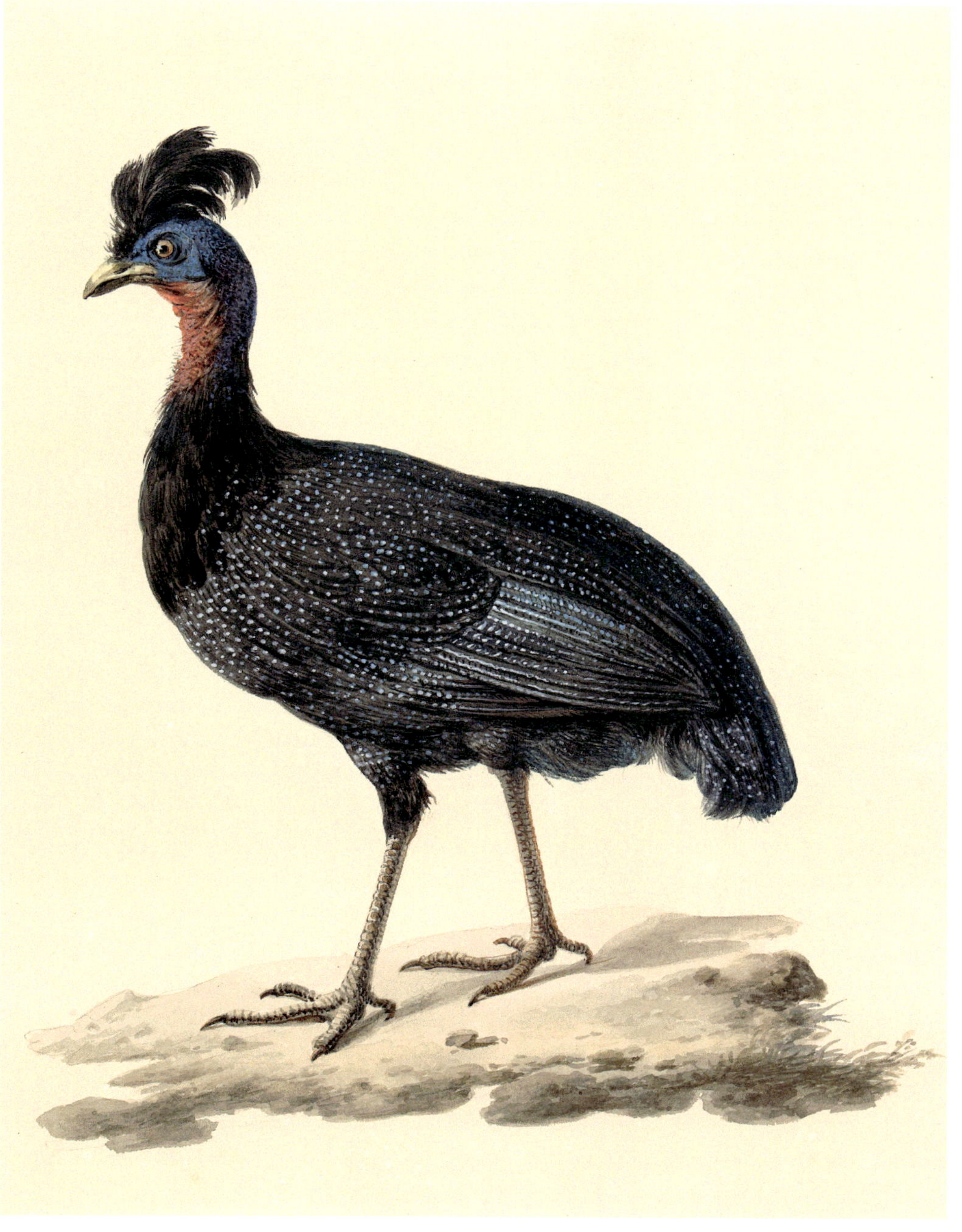

WESTERN CRESTED GUINEAFOWL
Guttera verreauxi

Pieter Pietersz Barbiers, watercolour on paper, *ca.* 1759–1842

The western crested guineafowl is commonly found in the forests and woodlands of West, Central and East Africa. It boasts a distinctive black crest on the top of its head that resembles a toupée. This particular bird was drawn by the Dutch artist Pieter Pietersz Barbiers. We can assume that the bird most likely came into the artist's possession by way of the colony in West Africa that was then called the Dutch Gold Coast.

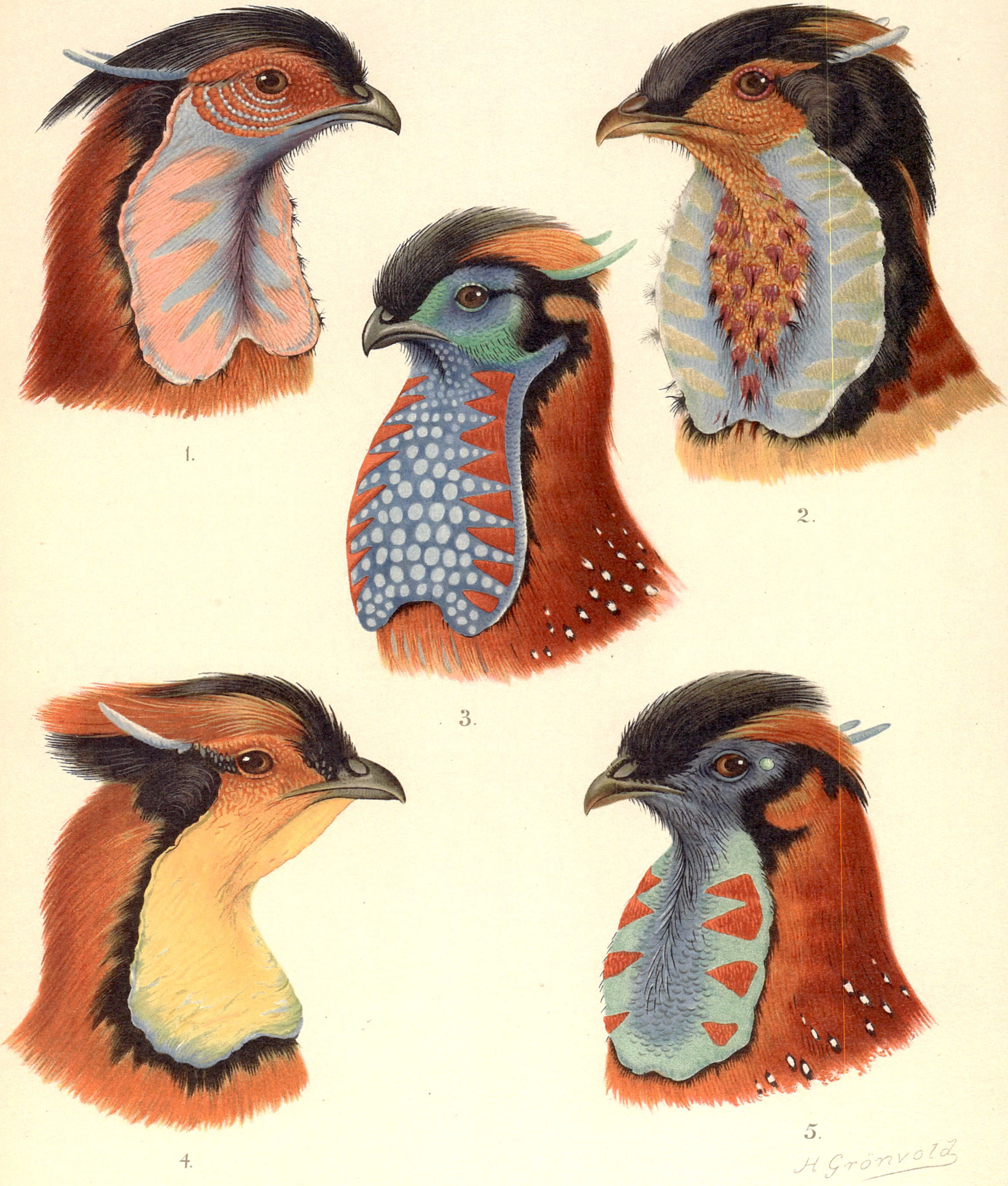

WATTLES OF COCK TRAGOPANS.

WATTLES OF COCK TRAGOPANS

1. WESTERN TRAGOPAN
   Tragopan melanocephalus
2. CABOT'S TRAGOPAN
   Tragopan caboti
3. TEMMINCK'S TRAGOPAN
   Tragopan temminckii
4. BLYTH'S TRAGOPAN
   Tragopan blythii
5. SATYR TRAGOPAN
   Tragopan satyra

Henrik Grönvold,
A Monograph of the Pheasants, 1918–22

Published between 1918 and 1922, *A Monograph of the Pheasants* was a monumental work that was produced by the American naturalist, ornithologist and explorer William Beebe. Remembered for his impressive research expeditions, this work was the result of a seventeen-month voyage across the globe which was financed by the businessman, philanthropist and pheasant enthusiast Anthony R. Kuser.

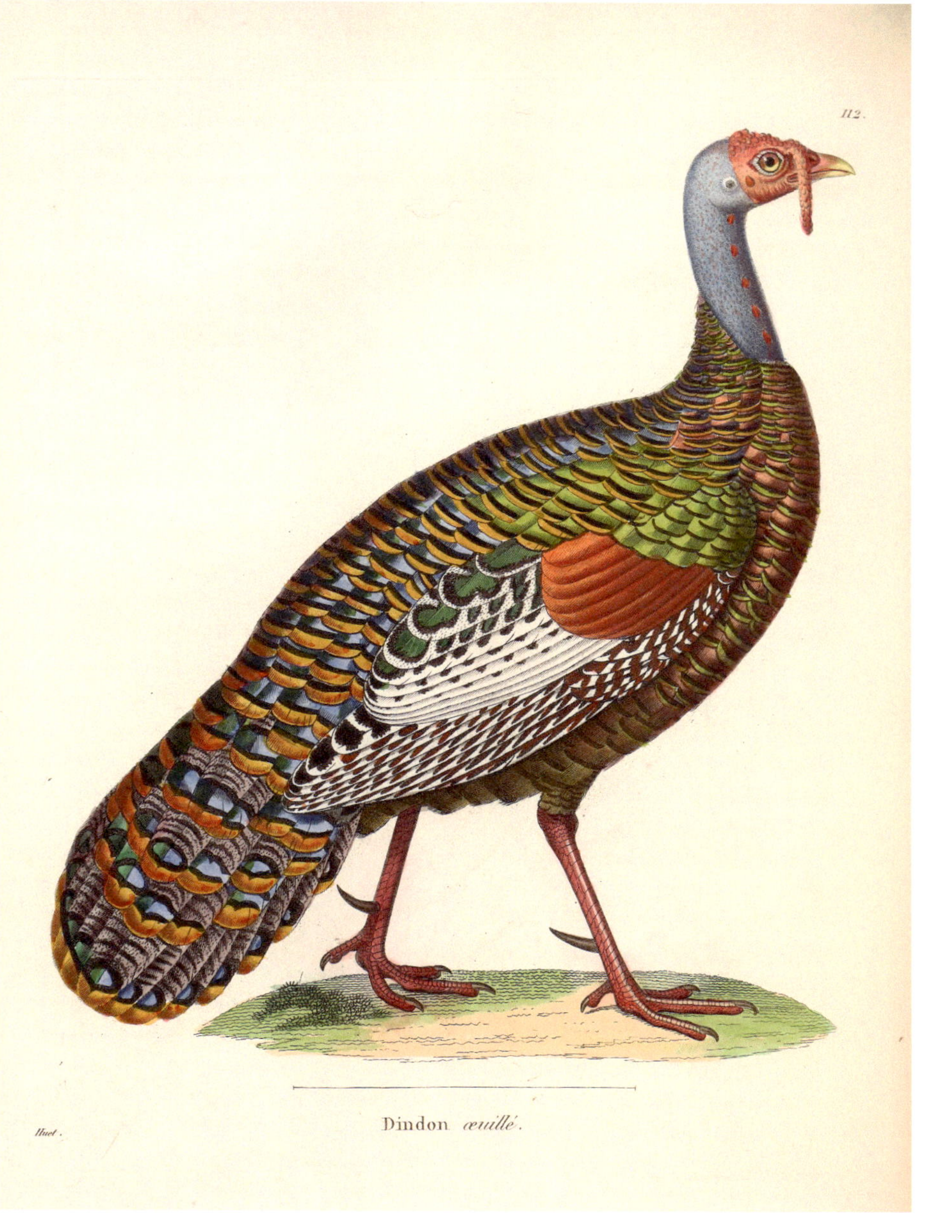

OCELLATED TURKEY
Meleagris ocellata

Nicolas Huet II, *Nouveau recueil de planches coloriées d'oiseaux*, 1820–39

The ocellated turkey is a species of turkey that is only found on the Yucatán Peninsula of Mexico, Belize and Guatemala.

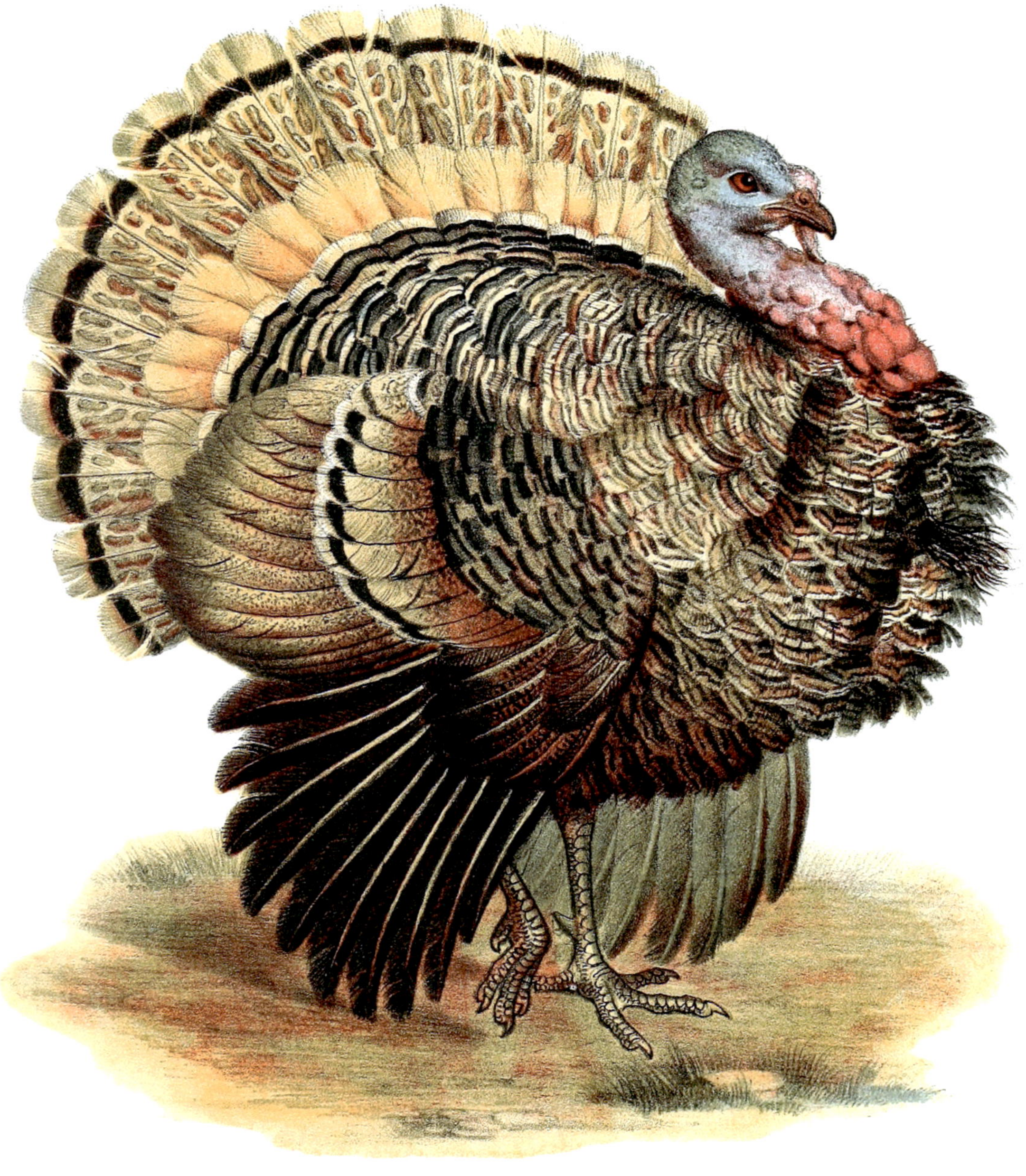

WILD TURKEY
Meleagris gallopavo

Uncredited, *Bilder-atlas*, 1860–4

Native to North America, the wild turkey is the heaviest member of the Galliformes. They are the same species as the domesticated turkey (*Meleagris gallopavo*), yet in the wild these birds are capable of flight. Despite this, they are reluctant fliers and, by the early 1900s, the turkey population of North America had plummeted through over-hunting. While the bird may be North American, its name is said to have come from Britain, where one theory suggests that these birds had originally been imported from merchants in Turkey.

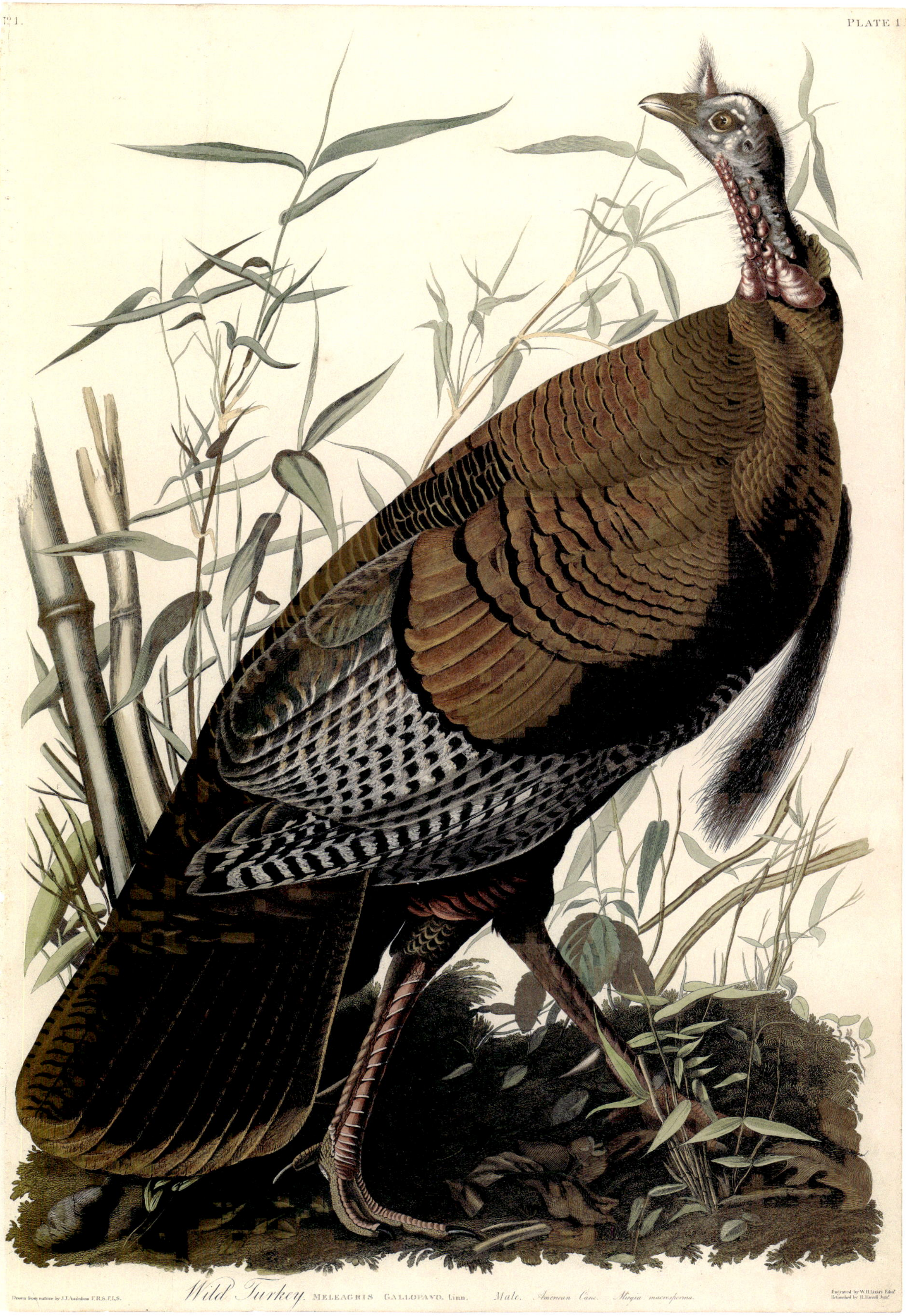

WILD TURKEY
Meleagris gallopavo

John James Audubon,
The Birds of America, 1827–38

This wild turkey is the first image to appear in Audubon's *Birds of America*. Like all birds featured in this work, Audubon depicts the bird at life size. The turkey was selected to emphasize the necessity for the book's impressive size. The original edition of *Birds of America* was published on massive 'double-elephant' sheaves of handmade paper. The final work is divided into four volumes, with the first three containing 100 illustrations, and the final volume containing 135.

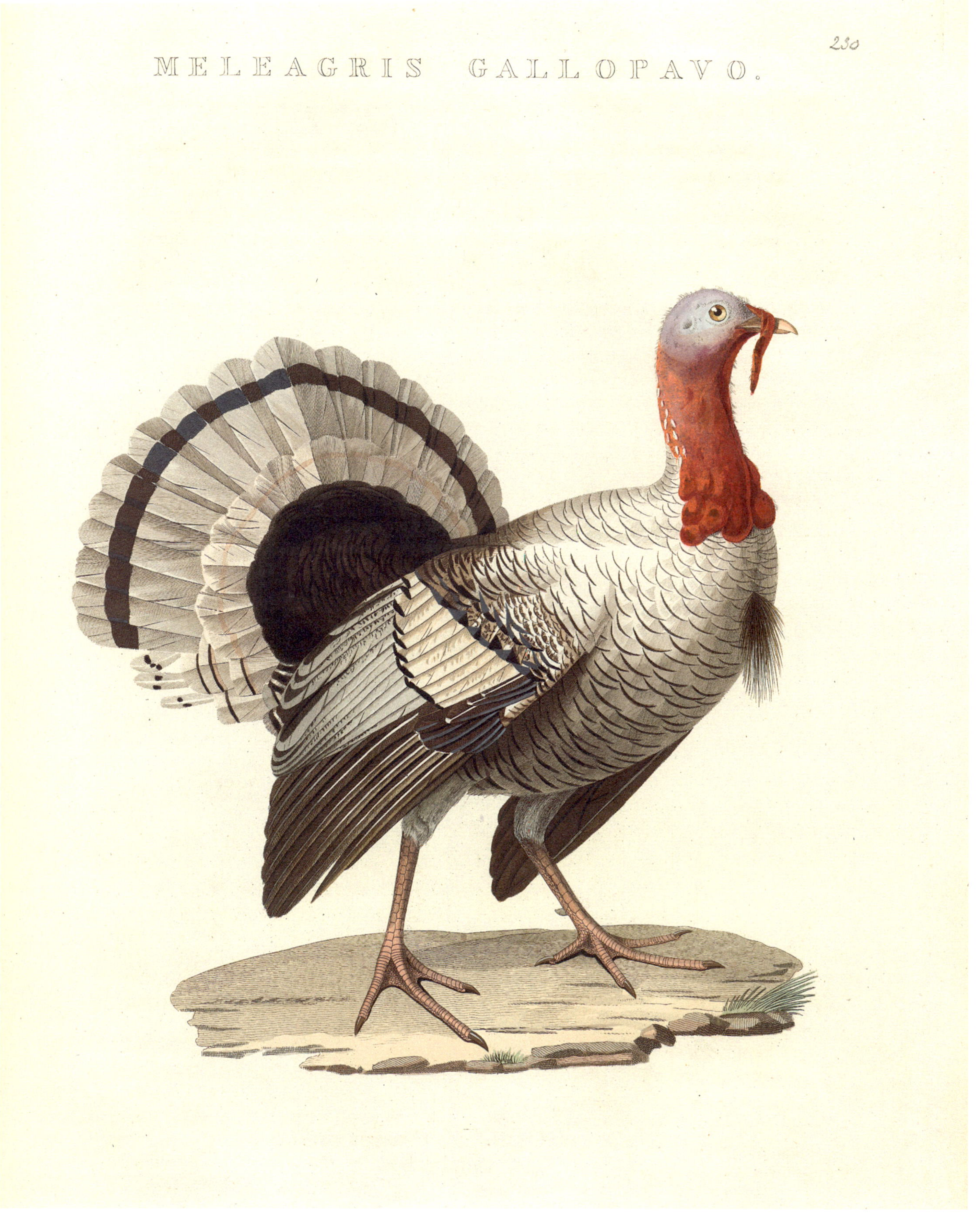

DOMESTIC TURKEY
Meleagris gallopavo

Jan Christiaan Sepp and son, *Nederlandsche Vogelen*, 1770–1829

Nederlandsche Vogelen was published between 1770 and 1829 and was the first comprehensive account of the birds of the Netherlands. Printed over five volumes, it contains 250 hand-coloured engravings and is one of the most expensive works in the history of Dutch printing. The initial engravings were made by Christiaan Sepp and later by his son Jan Christiaan Sepp and his grandson Jan Sepp. The plates were not signed, so it remains unclear who was responsible for producing each work.

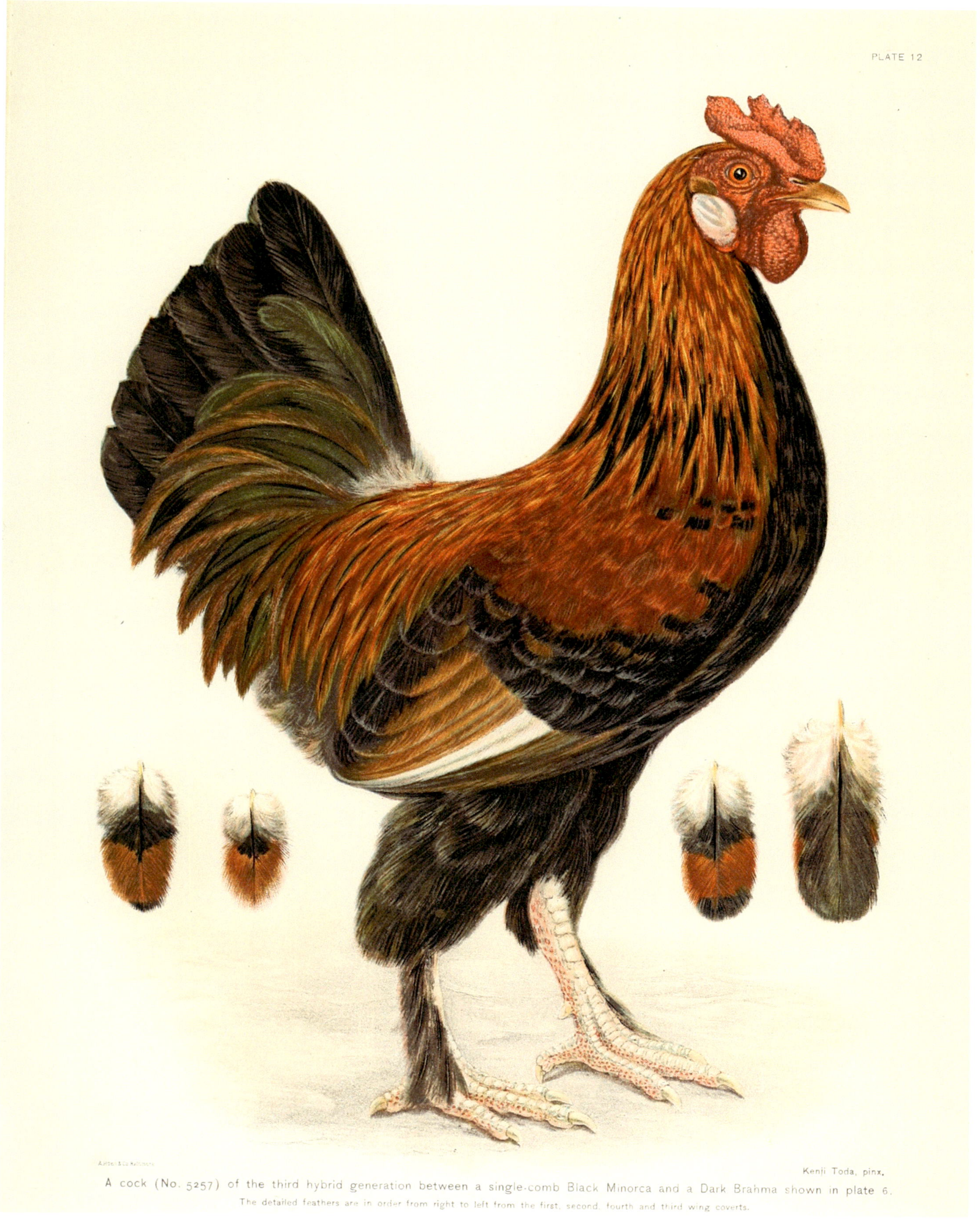

DOMESTIC CHICKEN (COCK)
Gallus gallus domesticus

Kenji Toda, *Inheritance of Characteristics in Domestic Fowl*, 1909

This illustration is taken from a guide to poultry genetics that was published in 1909 by the American biologist Charles Davenport. Much of Davenport's research focused on natural selection, variation and hybridization, and it led him to become one of the leading figures in the American eugenics movement. This highly controversial movement aimed at improving the genetic quality of the human population through compulsory sterilization and selective breeding. Toda's illustration examines the hybridization of a chicken.

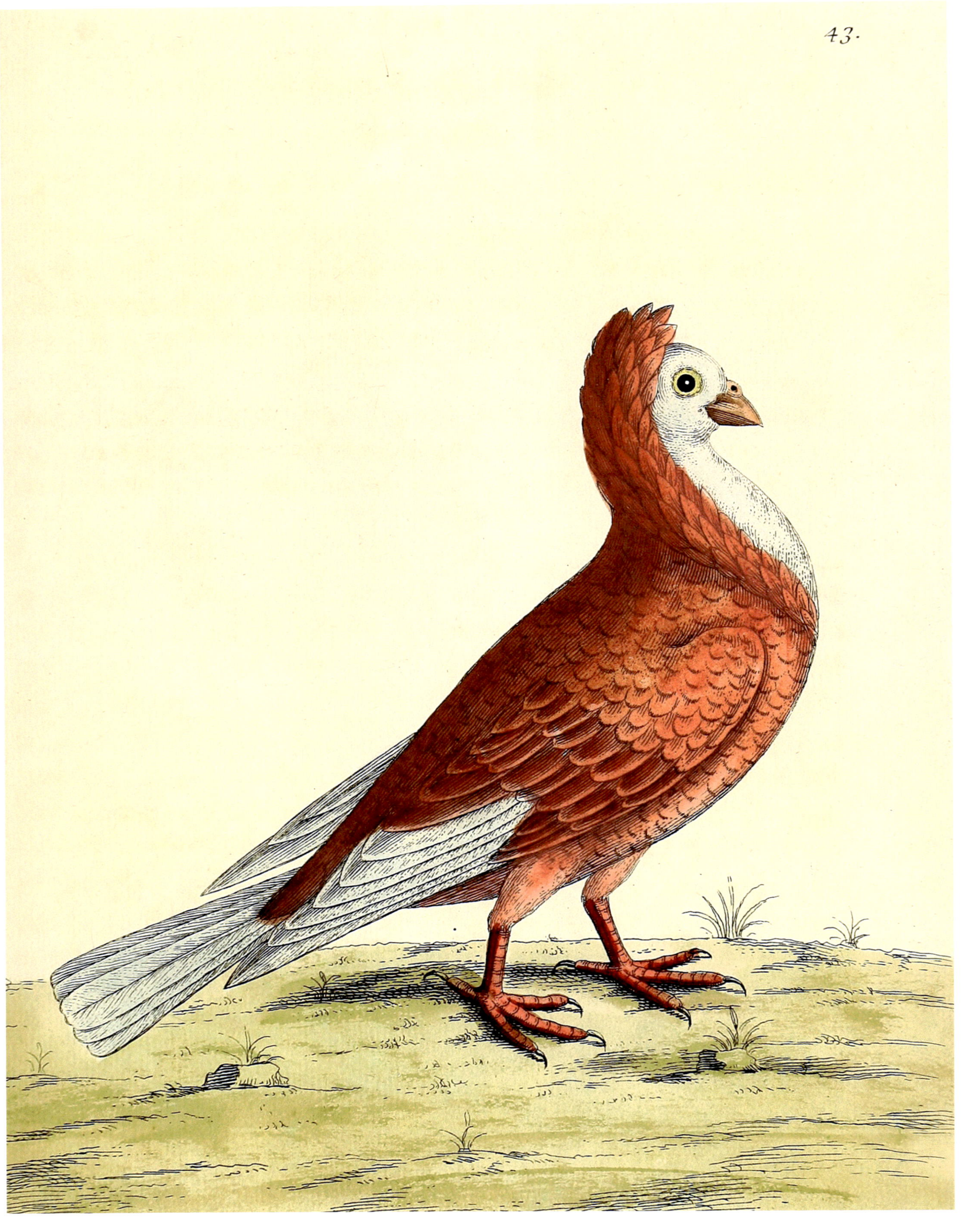

DOMESTIC PIGEON (JACOBIN)
Columba livia domestica

Eleazar Albin & Elizabeth Albin,
A Natural History of Birds, 1731–8

The practice of keeping and breeding pigeons has existed all over the world for thousands of years. This practice is typically referred to as 'pigeon fancying'. Over the course of many years, fanciers will selectively breed specific pigeons to alter their morphology in the hopes of creating a bird that will suit their own aesthetic sensibilities. The Jacobin pigeon is one such example, and is a common breed of fancy pigeon. It is noted for the feathered hood that surrounds its head.

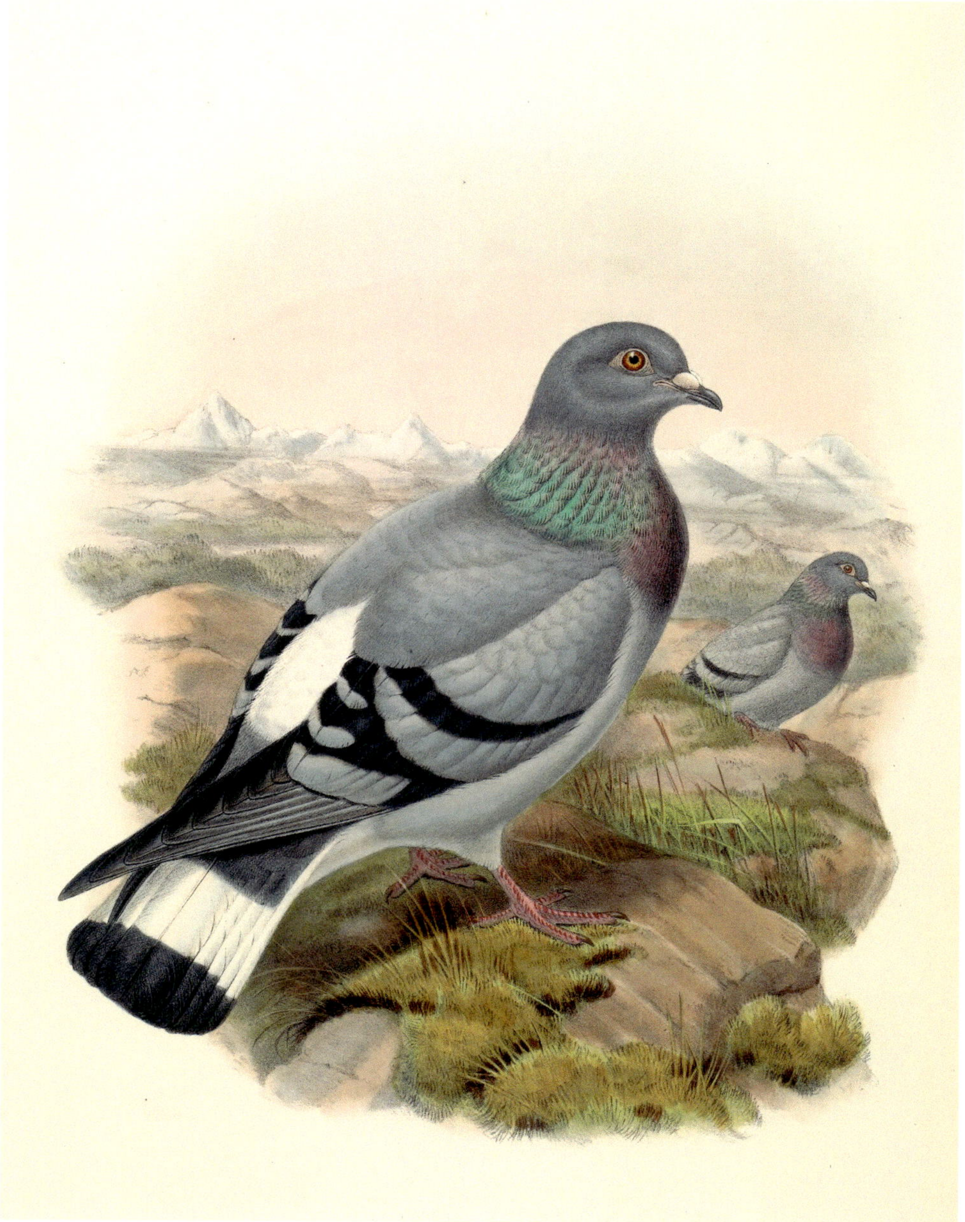

HILL PIGEON
Columba rupestris

John Gould & William Matthew Hart, *The Birds of Asia*, 1850–83

Pigeons (*Columba livia*) are said to have been domesticated in the Middle East at the end of the Stone Age, and today there are hundreds of established domestic breeds of this bird. The oldest of these is the rock pigeon, which is considered to be the oldest domesticated bird in the world. This species, the hill pigeon, tends to resemble this familiar bird, but it has a very distinctive white band on its tail. It is commonly found throughout Central and Eastern Asia.

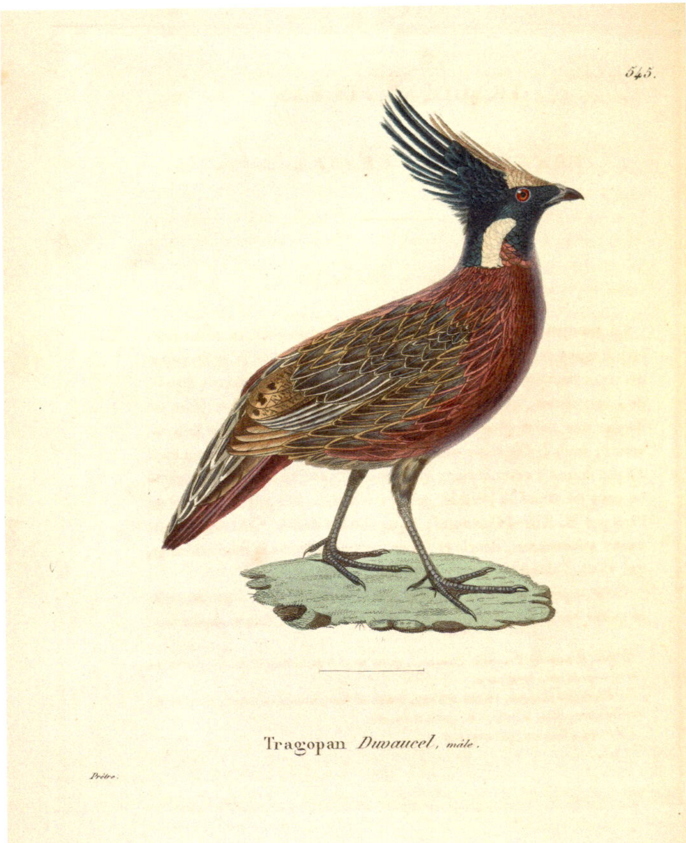

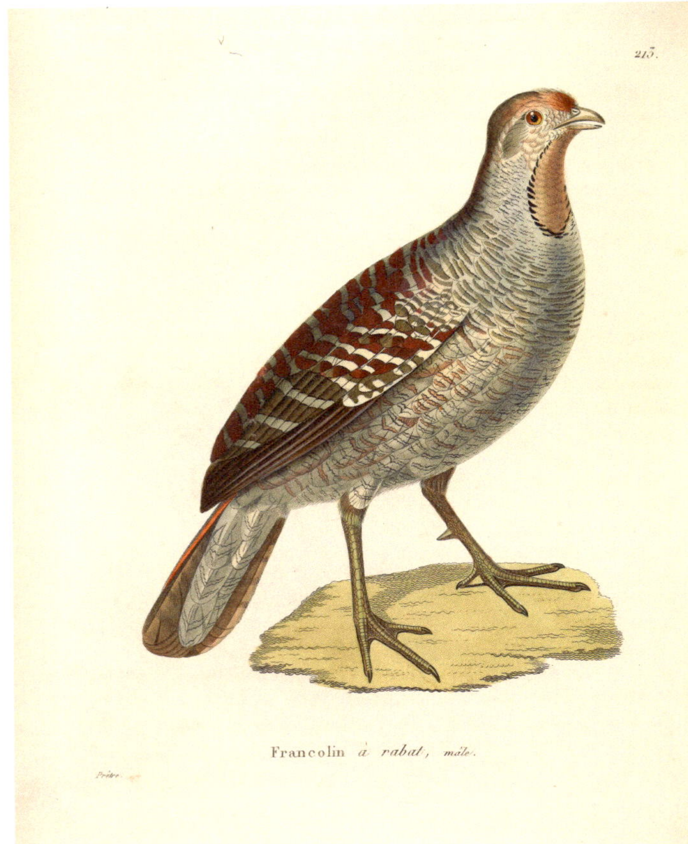

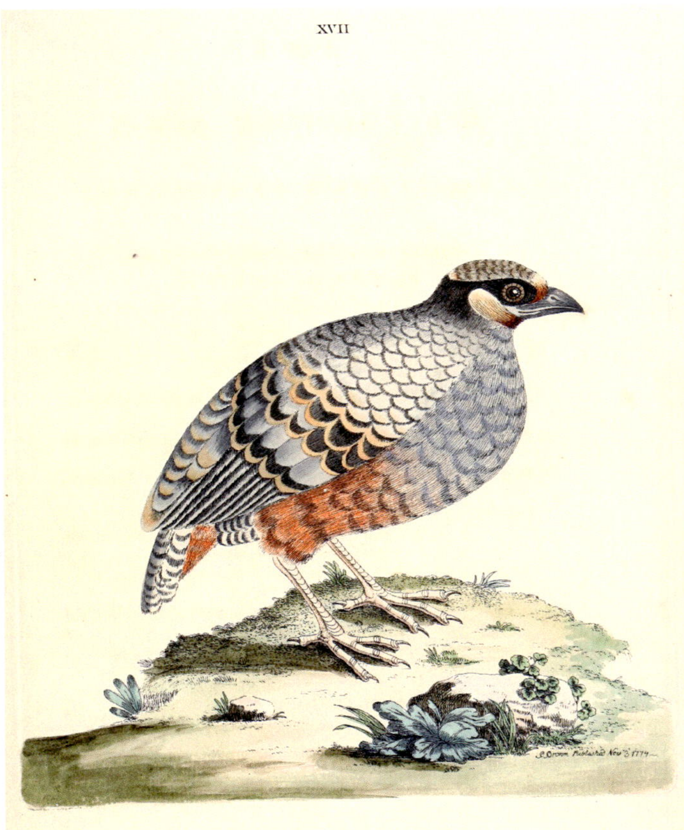

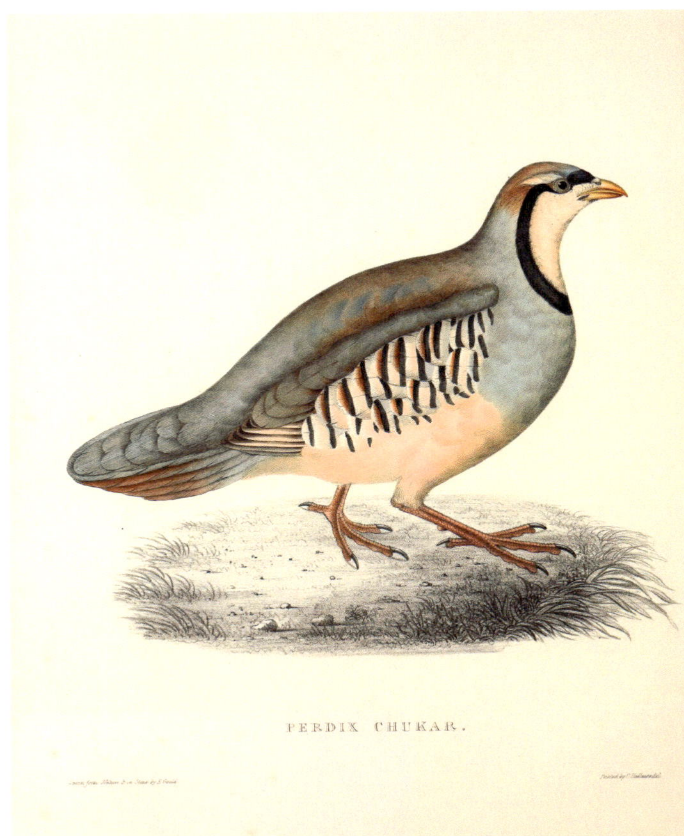

TOP LEFT:
KOKLASS PHEASANT
Pucrasia macrolopha

Jean-Gabriel Prêtre, *Nouveau recueil de planches coloriées d'oiseaux*, 1820–39

BOTTOM LEFT:
CHESTNUT-BELLIED PARTRIDGE
Arborophila javanica

Peter Brown, *New Illustrations of Zoology*, 1776

TOP RIGHT:
GREY FRANCOLIN
Francolinus pondicerianus

Jean-Gabriel Prêtre, *Nouveau recueil de planches coloriées d'oiseaux*, 1820–39

BOTTOM RIGHT:
CHUKAR
Alectoris chukar

Elizabeth Gould, *A Century of Birds from the Himalaya Mountains*, 1830–2

Tragopans, francolins and partridges are birds within the subfamily *Phasianinae*, of the pheasant family (*Phasianidae*). Other birds within this subfamily include junglefowl and peafowl.

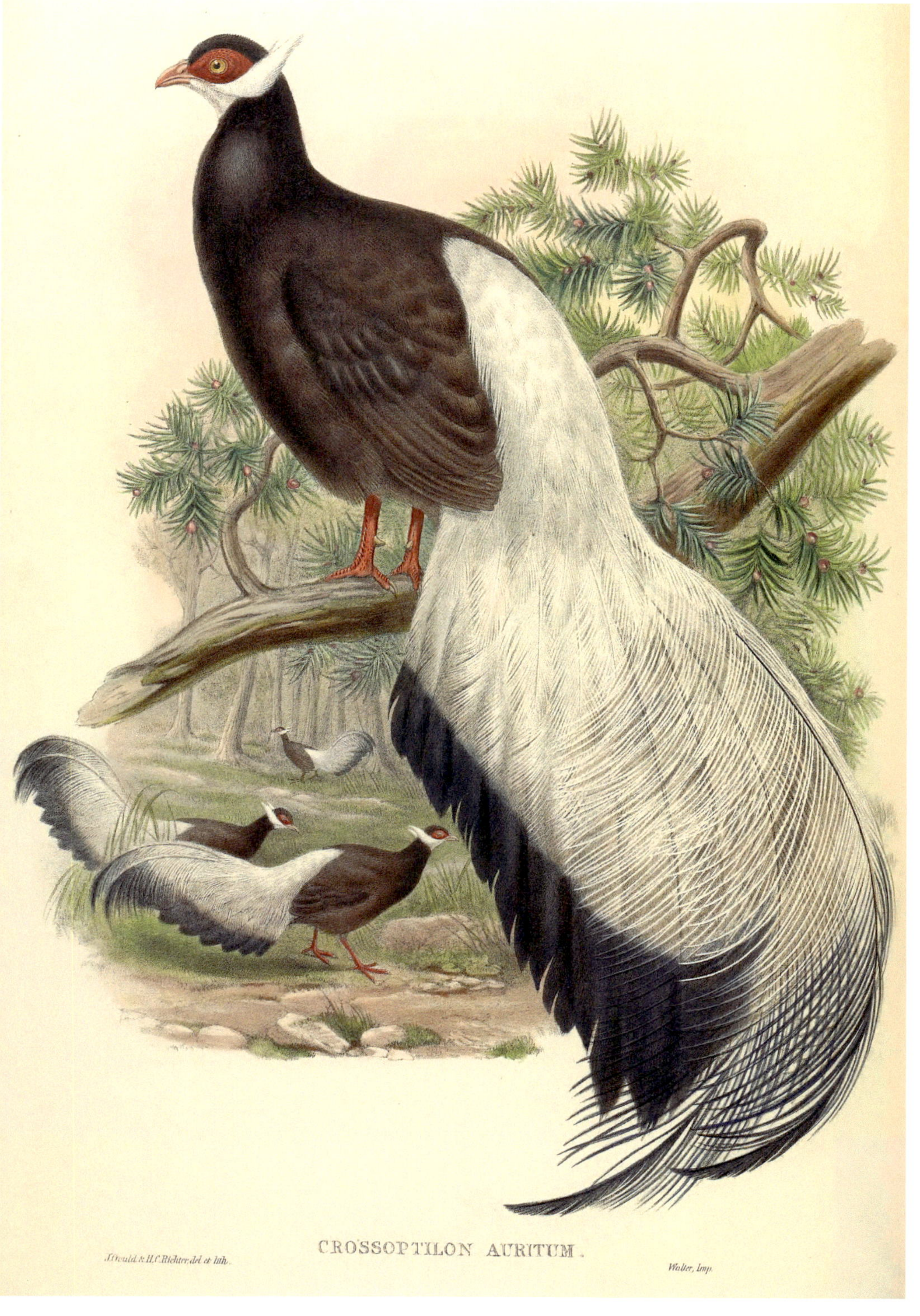

BROWN-EARED PHEASANT
Crossoptilon mantchuricum

John Gould & Henry Constantine Richter, *The Birds of Asia*, 1850–83

The brown-eared pheasant is found throughout the mountain forests of northeastern China. This plate appeared in the final volume of John Gould's *The Birds of Asia*. Spanning a total of seven volumes, this mammoth project began publication in 1850 and ran for 33 years. At the time it was the most comprehensive work on the ornithology of Asia, and contained a staggering 530 plates. When Gould died in 1881, the final parts of the work were completed by English ornithologist Richard Bowdler Sharpe.

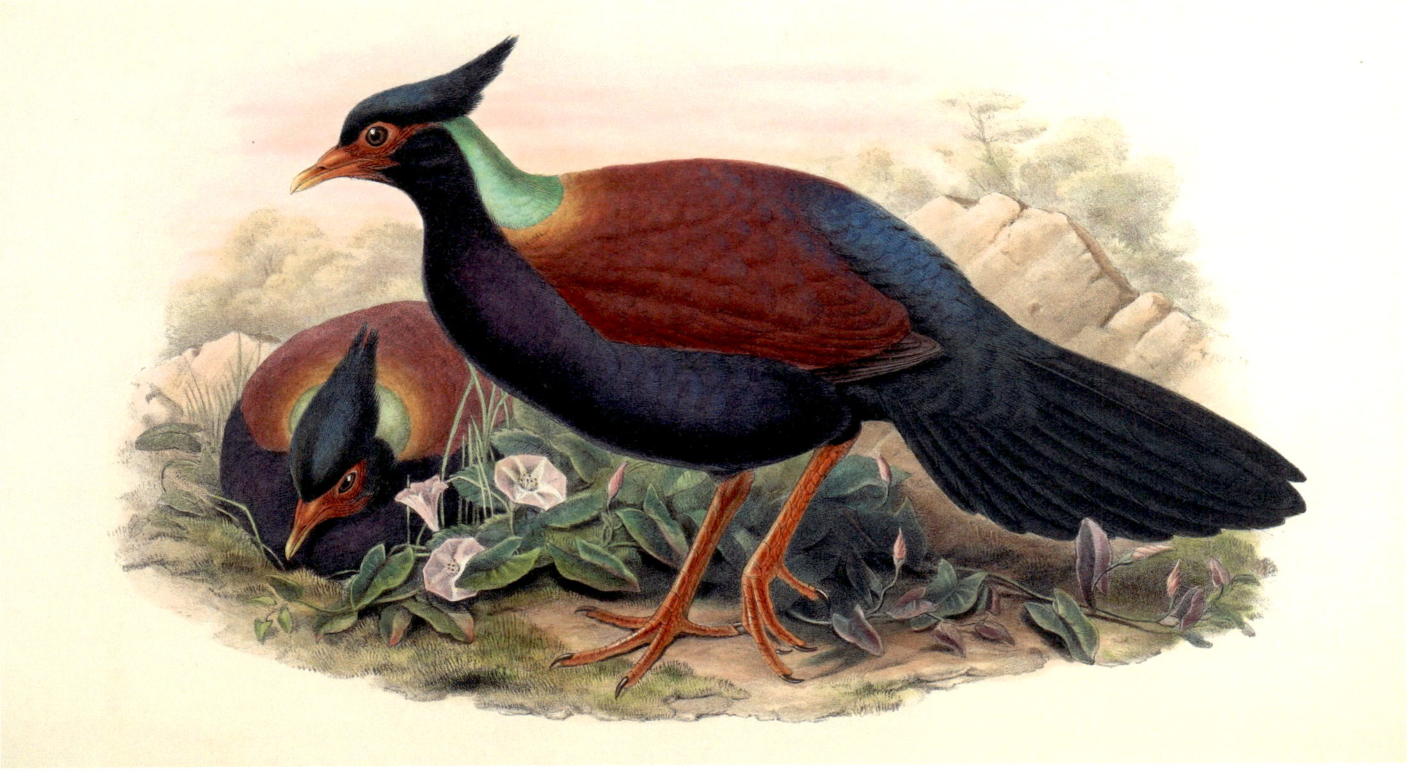

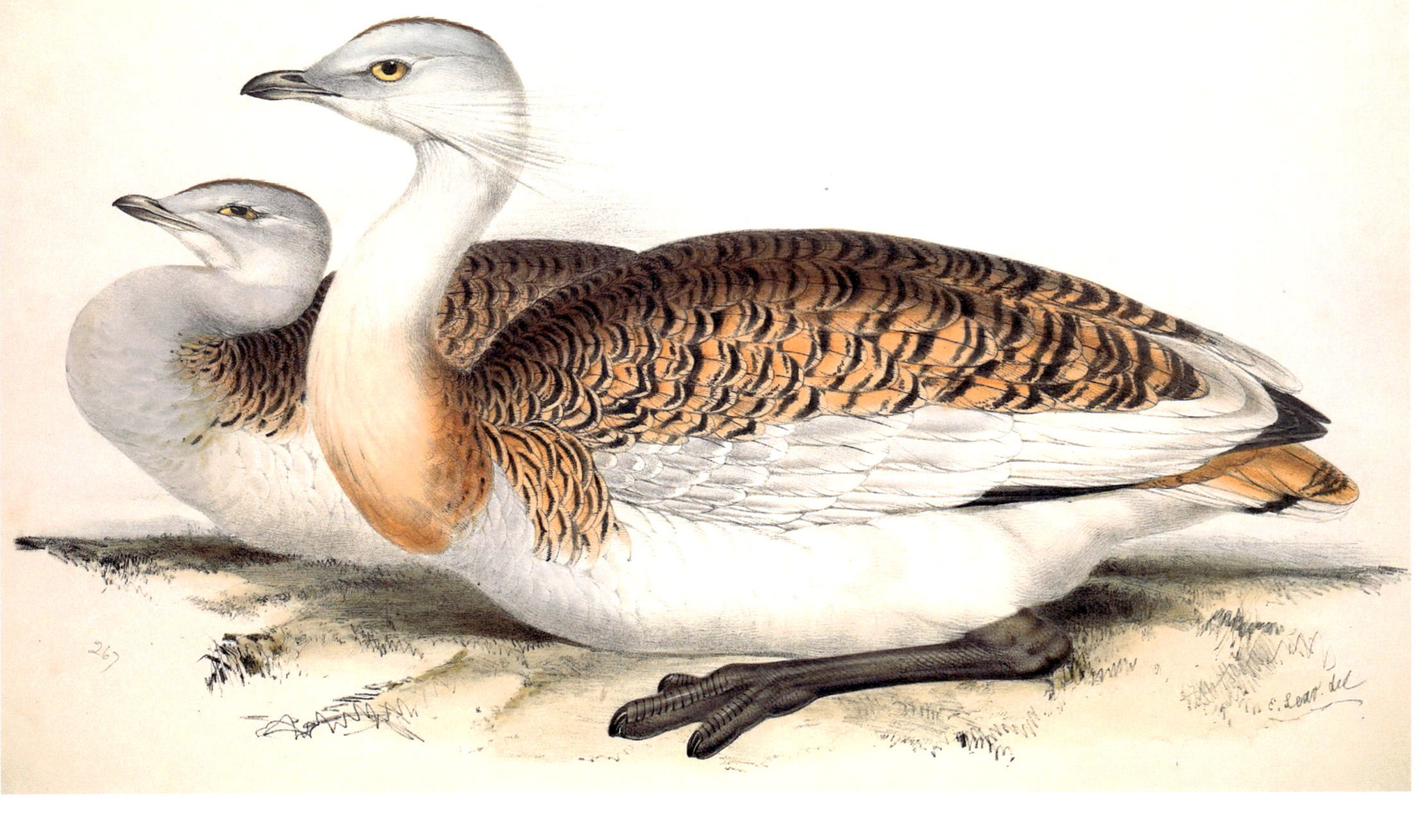

TOP:
GREEN-NAPED PHEASANT PIGEON
Otidiphaps nobilis

John Gould & Henry Constantine Richter, *The Birds of Asia*, 1850–83

BOTTOM:
GREAT BUSTARDS
Otis tarda

Edward Lear, *The Birds of Europe*, 1832–7

Both of these species highlight the challenges in categorizing birds. The pheasant pigeon, for example, may resemble a pheasant somewhat, but it is, in fact, a large terrestrial pigeon that belongs to the family *Columbidae*. Its genus name – *Otidiphaps nobilis* – is in reference to its similarity to the bustard family (*Otidae*). While bustards may resemble either gamebirds (Galliformes) or waterfowl (Anseriformes), they actually belong to a distinctive order all of their own known as Otidiformes.

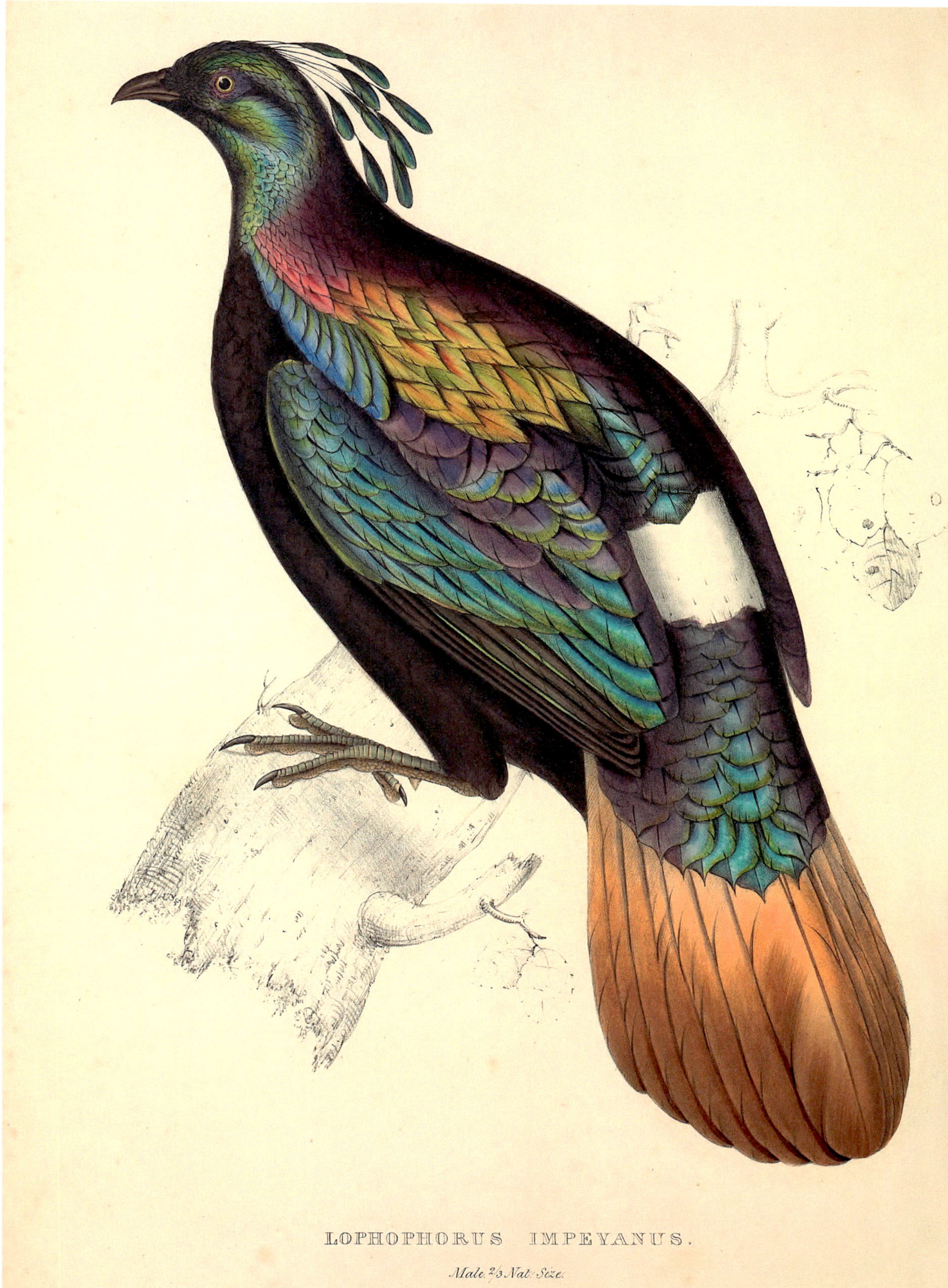

HIMALAYAN MONAL
Lophophorus impejanus

Elizabeth Gould, *A Century of Birds from the Himalaya Mountains*, 1830–2

At the age of just 23, John Gould became the first curator and preserver of the Zoological Society of London. In this role, he met many influential naturalists and academics. He was also often the first to see each new collection of skins that were brought into the society. One such collection came from the Himalayas and, in 1831, Gould published these in *A Century of Birds from the Himalaya Mountains*. The work was printed in stages and published in twenty monthly parts between 1830 and 1832.

PLATE IV

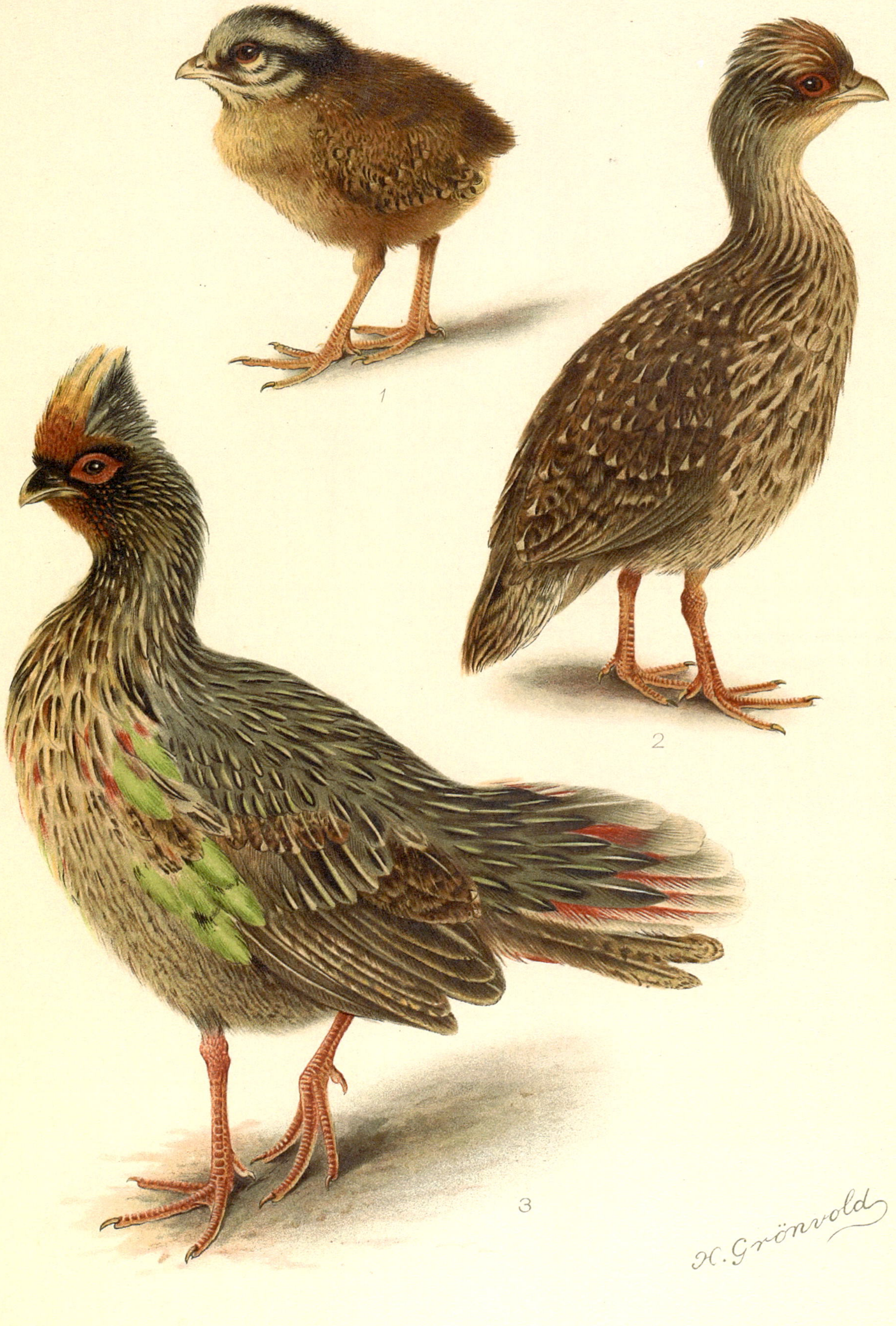

PLUMAGES OF THE HIMALAYAN BLOOD PARTRIDGE.

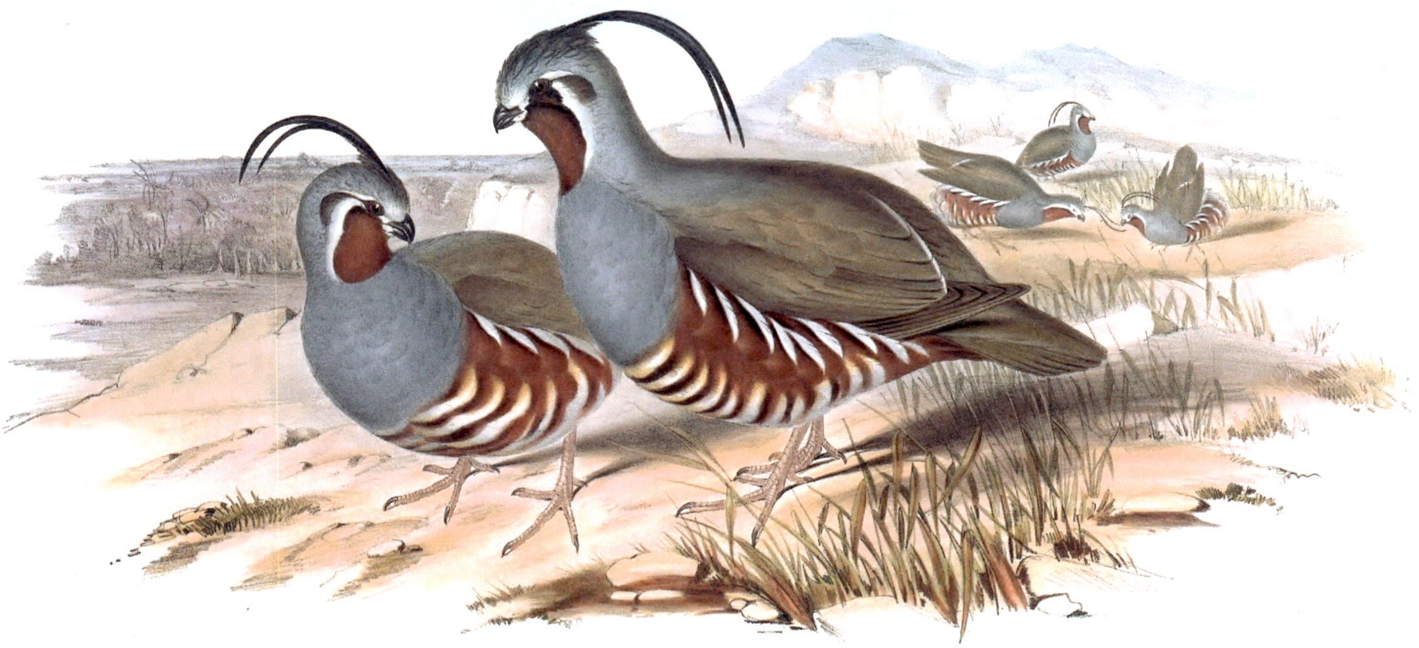

OPPOSITE
BLOOD PHEASANT
Ithaginis cruentus

Henrik Grönvold, *A Monograph of the Pheasants*, 1918–22

MOUNTAIN QUAILS
Oreortyx pictus

John Gould & Henry Constantine Richter, *A Monograph of the Odontophorinæ*, 1844–50

Throughout his career, Gould published a number of monographs. Typically these works were inspired by his own personal interest. His *Monograph of the Odontophorinæ, or Partridges of America* was his fourth monograph and it grew from a fondness he had for the American partridge family. The mountain quail seen here had been in his personal collection, but many of the other specimens came from several public and private collections throughout Europe.

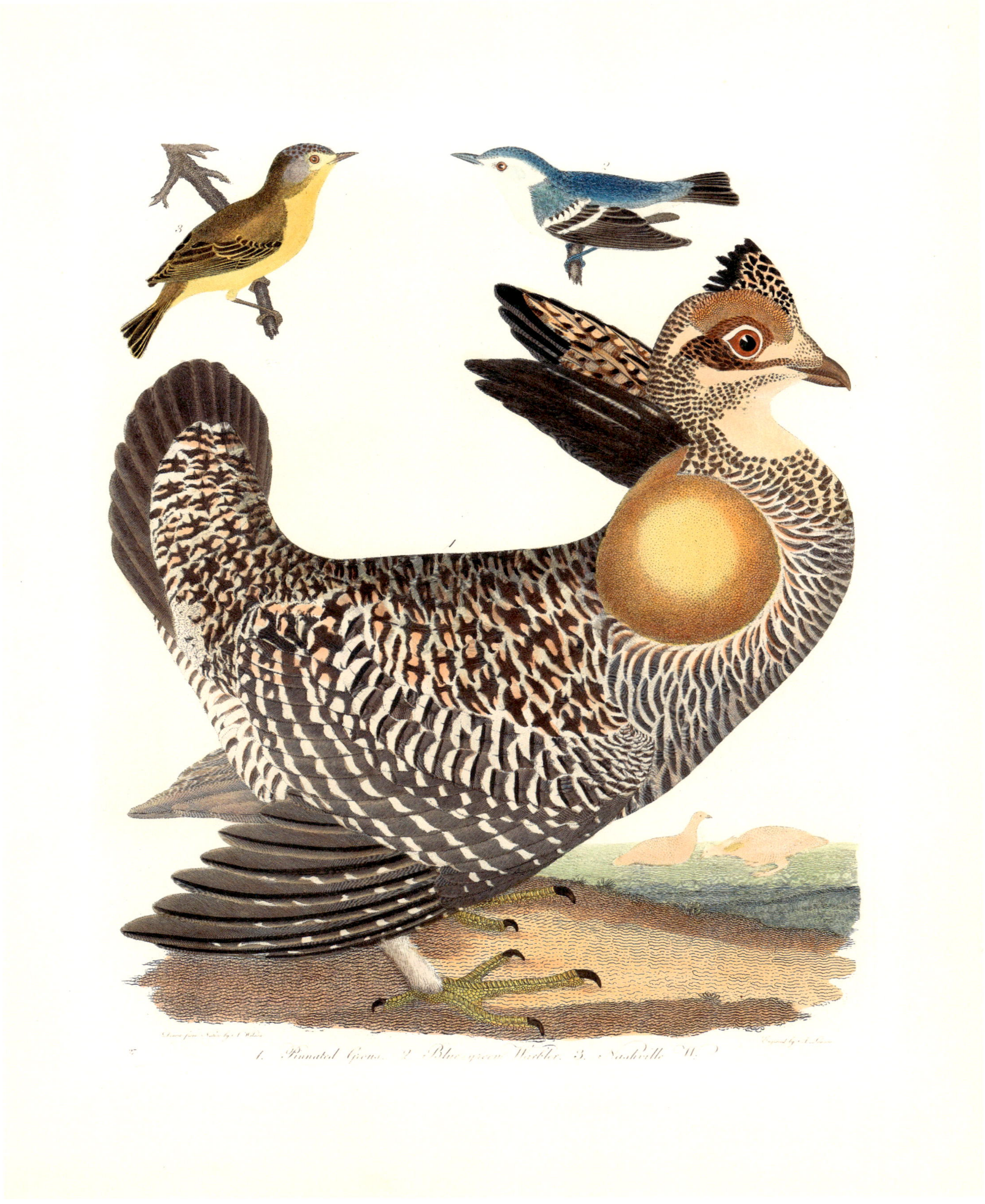

1. GREATER PRAIRIE CHICKEN
Tympanuchus cupido

2. CERULEAN WARBLER
Setophaga caerulea

3. NASHVILLE WARBLER
Leiothlypis ruficapilla

Alexander Wilson, *American Ornithology*, 1808–14

The distinctive greater prairie chicken is seen with two warblers in this work from Wilson's *American Ornithology*. Once abundant in North America, hunting and habitat loss has made these birds extremely rare. Often known as the pinnated grouse or *boomer*, the male of the species has a distinctive orange gular sac on its neck. This is used as part of its mating ritual, during which males produce loud booming calls, which they can amplify by inflating their neck sacs.

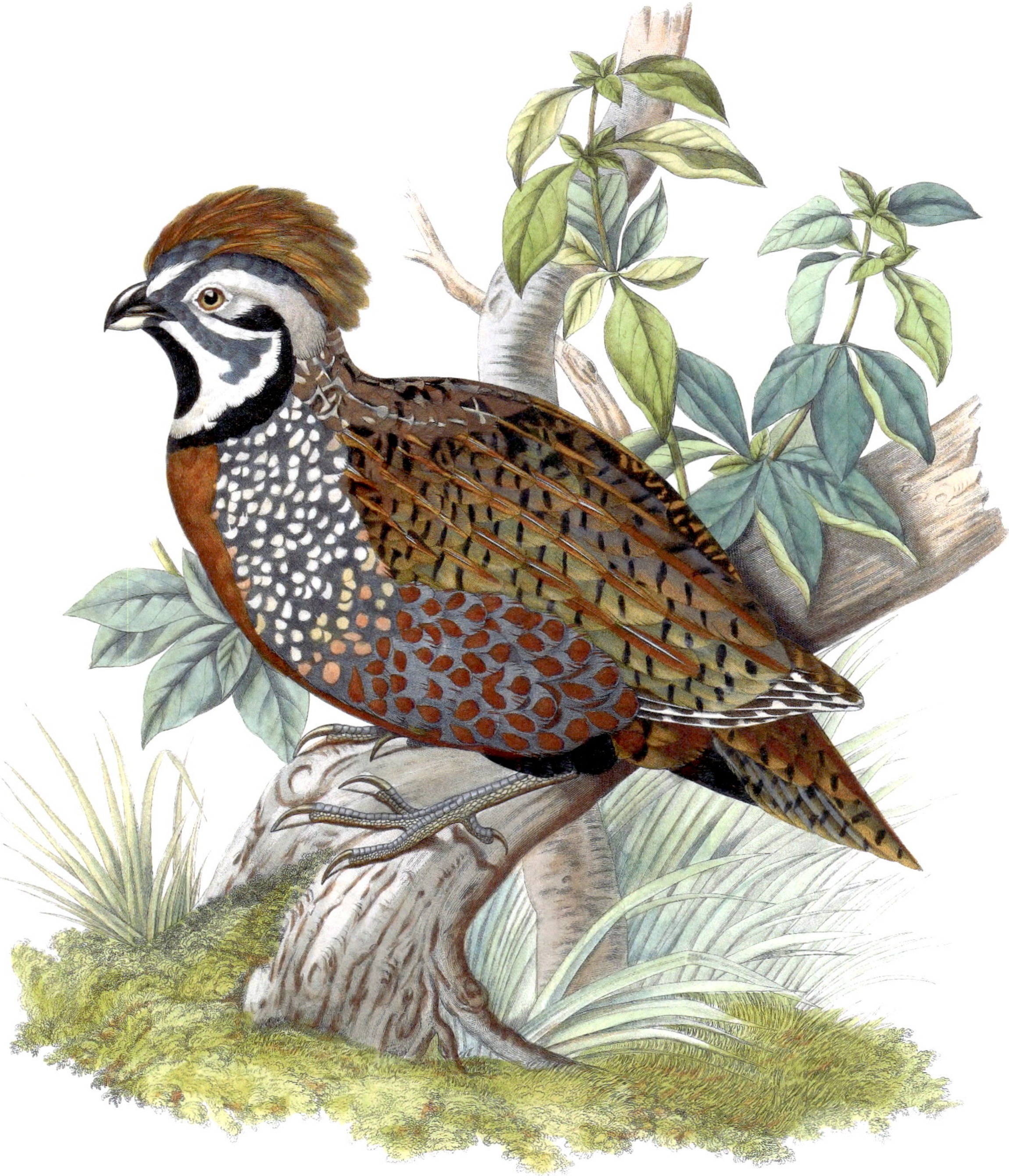

SPOT-BREASTED QUAIL
Cyrtonyx sallei

Paul Louis Oudart, *Arcana naturæ*, 1859

Found in the Mexican mountains as well as some parts of the southern United States, the spot-breasted quail is noted for its distinctive speckled plumage. This illustration is by French artist Paul Louis Oudart. Oudart was one of the first French artists to have his natural history paintings reproduced as lithographic prints and, later on in his career, he learnt the art of lithography himself. This work is taken from James Thomson's *Arcana naturæ*.

RUFFED GROUSE
Bonasas umbellus

Alexander Pope Jr., *Upland Game Birds and Water Fowl of the United States*, 1877–8

The ruffed grouse takes its name from the bird's distinctive ruffs on the side of its neck. This ring of feathers is found on both the male and female and both sexes appear similar. Found in Canada and northern parts of the United States, they are common game birds. In fact, their popularity led to some of North America's earliest game-management efforts. As far back as 1708, New York had a closed season on ruffed grouse, with no hunting allowed during specific times of the year.

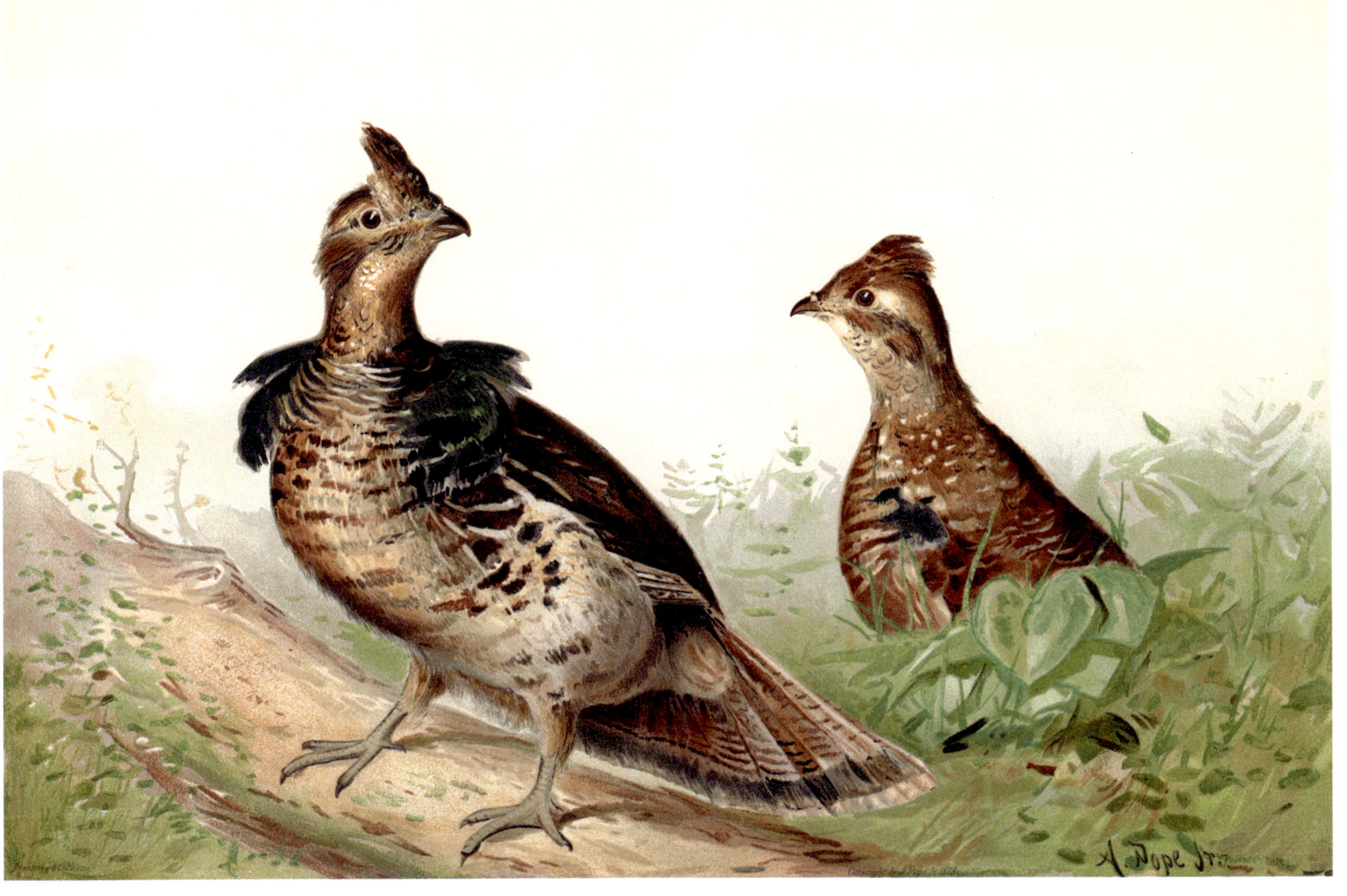

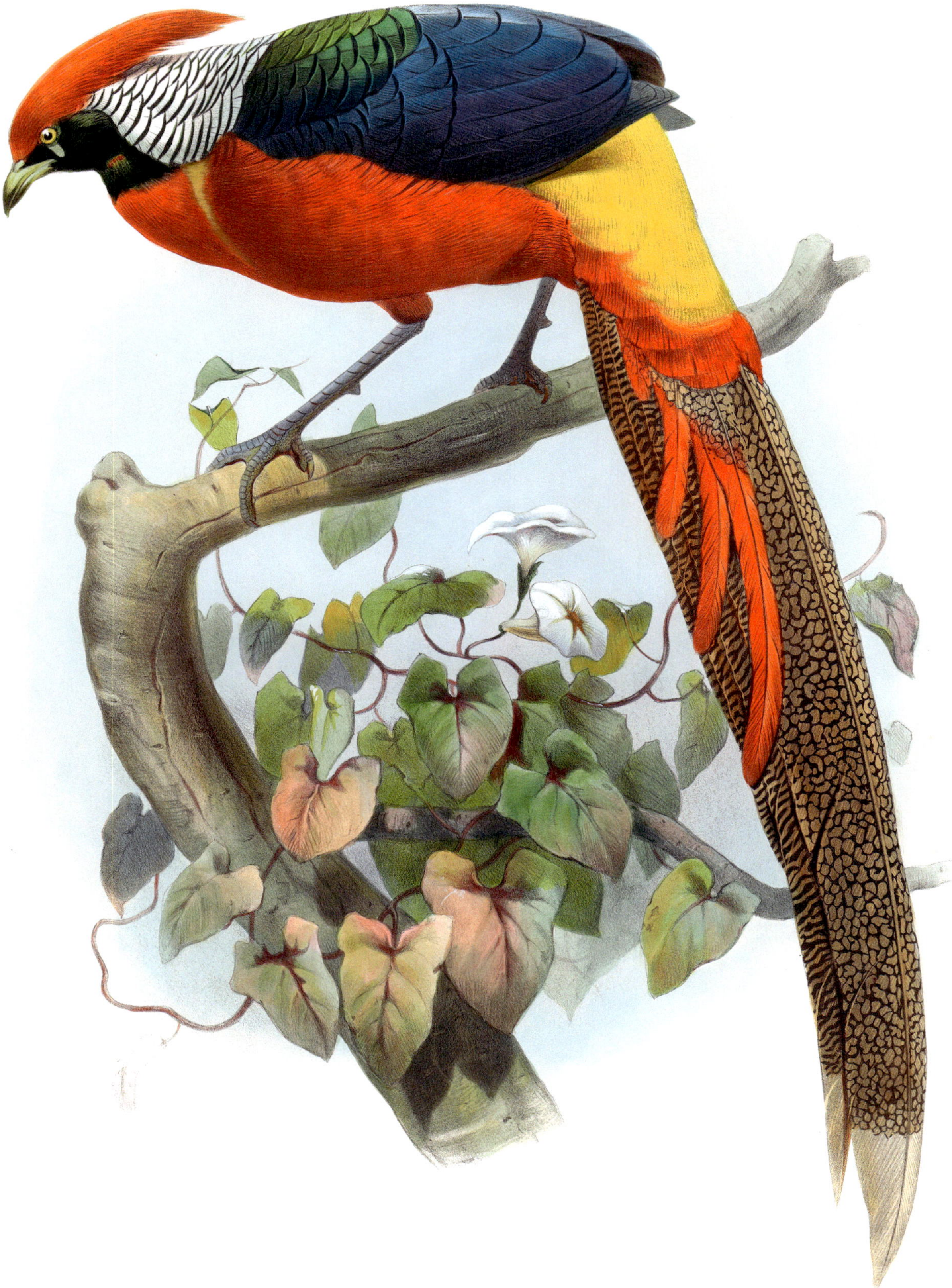

HYBRID PHEASANT

Joseph Wolf & Joseph Smit, *A Monograph of the Phasianidae*, 1870–2

This hybrid pheasant is the result of a golden pheasant (*Chrysolophus pictus*) breeding with a Lady Amherst's pheasant (*Chrysolophus amherstiae*). Several bird species hybridize with other birds and this can occur both naturally and, more commonly, through the intervention of humans. The fact that some birds can produce hybrids calls into question our notion of what a species is, and suggests that nature has a greater fluidity to it than our scientific constructs might suggest.

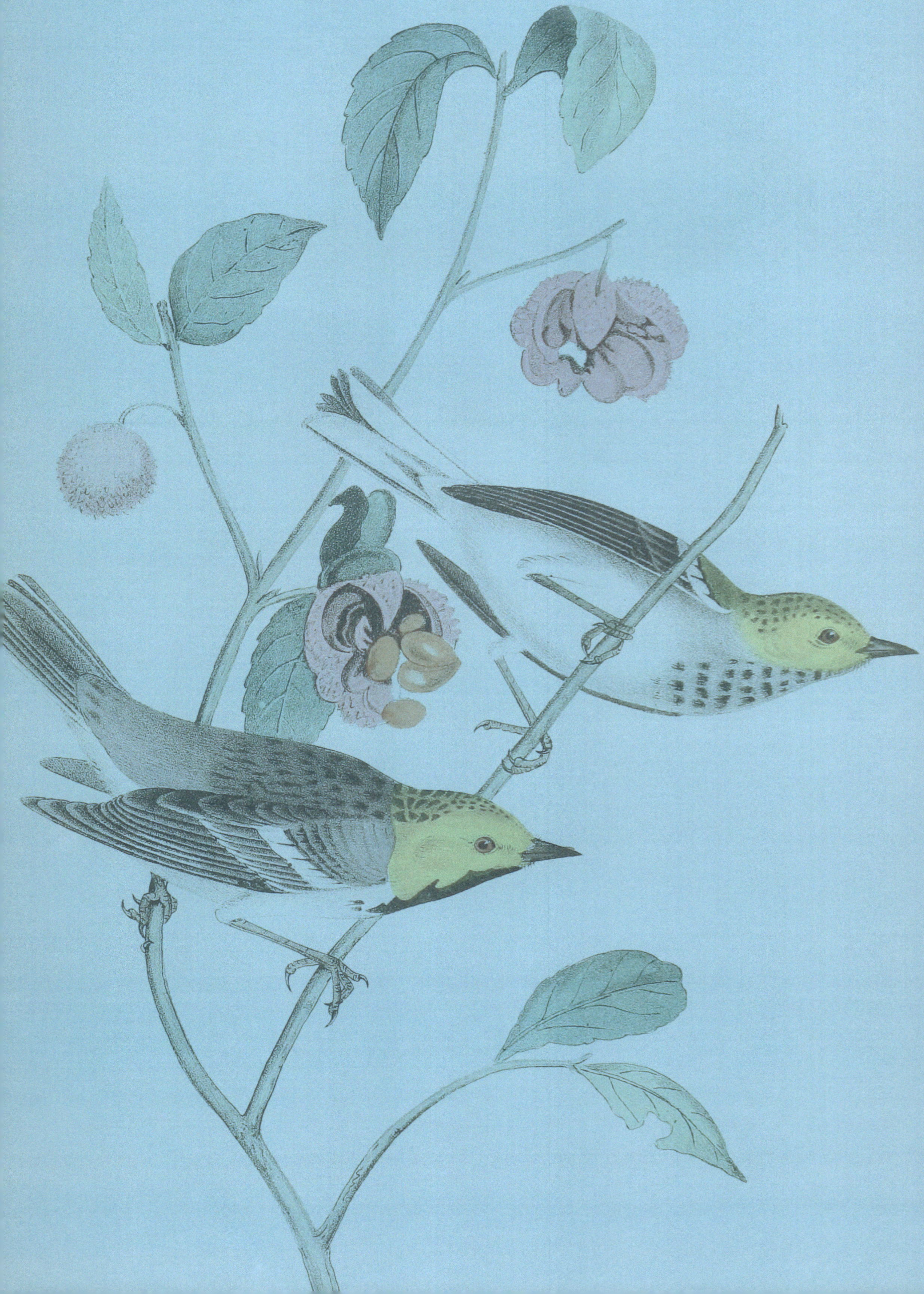

Songbirds

The characteristic cry of certain birds can sound musical to the human ear and, for this reason, their cries are often referred to as 'birdsong'. This ability for melodious vocalization exists in a grouping of birds that have a highly developed vocal organ. Their songs are uttered in characteristic bursts and melodies, and their unique feet allow them to deliver these songs while resting on perches.

Songbirds are, as their name might suggest, birds that have the ability to make sounds that seem musical. Often called 'perching birds' or 'passerines', these birds belong to the order Passeriformes. Apart from their vocal organs, what makes these birds distinct is their feet. Passerines have three toes in the front and one strong toe in the back. This allows them to easily perch on top of things like rocks, trees and cliffs. More than half of all birds have this ability – which makes Passeriformes the largest bird order in the world.

With roughly 100 families and about 6,000 species, Passeriformes are a diverse order of small to medium-sized birds. All of them have voice-boxes, but it is the suborder Passeri – or Oscine Passerines – that is distinguished by its highly developed vocal organ that is known as the syrinx. This allows these birds to communicate with other birds through a series of songs and calls.

There are different reasons for birdsong. Often these songs are used to attract mates, while at other times they are used to defend breeding territories. Sometimes, a bird will communicate simply through a short burst of sound known as a call. These calls can be a way of warning of danger – such as when a blackbird is startled and cries out in the hopes of chasing off an intruder. Sometimes a bird will use their call to locate their mate or offspring. Calls can even be used as a means to communicate with other birds while they are flying as a flock.

Different types of birds communicate through different melodies and so it is often possible to identify a bird solely by the sounds they produce. Some birds, such as the nightingale (*Luscinia megarhynchos*), can produce sounds that we often think of as beautiful, while others, such as the magpie (*Pica pica*), might be said to produce a harsh sound. Some songbirds sound incredibly musical while others do little or no singing.

An illustration might not be able to capture the sounds that these birds make, but their ability to perch has encouraged many illustrators to attempt to capture the environments that they inhabit. Mark Catesby's *Natural History of Carolina, Florida and the Bahama Islands* (1729–47; see page 44) was the first published account of the flora and fauna of North America and his illustrations tended to explore the relationship between birds and their habitats. Passionate about botany, he rarely drew his subjects in isolation and instead included the plant life and insects that he found in the field.

John James Audubon (see page 198) was another illustrator who spent time in the field and worked hard to give readers an idea of the habitats birds lived in. Before the publication of *Birds of America* (1827–38), similar works presented their subjects on blank backgrounds, but Audubon was determined to create images that could give readers a sense of each bird's scale and environment. Audubon and half a dozen assistants produced the original paintings for the book, with artists Maria Martin (1796–1863), George Lehman (*ca.* 1800–70) and Joseph Mason (1808–42) all producing impressive uncredited landscapes that truly help to elevate the work.

Audubon's elaborate scenic backgrounds clearly influenced the works of John Gould (see page 292). The illustrations in his work always featured backgrounds that closely related to his birds and occasionally even included details of where these birds nested. The illustrations in *Birds of Australia* (1840–8) are particularly noteworthy and this is no doubt in some part to do with Gould spending just under two years travelling and documenting the fauna of the country.

Illustrations from all these works are included in this chapter along with many other impressive depictions of Passerines. (A few songbirds, such as birds of paradise, are included in the next section.) As these form the largest order of birds, they are common throughout the world and their commonality is exemplified by the geographic variety that we can see throughout this chapter. Illustrations include a warbler from Australia (*Origma solitaria*), an American goldfinch (*Spinus tristis*) and a sparrow-lark (*Eremopterix australis*) from southern Africa.

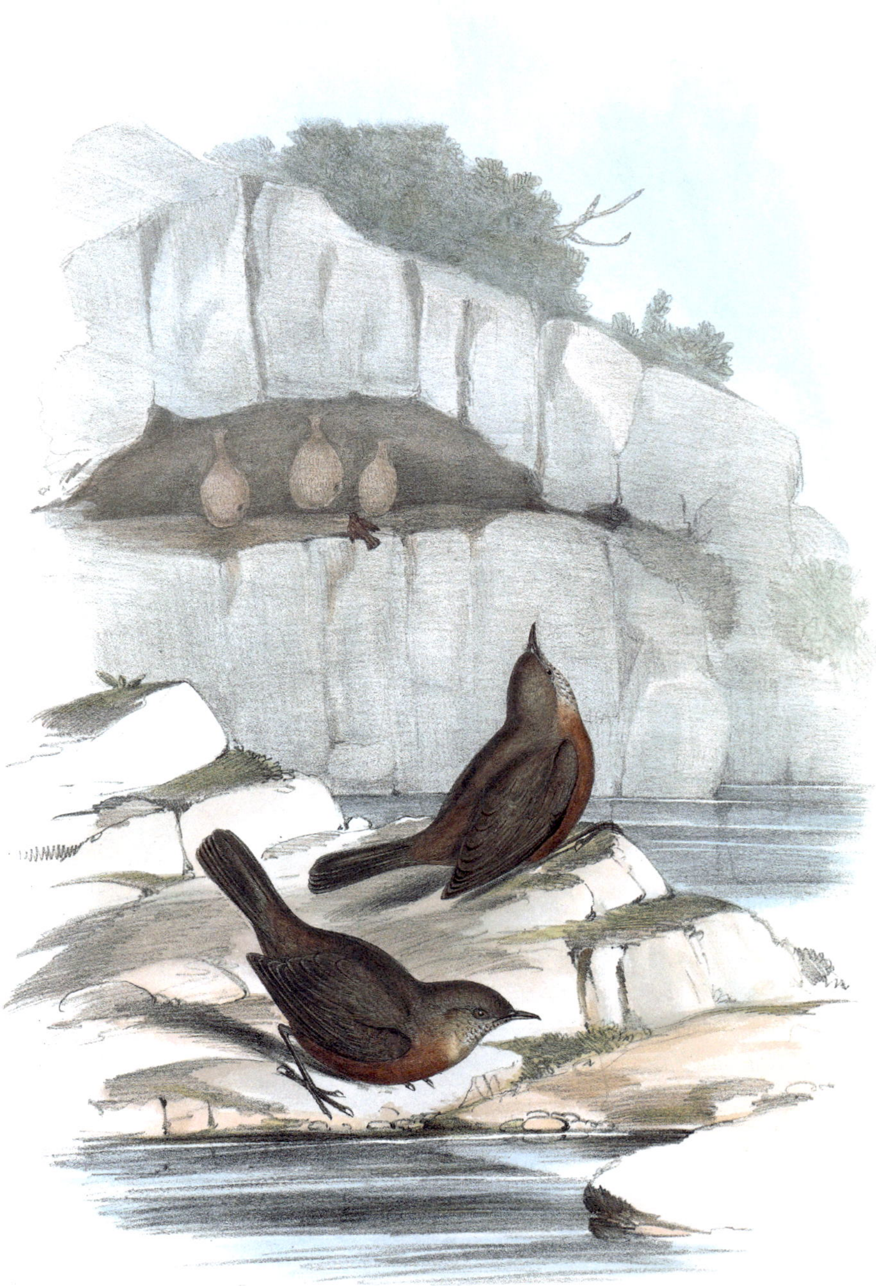

ROCKWARBLER
Origma solitaria

John Gould & Henry Constantine Richter, *The Birds of Australia*, 1840–8

OPPOSITE:
BROWN DIPPER
Cinclus pallasii

Diving Dipper and Other Birds, ca. 1610–15, Ustad Mansur, *The Shah Jahan Album*

Ustad Mansur was a Mughal painter and court artist who served Emperor Jahangir. Mansur painted many images of nature for the emperor and some of these can be seen in the *Shah Jahan Album*.

This impressive album provides a fascinating insight into the diverse interests of the era and contains studies from nature, portraits of important figures and illuminated folios of calligraphy by renowned calligraphers. Initiated by Jahangir, it was built upon over the generations.

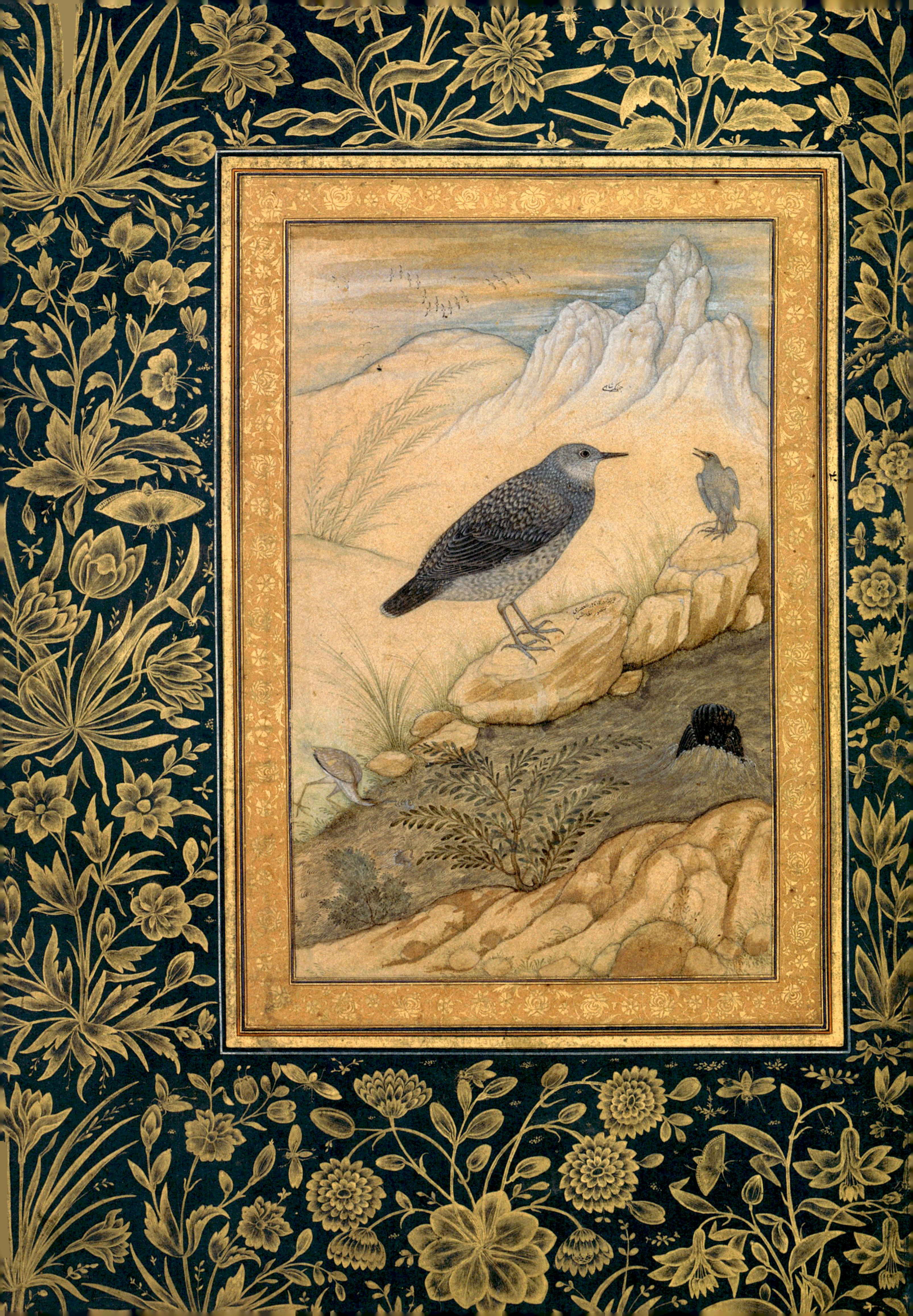

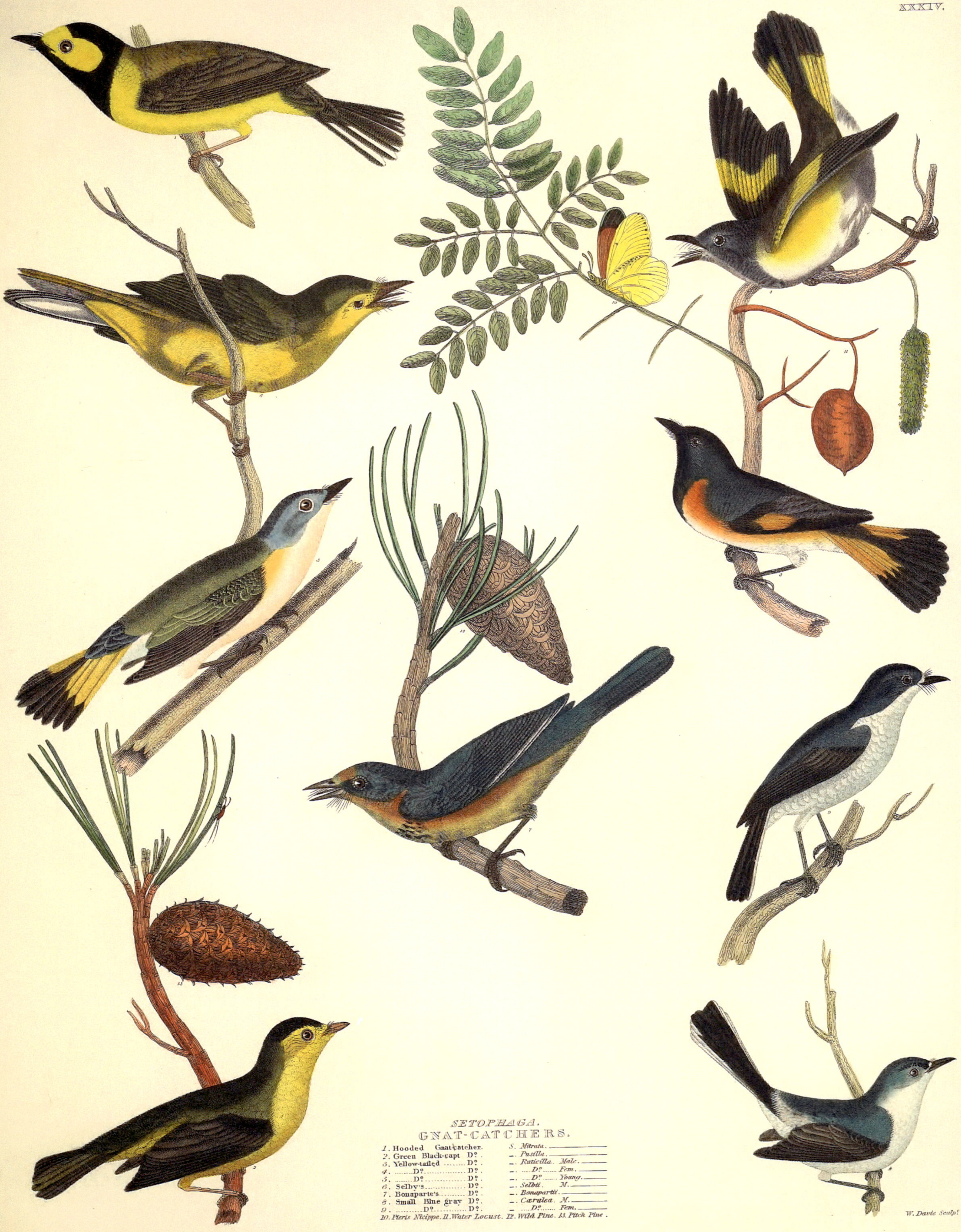

SETOPHAGA.

GNAT-CATCHERS.

1. Hooded Gnat-catcher.	*S. Mitrata.*
2. Green Black-capt D°.	*— Pusilla.*
3. Yellow-tailed .... D°.	*— Ruticilla. Male.*
4. ....D°................ D°.	*— .....D°...... Fem.*
5. ....D°................ D°.	*— ....D°...... Young.*
6. Selby's.................. D°.	*— Selbii. M.*
7. Bonaparte's.......... D°.	*— Bonapartii.*
8. Small Blue gray D°.	*— Cærulea. M.*
9. ....D°.............. D°.	*— ....D°...... Fem.*

10. Pieris Nicippe. 11. Water Locust. 12. Wild Pine. 13. Pitch Pine.

W. Davis Sculp.

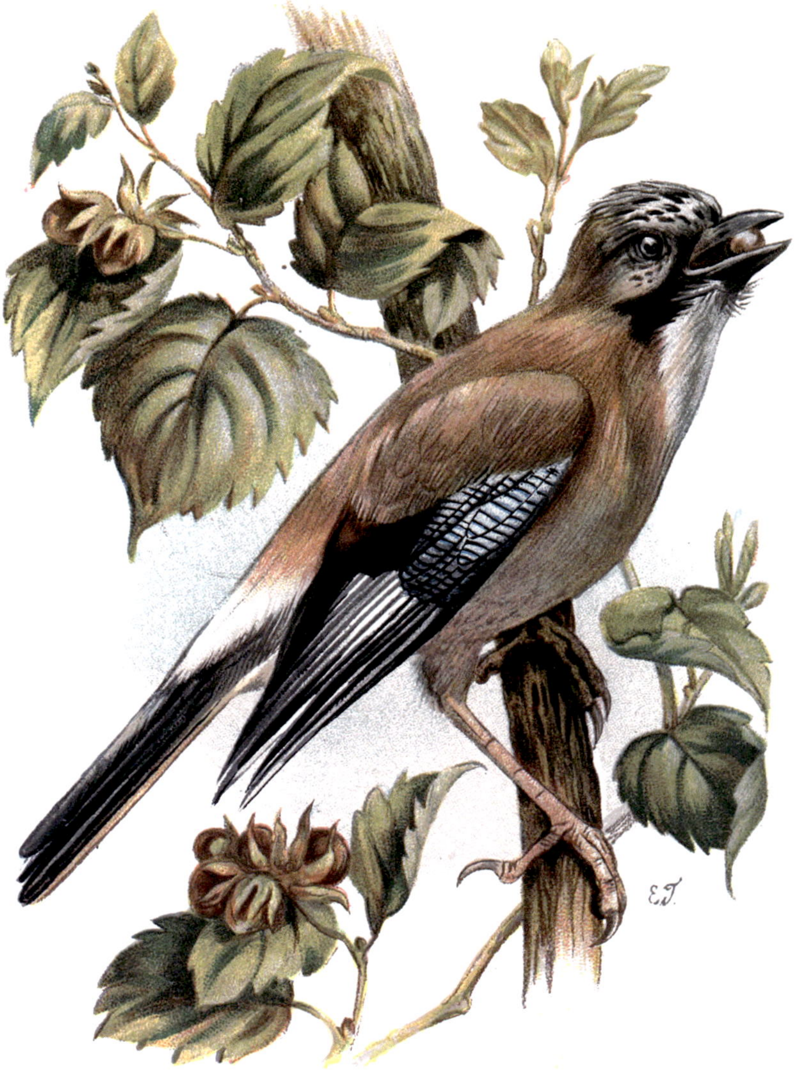

OPPOSITE, CLOCKWISE FROM TOP RIGHT:

AMERICAN REDSTART (FEMALE)
AMERICAN REDSTART
BLUE-GRAY GNATCATCHER
BLUE-GRAY GNATCATCHER
CANADA WARBLER
WILSON'S WARBLER (MALE)
AMERICAN REDSTART (JUVENILE)
'SELBY'S GNATCATCHER'
HOODED WARBLER (MALE)

Alexander Wilson, *Illustrations of the American Ornithology of Alexander Wilson and Charles Lucian Bonaparte*, 1835

Although labelled as Gnatcatchers (*Polioptilidae*), only one of the species on this plate (blue-gray gnatcatcher) remains in the family. The rest are New World warblers (*Parulidae*).

EURASIAN JAY
Garrulus glandarius

Eliza Turck, *Familiar Wild Birds*, 1883

Published in four volumes between 1883 and 1888, Walter Swaysland's *Familiar Wild Birds* featured illustrations by Eliza Turck and Archibald Thorburn. This Eurasian jay is a work by Turck. Known predominantly as a painter, she exhibited frequently at the Royal Academy and was known for her portraiture, landscape and genre paintings. Her contribution to *Familiar Wild Birds* seems to be her only work as a bird illustrator.

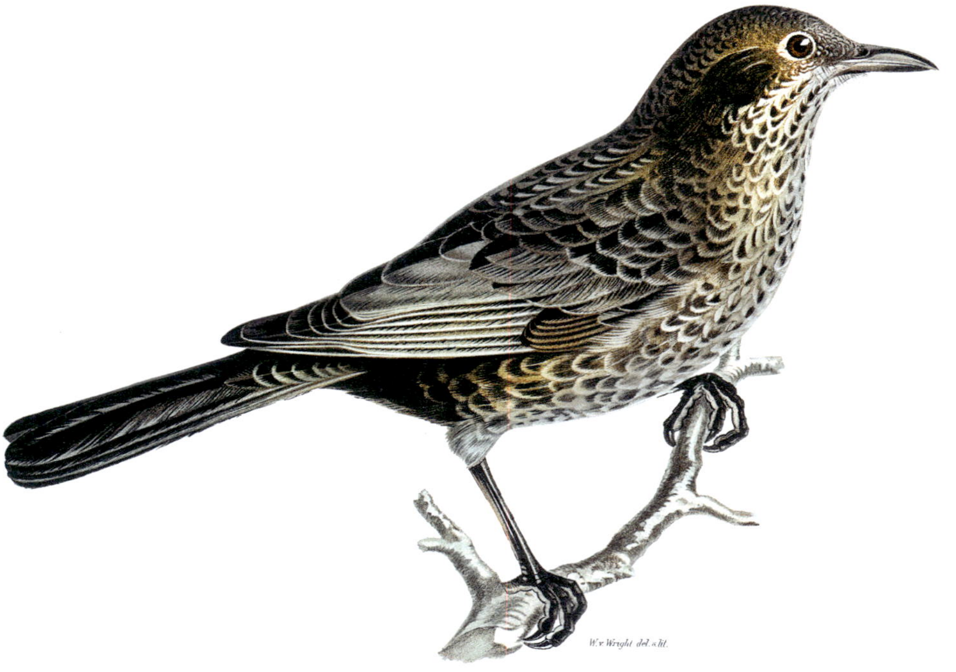

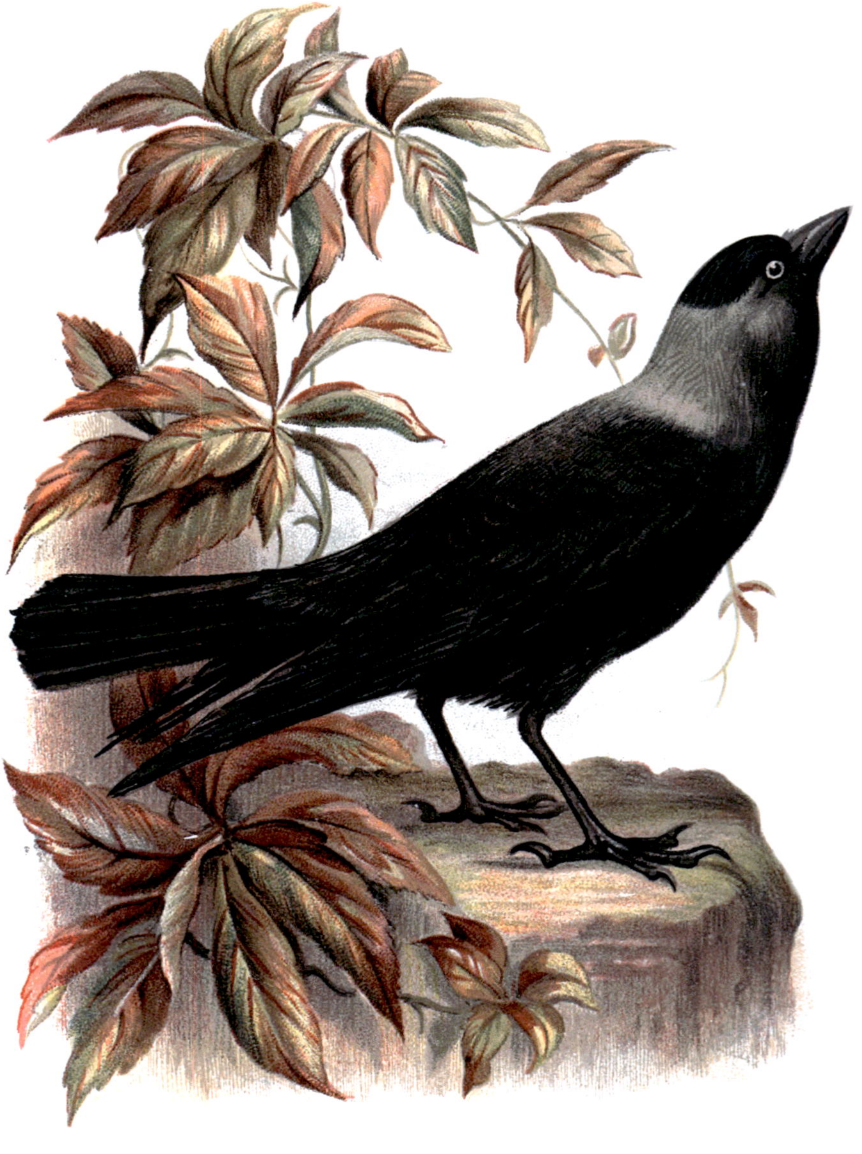

TOP:

EURASIAN BLACKBIRD (FEMALE)

Turdus merula

Wilhelm von Wright, *Svenska Fåglar*, 1828–38 (this edition: 1929)

BOTTOM:

EURASIAN JACKDAW (FEMALE)

Coloeus monedula

Archibald Thorburn, *Familiar Wild Birds*, 1883

This jackdaw also comes from Swaysland's *Familiar Wild Birds* but is the work of Thorburn. A Scottish bird artist, he is among the most famous of all British ornithological illustrators. His contribution to Swaysland's work marked his first major commission, and 144 of the book's 160 coloured plates were produced by him.

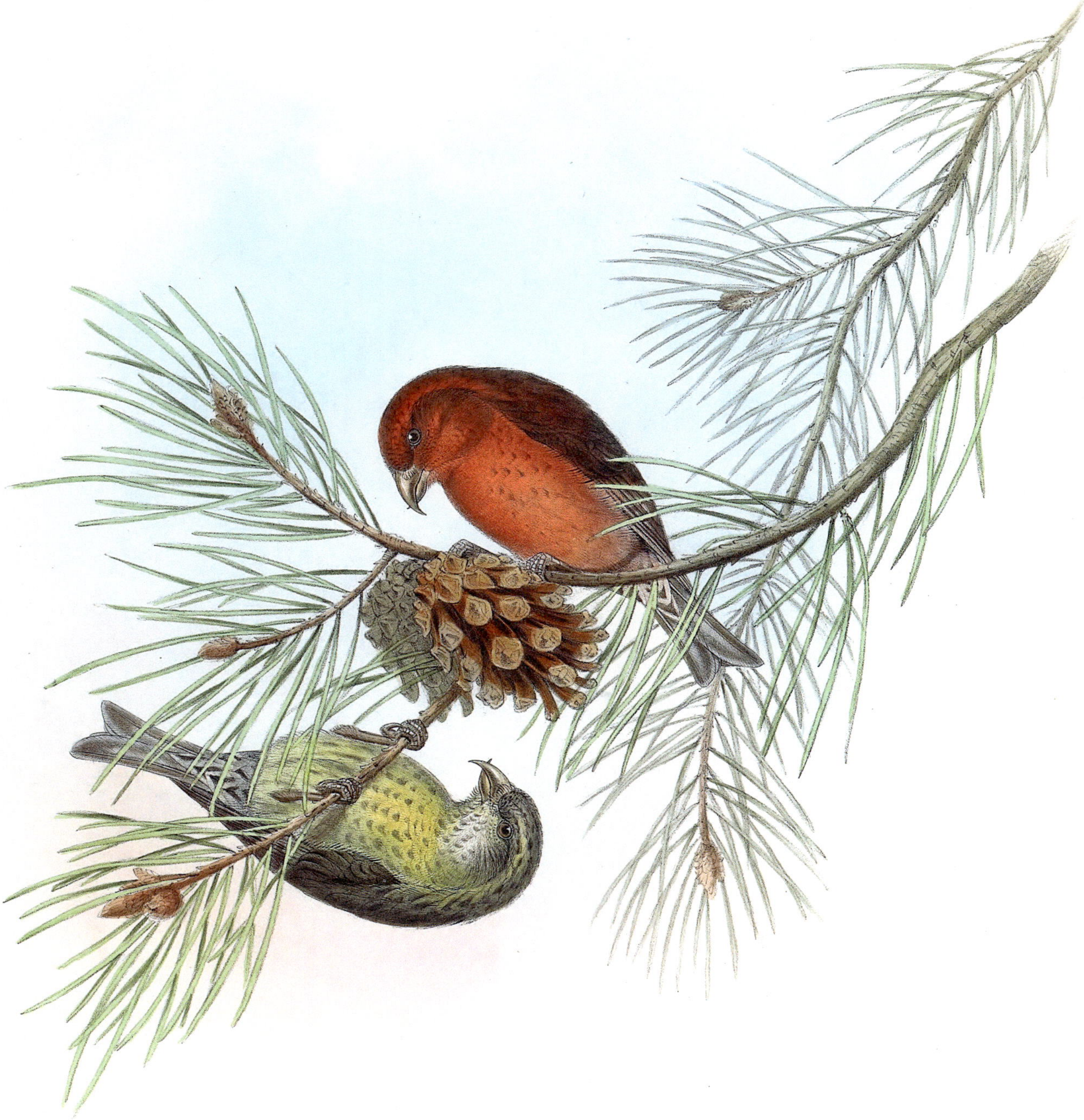

RED CROSSBILLS
Loxia curvirostra

John Gould & Henry Constantine Richter, *The Birds of Asia*, 1850–83

Crossbills are striking birds that belong to the finch family (*Fringillidae*). Characterized by their distinctive bills, their mandibles cross at the tip, allowing them to twist their upper and lower bill in opposite directions. Crossbills often feed on pinecones – as this illustration shows – and their unique bills make them excellent foragers. It allows them to prise open the cone and extract the seed inside. They are the only finch capable of doing this and their bills represent their unique evolution.

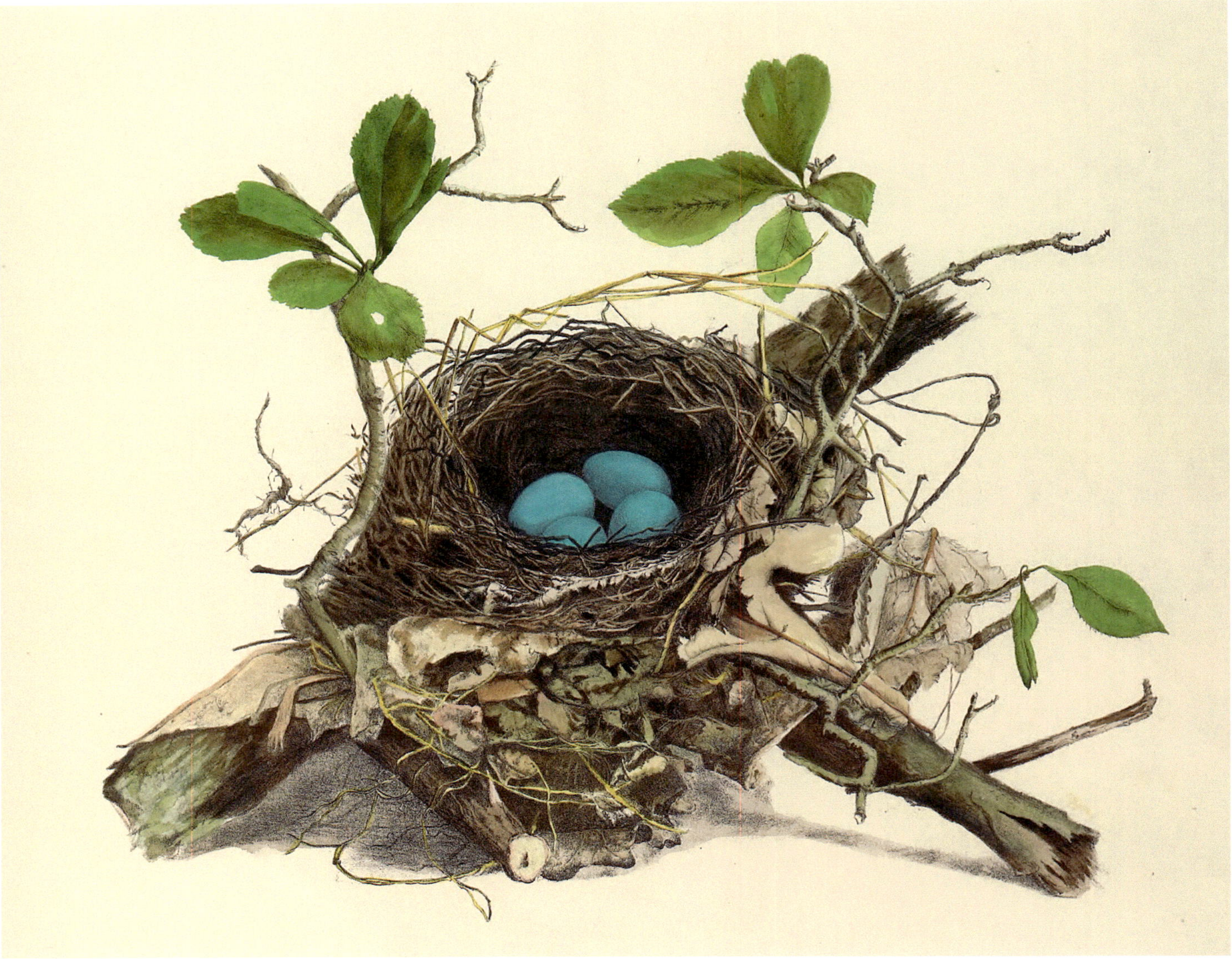

NEST OF A WOOD THRUSH
Hylocichla mustelina

Genevieve Estelle Jones, *Illustrations of the Nests and Eggs of Birds of Ohio*, 1879–86

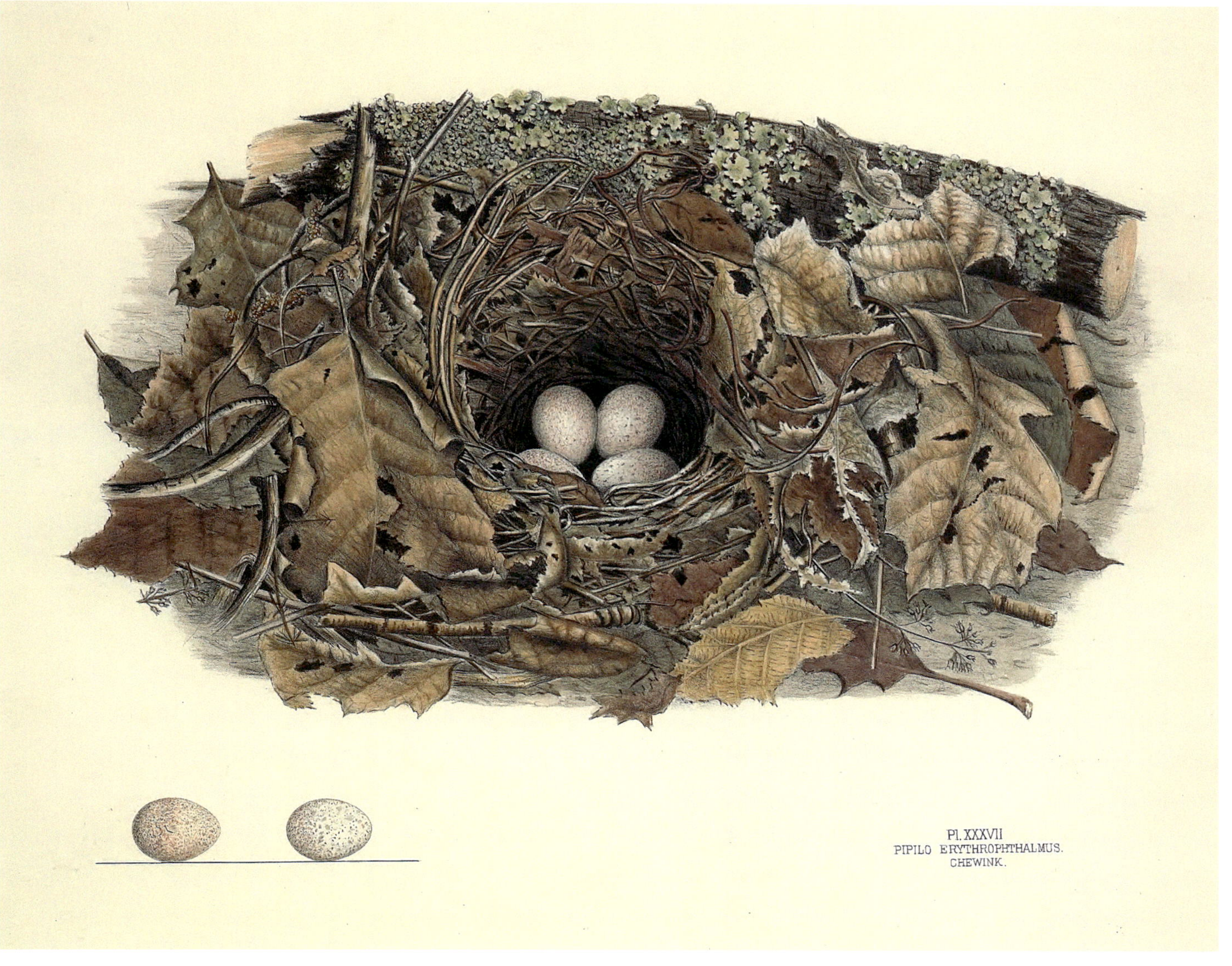

NEST OF AN EASTERN TOWHEE
Pipilo erythrophthalmus

Genevieve Estelle Jones, *Illustrations of the Nests and Eggs of Birds of Ohio*, 1879–86

In 1876, Genevieve Jones viewed John James Audubon's *The Birds of America* at the World's Fair in Philadelphia and was so inspired by the work that she was determined to produce a book of her own. Her ambition was to create a companion work to Audubon's that would focus on the nests and eggs of the birds seen in his original book. Persuaded to limit her workload to simply the birds of her home state, her *Illustrations of the Nests and Eggs of Birds of Ohio* documented the nests and eggs of some 130 species.

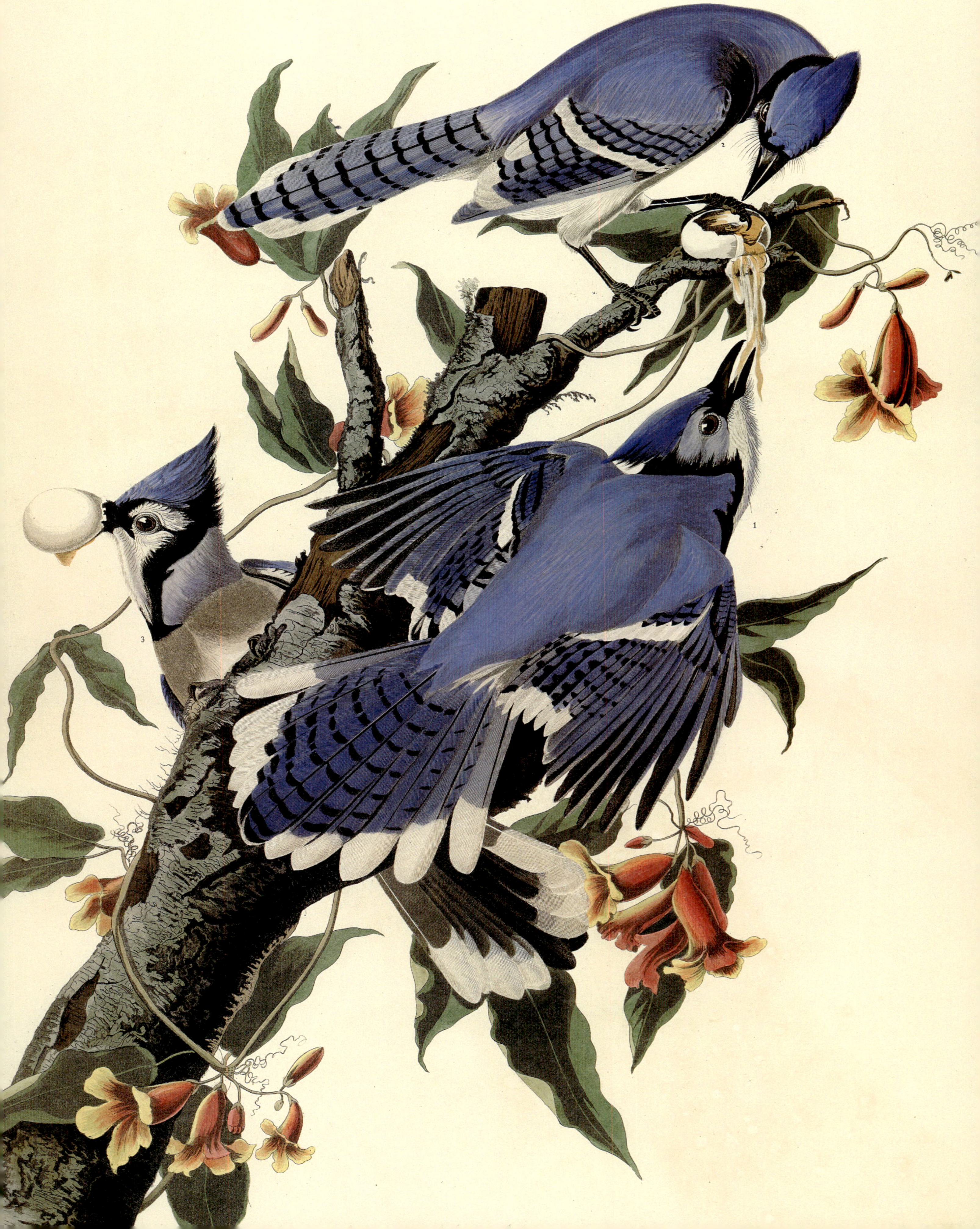
2
1
3

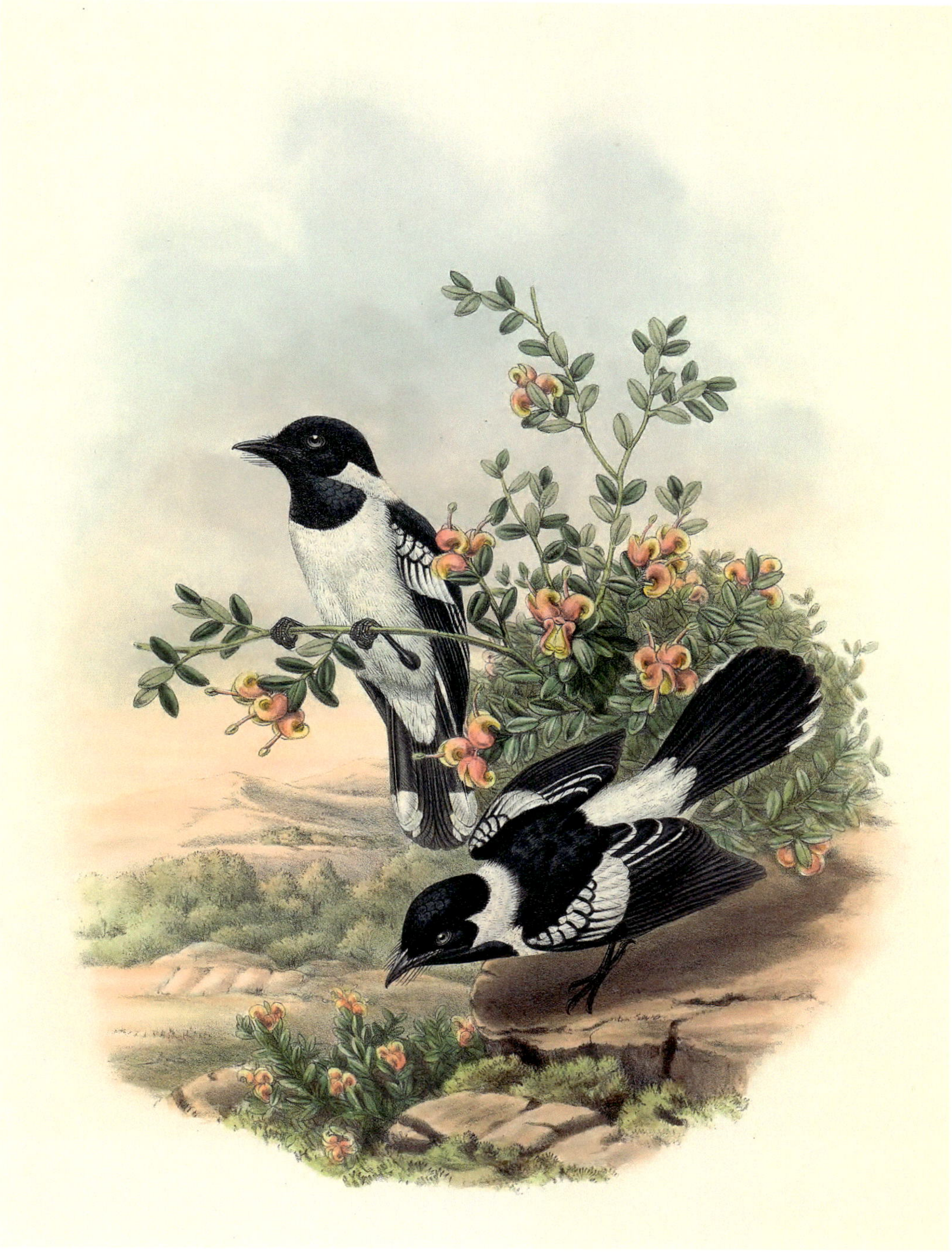

OPPOSITE:
BLUE JAY
Cyanocitta cristata

John James Audubon, *The Birds of America*, 1827–38

'Who could imagine that a form so graceful, arrayed by nature in a garb so resplendent, should harbour so much mischief.' These, the words of Audubon, upon seeing a blue jay that was feasting on an egg it had pilfered from another bird's nest. The blue jay is known for raiding nests, stealing eggs and attacking smaller birds, yet this behaviour is far less common than their reputation might suggest.

WHITE-COLLARED MONARCHS
Symposiachrus vidua

William Matthew Hart, *The Birds of New Guinea and the Adjacent Papuan Islands*, 1875–88

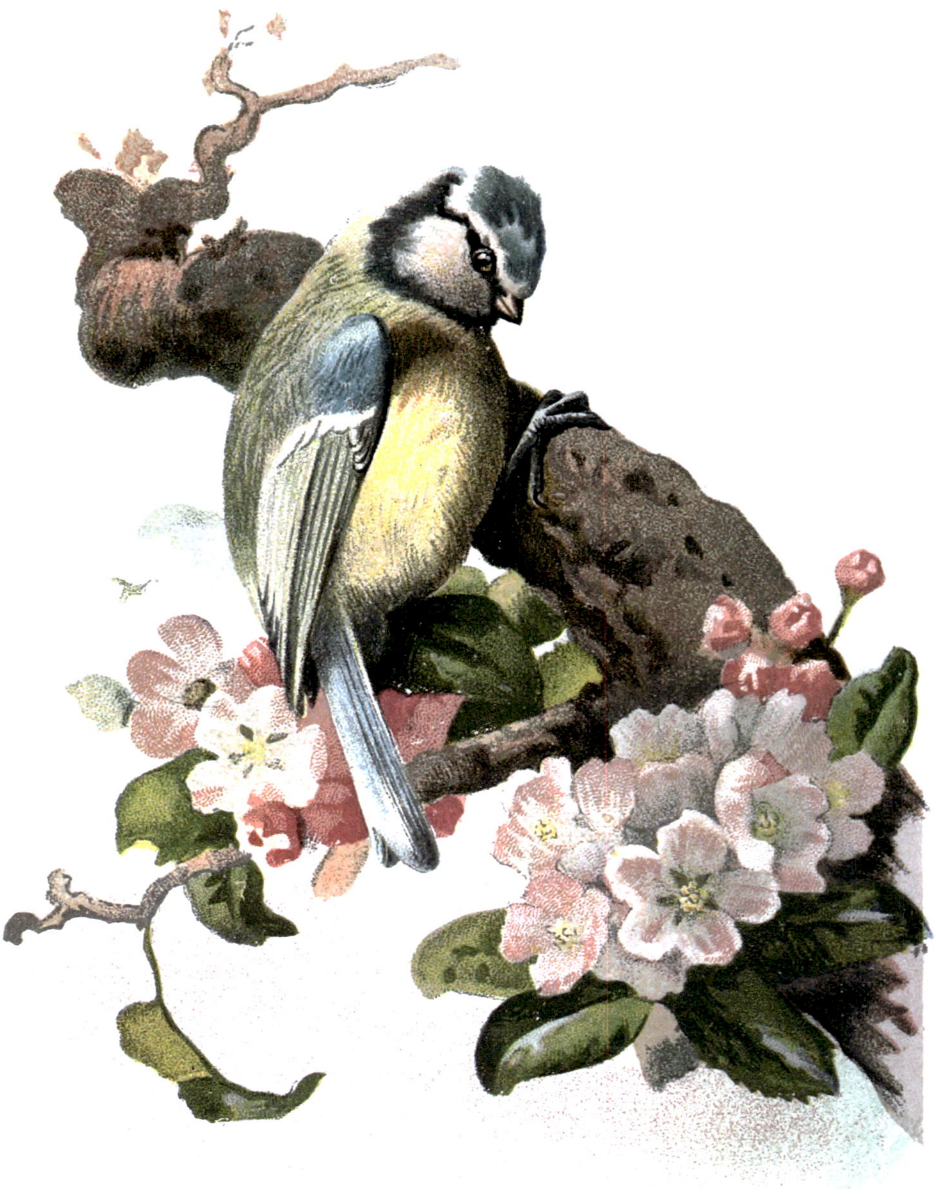

EURASIAN BLUE TIT
Cyanistes caeruleus

Archibald Thorburn,
Familiar Wild Birds, 1883

Tits are a widespread and common family of birds that can be found throughout much of Europe, North America, Asia and Africa. They are small birds that can have either plain or colourful plumages. The blue tit is a particularly attractive species, with a colourful mix of blue and yellow plumage. Originally a woodland bird, they have adapted well to man-made habitats and are regular garden visitors throughout much of Europe.

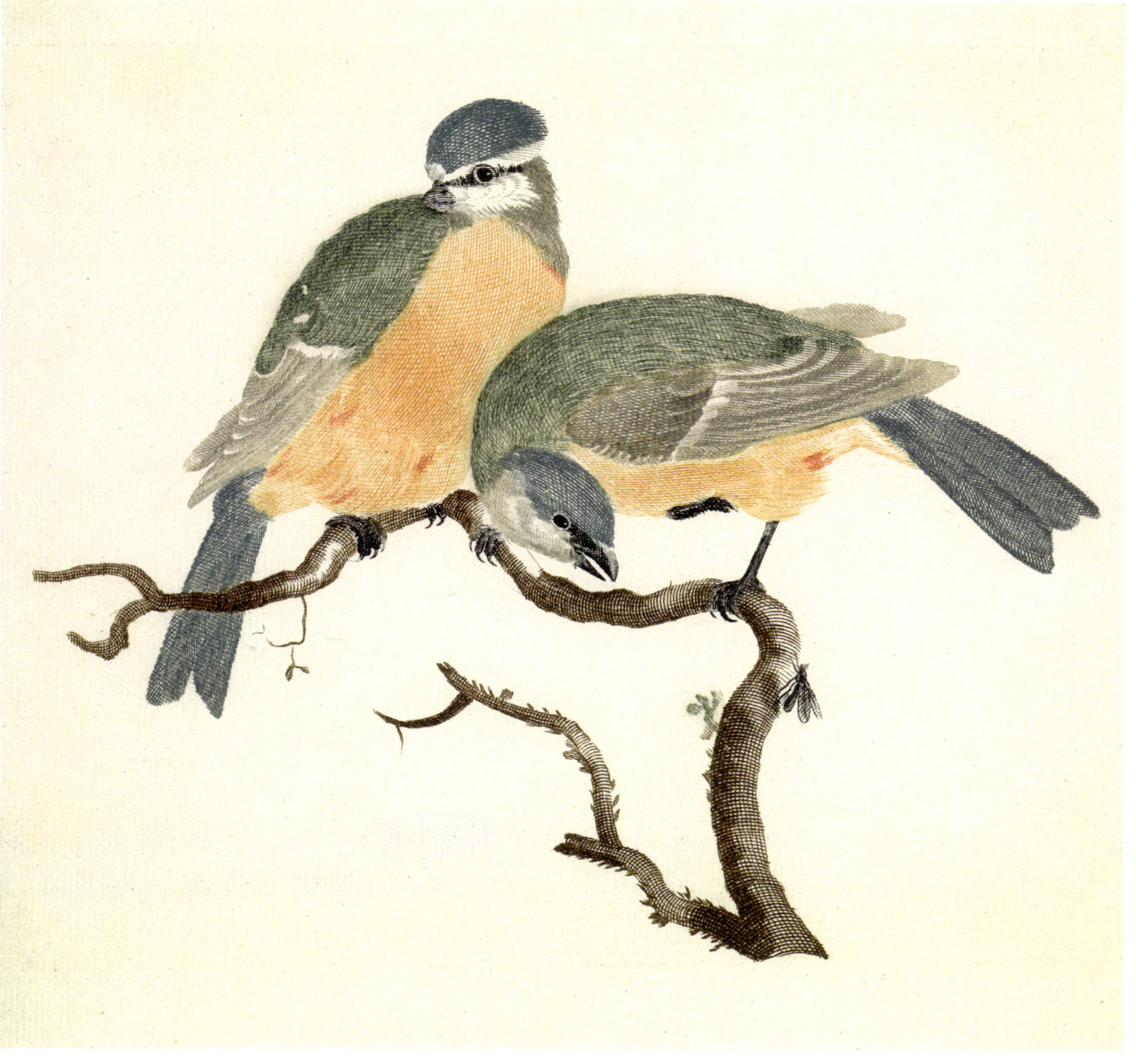

EURASIAN BLUE TIT & GREAT TIT
Cyanistes caeruleus & Parus major

Een Pimpelmees en een Koolmees op een tak, Johan Teyler, engraving, 1688–98

Johan Teyler was a Dutch Golden Age painter and engraver. He was a pioneer of print and, around 1680, he developed a colour printing process that allowed for full-colour impressions. Intaglio engraving had been the standard method for printing in Europe since the mid-fifteenth century, with images being printed solely in black-and-white. Teyler's method used coloured inks when preparing the intaglio plate. Unfortunately, the process was not commercially viable, despite the results being artistically beautiful.

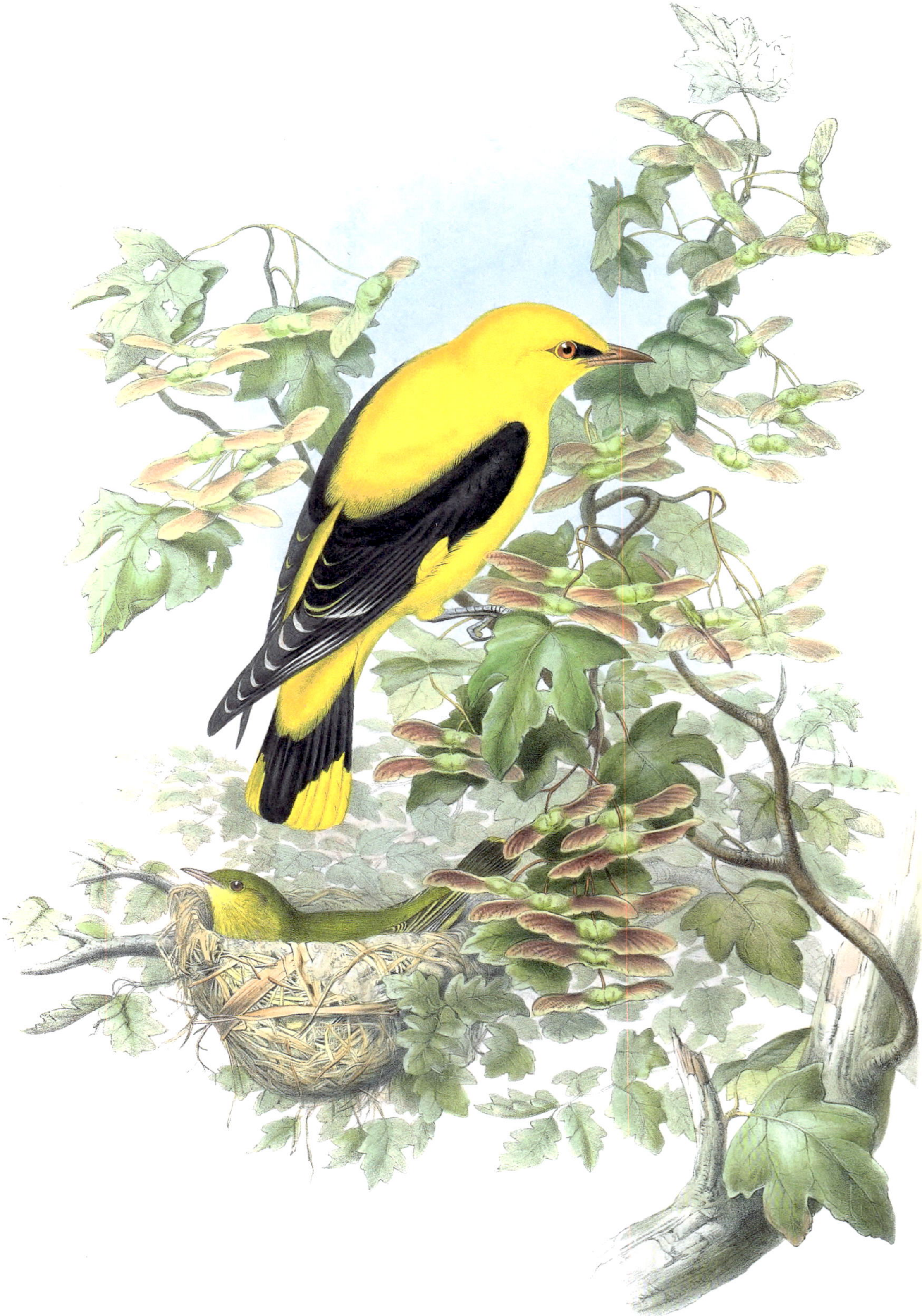

EURASIAN GOLDEN ORIOLE
Oriolus oriolus

John Gould & Henry Constantine Richter, *The Birds of Great Britain*, 1862–73

With its fluting song and brilliant yellow plumage, the golden oriole seems more like a bird found in the tropics than in the coastal plains of eastern England. Indeed, when Gould included it in his *Birds of Great Britain*, he wrote that some readers may think he was adding a species which 'does not belong to our avifauna'. In truth, this oriole is an extremely rare bird in Britain, but sightings are not impossible. A number of them began breeding in East Anglia in the mid-1960s when the match company Bryant & May began growing a forest of poplars.

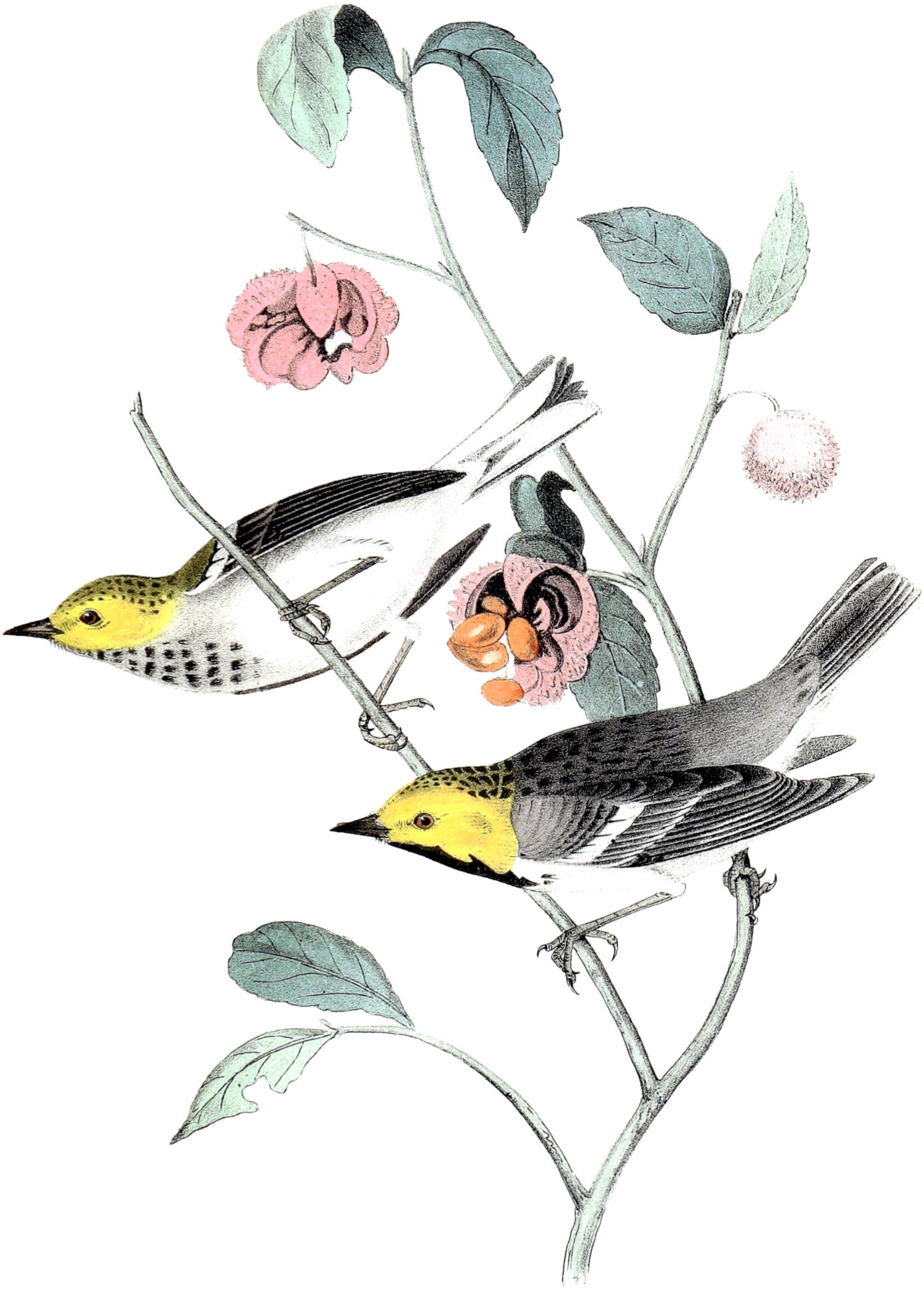

HERMIT WARBLERS
Setophaga occidentalis

John James Audubon, *The Birds of America*, 1827–38 (this edition: 1840–4)

This hermit warbler comes from the Royal Octavo Edition of Audubon's *Birds of America*. Published over seven volumes between 1840 and 1844, it was a smaller and more affordable edition of his masterwork. While the original had been published in London, Audubon kept things in America for this edition, employing the talents of J.T. Bowen and his team who lithographed and hand coloured the work in Philadelphia and New York.

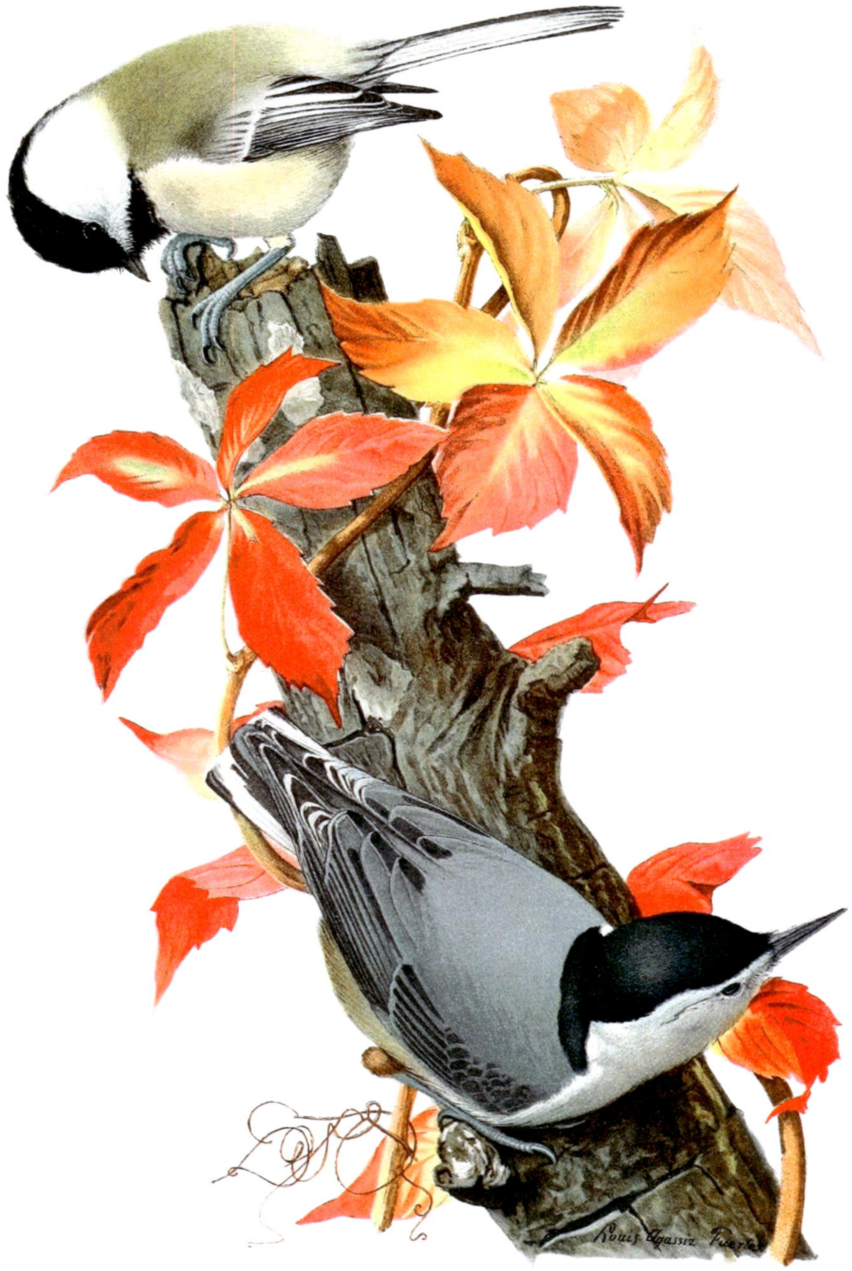

BLACK-CAPPED CHICKADEE &
WHITE-BREASTED NUTHATCH
Poecile atricapillus
& Sitta carolinensis

Louis Agassiz Fuertes, *The Economic Value of Birds to the State*, 1903

Louis Agassiz Fuertes was one of America's most widely acclaimed ornithological illustrators at the turn of the twentieth century. Inspired by the work of Audubon, he often travelled in pursuit of birds and visited many parts of the United States as well as parts of Canada, Mexico, Colombia and Jamaica. His work was based on studies of living birds and his illustrations are characterized by their sense of vitality and liveliness.

OPPOSITE, CLOCKWISE FROM TOP RIGHT:

PURPLE FINCH (FEMALE)
Haemorhous purpureus

BLUE GROSBEAK (FEMALE), *POSSIBLY* INDIGO BUNTING (FEMALE)
Passerina cyanea or *P. caerulea*

NORTHERN CARDINAL (FEMALE)
Cardinalis cardinalis

NORTHERN CARDINAL (MALE)
Cardinalis cardinalis

BLUE GROSBEAK (MALE), *POSSIBLY* INDIGO BUNTING (MALE)
Passerina cyanea or *P. caerulea*

DICKCISSEL (MALE)
Spiza americana

PURPLE FINCH (MALE)
Haemorhous purpureus

Alexander Wilson, *Illustrations of the American Ornithology of Alexander Wilson and Charles Lucian Bonaparte*, 1835

In 1835 all of these birds were identified as finches, but today only the purple finch is still considered a member of the Finch family (*Fringillidae*). The rest on the plate are now in the Cardinal family (*Cardinalidae*). There may be extra confusion here: a bird is depicted that resembles a male blue grosbeak next to a female that suspiciously resembles a streaky-breasted female indigo bunting.

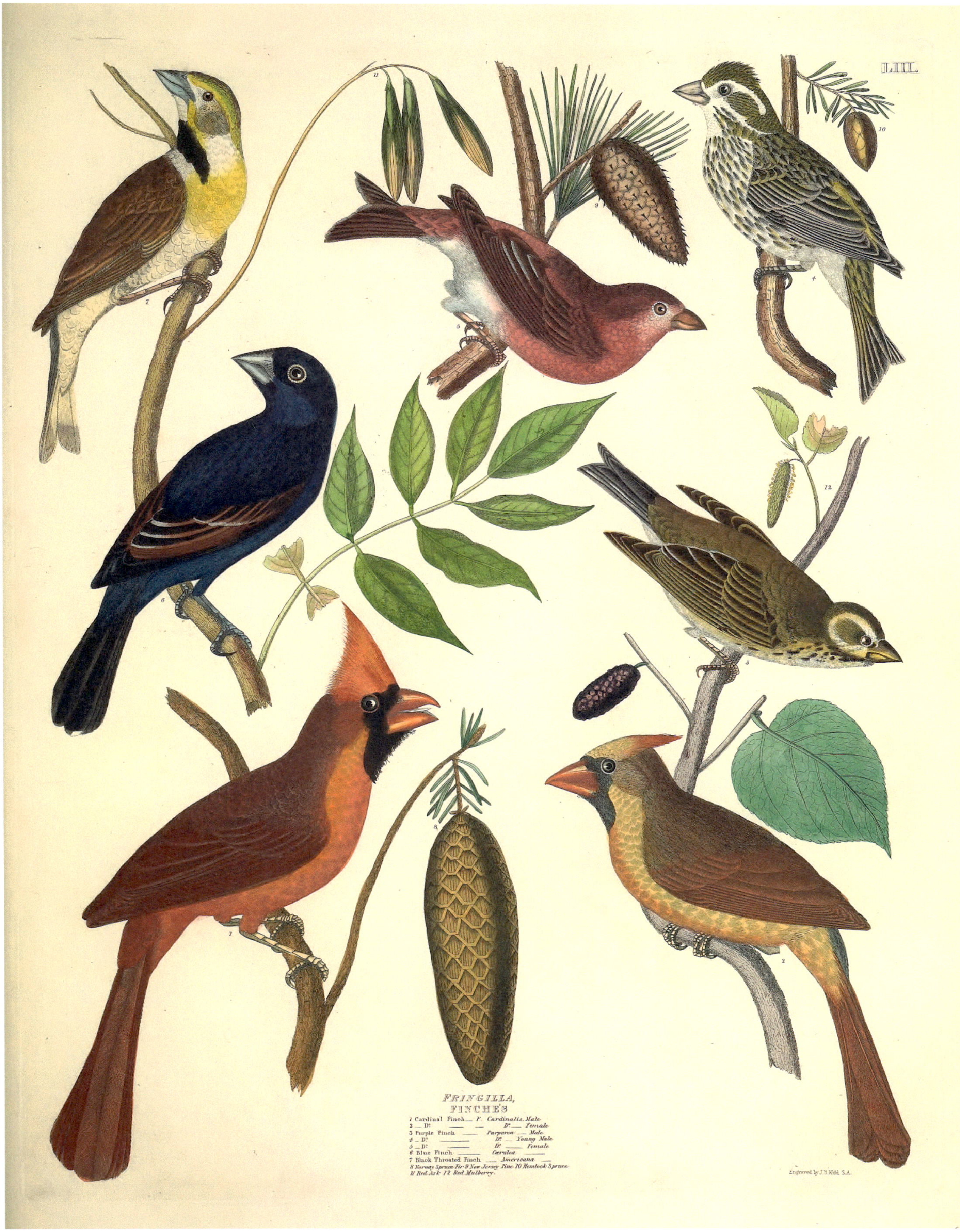
LIII.
FRINGILLA,
FINCHE'S
1 Cardinal Finch — F. Cardinalis. Male
2 — Do. — — Do. — Female
3 Purple Finch — Purpurea — Male
4 — Do. — Do. — Young Male
5 — Do. — Do. — Female
6 Blue Finch — Cærulea —
7 Black Throated Finch — Americana —
8 Norway Spruce Fir 9 New Jersey Pine 10 Hemlock Spruce
11 Red Ash 12 Red Mulberry.
Engraved by J. B. Kidd. S.A.

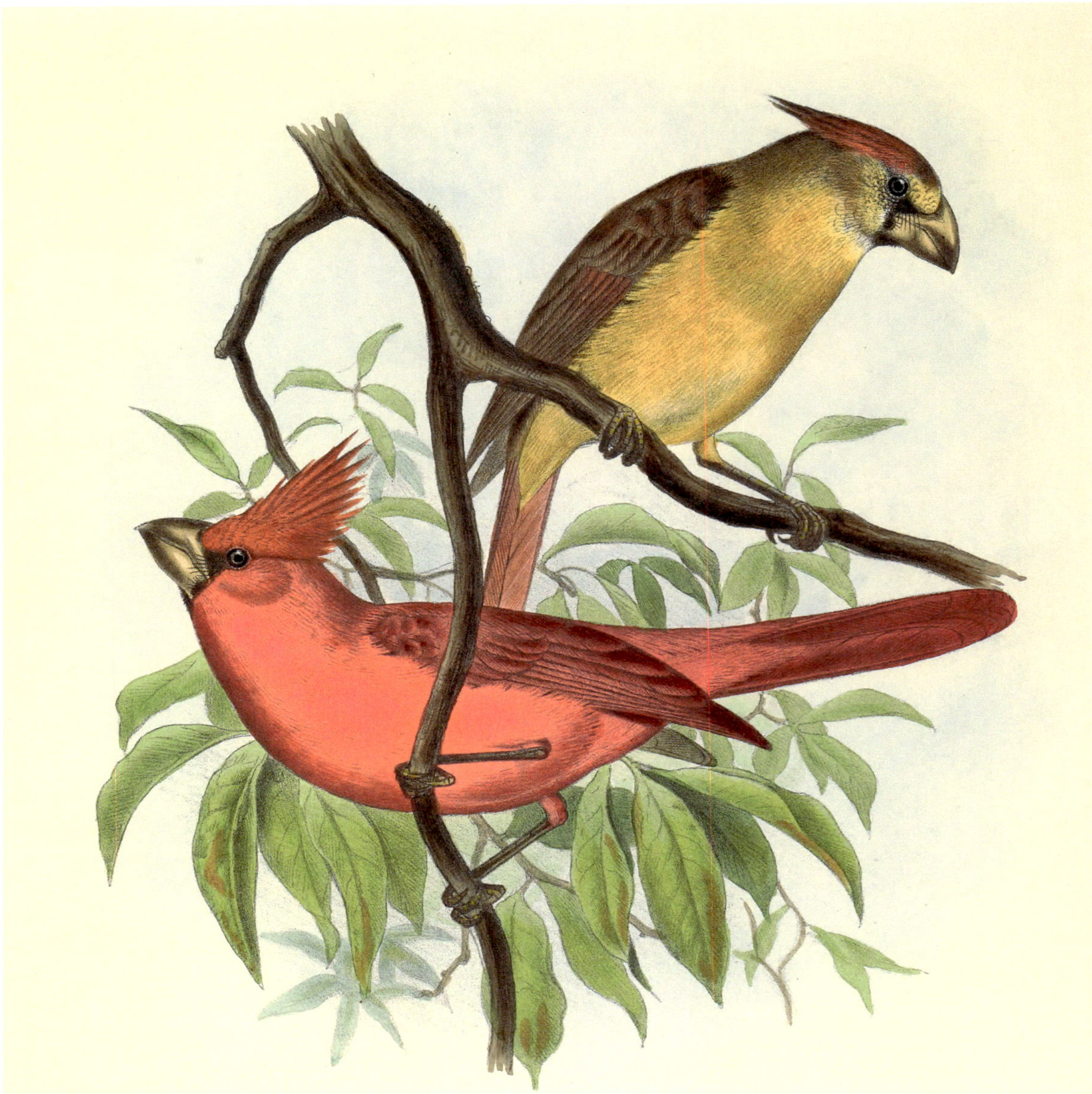

VERMILION CARDINALS
Cardinalis phoeniceus

Joseph Smit, *Exotic Ornithology*, 1869

The cardinals (*Cardinalis*) are a small group of thick-billed birds found in North and South America. They have stout bills and males are predominantly red in colour. This is particularly pronounced in the vermillion cardinal, a bird found in parts of Colombia and Venezuela. This study of both a male and female comes from Sclater's *Exotic Ornithology*, which describes many new and rare birds in both Central and South America.

OPPOSITE:
BLACK-CAPPED DONACOBIUS & NORTHERN CARDINAL
Donacobius atricapilla & Cardinalis cardinalis

Jan Hoynck van Papendrecht, pen and brush on paper, 1895

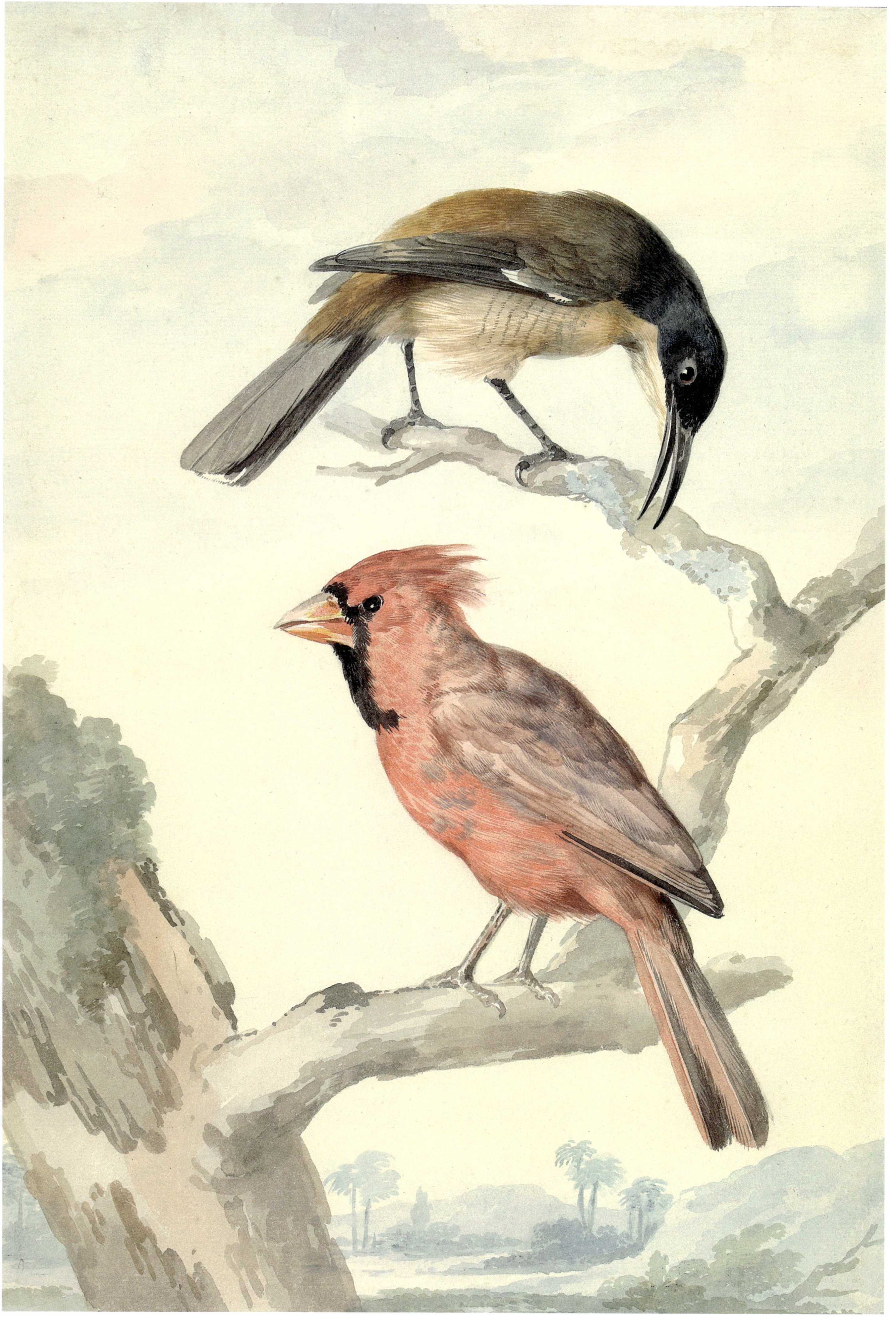

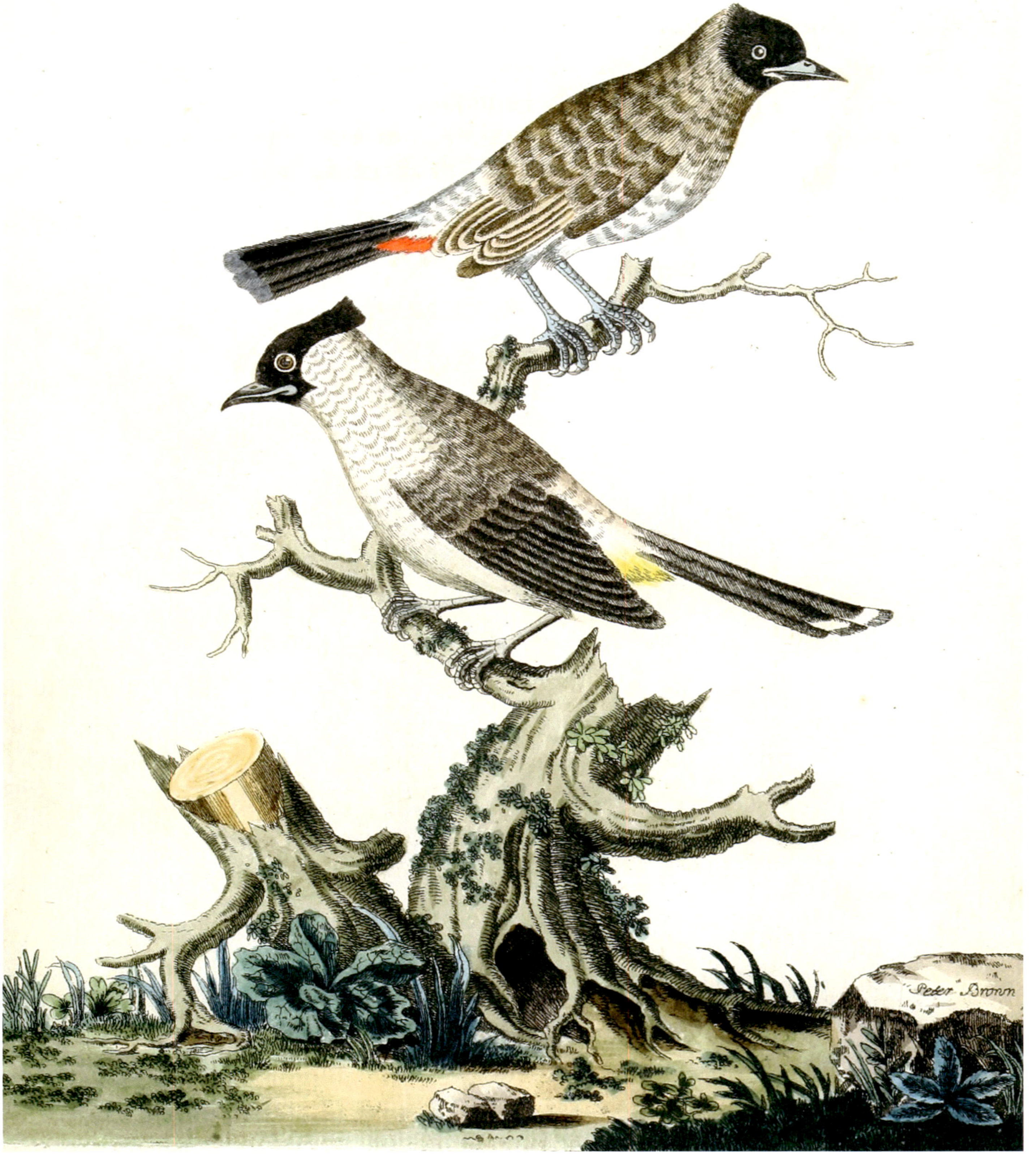

RED-VENTED BULBUL & SOOTY-HEADED BULBUL
Pycnonotus cafer & Pycnonotus aurigaster

Peter Brown, *New Illustrations Of Zoology*, 1776

Peter Brown was a Danish-born British natural history illustrator who is best remembered for his botanical illustrations. Appointed botanical painter to the Prince of Wales in 1783, he demonstrated a great skill for the botanic arts. His bird studies for *New Illustrations Of Zoology* are somewhat less accomplished, but despite their somewhat stilted nature, they still hold some appeal. The text, as well as many of the book's specimens, were provided by Welsh naturalist and traveller Thomas Pennant.

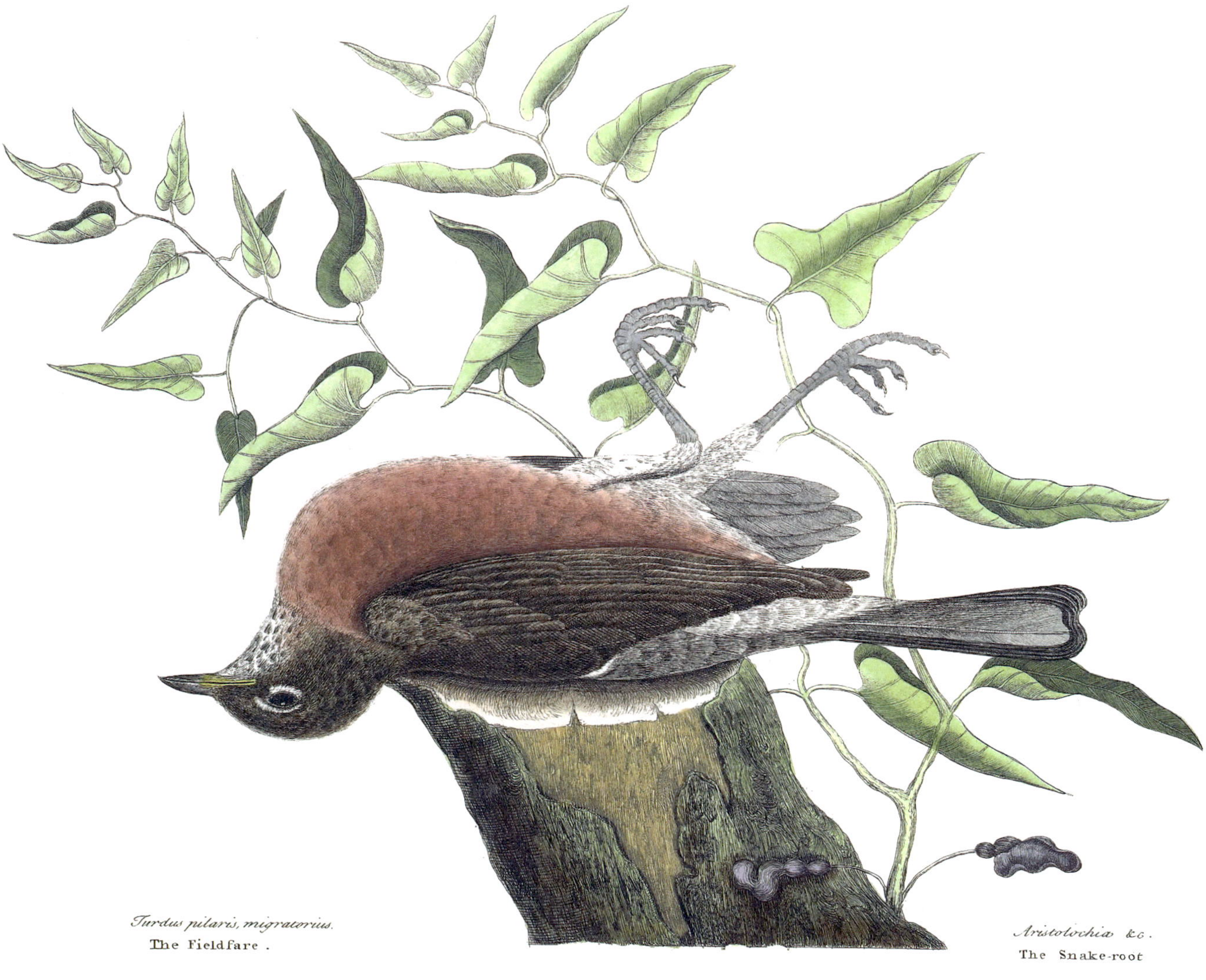

AMERICAN ROBIN
Turdus migratorius

Mark Catesby, *The Natural History of Carolina, Florida and the Bahama Islands*, 1729–47

Mark Catesby's seminal account of North America's flora and fauna was not without humour. In this curious illustration from his *Natural History*, we see an American robin lying dead on a tree stump. The bird is surrounded by leaves of Virginia snakeroot (*Aristolochia serpentaria*). Many had claimed that these leaves were an antidote to the venomous bites of snakes, but Catesby held a more sceptical view – reflected in his inclusion of a dead robin.

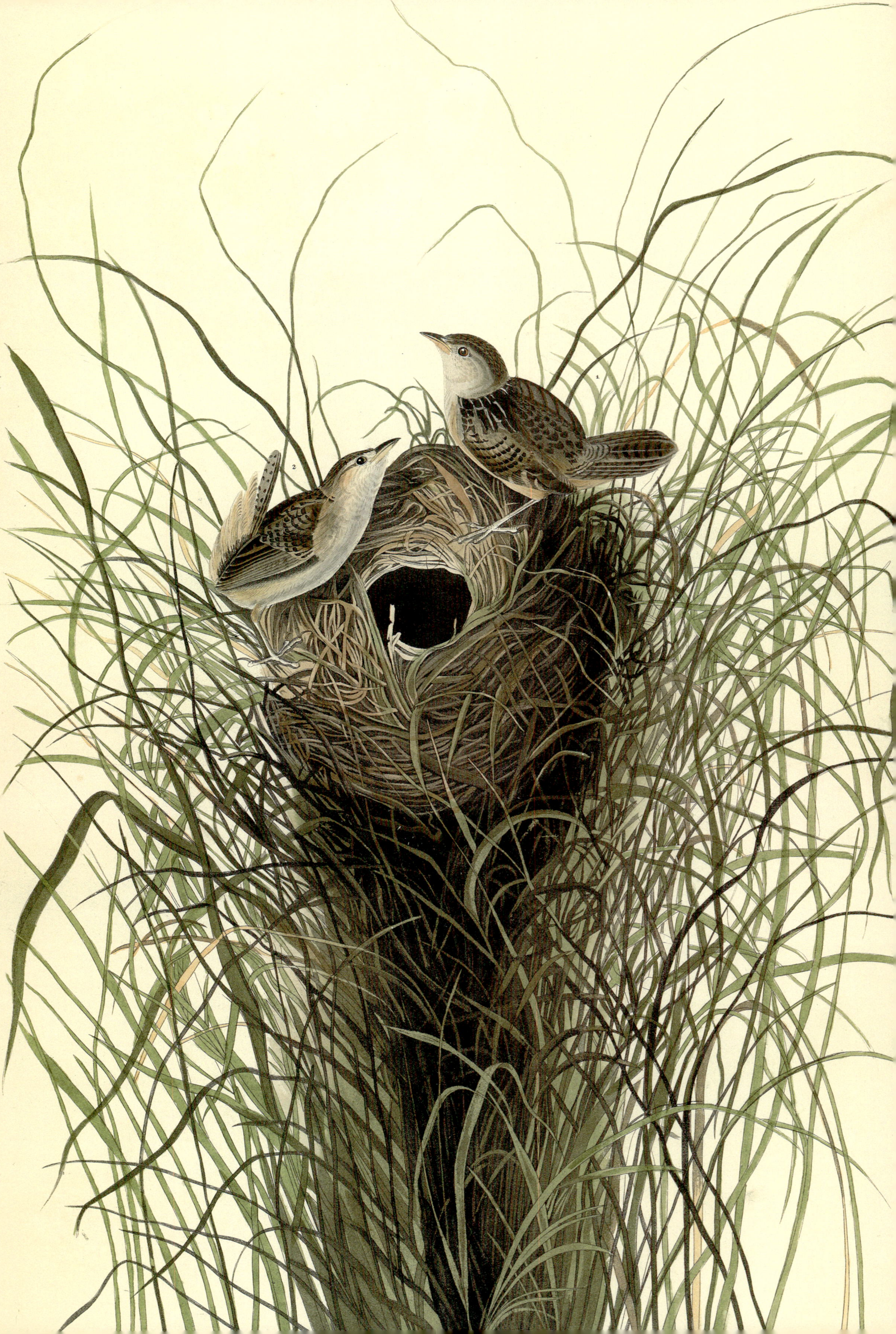

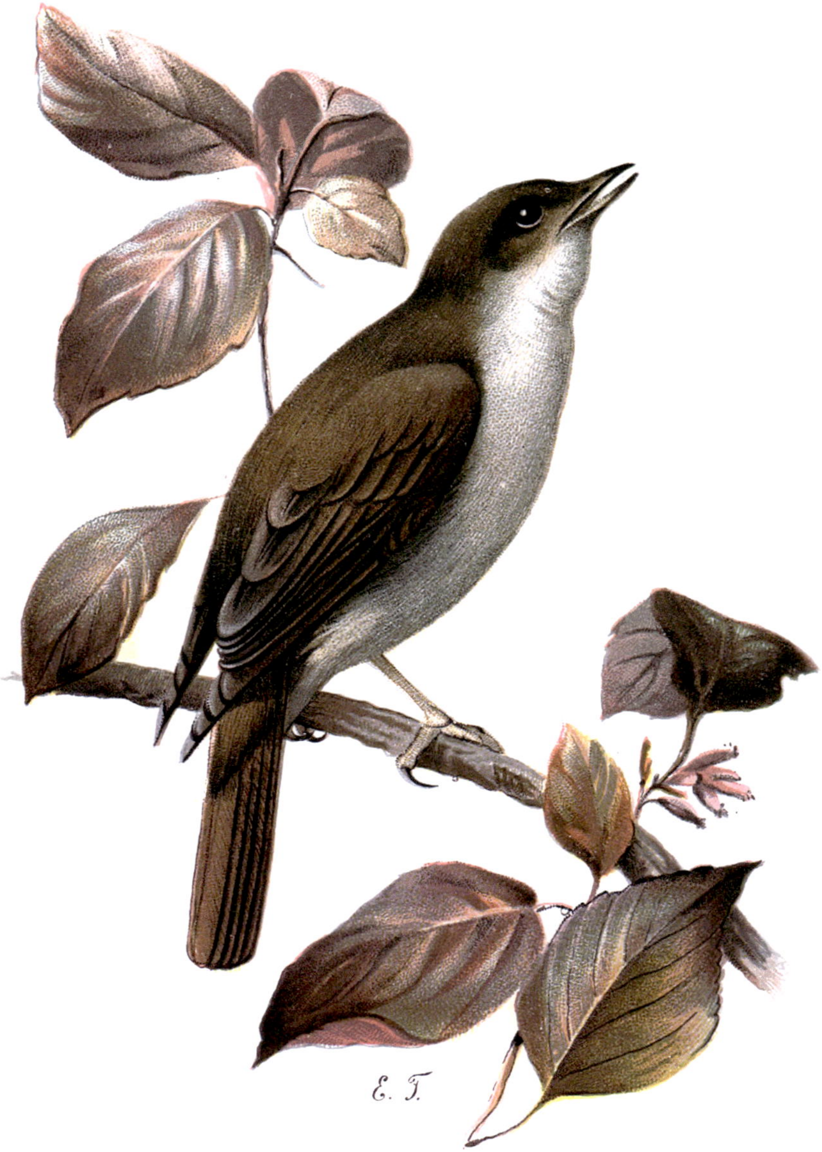

OPPOSITE:
SEDGE WREN
Cistothorus stellaris

John James Audubon,
The Birds of America, 1827–38

COMMON NIGHTINGALE
Luscinia megarhynchos

Eliza Turck, *Familiar Wild Birds*, 1883

For many, the nightingale is the ultimate songbird. Frequently heard singing at night, their name is derived from the Old English *galan*, meaning to sing. While plain in appearance, their voice is sweet and melodic, and this led many birdcatchers to sell these birds in cages. Under these conditions, the nightingale sang infrequently and captivity greatly shortened their lifespan. Those that made it to autumn would often kill themselves against the bars of their cage when their migratory impulse took hold.

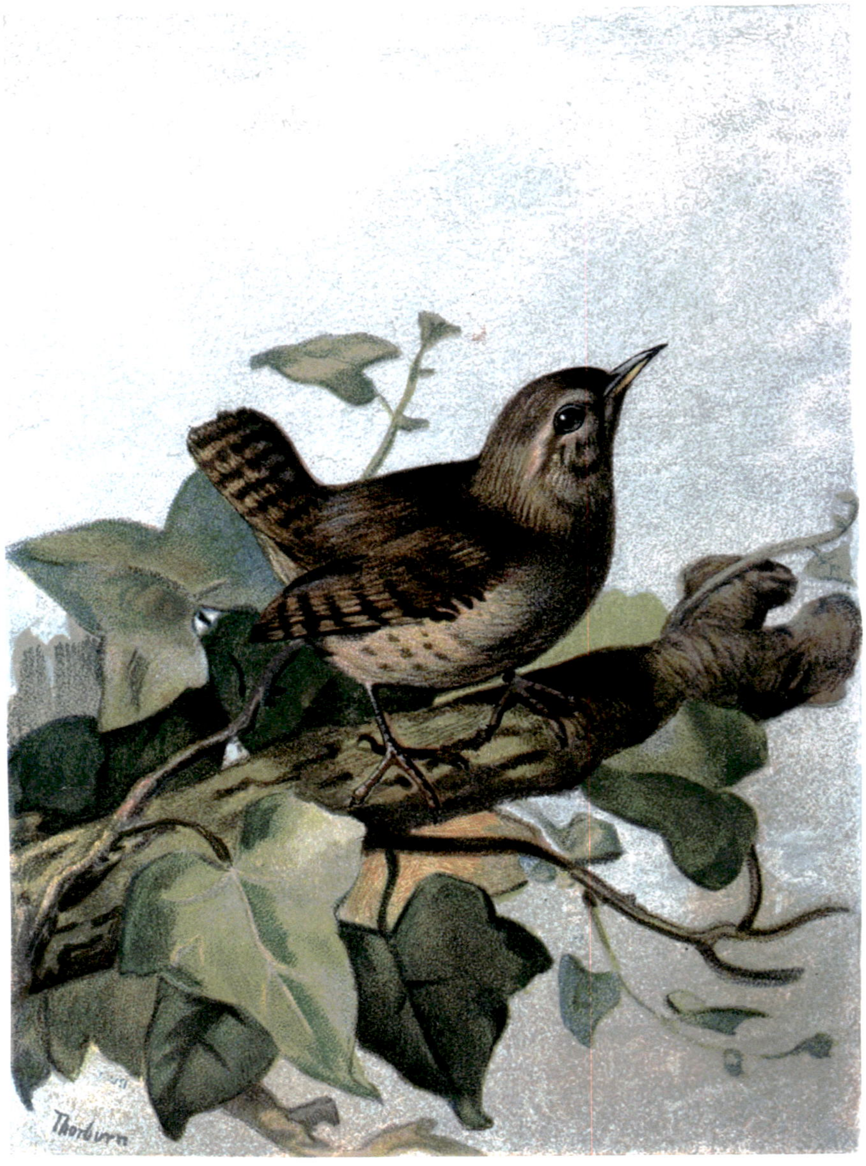

NORTHERN WREN
Troglodytes troglodytes

Archibald Thorburn,
Familiar Wild Birds, 1883

Despite the large size of the wren family (*Troglodytidae*), the northern wren (*Troglodyres troglodytes*) is the only species of the bird that can be found in Eurasia. There, it is commonly known simply as the wren. Despite their tiny size, these birds are fiercely territorial and are quick to quarrel and fight. Their scientific name, *Troglodytidae*, has its origins in Greek and translates as 'cave-dweller'. This refers to the bird's habit of disappearing into crevices, ledges and other cavities.

OPPOSITE:
BUFF-THROATED PURPLETUFT
& WHITE-BROWED PURPLETUFT
Iodopleura pipra & Iodopleura isabellae

Paul Louis Oudart, *Iconographie ornithologique*, 1845–9

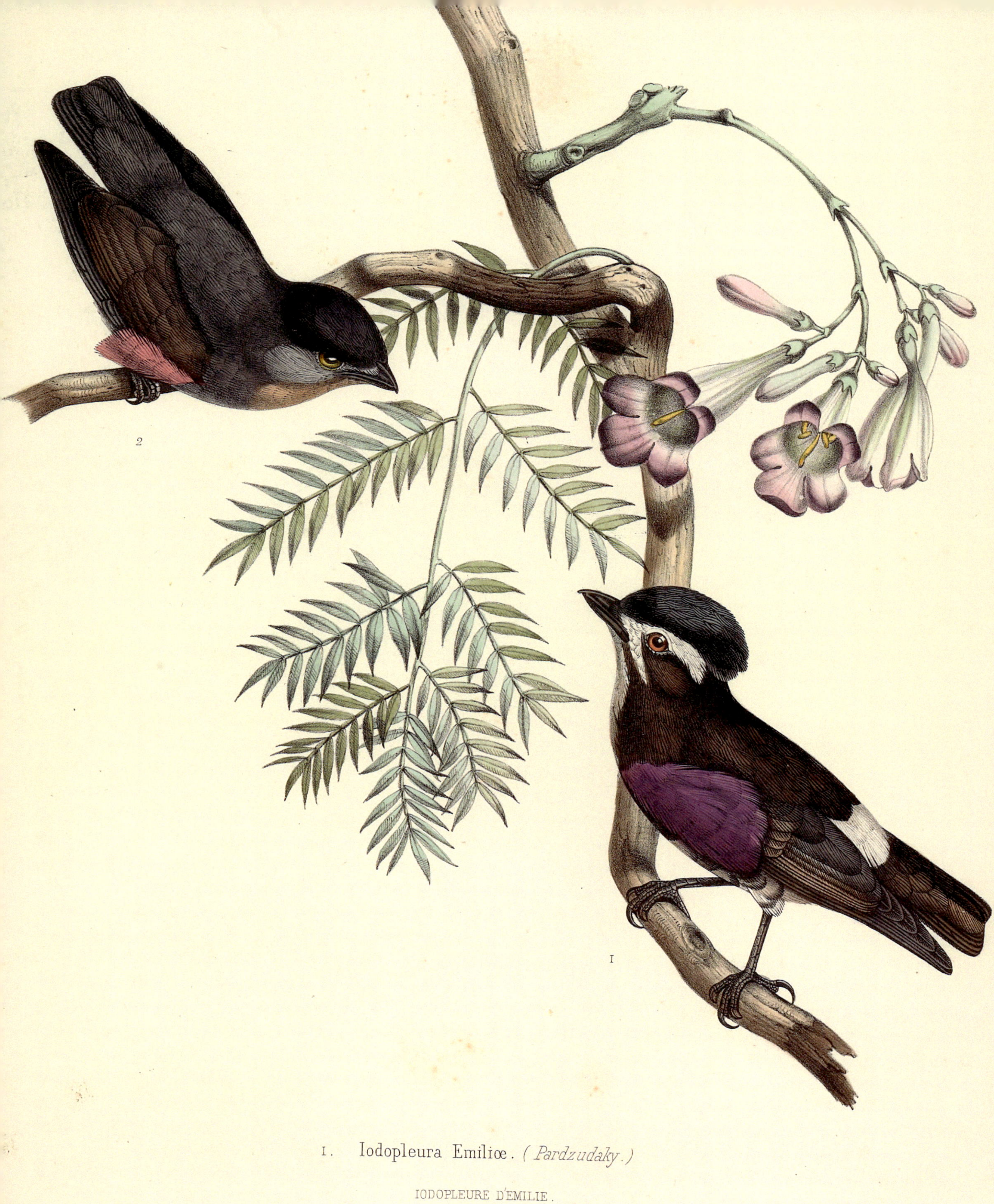

1. Iodopleura Emiliœ. *(Pardzudaky.)*

IODOPLEURE D'EMILIE.

2. Iodopleura pipra.

IODOPLEURE MANAKIN.

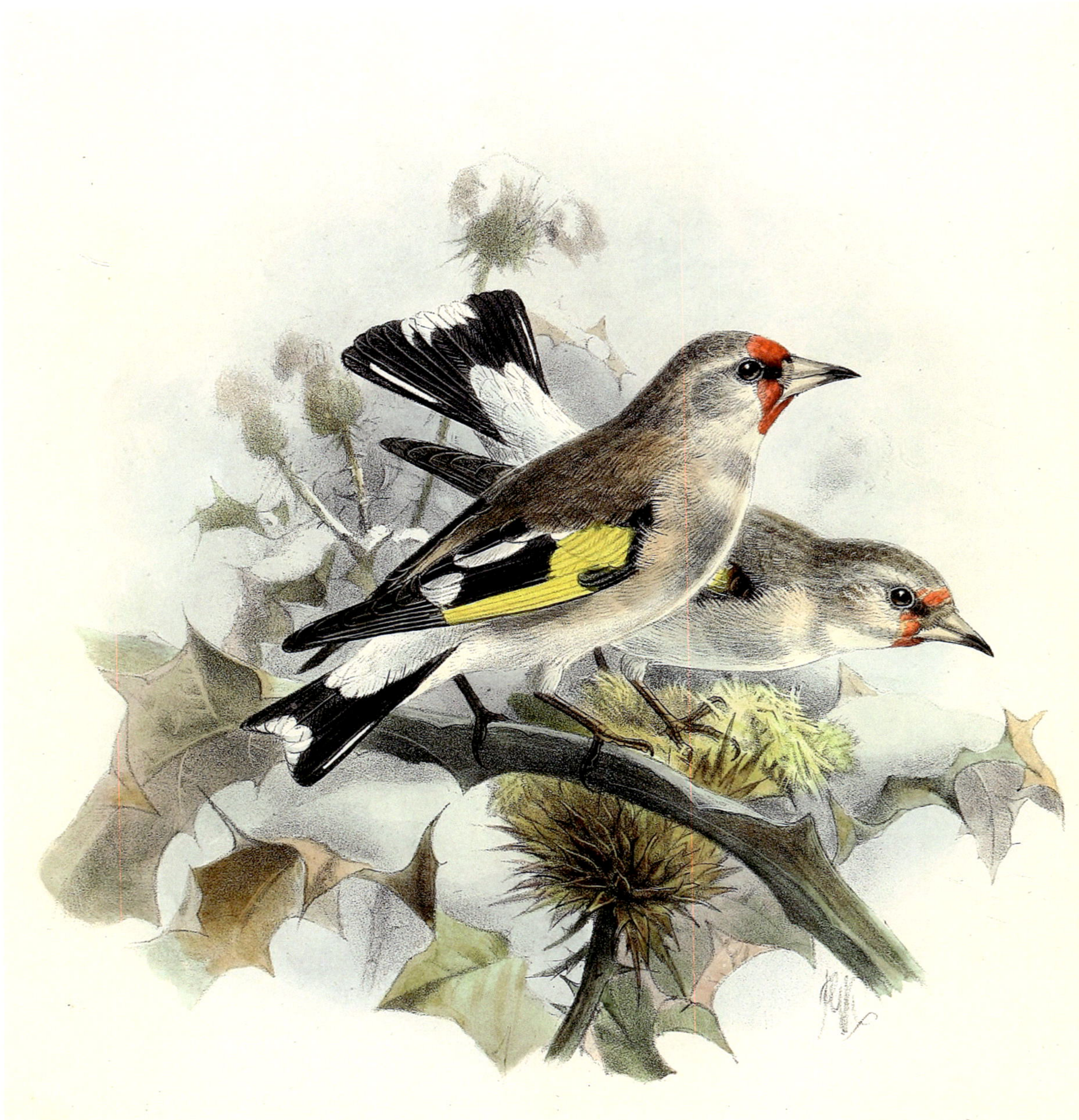

EASTERN GOLDFINCH
Carduelis caniceps

John Gerrard Keulemans, *A History of the Birds of Europe*, 1871–81

Celebrated for its playful song, the goldfinch is a small and charming bird that is found throughout much of Europe, North Africa and Western Asia. The Italian composer Vivaldi was so taken by the bird's pleasant song that he wrote Il Gardellino – a concerto for flute that imitates the call of the goldfinch. The bird found in the Himalayas is now considered to be a separate species.

OPPOSITE:
WAGTAILS
Motacilla

From top to bottom:
WHITE WAGTAIL
Motacilla alba

GREY WAGTAIL
Motacilla cinerea

GREY-HEADED WAGTAIL
Motacilla flava thunbergi

PIED WAGTAIL
Motacilla alba yarrellii

YELLOW WAGTAIL
Motacilla flava flavissima

Alexander Francis Lydon, *The Smaller British Birds*, 1894

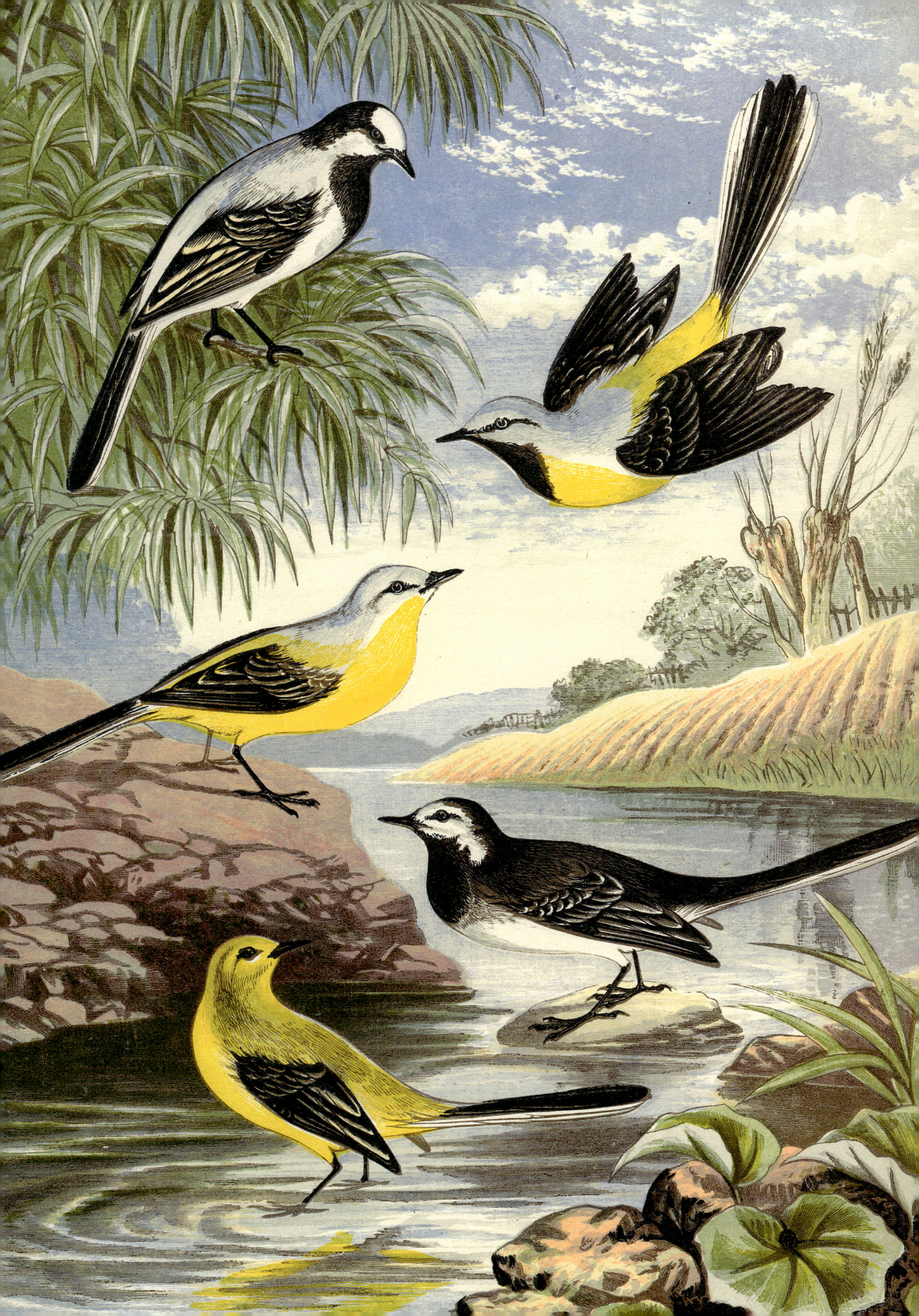

No 5. PLATE XXI.

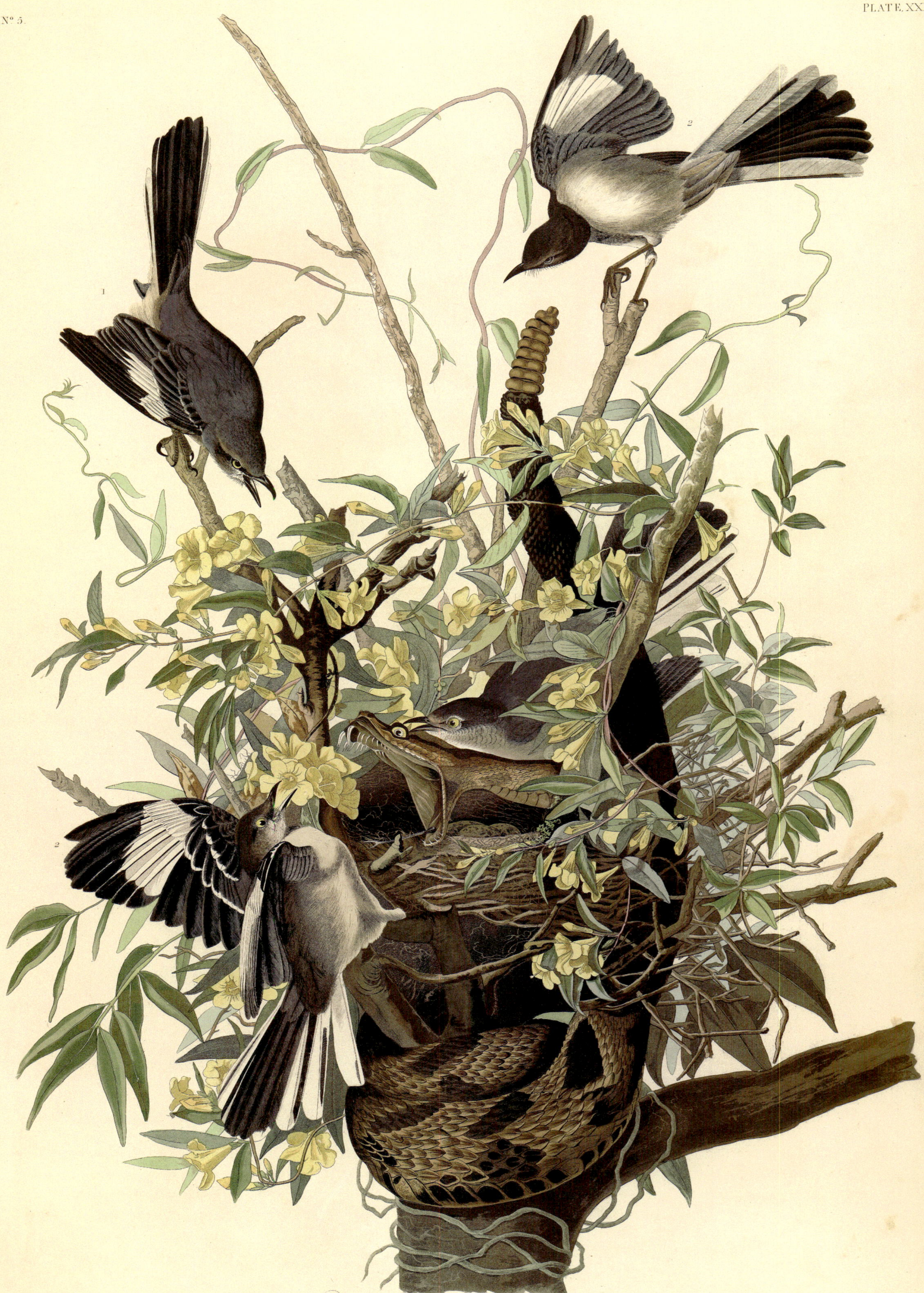

Drawn from Nature and Published by John J. Audubon, F.R.S.F.L.S.

Engraved, Printed and Coloured by R Havell

Mocking Bird, TURDUS POLYGLOTTUS, Linn. *Males, 1. Females, 2. Florida Jessamine Gelseminum nitidum*

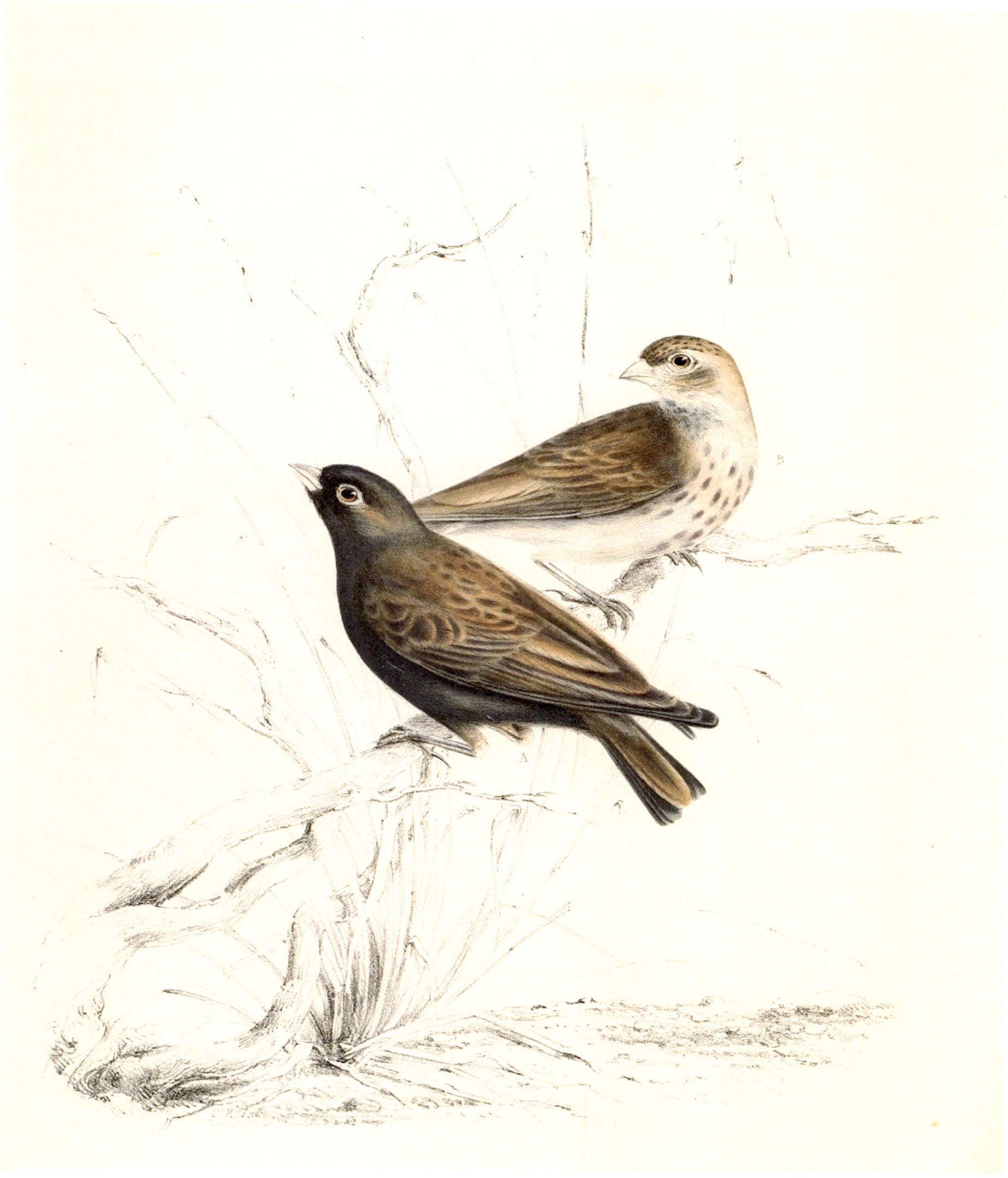

OPPOSITE:
NORTHERN MOCKINGBIRDS
Mimus polyglottos

John James Audubon, *The Birds of America*, 1827–38

This work by Audubon demonstrates that artist's unparalleled skill for depicting scenes of birds in motion. This particular moment is based on instances where Audubon had witnessed snakes stealing both eggs and young from the nests of mockingbirds. He also saw the mobbing tactics of the bird which would sometimes successfully drive the snakes away.

BLACK-EARED SPARROW-LARK (LEFT: MALE, RIGHT: FEMALE)
Eremopterix australis

George Henry Ford, *Illustrations of the Zoology of South Africa*, 1838–49

These black-eared sparrow-larks are found in parts of Namibia, Botswana and South Africa. They appear here in an illustration from Sir Andrew Smith's *Illustrations of the Zoology of South Africa*. Many consider Smith to be the father of South African zoology. The illustration was produced by George Henry Ford who was born in South Africa and worked at the British Museum.

John James Audubon
1785–1851

An unparalleled influence on ornithology and natural history, John James Audubon was the author and artist behind *The Birds of America*, now one of the most prized illustrated books in the world.

One of the most important naturalists of his era, John James Audubon (1785–1851) is responsible for producing a book that many consider to be the ornithological masterpiece of the nineteenth century. A work of tremendous artistic merit, *The Birds of America* (1827–38) was also of great cultural and artistic importance. As extensive as it was ambitious, the work took the artist most of his life to complete and today it is currently the most expensive printed work to exist in the world.

Born in Haiti in 1785, Audubon was the illegitimate son of an affluent French sea captain and plantation owner and a young chambermaid who died soon after his birth. Named Jean-Jacques, he spent his childhood in Nante where he was often seen drawing pictures of birds, their nests and their eggs. In 1803 he was sent by his father to work on a family property in Pennsylvania as a way to avoid conscription in the Napoleonic Wars. His father had hoped that this move would encourage his son to gain business experience, but Pennsylvania offered a wealth of opportunities for anyone who was passionate about fishing, hunting, exploration and drawing. Audubon's successes as a businessman were variable at best and by 1819, he found himself bankrupt and in a debtors' prison. This situation made him reconsider his life and he realized that his true calling was nature. He soon abandoned any further business pursuits and decided to direct all his energies towards his passion.

In 1820, he embarked on his ambitious project. Despite having a young family at home, he would travel for months at a time, often into unexplored territories, in hopes of finding new specimens. He travelled mainly on foot but also on horseback, canoe, schooner and steamboat. While he worked from live specimens, he also killed a lot of birds and while this practice seems savage to many today, it was the norm in the eighteenth and nineteenth centuries. The birds that he shot would then become specimens which he would pin into position using wooden frames and wire. This allowed him to continue his precise and delicate work for several days and allowed him to replicate the natural poses of the birds he observed.

Unlike other natural history illustrators, Audubon's ambition was not simply to illustrate the birds that he saw – his ambition was to make art. This was something that caused a challenge when he began to look for subscribers among the science communities of Cincinnati and New Orleans. Not only did these men want to stay faithful to the esteemed ornithologist Alexander Wilson – whose book, at the time, was considered to be the definitive work of American ornithology – but they also saw Audubon's work as being too dramatic and overly romantic. Despite his brilliance, the reality of the situation was that Audubon was little more than a self-taught artist who, at almost forty years of age, had no real credentials in either the art world or the world of science. Undeterred, he set his sights on Britain in the hopes of finding subscribers and patrons there.

In 1828 he crossed the Atlantic and found a much warmer reception for his work. A formidable entrepreneur, he played up on his frontier reputation, dressing in pioneer attire and sharing tales of American exoticism with the upper classes and members of the scientific establishment. In Edinburgh, he found Selby's engraver, W.H. Lizars, who produced a number of prints for him. When Lizars' colourists went on strike, Audubon was forced to move his production to London, where he worked with Robert Havell Jr. (1793–1878). Havell was a hugely talented engraver, and his work with Audubon on *The Birds of America* was an artistic achievement all of its own. The pairing was perfect, and when Audubon returned to America he had secured the majority of his subscribers and patrons.

Production on *The Birds of America* was ambitious, to say the least. Printing began in 1827 and eventually ended eleven years later. The work was issued over eighty parts, with patrons receiving tin boxes that would contain a set of five plates including one large composition, one medium composition and three smaller ones. Each set depicted a species new to science. This was a smart spot of marketing that was designed to keep subscribers engaged over the long duration of the project's release. Upon completion, the work contained some 435 hand-coloured, life-size prints that were printed on the largest sheets of paper then available. Each print showed the bird in its natural environment, often with an exquisitely detailed backdrop. The popularity of the work led to a smaller octavo edition being published in America between 1840–44. This work also enjoyed great success and both editions brought Audubon fame and fortune in his lifetime. Not one to rest easy, he continued to travel, making excursions out West in pursuit of his next project: *Viviparous Quadrupeds of North America*. Sadly, he began to suffer signs of dementia and his work on the project was cut short. He passed away in 1851 at the age of sixty-five. His son, John Woodhouse Audubon, took over publishing of his final work.

Audubon's influence on ornithology and natural history is unparalleled. His artistry influenced the ornithological books that followed, while his writings on birds played a significant role in helping us to understand their anatomy and behaviour. In 1905, the Audubon Society was named in his honour and it continues working towards its original goal of promoting bird conservation in the United States. Today, *The Birds of America* is considered to be one of the great masterpieces of book art and is now one of the most prized illustrated books in the world.

AMERICAN TREE SPARROW
Spizelloides arborea

John James Audubon,
The Birds of America, 1827–38

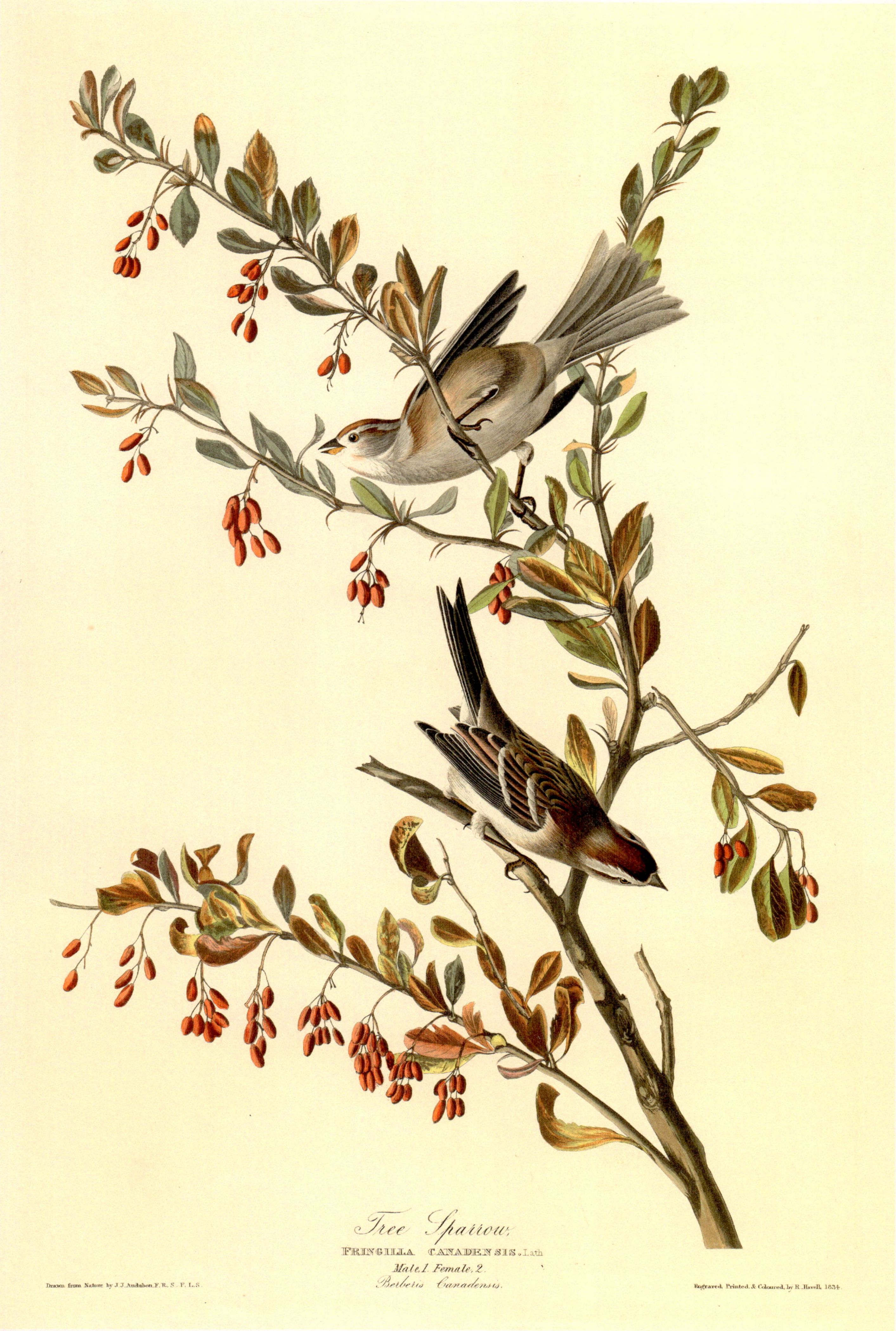
Tree Sparrow,
FRINGILLA CANADENSIS. Lath
Male, 1. Female, 2.
Berberis Canadensis.
Drawn from Nature by J. J. Audubon, F. R. S. F. L. S.
Engraved, Printed & Coloured, by R. Havell, 1834.

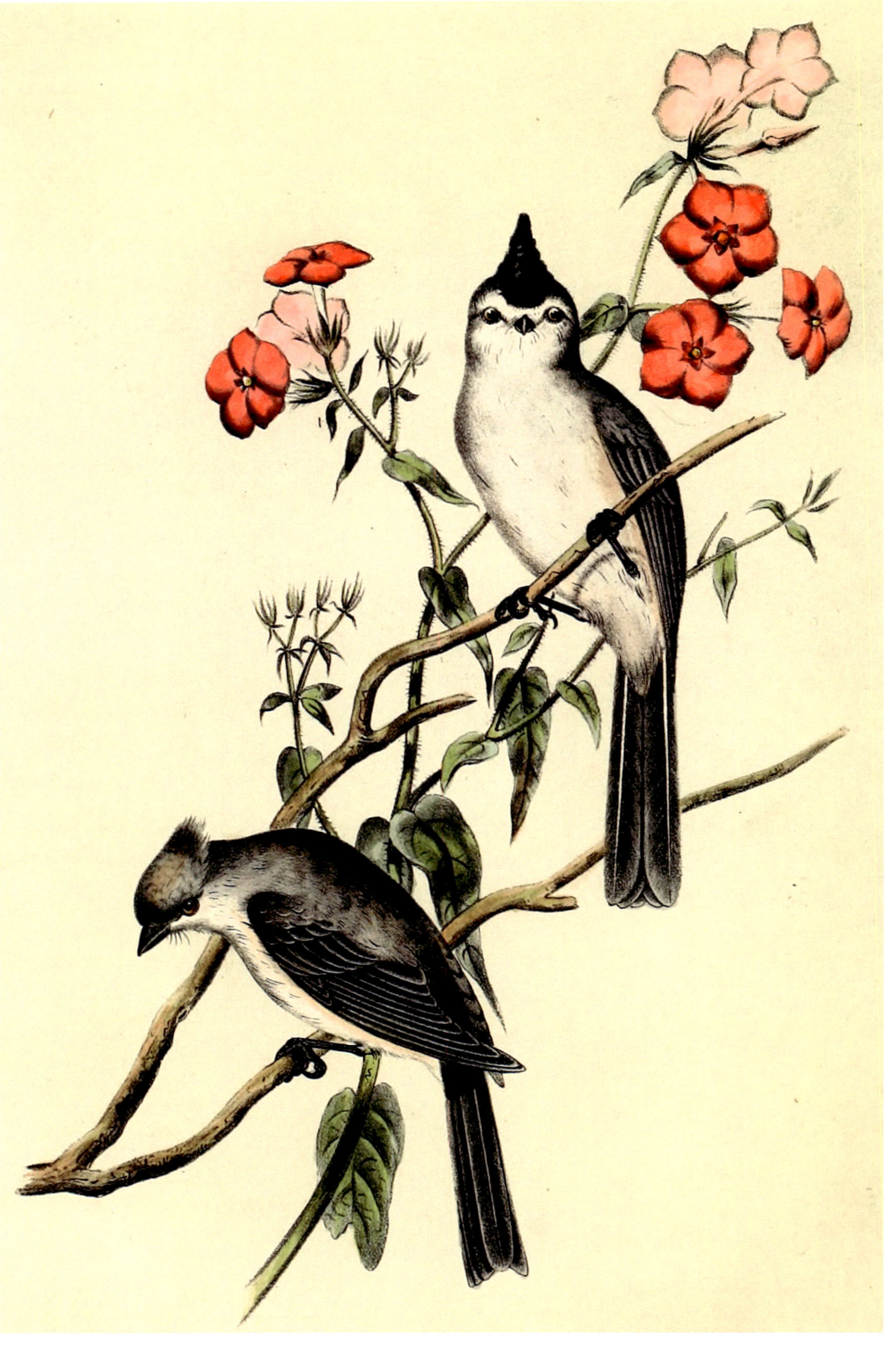

BLACK-CRESTED TITMOUSE (ABOVE) & TUFTED TITMOUSE
Baeolophus atricristatus & *Baeolophus bicolor*

George G. White, *Illustrations of the Birds of California, Texas, Oregon, British and Russian America*, 1853–6 (this edition: 1862)

This illustration by the relatively unknown George G. White is taken from the book *Illustrations of the Birds of California, Texas, Oregon, British and Russian America*. Written by American ornithologist John Cassin, the work aimed to act as a supplement to Audubon's *Birds of America* and depicts examples of North American birds that had not yet been illustrated by other authors.

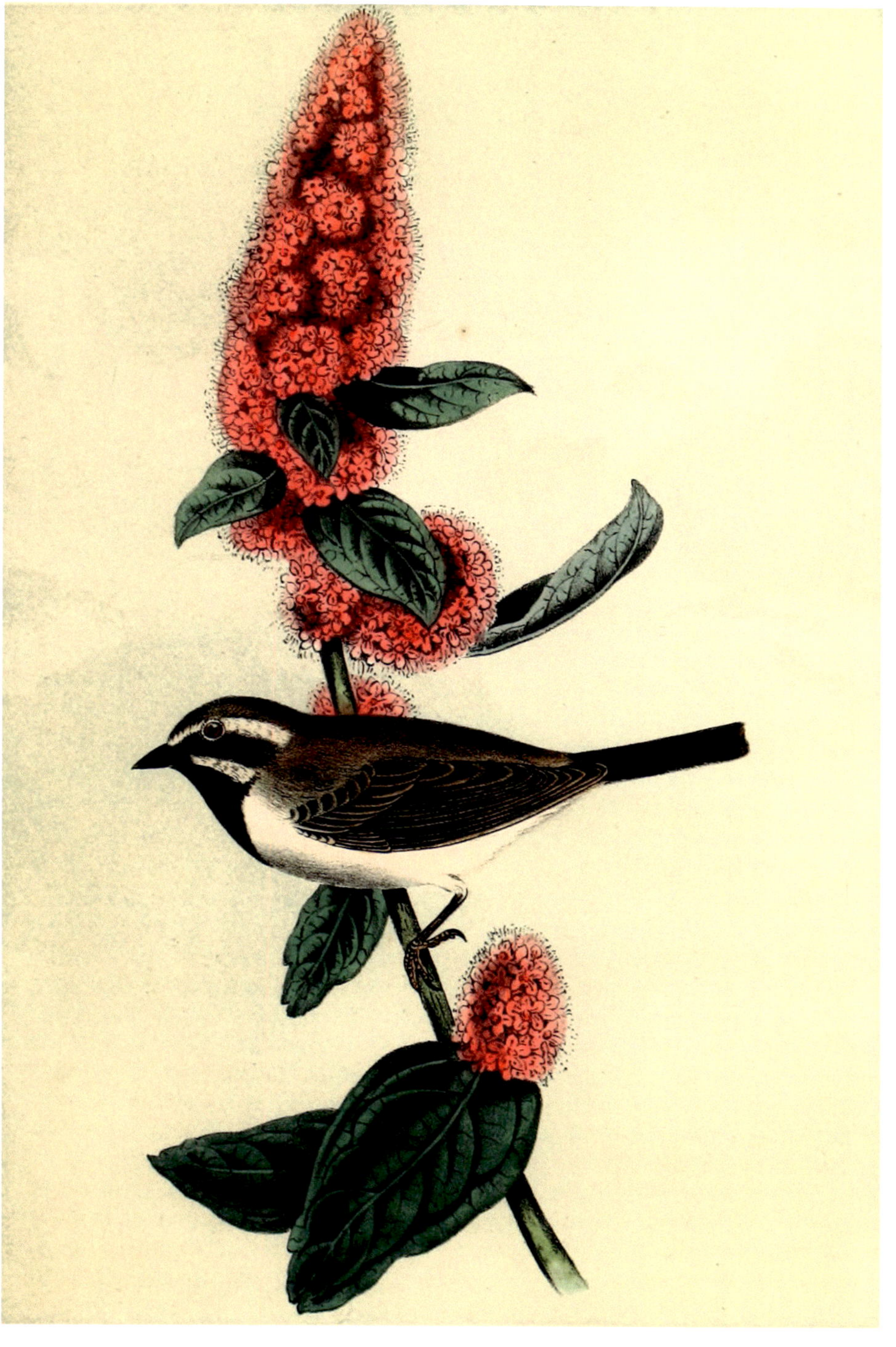

BLACK-THROATED SPARROW
Amphispiza bilineata

George G. White, *Illustrations of the Birds of California, Texas, Oregon, British and Russian America*, 1853–6 (this edition: 1862)

Cassin's work is considered to be hugely important within American ornithology, and with this book he was eager to present his findings as a worthy continuation of Audubon's work. Indeed, he even suggested to Audubon's sons that his work be published as part of an extended edition to their father's octavo edition. While a disagreement over credits in the work led to the collaboration falling through, Cassin continued to pay tribute by having his book published by Audubon's lithographer, J.T. Bowen.

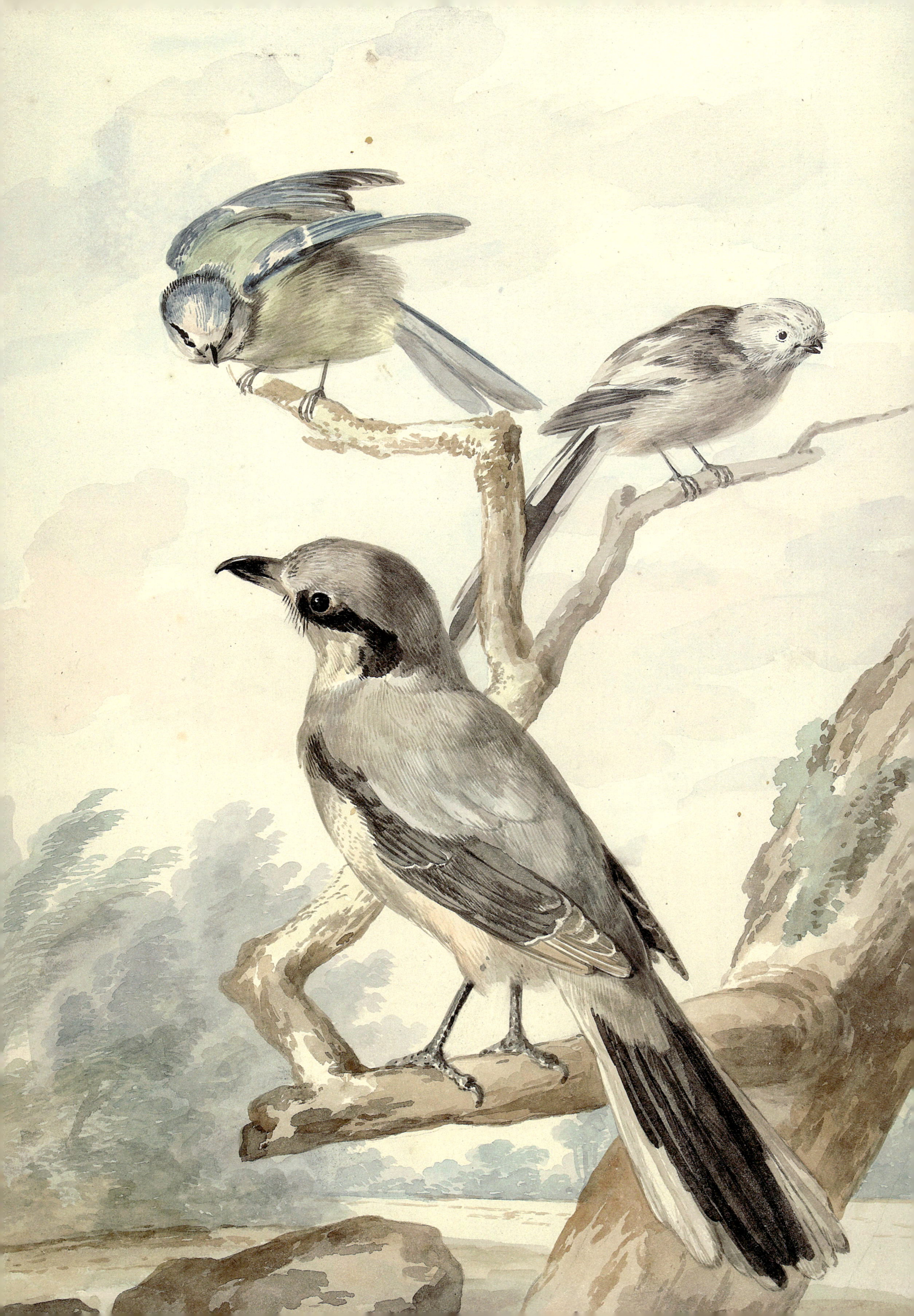

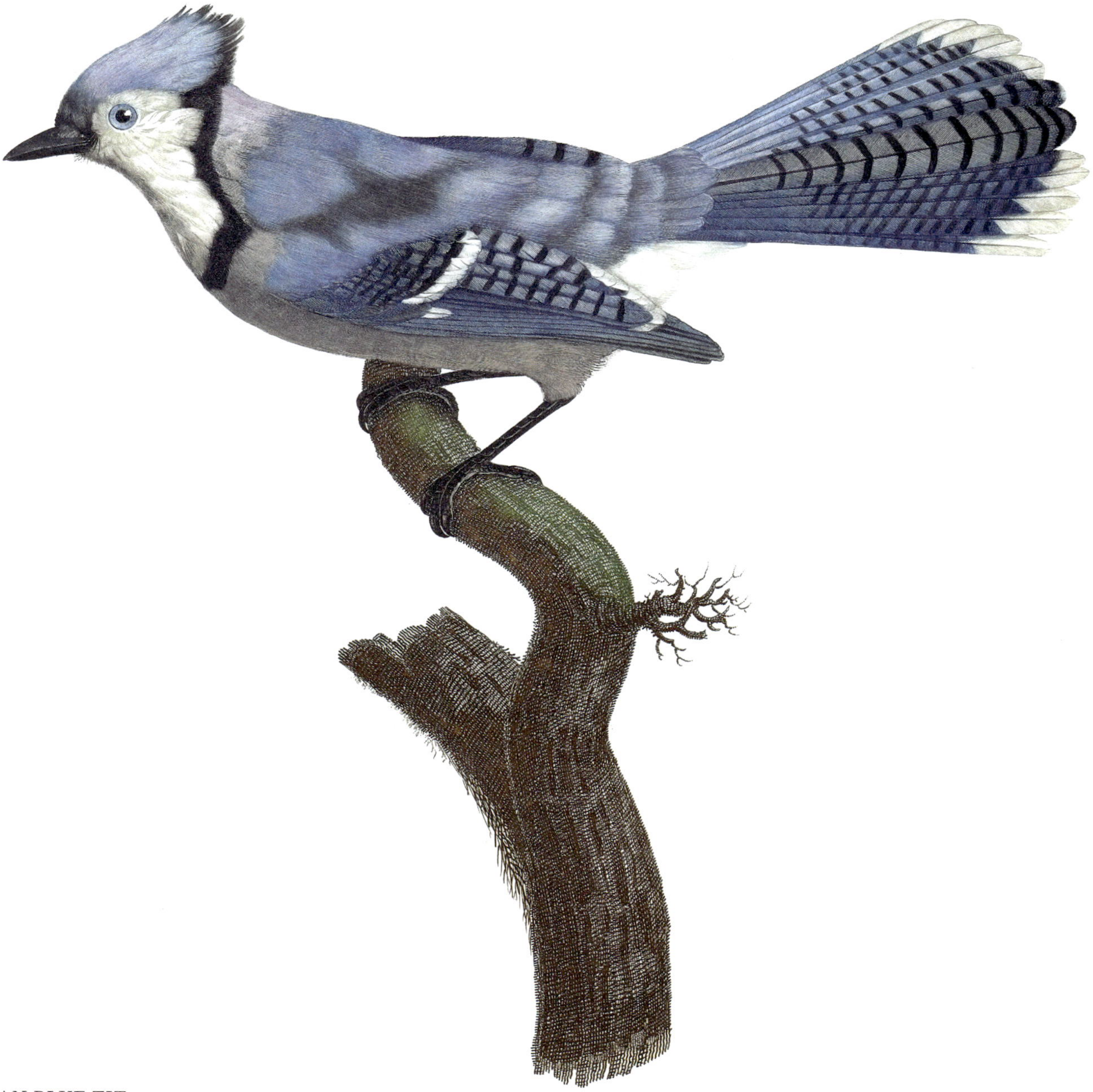

OPPOSITE:

EURASIAN BLUE TIT,
LONG-TAILED TIT
& GREAT GREY SHRIKE

Cyanistes caeruleus, Aegithalos caudatus & Lanius excubitor

Drie vogels: een Staartmees, een Pimpelmees en een Klapekster of Grote Klauwier, Aert Schouman, watercolour, 1756

Aert Schouman was a talented Dutch artist who painted many works featuring birds. These works were drawn from life: Schouman studied the birds that were kept on the estate of William IV. At the time, William was Prince of Orange and he had constructed a menagerie on his estate at Voorburg, one kilometre east of The Hague.

BLUE JAY

Cyanocitta cristata

Jacques Barraband, *Histoire naturelle des oiseaux de paradis et des rolliers*, 1806

With its striking blue and white plumage, the blue jay is arguably the most elegant member of the crow family. Jays tend to eat seasonal nuts and seeds and so will often bury their food to eat it later. A jay can bury thousands of acorns over a period of a month and, when winter arrives, it will remember exactly where most of these seeds are buried.

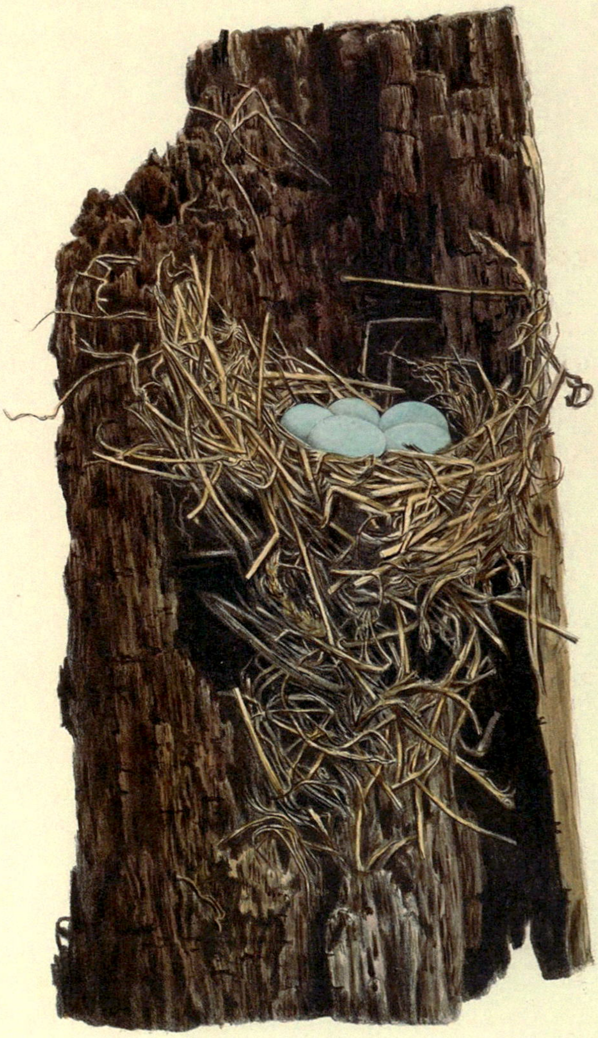

Pl. XII.
SIALIA SIALIS.
EASTERN BLUEBIRD.

NEST OF THE EASTERN BLUEBIRD
Sialia sialis

Genevieve Estelle Jones, *Illustrations of the Nests and Eggs of Birds of Ohio*, 1879–86

Jones released the first instalment of her work in 1879, and a month later she fell desperately ill. Stricken with typhoid fever, she passed away three weeks later at the age of only thirty-two. Despite these tragic circumstances, Jones was tenacious to the end. She instructed her brother Howard to continue the project, with her mother Virginia handling the remainder of the illustrations; her friend Eliza Shulze taught Virginia the lithographic technique. The book was eventually completed in 1886 and, despite its modest sales, it was hailed as an extraordinary achievement.

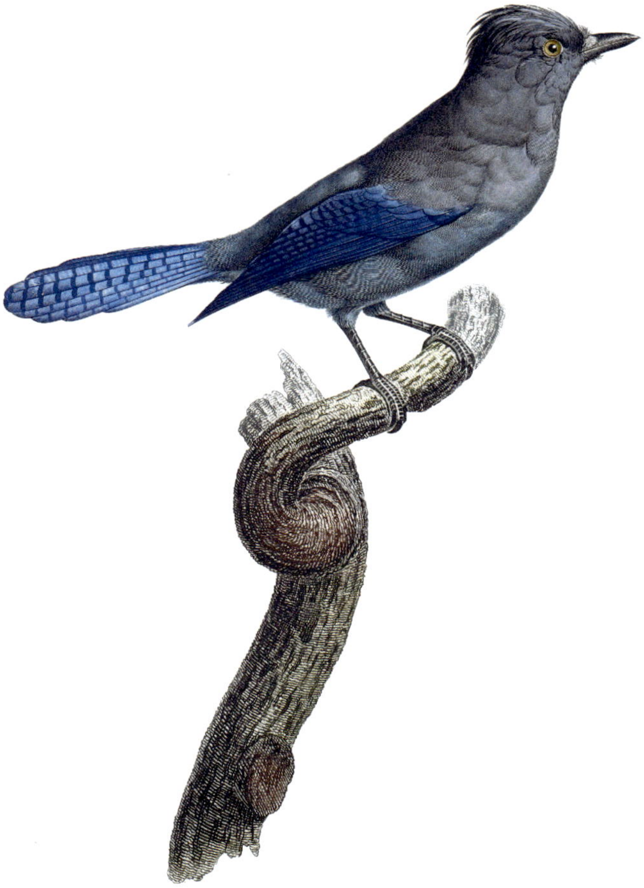

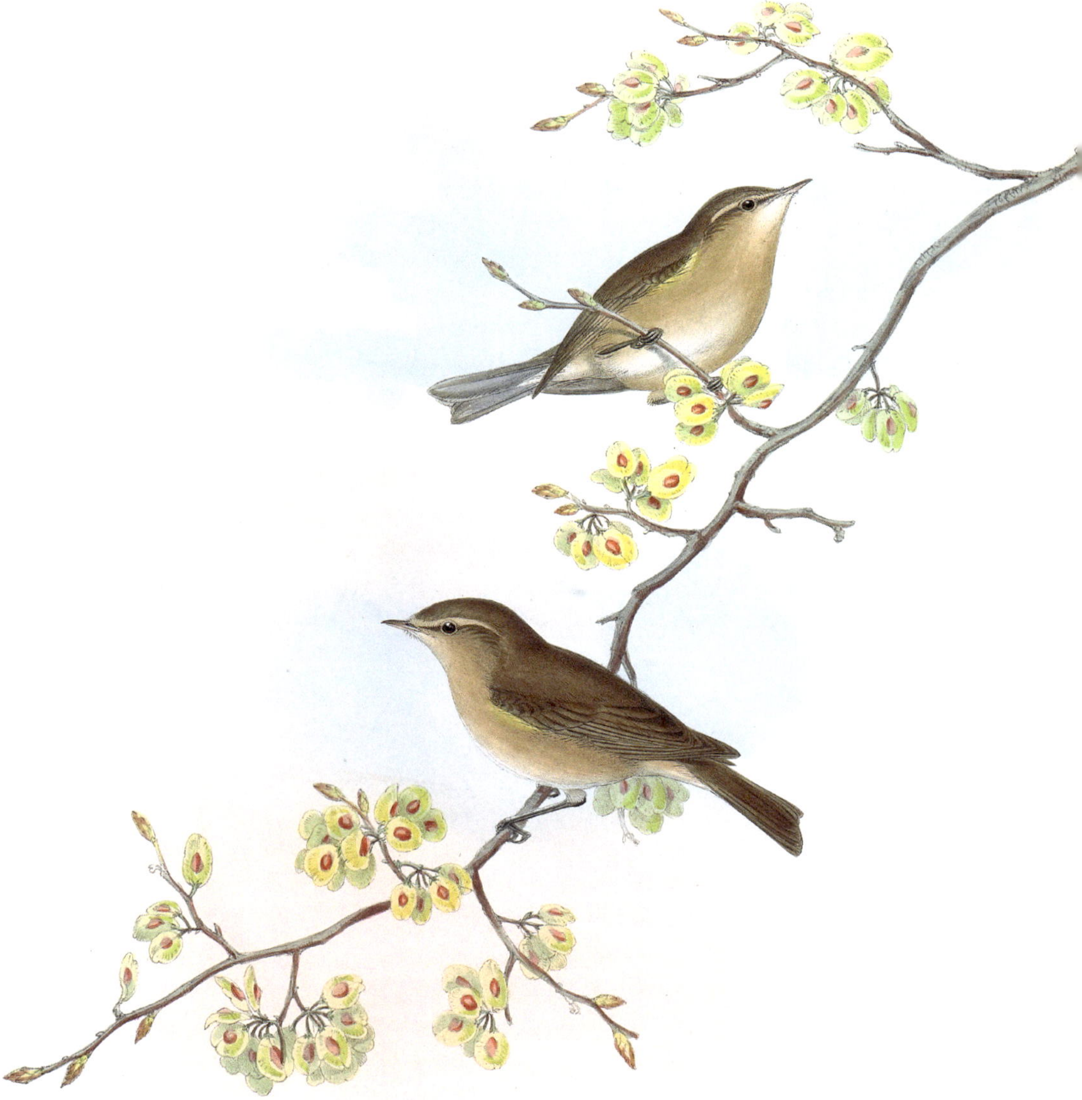

ABOVE:
STELLER'S JAY
Cyanocitta stelleri

Jacques Barraband, *Histoire naturelle des oiseaux de paradis et des rolliers*, 1806

RIGHT:
SIBERIAN CHIFFCHAFFS
Phylloscopus tristis

John Gould & Henry Constantine Richter, *The Birds of Asia*, 1850–83

The Siberian chiffchaff breeds in Siberia and migrates south to the lower Himalayas in the winter. It is shown here in an illustration from Gould's *The Birds of Asia*. In his text, Gould identifies it by the name *Phyllopneuste tristis* and compares it to the common chiffchaff. He notes that this bird is slightly larger and has slightly different colouring and notes that the zoologist Edward Blyth believed both birds to be identical. Today, it remains unclear if the Siberian chiffchaff is a subspecies of chiffchaff or if the bird is a species in its own right.

OVERLEAF:
SAND MARTIN
(BANK SWALLOW)
Riparia riparia

John Gould & William Matthew Hart, *The Birds of Great Britain*, 1862–73

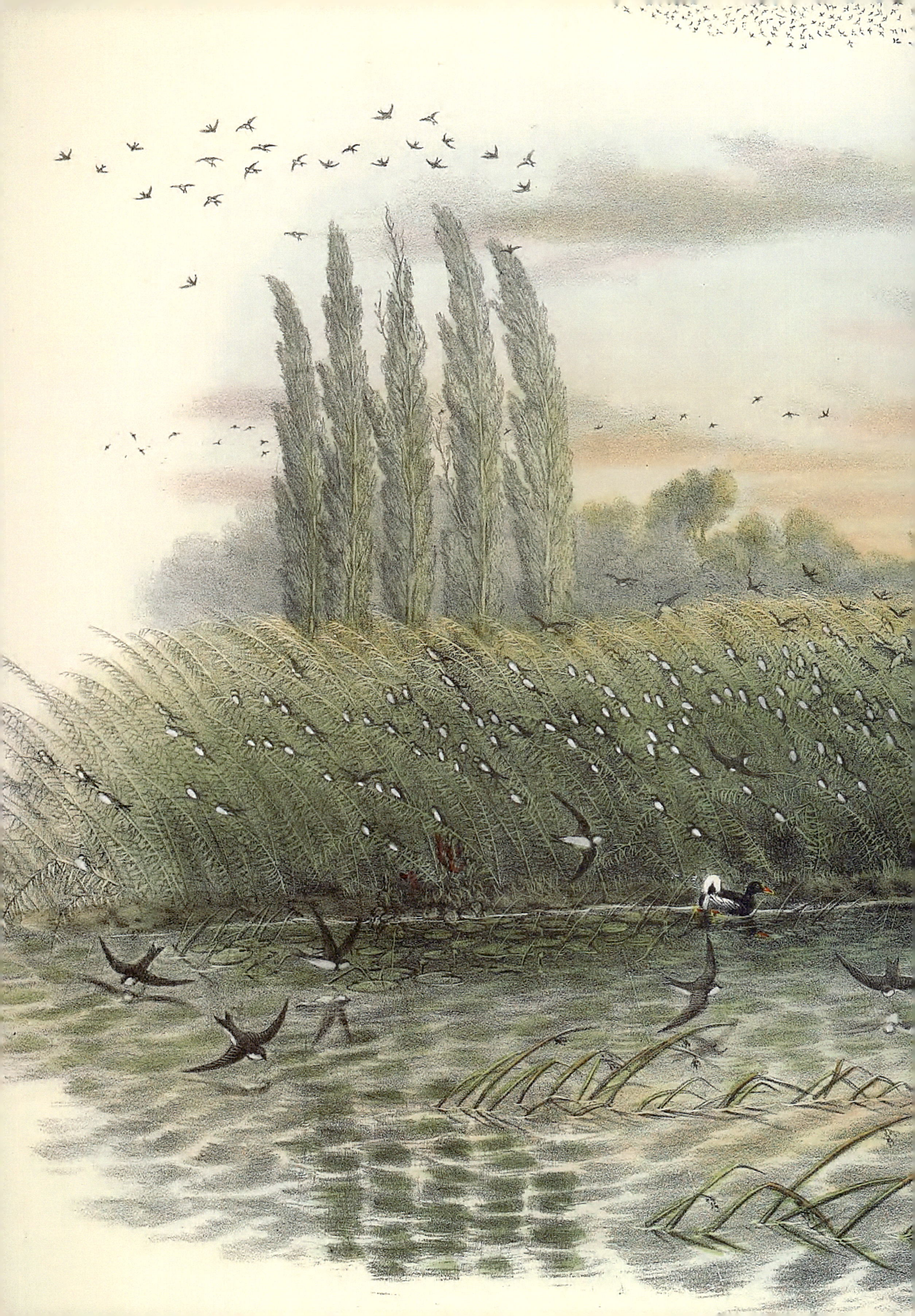

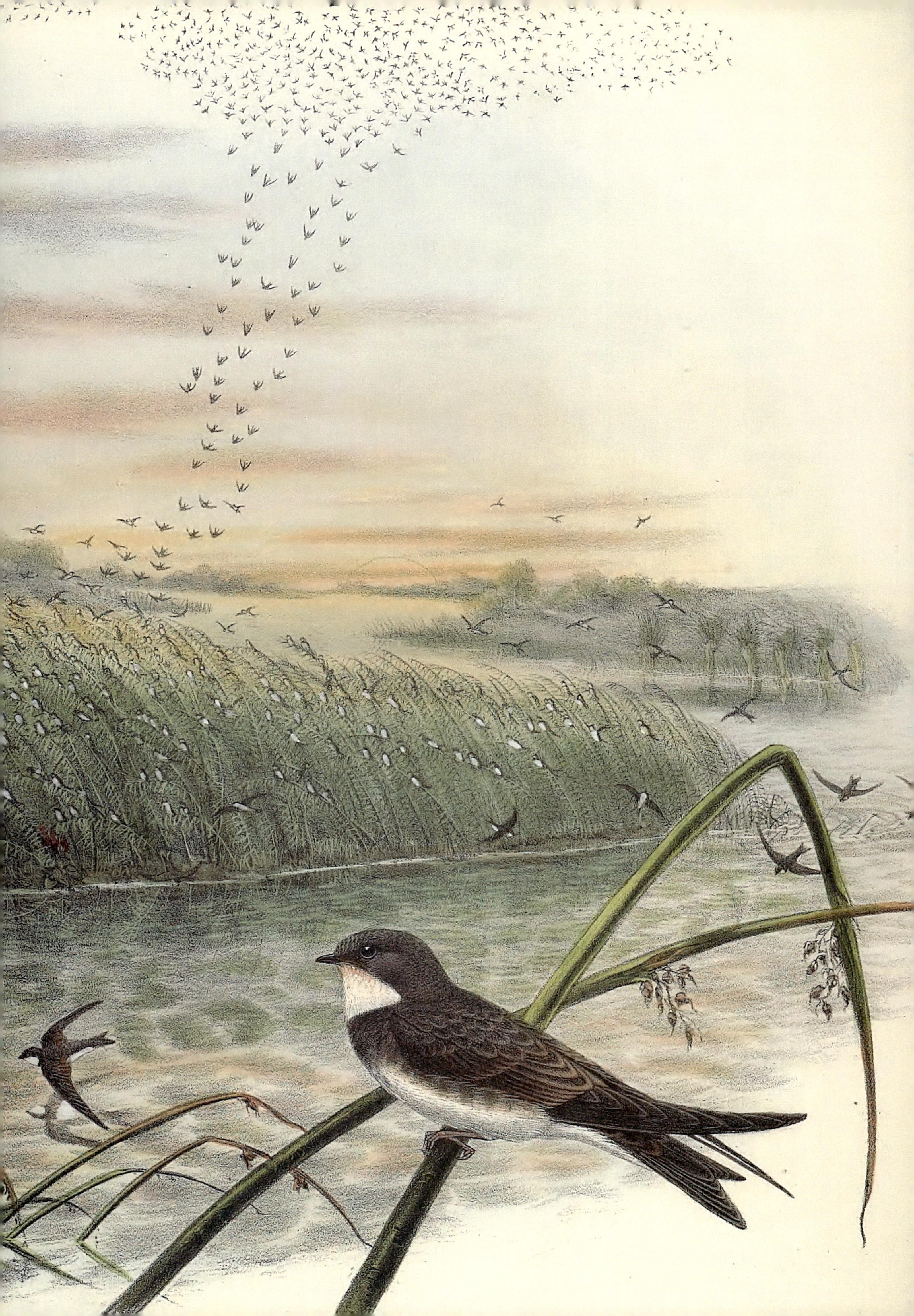

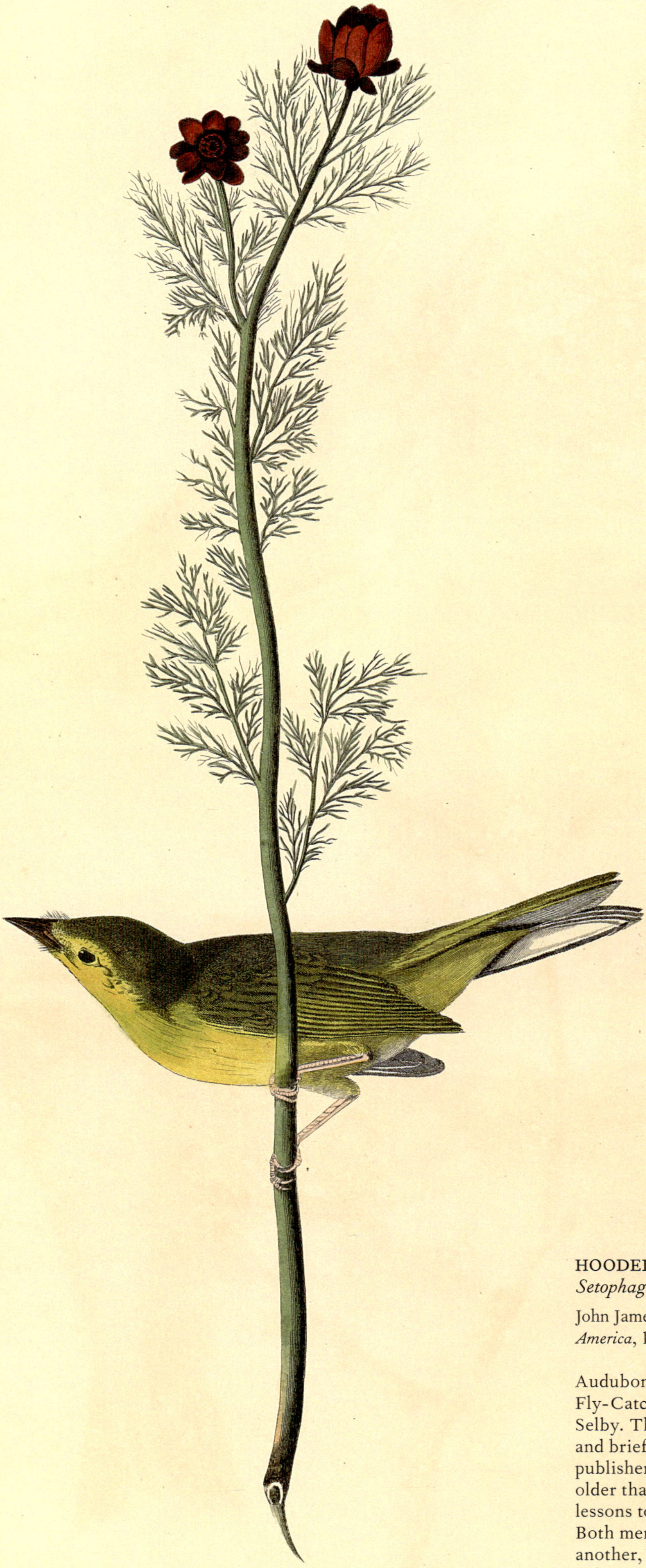

HOODED WARBLER
Setophaga citrina

John James Audubon, *The Birds of America*, 1827–38

Audubon named this bird Selby's Fly-Catcher after Prideaux John Selby. The pair had met in Scotland and briefly shared the same publisher. Audubon was three years older than Selby and gave drawing lessons to the celebrated naturalist. Both men were admirers of one another, and today, Selby is often referred to as the 'British Audubon'.

PINE BUNTINGS
AND MEADOW BUNTINGS
Emberiza leucocephala
and *Emberiza cioides*

Takeuchi Seihō (after Tsuchida Eishō),
Eishō hyakuchō gafu, 1913

Translated as *Pictures of One Hundred Birds*, this folio of woodcut illustrations was made by Takeuchi Seihō and was based on drawings originally created by Tsuchida Eishō. The folio contains a series of illustrations that examine a variety of different birds. Many of the illustrations present detailed renderings of specific attributes such as wings and feet, while small notes are often written directly beside each illustration.

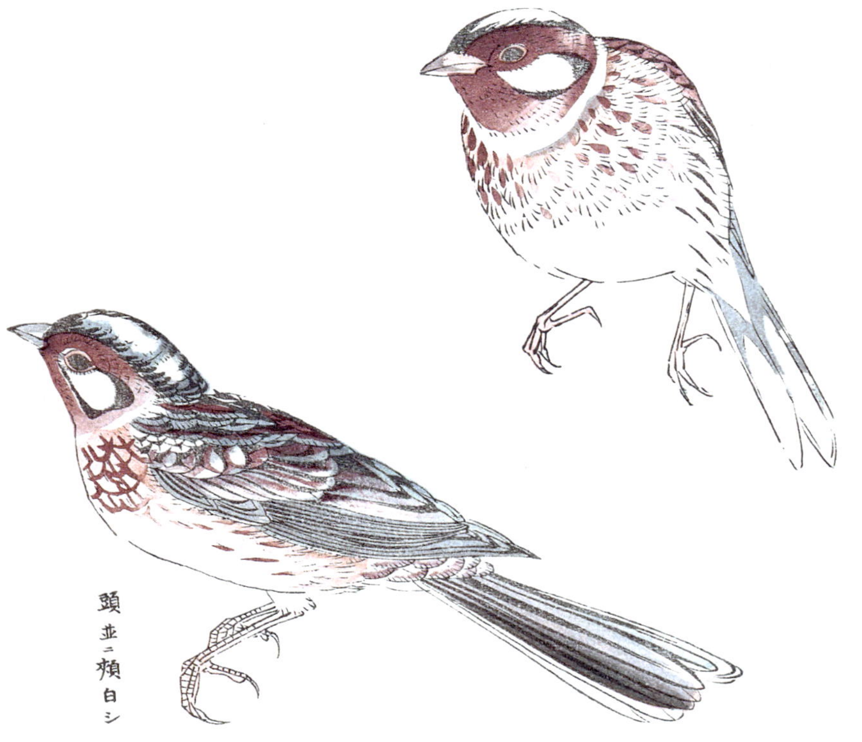

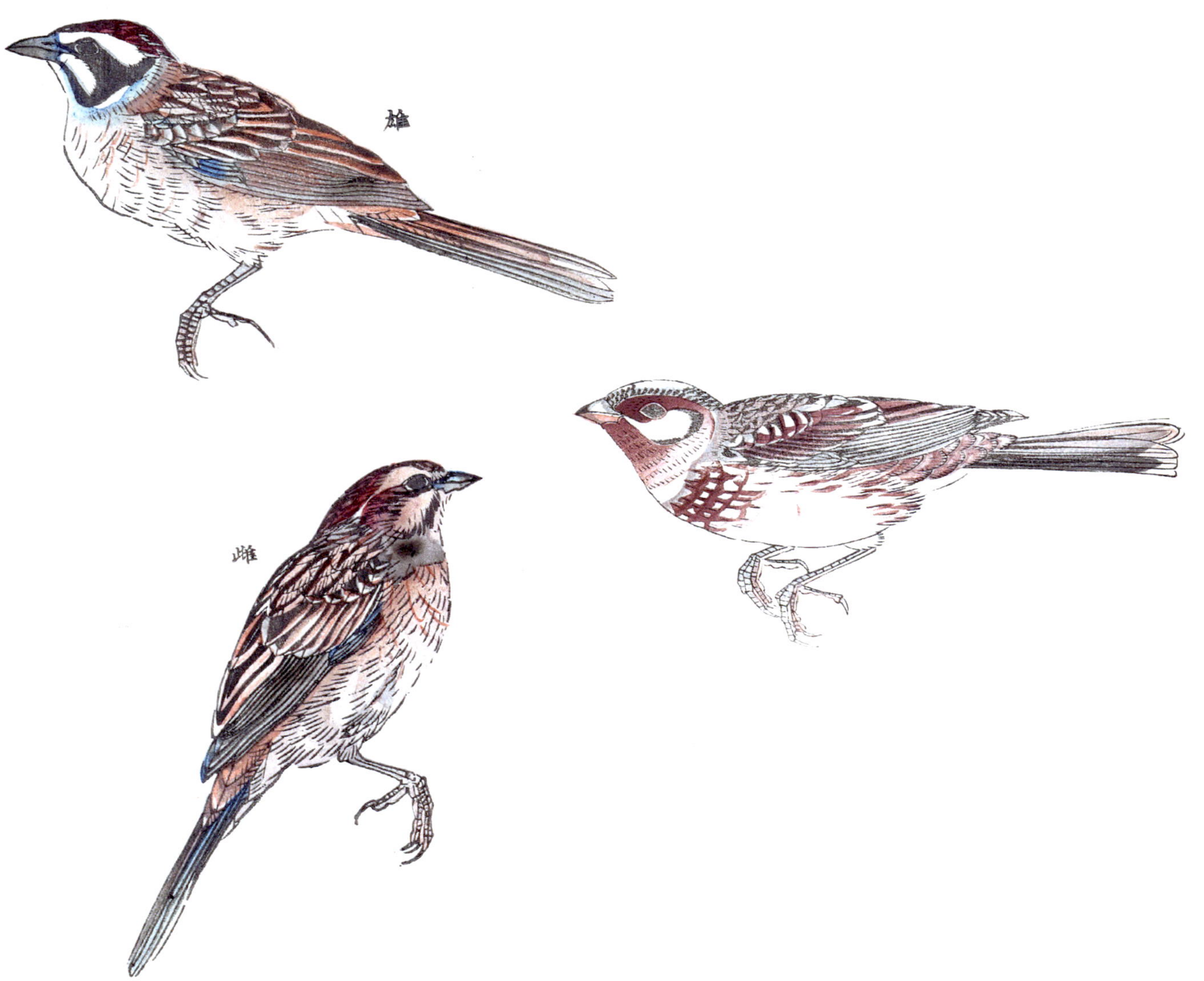

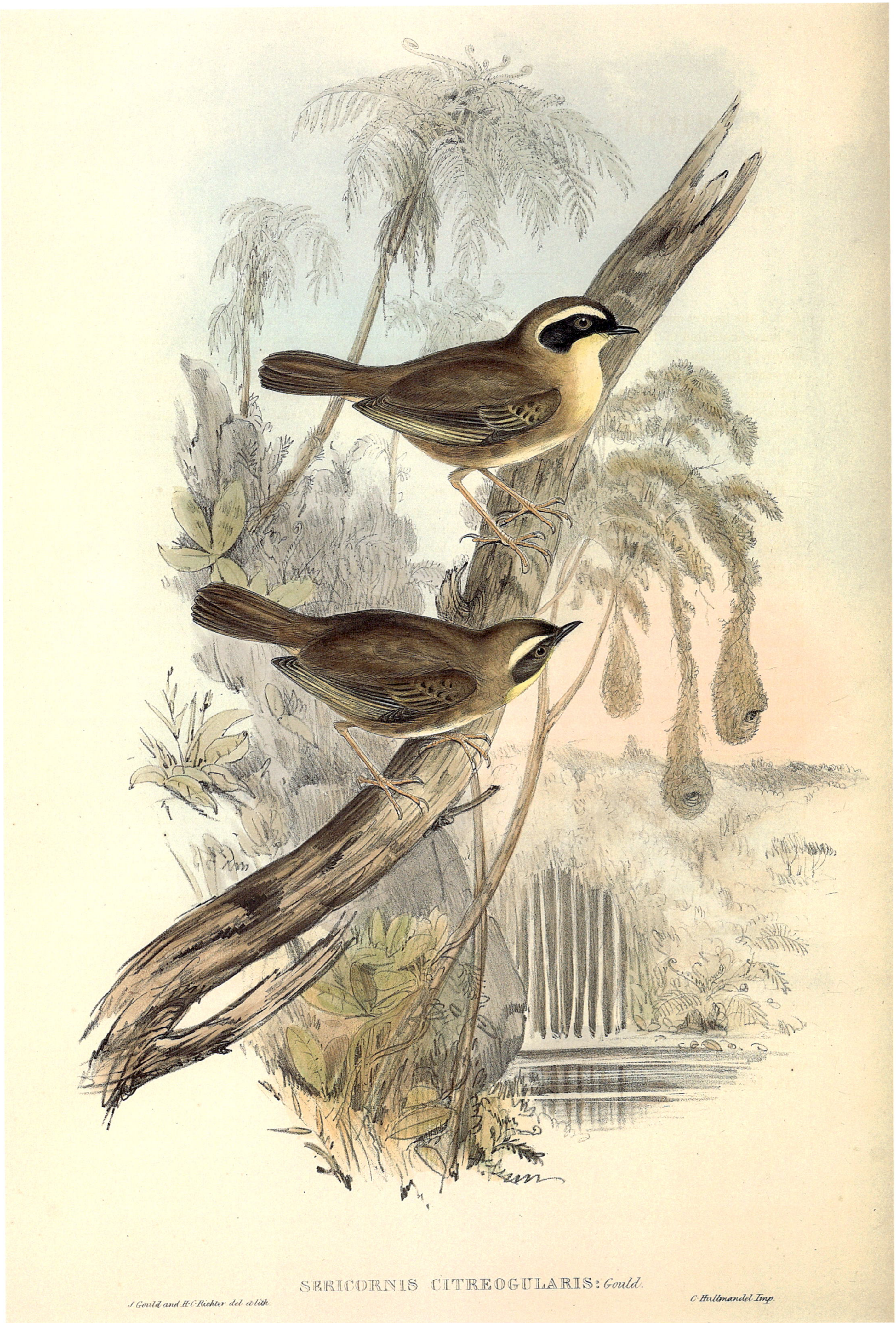
SERICORNIS CITREOGULARIS: Gould.
J. Gould and H. C. Richter del. et lith.
C. Hullmandel Imp.

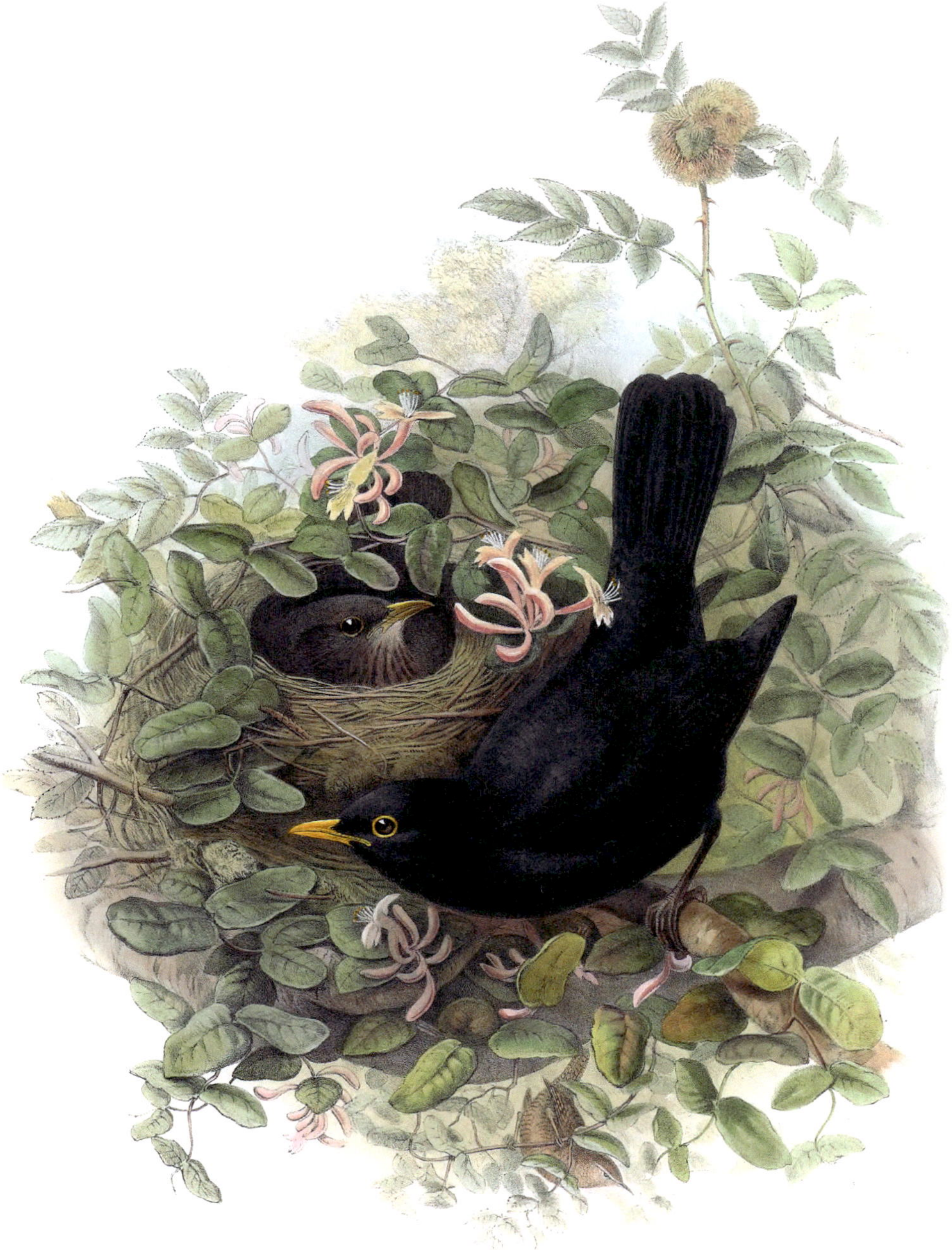

OPPOSITE:
YELLOW-THROATED SCRUBWRENS
Sericornis citreogularis

John Gould & Henry Constantine Richter, *The Birds of Australia*, 1840–8

EURASIAN BLACKBIRD
Turdus merula

John Gould & Henry Constantine Richter, *The Birds of Great Britain*, 1862–73

Blackbirds can build their nests from twigs, grass, leaves, straw, moss and mud. It is the female that takes responsibility for building duties and will tend to build them in spaces that can provide cover, such as trees, shrubs and bushes. Some build their nests inside buildings, garden sheds and even occasionally on the ground. The process takes approximately two weeks to complete, and the incubation of eggs can last for between eleven and seventeen days.

PIED CURRAWONG
Strepera graculina

Sarah Stone, watercolour, *ca.* 1790

This pied currawong is a watercolour by Sarah Stone which comes from an album of thirty-one works she produced on the fauna of New South Wales. This work was later translated into a copper plate engraving, which can be seen in John White's *Journal of a Voyage to New South Wales*. Often her drawings were the very first studies of these birds, and this is true in the case of this pied currawong. These birds are native to Eastern Australia and to the tiny Australian island of Lord Howe.

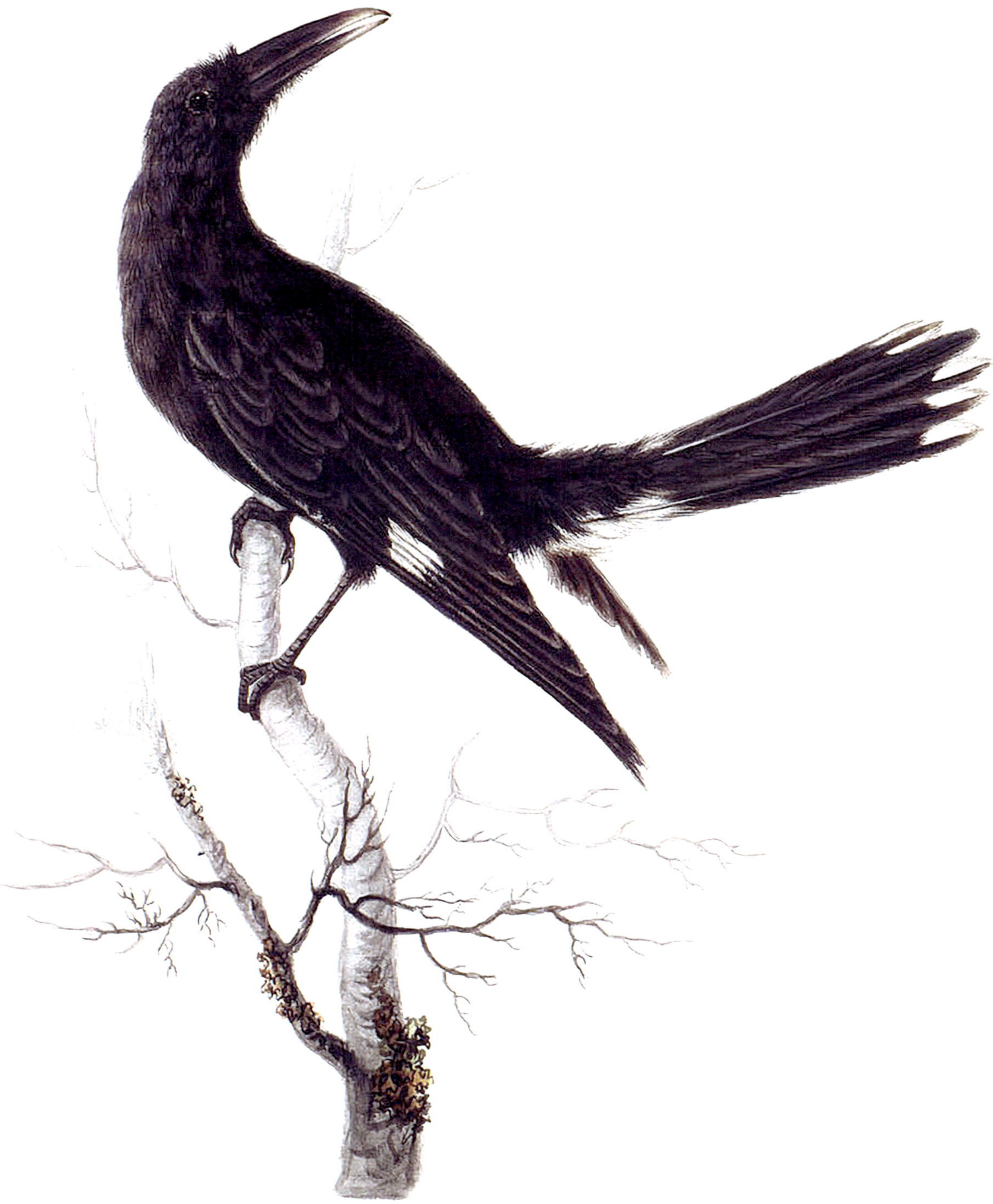

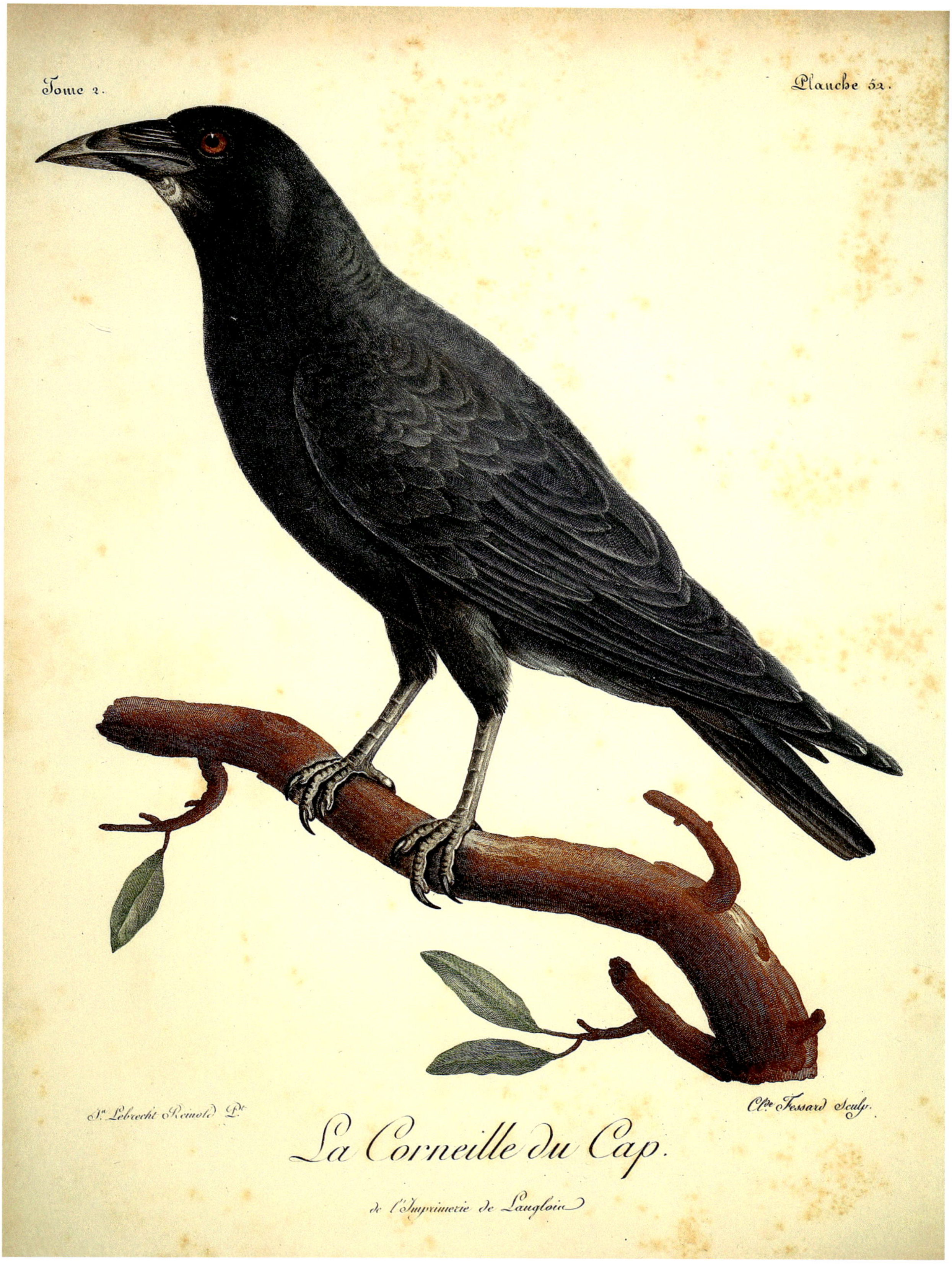

CAPE CROW
Corvus capensis

Johann Lebrecht Reinold, *Histoire naturelle des oiseaux d'Afrique*, 1796–1812

The cape crow is a common bird in parts of south and east Africa and it is represented here in an illustration from François Levaillant's comprehensive account of the birds of the continent. Levaillant was one of the first major figures in the field of modern ornithology. He gained prominence towards the end of the eighteenth century through a practice that focused on the collection, recording and studying of birds. His six-volume account on the birds of Africa was a major undertaking and was based on his own personal account of exploring the region.

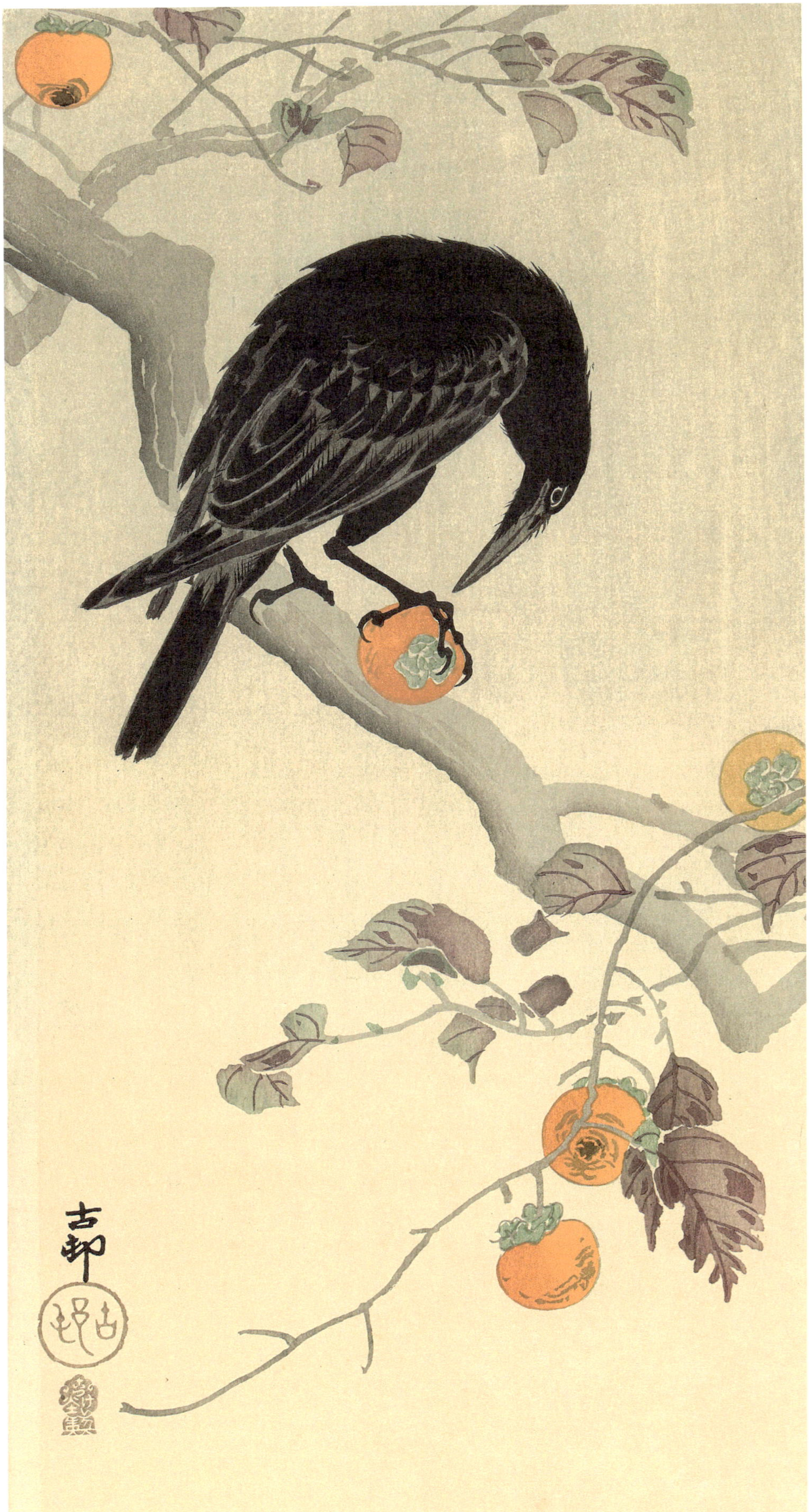

CARRION CROW

Corvus corone

Crow with kaki fruit, Ohara Koson, colour woodcut, *ca.* 1900–10

Koson's depiction of a crow shows the bird feeding on a small kaki fruit. A popular fruit in Japan, this is sometimes known as a Japanese persimmon. Crows are smart and adaptable birds that have adjusted well to living alongside humans, often finding ways to prosper in the human environment. While the crow has had a long tradition of being a symbol of power in Japanese mythology, today many Japanese people consider them to be a pest following a dramatic rise in their numbers.

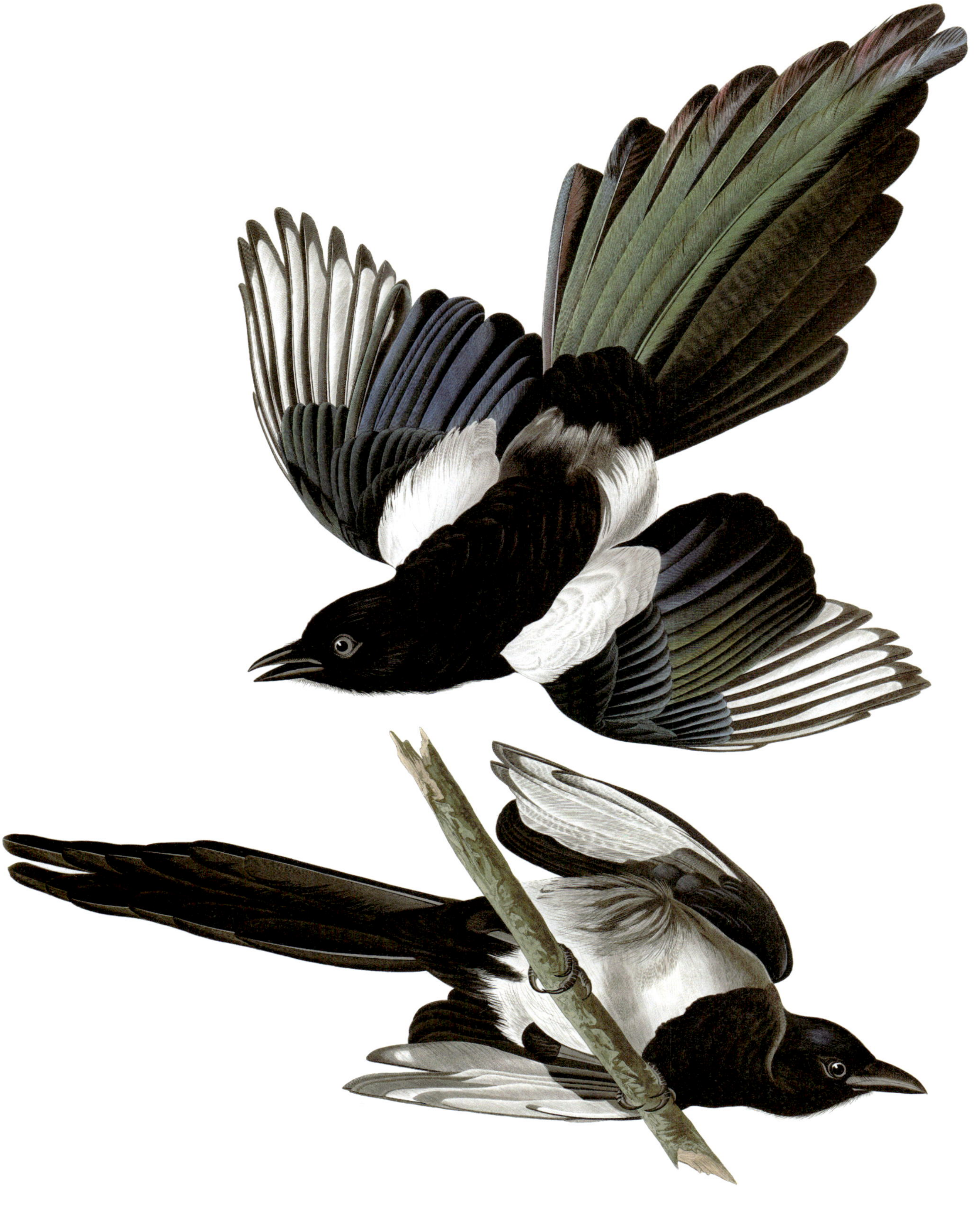

BLACK-BILLED MAGPIES
Pica hudsonia

John James Audubon,
The Birds of America, 1827–38

The American magpie – or black-billed magpie – looks almost identical to the European magpie (*Pica pica*) and Audubon is keen to point to this similarity in his text. Today, opinions remain split on whether or not it should be classed as a separate species.

Audubon's illustration manages to capture the fantastic colour of this bird's plumage which, despite its darkness, tends to have an impressive iridescent sheen that takes on colourful hues of purple, blue and green.

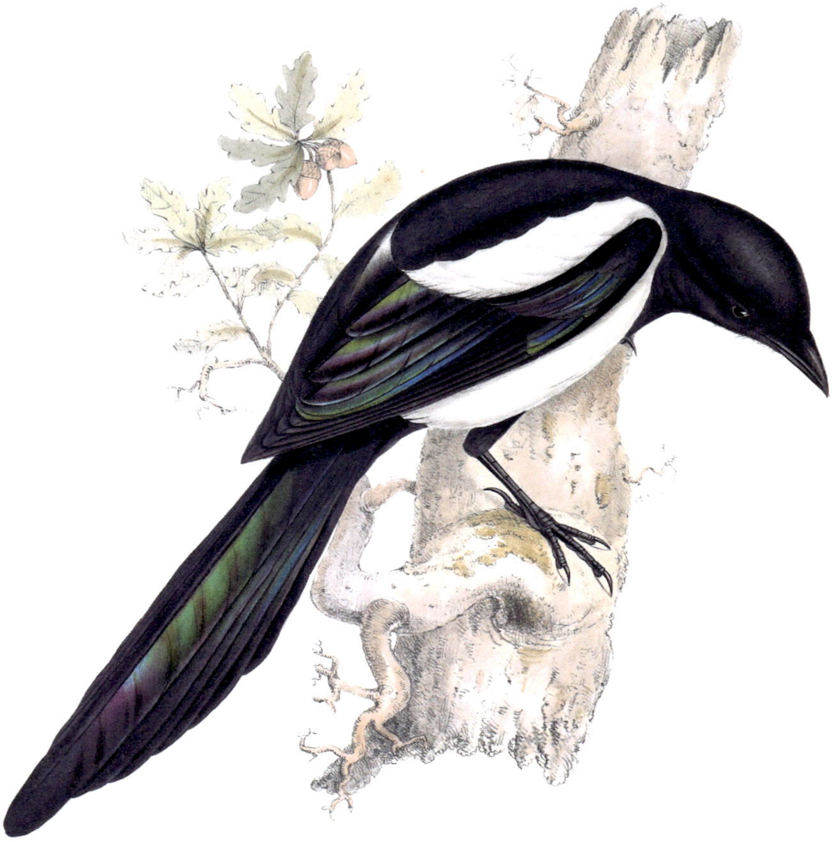

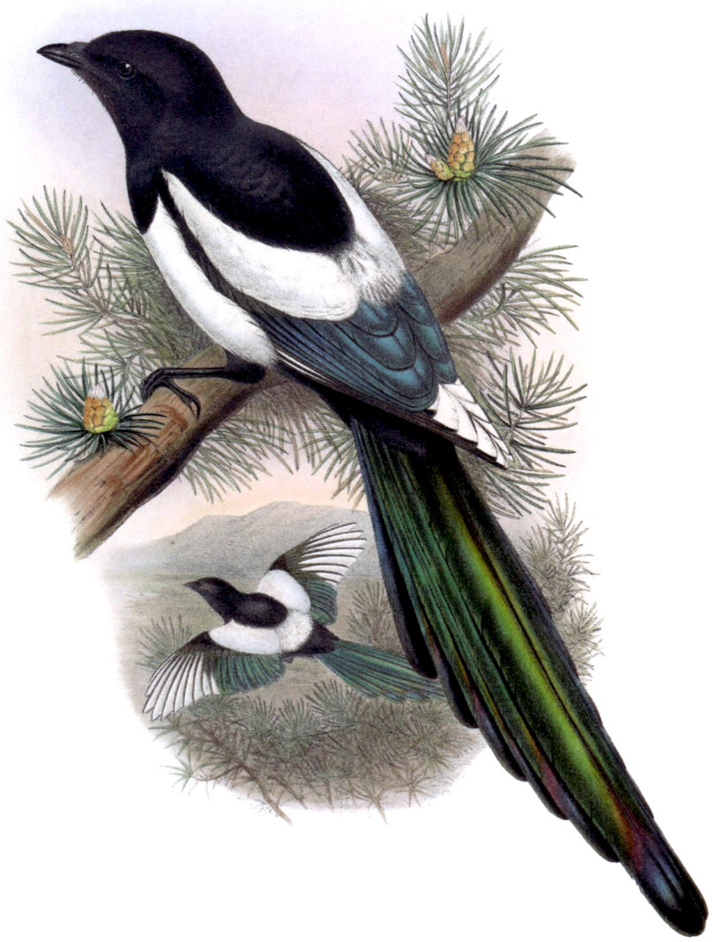

TOP:
EURASIAN MAGPIE
Pica pica

John Gould & Elizabeth Gould, *The Birds of Europe*, 1832–7

BOTTOM:
ORIENTAL MAGPIE
Pica serica

John Gould & Henry Constantine Richter, *The Birds of Asia*, 1850–83

These magpies come from Gould's *The Birds of Europe* and *The Birds of Asia*. Similar to the American magpie, the oriental magpie may, in fact, be the same species as the European species. Yet, despite having similarities, the Oriental magpie is a slightly larger bird with a shorter tail, and has a strong purplish-blue iridescence to its plumage.

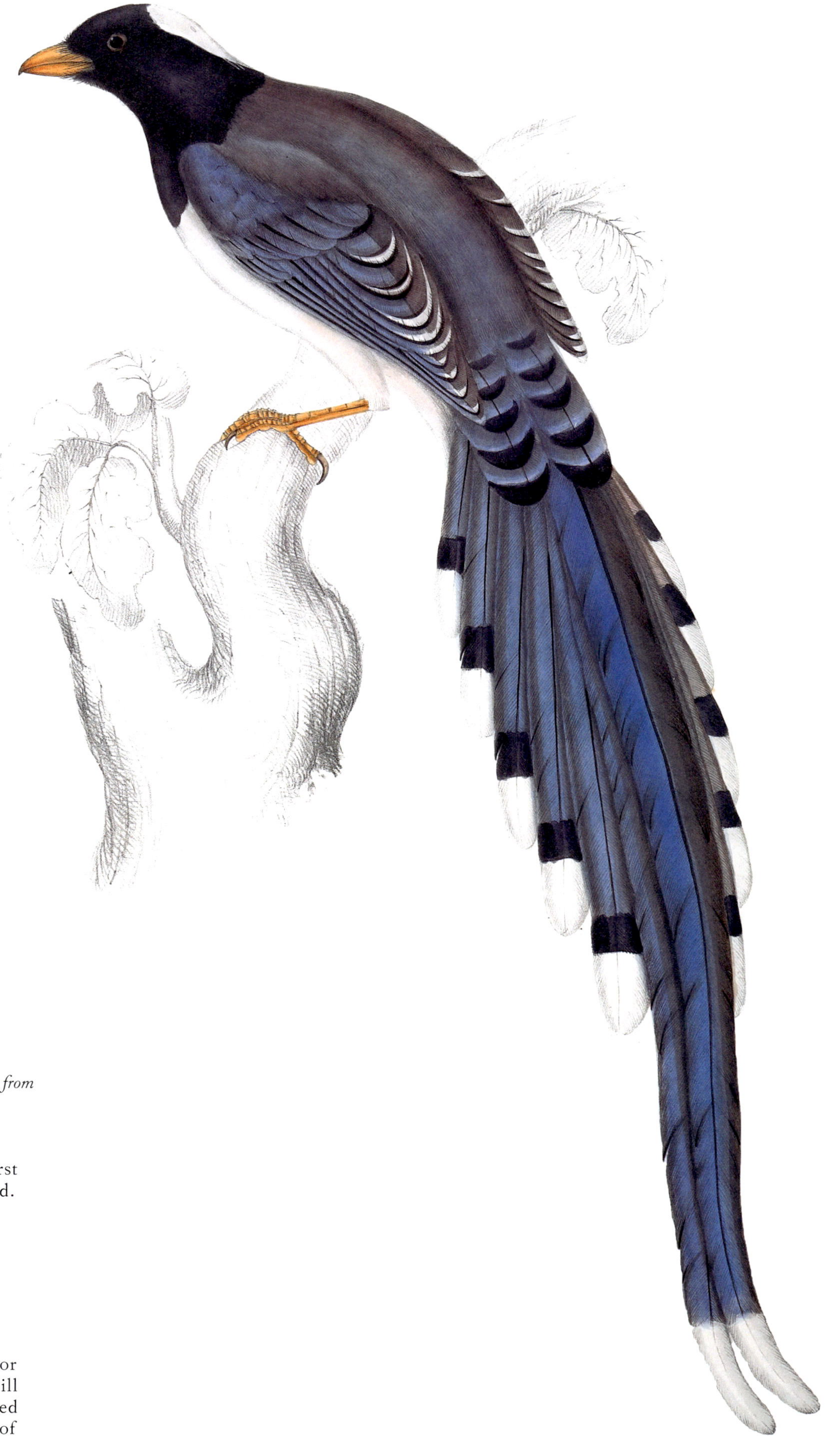

RED-BILLED BLUE MAGPIE
Urocissa erythroryncha

Elizabeth Gould, *A Century of Birds from the Himalaya Mountains*, 1830–2

A Century of Birds from the Himalaya Mountains was the first work that John Gould published. Elizabeth Gould provided the illustrations and the text was written by the Irish zoologist Nicholas Aylward Vigors. In 1826, Vigors co-founded the Zoological Society of London and, the following year, Gould became the society's first curator and preserver. The society is still in operation today and is devoted to the worldwide conservation of animals and their habitats.

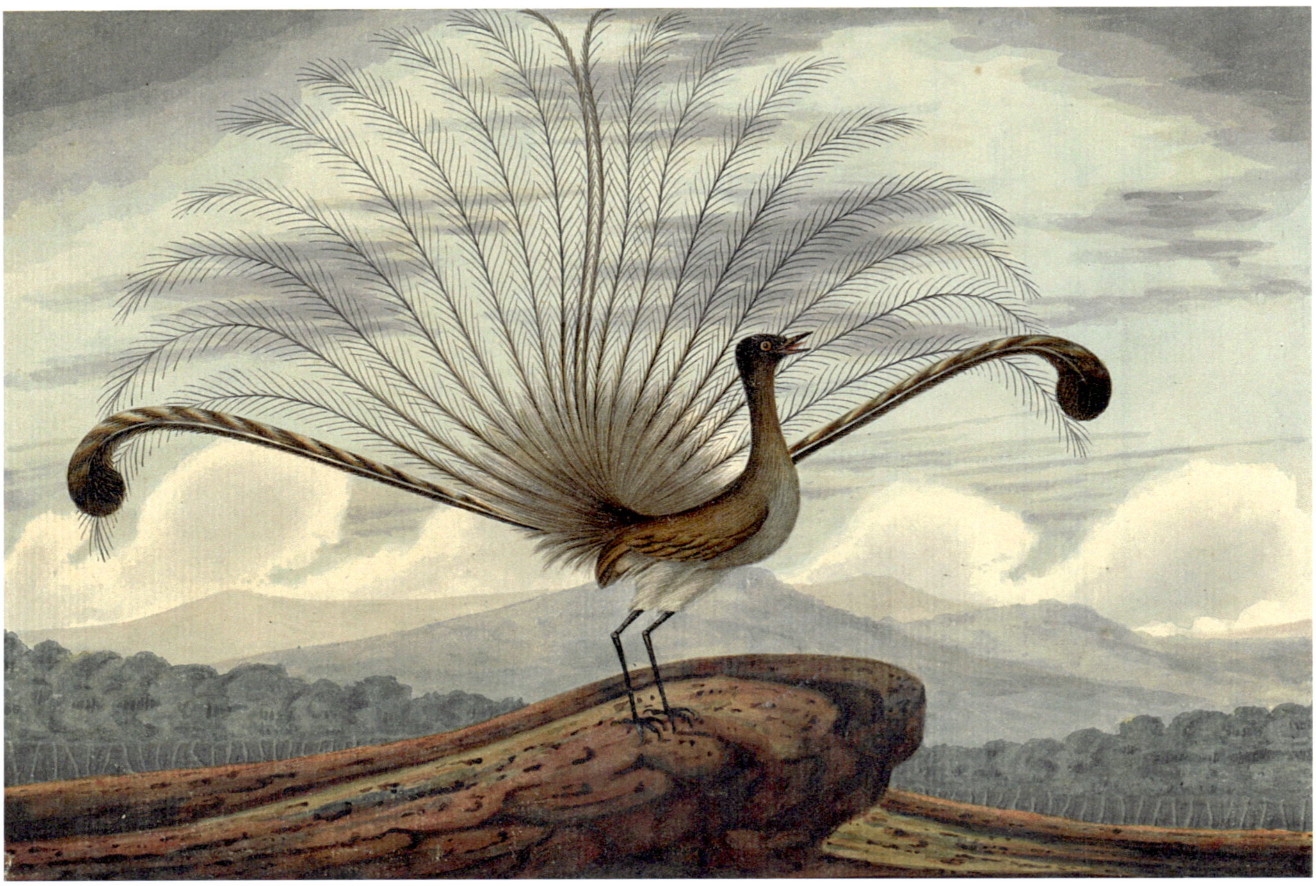

SUPERB LYREBIRD
Menura novaehollandiae

Richard Browne, *Select Specimens From Nature of the Birds & Animals of New South Wales*, 1813

Richard Browne was an Irish convict artist who was sent to New South Wales in 1810, possibly for the crime of forgery. After his sentence was complete, he moved to Sydney but quickly reoffended and was sent to the Newcastle penal colony. There he began producing artworks, including this illustration of a lyrebird.

It appears in the work *Select Specimens From Nature of the Birds and Animals of New South Wales* (1813), written by the commandant of Newcastle, Lieutenant Thomas Skottowe. It features watercolours of Aboriginal Australians as well as depictions and descriptions of mammals, insects and birds.

OPPOSITE:
ALBERT'S LYREBIRD
Menura alberti

John Gould & Henry Constantine Richter, *Supplement to the Birds of Australia*, 1851–69

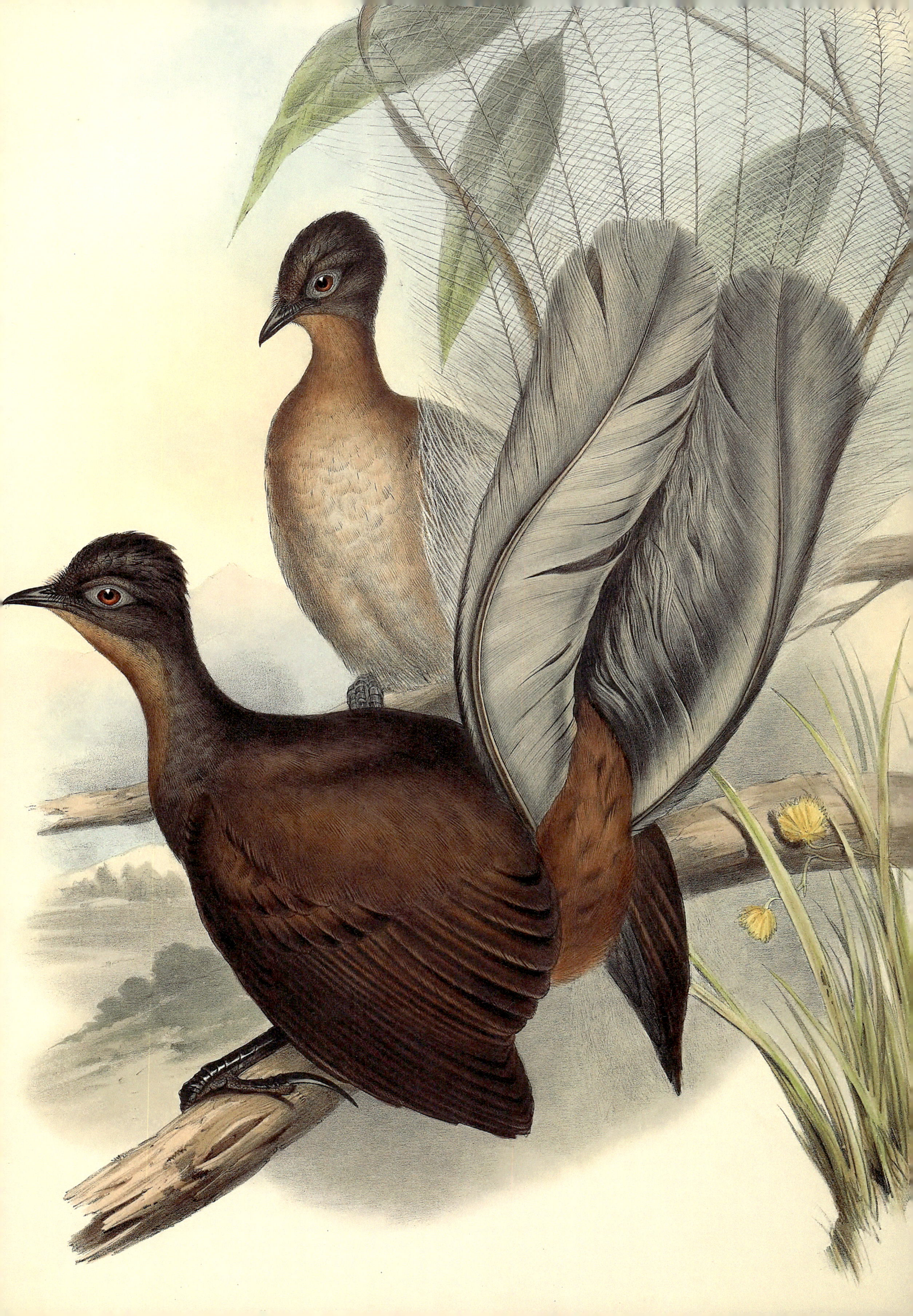

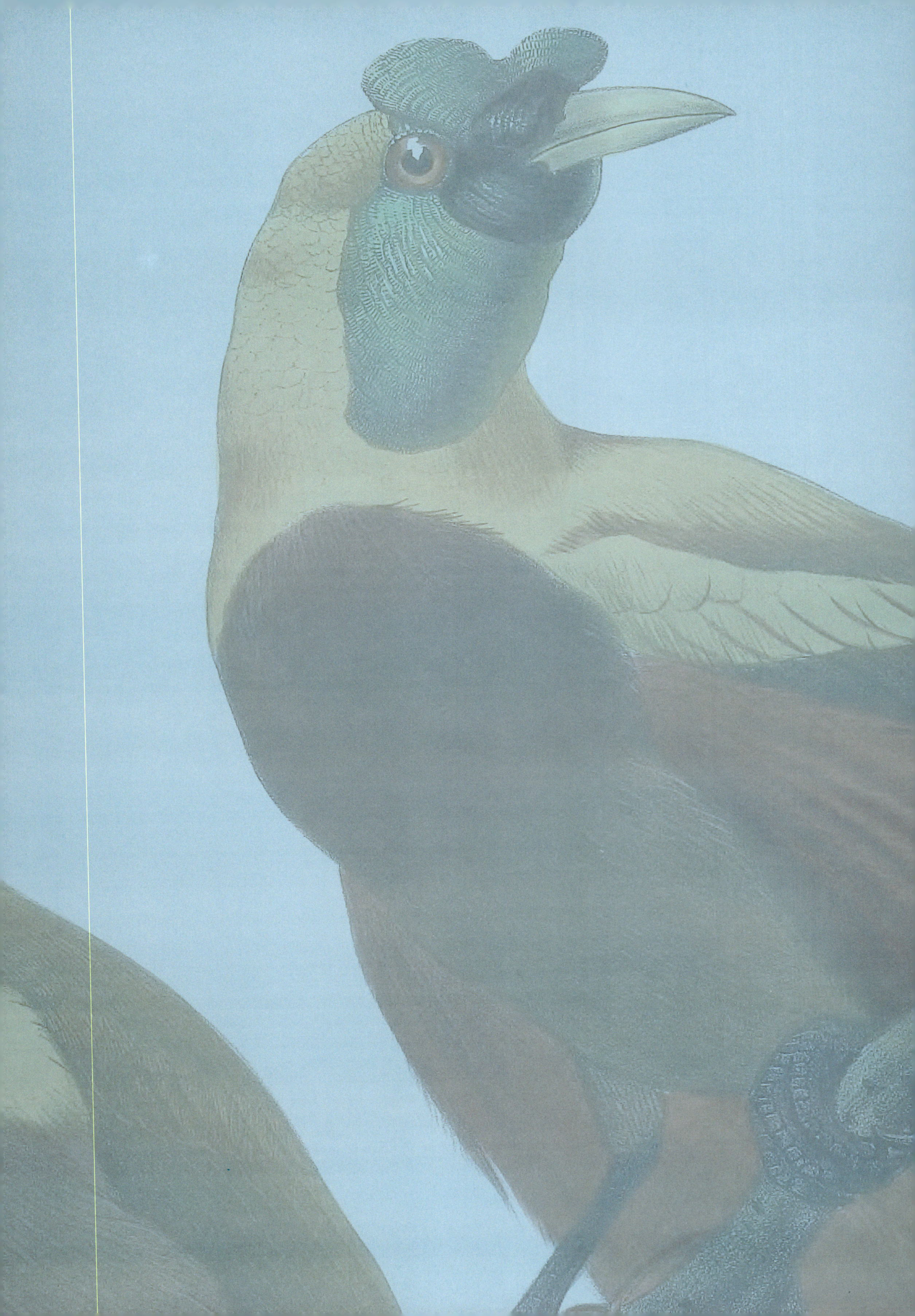

Parrots & Other 'Exotic' Birds

Birds-of-paradise, macaws, cockatoos, hornbills and hummingbirds are just some of the birds that attracted the interests of wealthy collectors during the era of European colonialism. Brought back from foreign lands, these 'exotic' birds became symbols of power, status, wealth and global reach. Often vibrantly coloured, they ranked among some of the most popular birds of the day and quickly became the subject of many ornithological monographs.

The Age of Discovery is said to have opened the world up to Europe. When the Magellan–Elcano expedition completed the first circumnavigation of the globe, it returned to Spain in 1522 with the skins of several greater birds-of-paradise (*Paradisaea apoda*). These skins had come from the Maluku Islands, where the native Moluccans had prepared each bird by removing their feet, wings and internal organs. Their skills in taxidermy were so impressive that when the specimens eventually reached Europe they created a huge stir of sensation.

Described as 'birds of the Gods', these elaborate skins took on a mythical status all of their their own. People believed that these strange creatures had descended directly from heaven and instead of using wings to fly, they could float through the air solely by using their elaborate plumage. These mythical creatures never needed to land for food, and instead they survived solely on the dew in the air; the only time a bird-of-paradise would touch the ground, people thought, was when it died.

This myth became so powerful that it continued to exist in Europe for well over 150 years. As more specimens reached Europe, the plumage of these birds became so popular that the birds started to become endangered. As European colonialism expanded, so too did an interest in the foreign and the exotic. Naturalists eagerly headed to foreign lands in the hopes of finding even more elaborate specimens – often competing with one another to be the first to discover a new species. Ideas and styles from distant regions influenced art and design and, as the Enlightenment broke, so too did a new interest in exoticism. Some of the most popular bird illustrations of the era were of 'exotic' birds.

By the nineteenth century, parrots had become one of the most sought-after birds in Europe. Found in warm climates, most species come from South America and Australasia. These birds were bought and sold for large sums of money; their bright colours, beautiful plumage and talents for mimicry made them a must-have for aristocrats and wealthy bird lovers around the continent. Owning exotic birds and animals was a way of displaying power, wealth and status.

Early zoos were also a means of displaying wealth and status, with monarchs constructing imperial menageries as a way to symbolize their power. Modern zoos emerged during the Enlightenment, and while they allowed for displays of colonial might, they also provided a space for scholarly learning. London Zoo opened as a space for scientific study in 1828, with the Zoological Society collecting, studying and documenting an extensive collection of animals from all around the world.

In 1829 an ornate ironwork aviary was opened that served as a summer residence for the zoo's macaws. It was here, the following year, that a young Edward Lear applied to the Society to request permission to draw their collection of parrots. Lear was the first major ornithological illustrator to draw birds directly from real life, and the work that he eventually published – *Illustrations of the Family of Psittacidae, or Parrots* (1832) – became the first monograph produced in England that focused solely on a single family of birds.

Images of cockatoos, parrots, hornbills and other tropical birds were in huge demand at this time, and publishers responded by producing beautiful books that captured their vibrancy and elegance. Some of the finest ornithological illustrations of this time were of these 'exotic' birds, as the illustrations in this chapter will demonstrate.

In addition to including many of Lear's remarkable parrots, this chapter is rich with work from other monographs, including illustrations from Daniel Giraud Elliot's stunning 1873 monograph on birds-of-paradise – arguably one of the most admired bird books ever produced. It also includes illustrations by the renowned French illustrator Jacques Barraband (1767–1809), whose illustrations for François Levaillant's *Histoire naturelle des perroquets* may rank as the most realistic and life-like illustrations of their era.

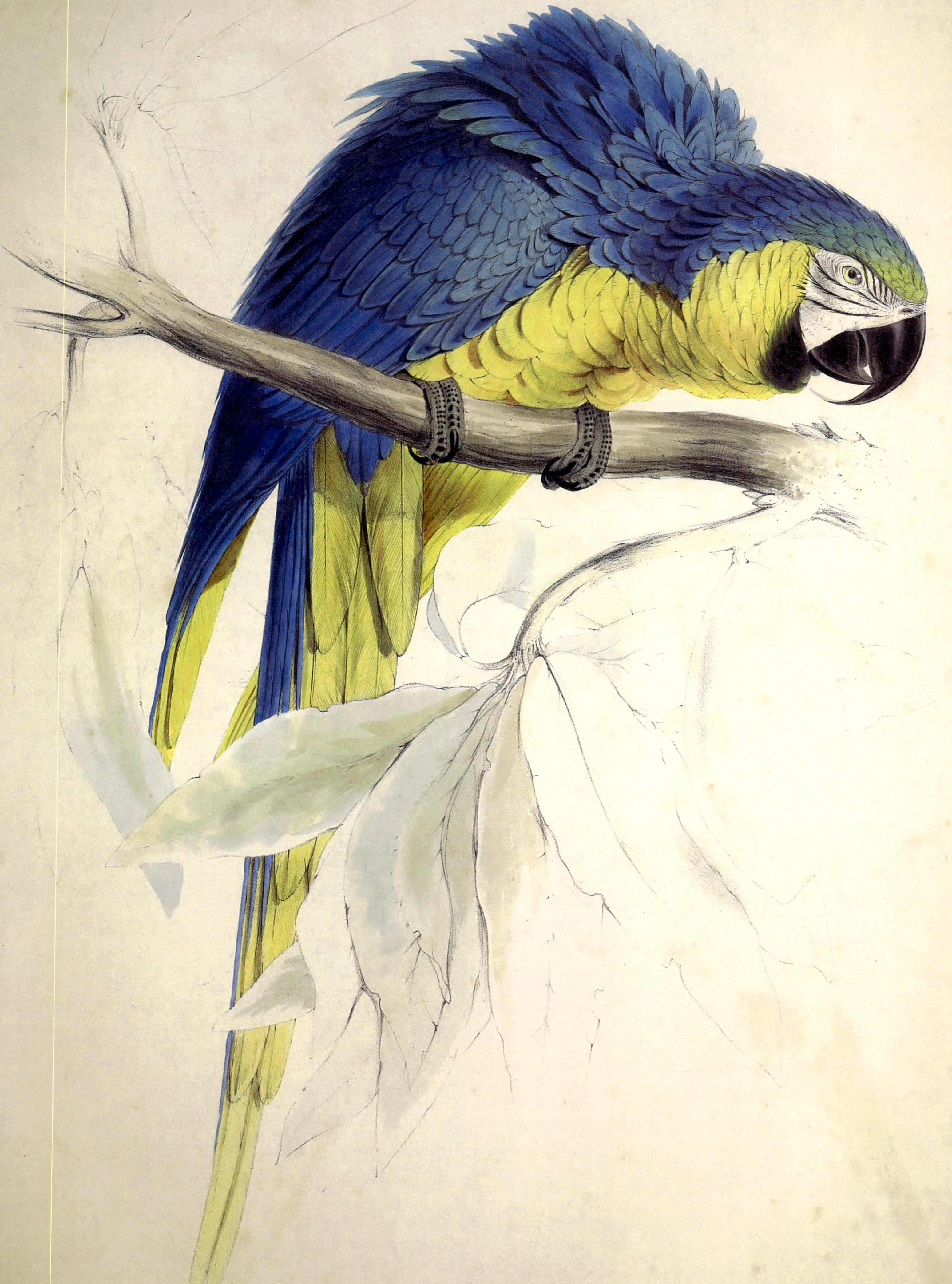

MACROCERCUS ARARAUNA.

Blue & Yellow Maccaw.

⅔ Nat. Size.

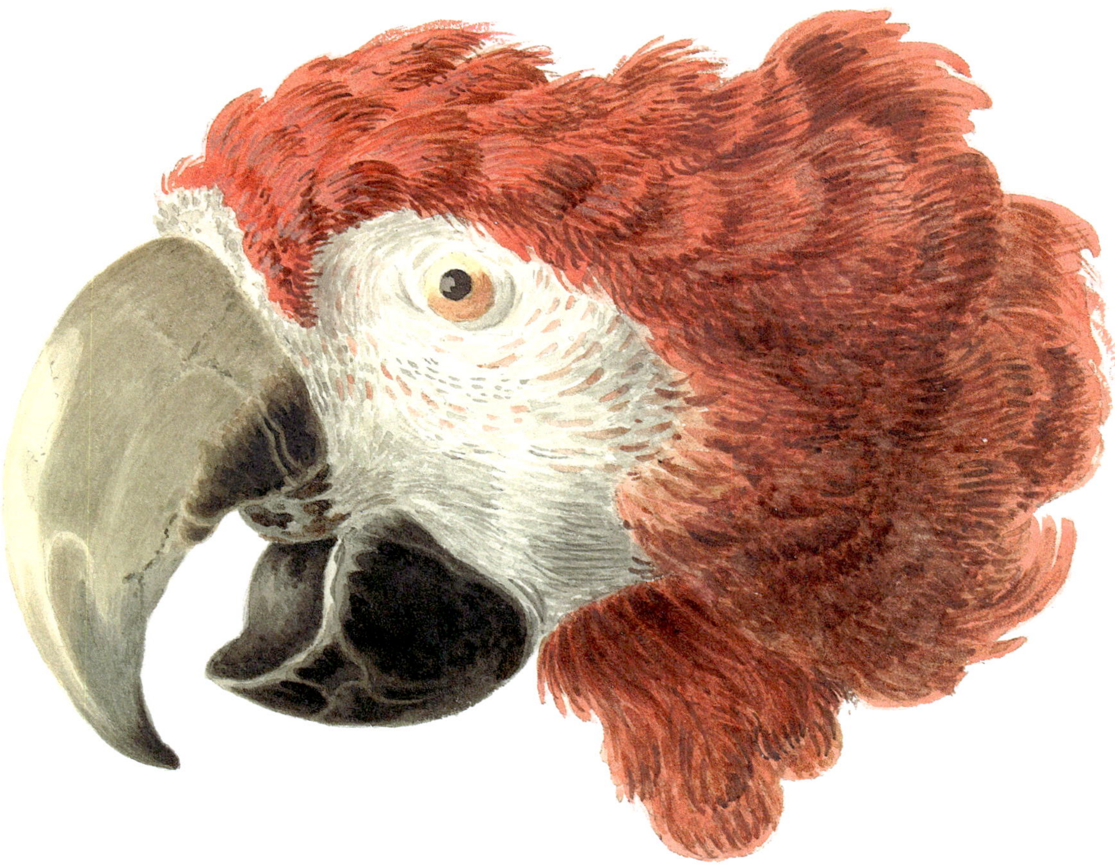

OPPOSITE:

BLUE-AND-YELLOW MACAW
Ara ararauna

Edward Lear, *Illustrations of the Family of Psittacidae*, 1832

This blue-and-yellow macaw is one of 42 hand-coloured lithographs that make up Edward Lear's *Illustrations of the Family of Psittacidae, or Parrots*. Published in 1832, it was the first monograph produced in England to focus on a single family of birds. Lear began the project when he was just eighteen years of age, producing it with no formal training or any institutional support. It quickly established him as a celebrated illustrator and his drawings of parrots remain unsurpassed even to this day.

SCARLET MACAW
Ara macao

Aert Schouman, drawing on paper, *ca.* 1725–92

Parrots are found mainly in tropical and subtropical regions and they represent a large family of birds that belong to the order Psittaciformes. They rank among some of the most intelligent birds, and their ability to be trained and also imitate the human voice has made them popular as pets. Above is a scarlet macaw. This bird is found in Central and South America, and it is also one of the most common captively bred parrots in the United States. Owning a scarlet macaw can be a life-long commitment, with some birds living for up to eighty years.

OVERLEAF:

HYACINTH MACAW
Anodorhynchus hyacinthinus

Charles Reuben Ryley, *Museum Leverianum*, 1792–6

The hyacinth macaw is the largest of the macaws. Vibrant, curious and playful, these birds have become popular pets, but heavy illegal trade has led to them becoming an endangered species. Found in parts of Brazil, Bolivia and Paraguay, habitat loss has also brought about declining numbers as cattle ranching and agriculture makes the land incapable of supporting this species.

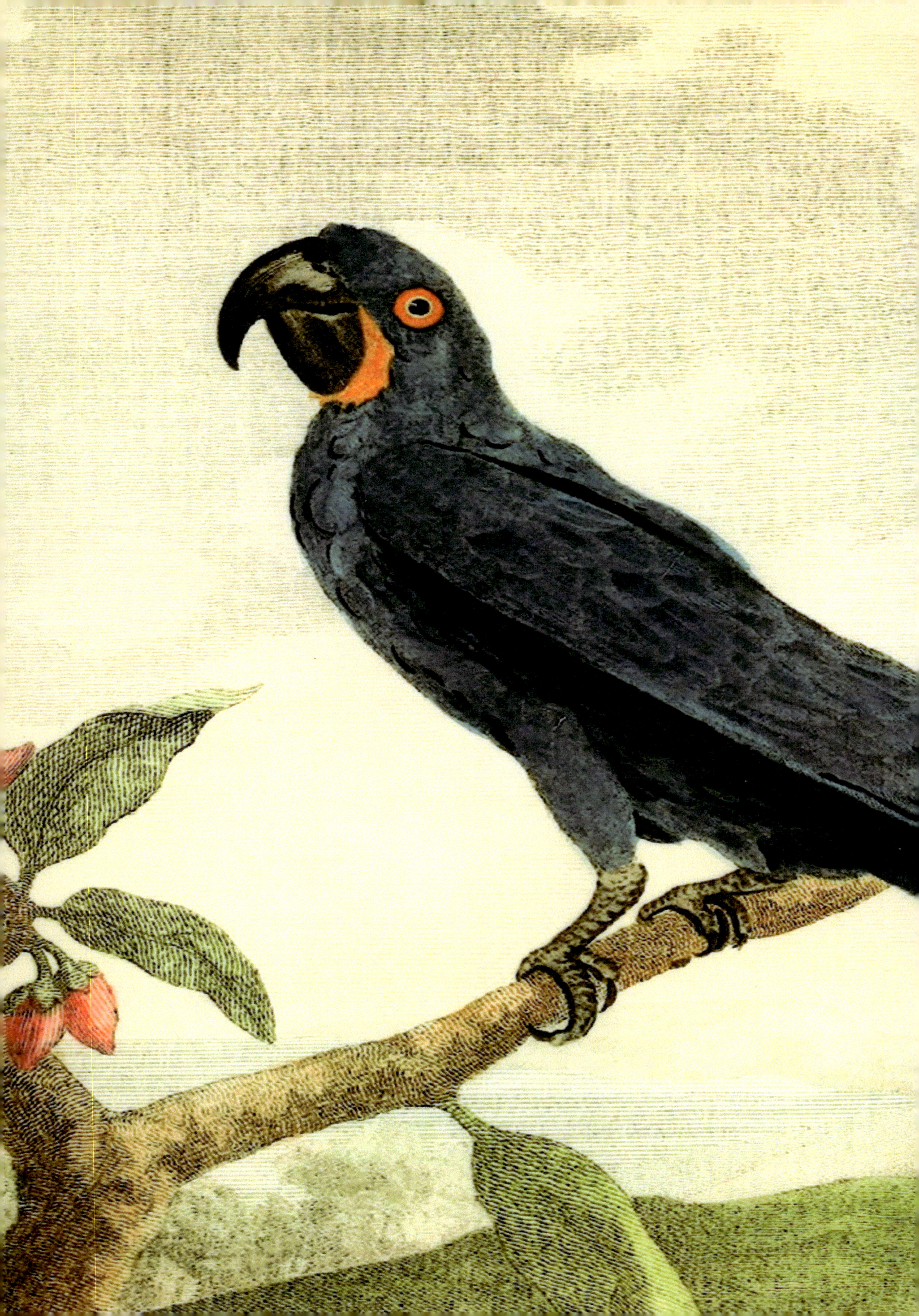

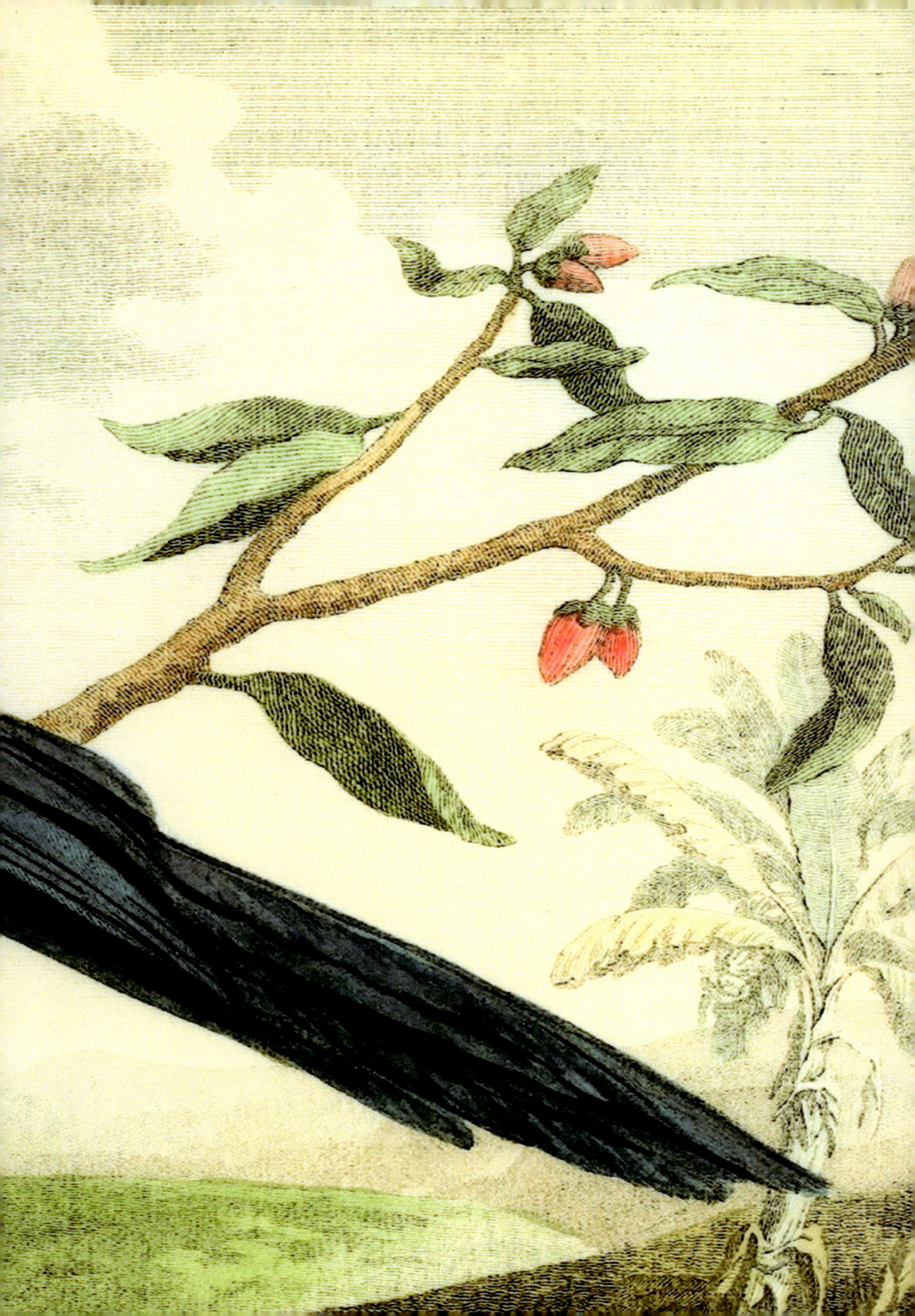

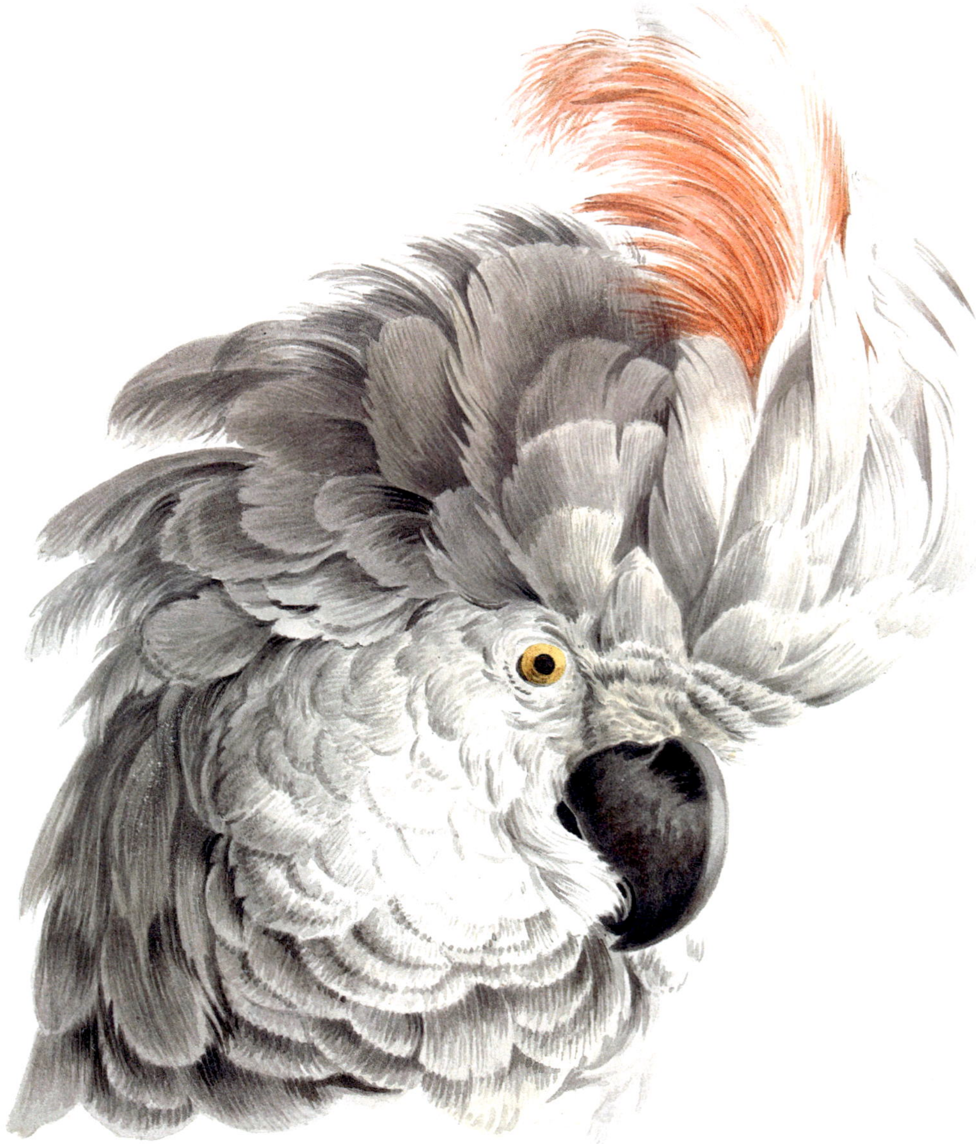

SALMON-CRESTED COCKATOO
Cacatua moluccensis

Aert Schouman, drawing on paper, *ca.* 1725–92

Native to Australasia and Indonesia, the cockatoo belongs to a family of parrots that are typically plain in colour and have thick curved beaks and a flamboyant crest atop their heads. The earliest depiction of the bird in Europe can be seen in *De Arte Venandi cum Avibus*. This manuscript was produced in the 1240s and was written by the Holy Roman Emperor Frederick II. The bird seen here is drawn by Schouman and the specimen would have presumably come from a Dutch colony in the Indonesian Archipelago.

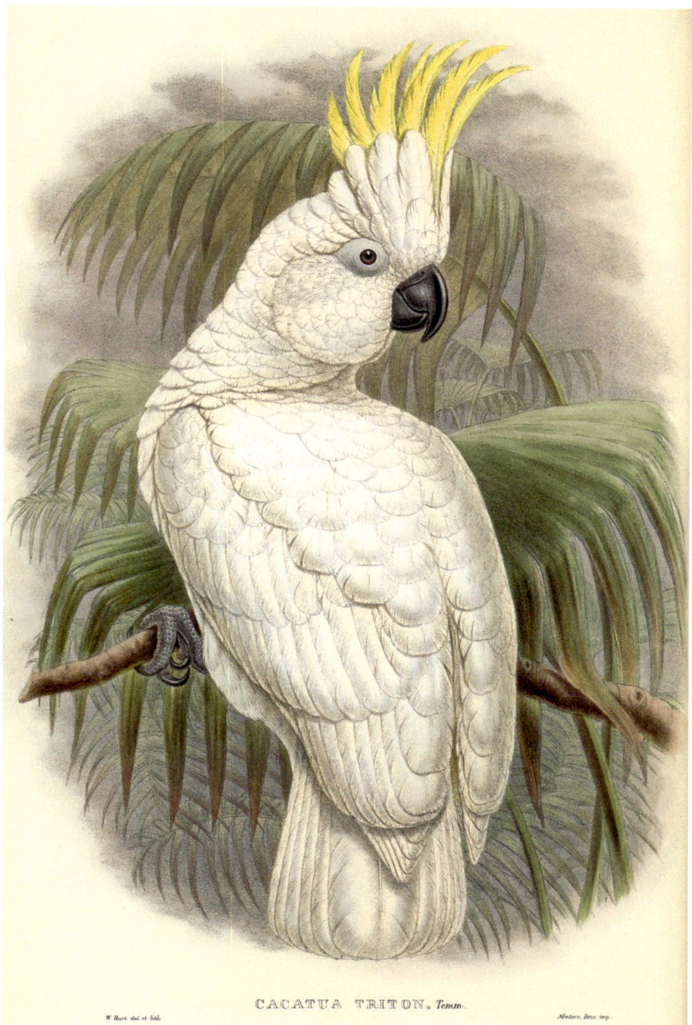

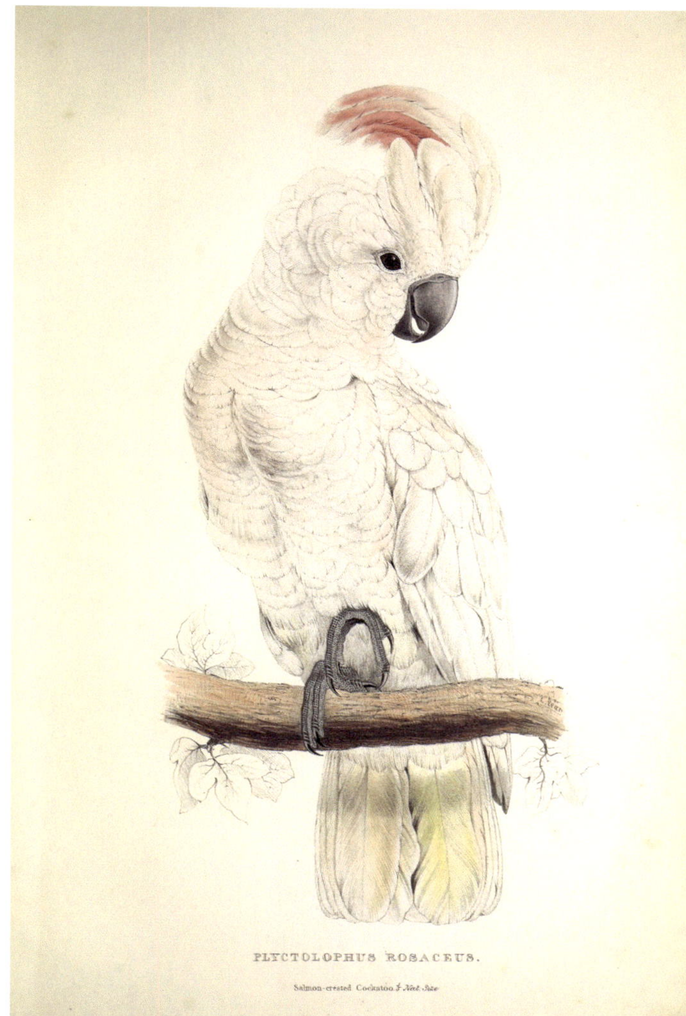

LEFT:

SULPHUR-CRESTED COCKATOO
Cacatua galerita

William Matthew Hart, *The Birds of New Guinea and the Adjacent Papuan Islands*, 1875–88

This cockatoo is noted for its large yellow crest which it fans out when excited. These birds also have a raucous call which is both distinctive and very loud. In captivity, they can live upwards of 80 years. In New South Wales, a cockatoo known as Cocky Bennett lived for well over a century. Born in the 1790s, Cocky spent his first 78 years sailing the South Sea Islands with his owner Captain George Ellis, and had lost all of his feathers by the time he died in 1916.

RIGHT:

SALMON-CRESTED COCKATOO
Cacatua moluccensis

Edward Lear, *Illustrations of the Family of Psittacidae*, 1832

Sometimes known as Moluccan cockatoos, these birds are native to the Maluku Islands in eastern Indonesia. Like all cockatoos, the salmon-crested cockatoo mates for life. The male and female will groom each other, forage together and sleep next to each other. During the breeding season, they will separate from the flock. When a cockatoo loses its mate, it is said to get depressed and very often the bird will stay alone for the rest of its life.

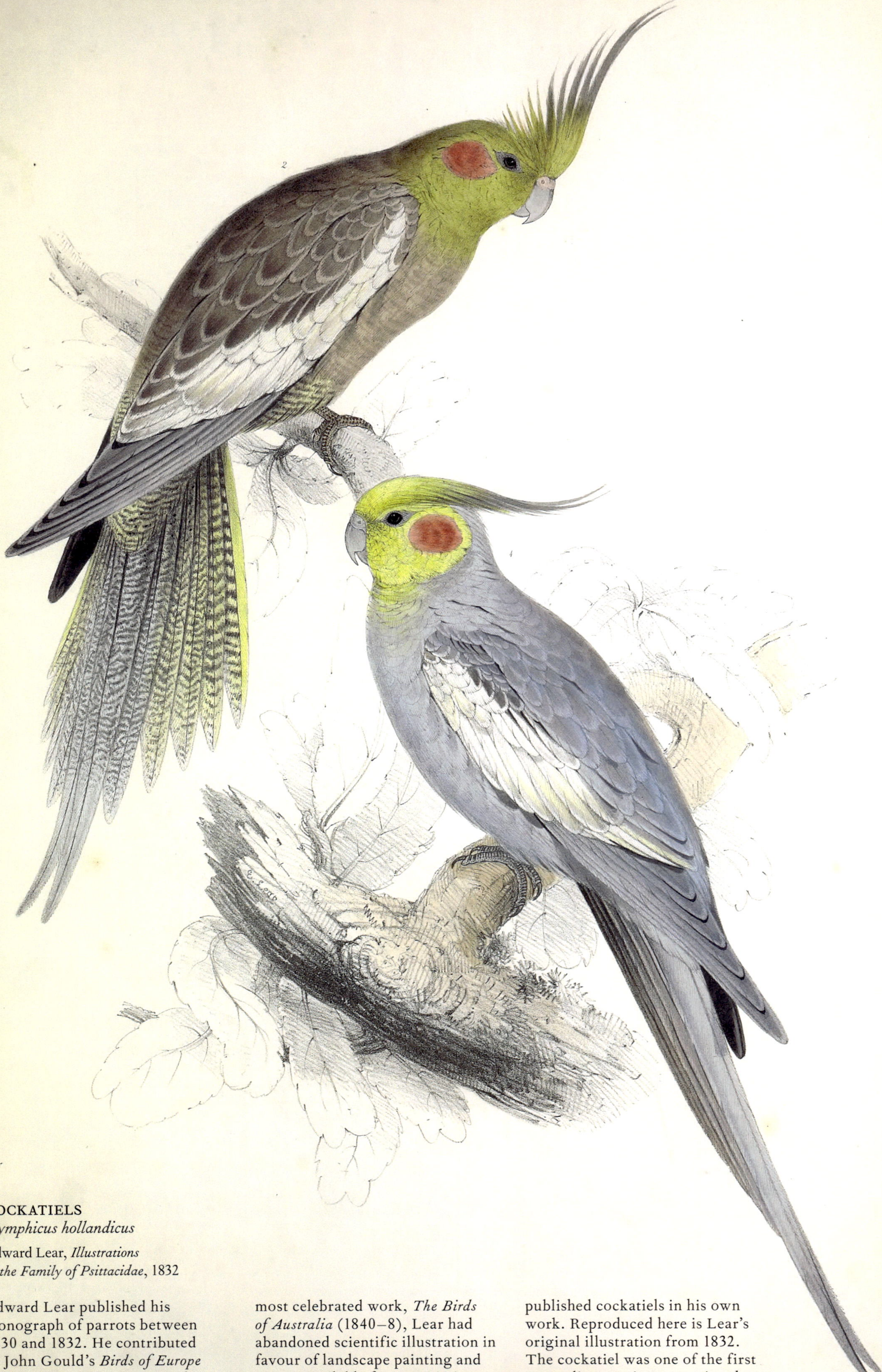

COCKATIELS
Nymphicus hollandicus

Edward Lear, *Illustrations of the Family of Psittacidae*, 1832

Edward Lear published his monograph of parrots between 1830 and 1832. He contributed to John Gould's *Birds of Europe* (1832–37) and his monographs on toucans and trogons. By the time Gould came to publish his most celebrated work, *The Birds of Australia* (1840–8), Lear had abandoned scientific illustration in favour of landscape painting and was unavailable for commission. This did not stop Gould from copying Lear's previously published cockatiels in his own work. Reproduced here is Lear's original illustration from 1832. The cockatiel was one of the first Australian species to survive the long voyage back to London.

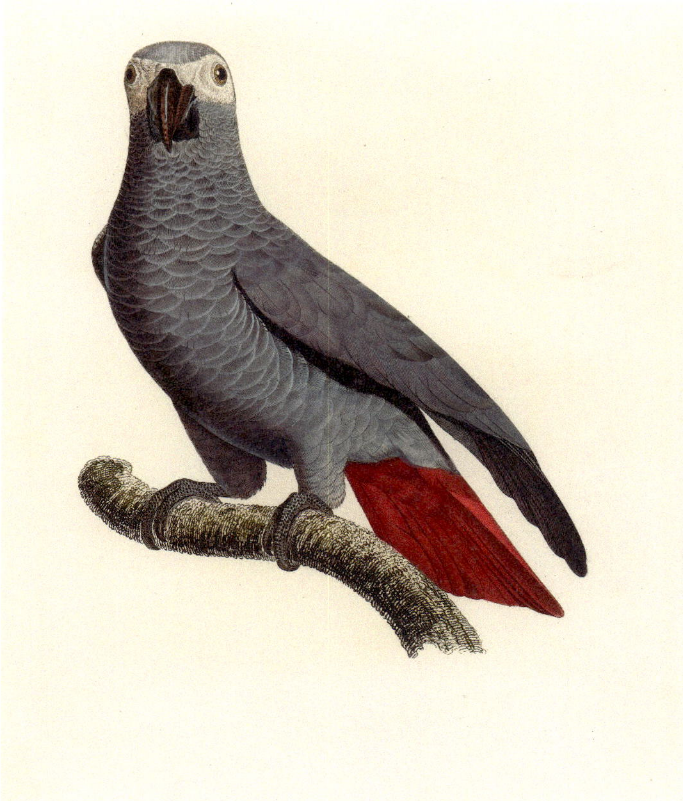

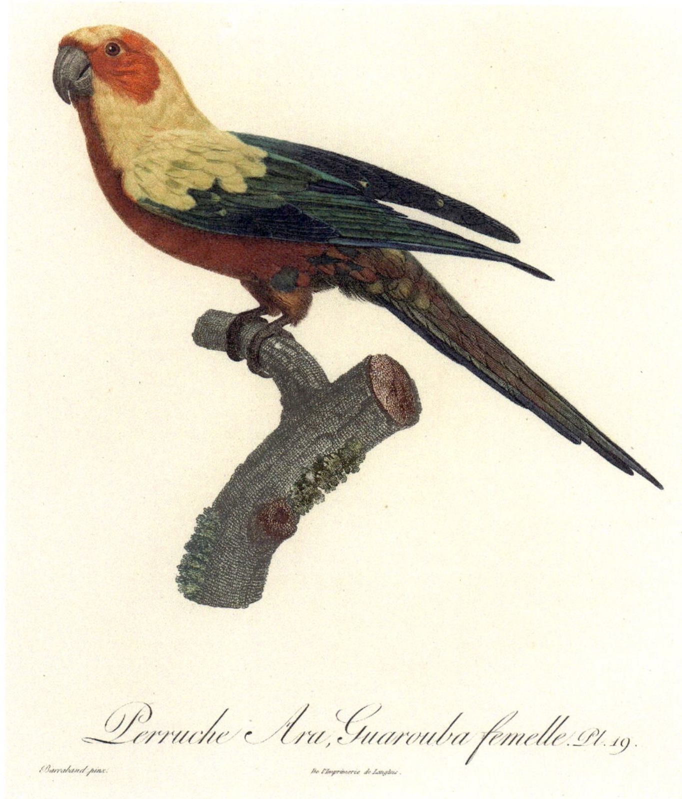

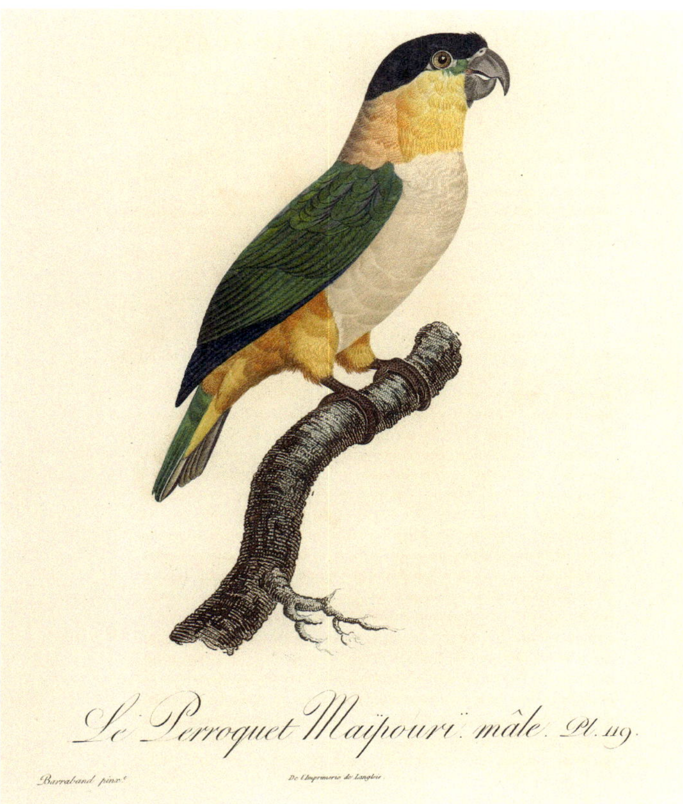

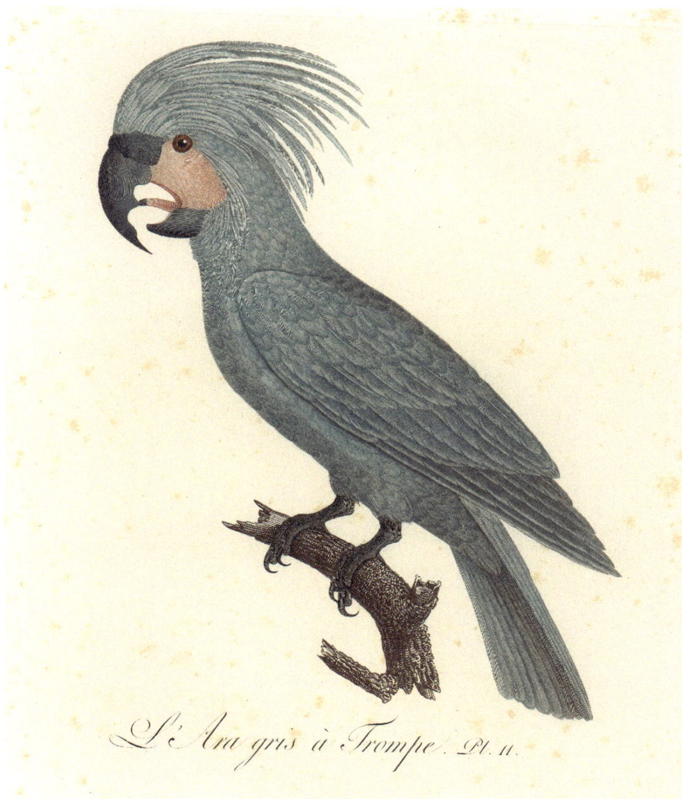

TOP LEFT:
GREY PARROT
Psittacus erithacus

Jacques Barraband, *Histoire naturelle des perroquets*, 1801–5

TOP RIGHT:
SUN PARAKEET
Aratinga solstitialis

Jacques Barraband, *Histoire naturelle des perroquets*, 1801–5

BOTTOM LEFT:
BLACK-HEADED PARROT
Pionites melanocephalus

Jacques Barraband, *Histoire naturelle des perroquets*, 1801–5

BOTTOM RIGHT:
PALM COCKATOO
Probosciger aterrimus

Jacques Barraband, *Histoire naturelle des perroquets*, 1801–5

François Levaillant's *Histoire naturelle des perroquets* was a monumental two-volume work published from 1801 to 1805 and illustrated by Jacques Barraband. A hugely talented artist, Barraband introduced the colour-printed copper plate technique to France and used this technique to create some of the era's most lifelike renderings of tropical birds.

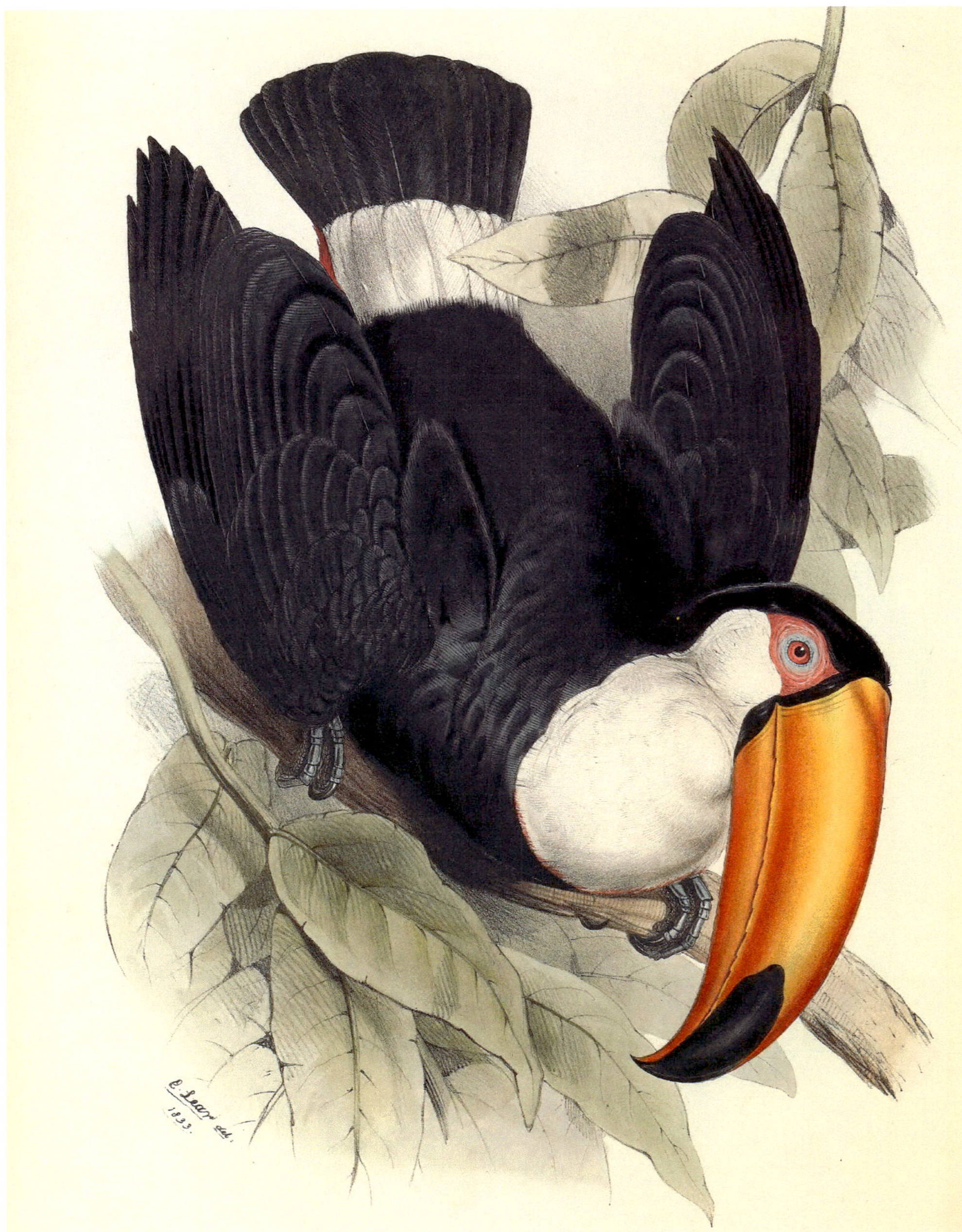

TOCO TOUCAN
Ramphastos toco

Edward Lear, *A Monograph of the Ramphastidae*, 1833–5

John Gould's inspiration for his monograph of the toucan family (*Ramphastidae*) came while he was researching his *Birds of Europe* and became fascinated by the toucan collection of a fellow ornithologist. He then made several trips to Europe, where he saw skins at museums and a small number of living birds in aviaries. Despite his limited opportunities to encounter these birds, he still managed to produce a rather splendid and detailed study of the family.

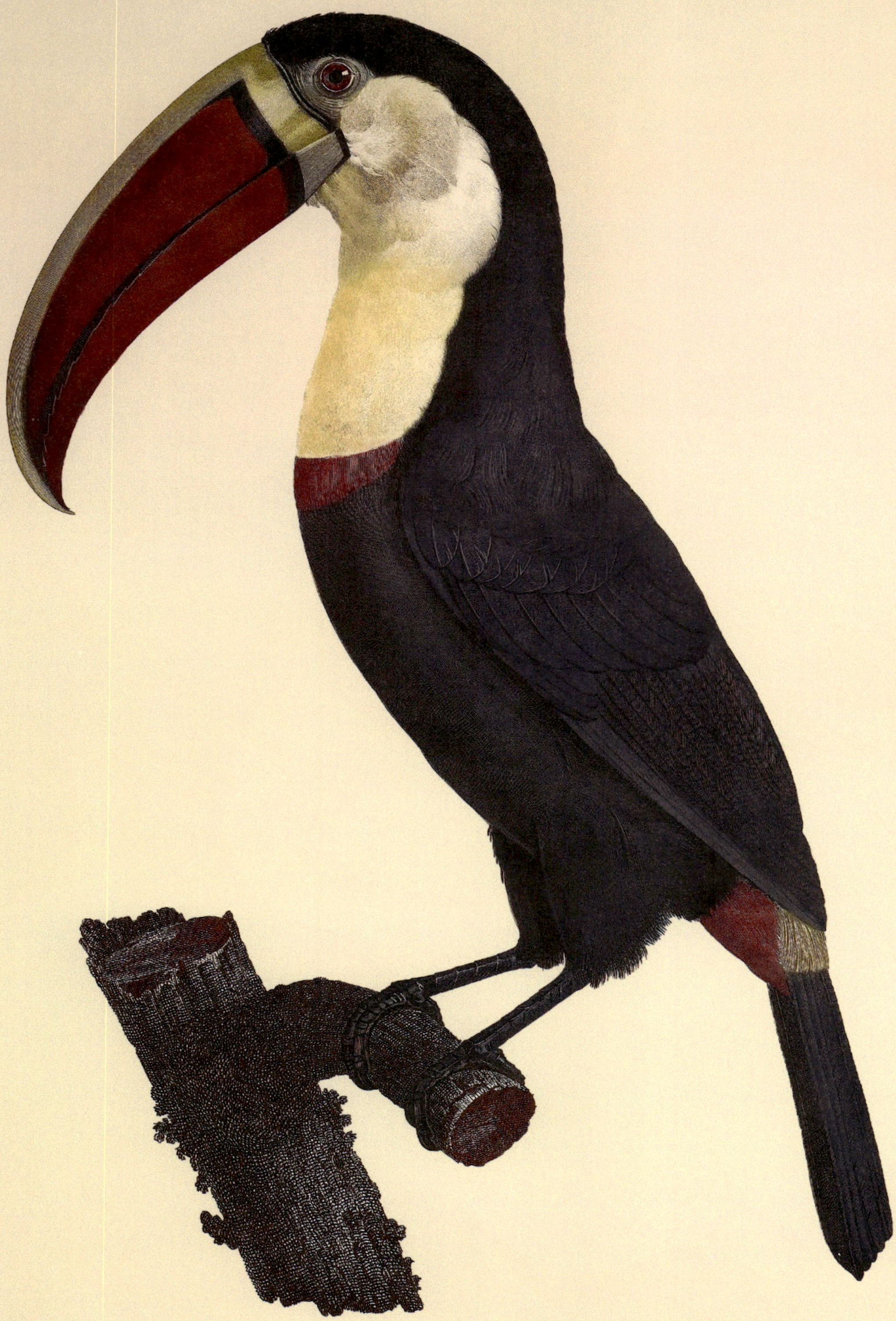

WHITE-THROATED TOUCAN
Ramphastos tucanus

Jacques Barraband, *Histoire naturelle des oiseaux de paradis et des rolliers*, 1806

Toucans are characterized by their large bills. These are lighter than they look and are mainly used to pluck fruit. The length of the bill is also the ideal size for stealing eggs from the nests of other birds. In some species, the bill can be up to half the length of the bird itself. This white-throated toucan appears in François Levaillant's final work *Histoire naturelle des oiseaux de paradis et des rolliers* – a stunning two-volume publication devoted to birds-of-paradise, rollers, toucans and barbus.

Edward Lear

1812–1888

Best known for writing nonsense verse and popularizing limericks, Edward Lear was also a gifted draftsman who innovated in ornithology by working in lithography and drawing from living specimens at the London Zoo.

A poet, a painter, an illustrator, a musician and an author, Edward Lear was a prolific Victorian polymath who is best known today for his absurdist wit, whimsical verse and for popularizing the limerick. Yet, having spent his early years as an ornithological illustrator, Lear is also revered in science circles for his natural history work. While his illustrated book of verse, *The Book of Nonsense* (1846), might be his best-remembered creation, it is his monograph devoted to one bird family, the *Psittacidae*, that is his most important.

The twentieth child of a bankrupt and widowed businessman, Lear was largely raised by his sister Ann, who was twenty-one years his senior. Shy, bespectacled, asthmatic and epileptic, the young Lear most likely developed an early passion for art through his sister, who – like many educated ladies of the day – was a competent watercolourist. With family finances in turmoil, Lear was forced to earn his living from a young age. He did so by painting decorations on ladies' fans and also producing visual records of diseases and afflictions for doctors and hospitals. Eventually, his talents took him to the Zoological Society where, aged sixteen, he was employed as an ornithological draughtsman. There he worked as an assistant for Prideaux John Selby, a connection that opened many opportunities for him. Two years later, aged a mere eighteen, he set about producing his own illustrated monograph.

Illustrations of the Family of Psittacidae, or Parrots (1832) was the first monograph produced in England to focus on a single family of birds. The process of publishing, illustrating and distributing such a work on his own was a huge undertaking at such a young age, not to mention a huge expense. To produce this work, he chose to adopt lithography as the medium for his illustrations. At this point in time, producing illustrations on lithographic stone was relatively new and only a few lithographic works had been produced before, most notably William Swainson's *Zoological Illustrations* (1820–3). Previously, an illustrator would have needed to hire an engraver to reproduce their drawing on a plate, but if Lear produced his works onto lithographic stone, he would not risk the possibility of his images being compromised. Perhaps more importantly, he could cut out the added expense of paying for a middleman and make his production costs far more affordable.

In many ways, this risk paid off. Lear was an expert draftsman and his illustrations are among some of the best parrot illustrations ever produced. This was also partly down to his ability to capture the personalities and individuality of each of these birds. Working from live parrots observed at London's Zoological Gardens, he was the first major bird artist to move away from simply drawing skins to taking the time to draw directly from life. His illustrations capture a sense of movement, while his use of colour demonstrates the splendour of many of the birds found within this family.

Released in parts between 1830 and 1832, the book contained forty-two hand-coloured lithographs and was printed at Imperial folio size (59 × 42cm). This made it considerably smaller than – say – Audubon's work, but it suited the size of the birds and allowed Lear to print many of them at lifesize. The work caught the attention of John Gould, who at the time was the curator and preserver at the museum of the Zoological Society of London. Almost eight years Lear's senior, he saw the commercial possibilities in the lithographic method and began producing a lithographic work of his own based on a large collection of Himalayan bird specimens that had arrived at the Zoological Society's museum. Gould was not a great artist, but he was a shrewd businessman. His wife Elizabeth, much like Lear's sister, was a proficient watercolourist, and so Gould hired Lear to teach her his methods. He then managed to draw on his contacts at the society to enlist some 298 subscribers for his book: almost twice as many as Lear had for his monograph. By the time Lear issued a second instalment of his parrot plates, the first section of Gould's *A Century of Birds from the Himalaya Mountains* was published. Lear's work may have been a great artistic success, but as a financial enterprise, it was a failure.

Unlike Gould, Lear was a relatively inexperienced businessman. He had printed 175 copies, which had not been enough to even cover his own expenses. To make matters worse, he sold only 125 of these and payments were often late or lacking. Finding himself in financial difficulty, Gould stepped in and bought the remaining prints, but he also purchased the copyright, supposedly aiming to finish the work himself. The work was never completed, and it is unclear if Gould's act was a means to support his employee or simply as a way to eradicate the competition.

Lear worked with Gould on two further projects after this. In *The Birds of Europe* (1832–7) Lear illustrated some of the more spectacular birds – the buzzards, the pelicans, the owls – and then worked on illustrating Gould's monograph of the toucan (1833–5), a work based on a single family of birds that was not vastly dissimilar to Lear's own monograph of parrots. Eventually, Lear abandoned his employment with Gould by citing his deteriorating eyesight. In 1837, aged twenty-five, he abandoned the world of birds and set off for Italy to start a new career as a landscape painter. His ornithology days may have been behind him, but a life of nonsense, music and adventure was just about to begin.

LEAR'S MACAW
Anodorhynchus leari

Edward Lear, *Illustrations of the Family of Psittacidae*, 1832

Initially misidentified as a Hyacinthine Macaw (*Anodorhynchus hyacinthinus*), this striking indigo bird was fittingly named Lear's macaw (*Anodorhynchus leari*) in 1856 in tribute to the artist's love for parrots.

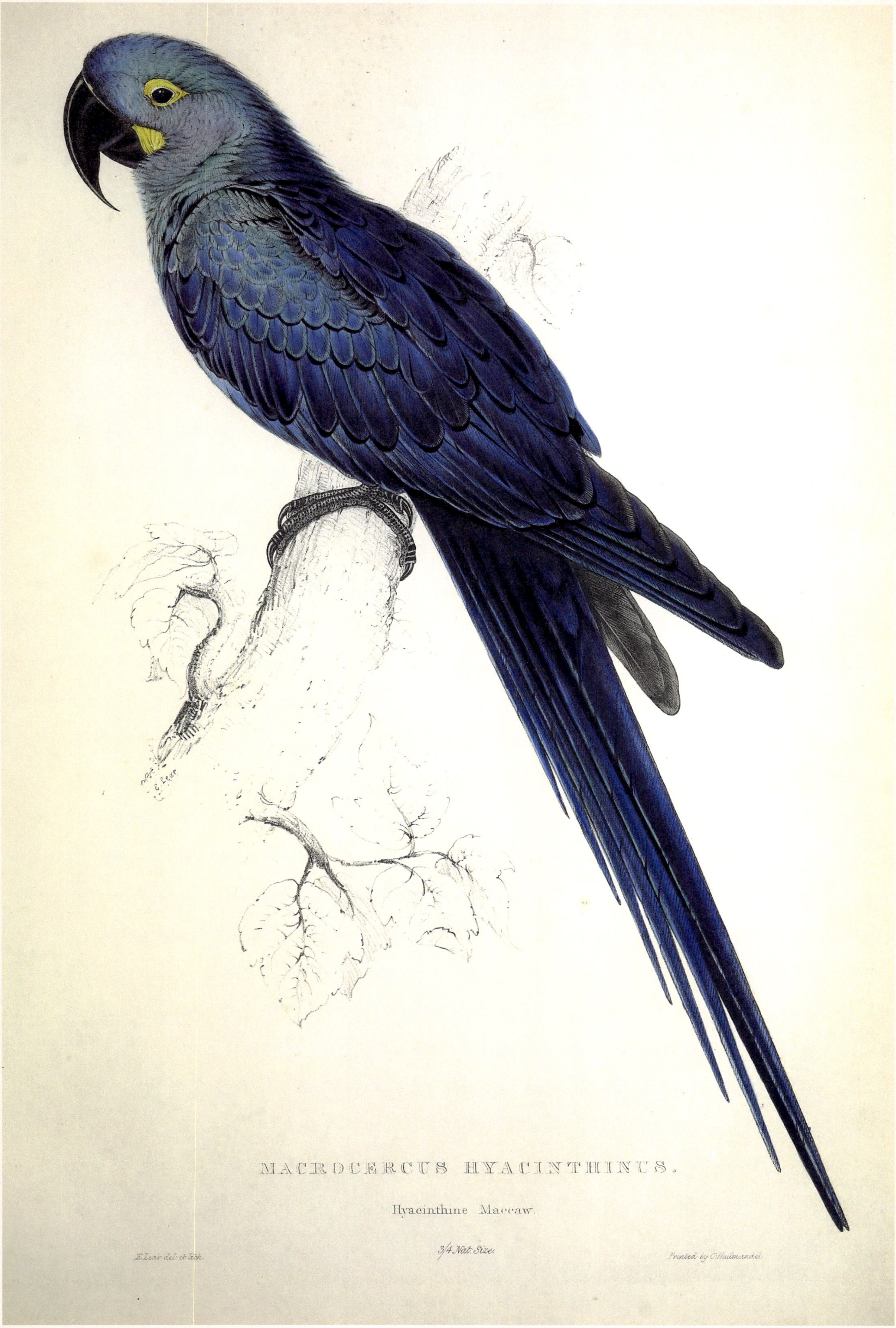

MACROCERCUS HYACINTHINUS.
Hyacinthine Maccaw.
E Lear del: et lith.
3/4 Nat: Size.
Printed by C Hullmandel.

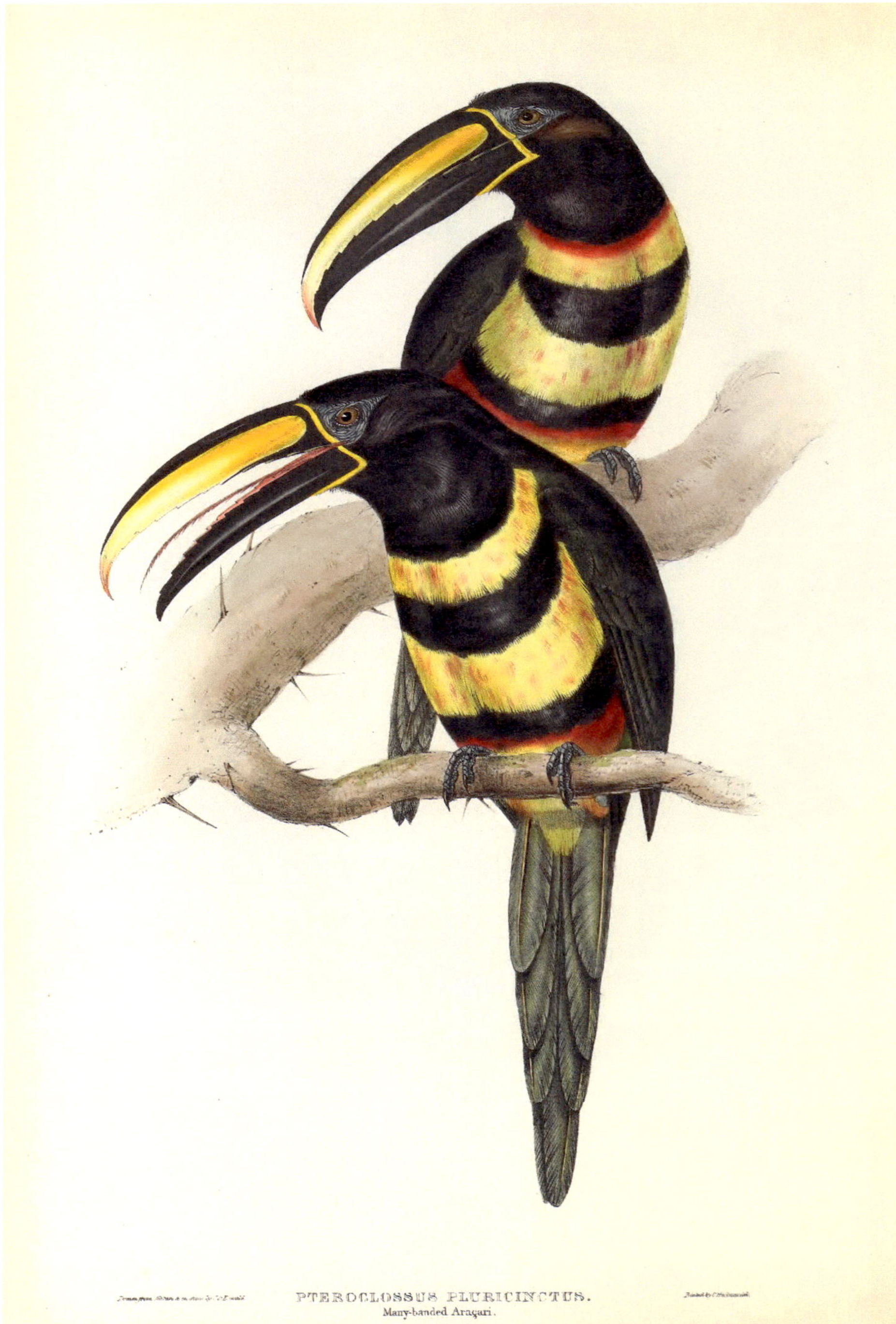

MANY-BANDED ARAÇARI
Pteroglossus pluricinctus

John Gould & Elizabeth Gould,
A Monograph of the Ramphastidae,
1833–5

Gould's monograph of the toucan was the first attempt to document the whole toucan family through descriptions and pictures. These birds are found in southern Mexico, Central America, parts of South America and the West Indies. The term 'toucan' is used as the common name for a number of species in this family, to which araçaris and toucanets also belong. The araçaris seen here are known as many-banded araçari on account of their distinctively banded underpart. This bird is found throughout Brazil, Colombia, Ecuador, Peru and Venezuela.

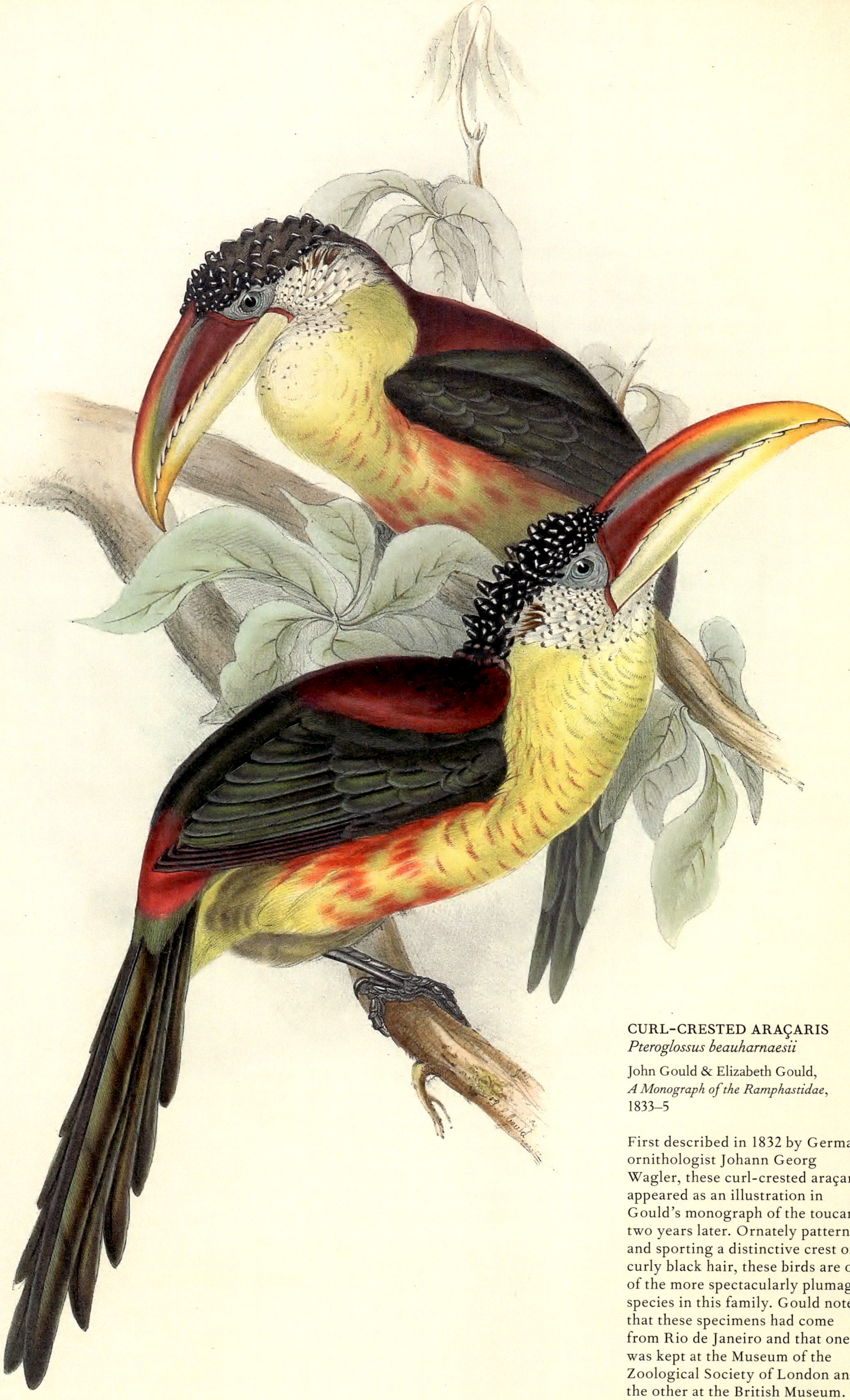

CURL-CRESTED ARAÇARIS
Pteroglossus beauharnaesii

John Gould & Elizabeth Gould,
A Monograph of the Ramphastidae,
1833–5

First described in 1832 by German ornithologist Johann Georg Wagler, these curl-crested araçaris appeared as an illustration in Gould's monograph of the toucan two years later. Ornately patterned and sporting a distinctive crest of curly black hair, these birds are one of the more spectacularly plumaged species in this family. Gould notes that these specimens had come from Rio de Janeiro and that one was kept at the Museum of the Zoological Society of London and the other at the British Museum.

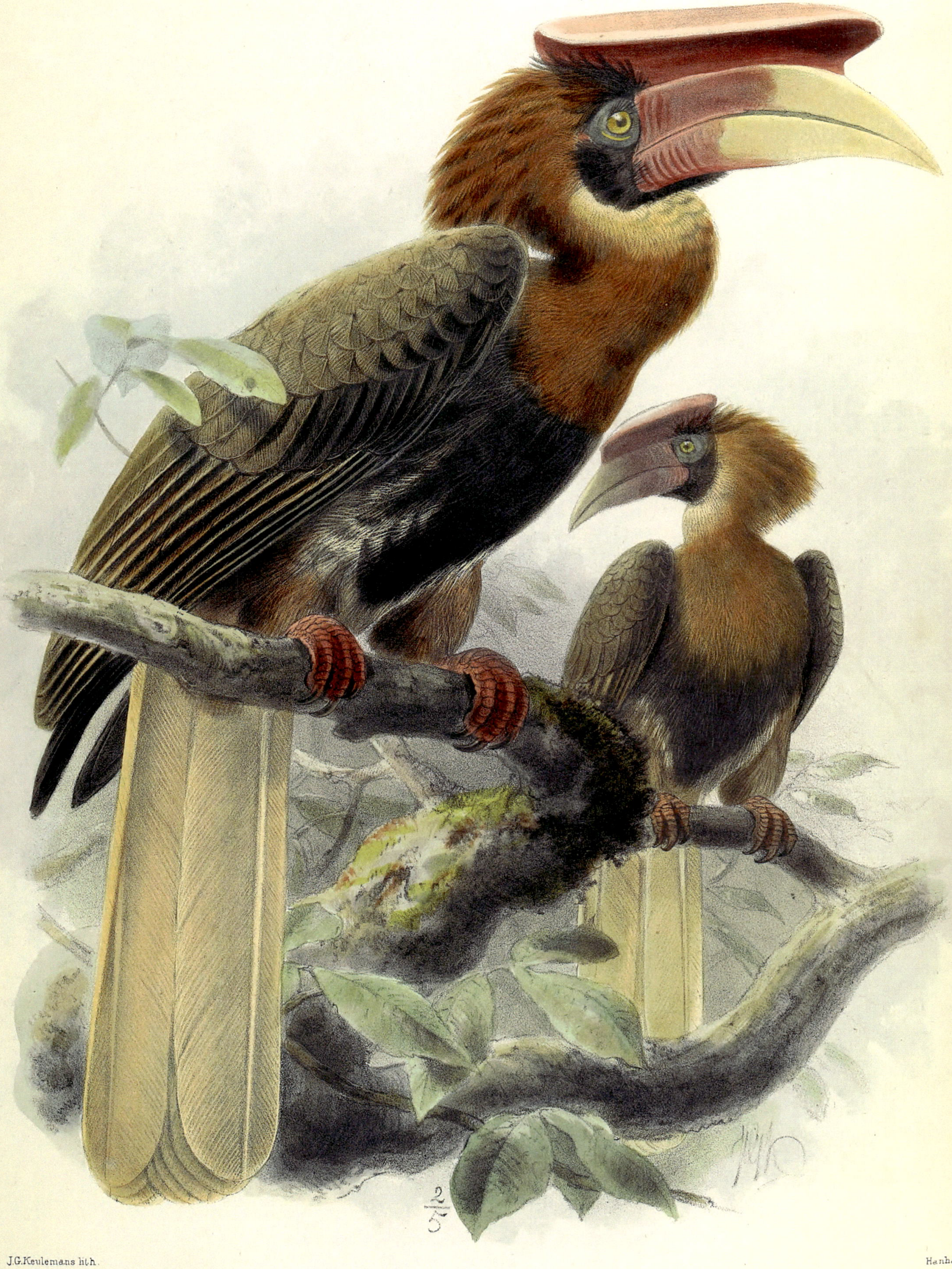

J.G.Keulemans lith. Hanhart imp.

HYDROCORAX MINDANENSIS.

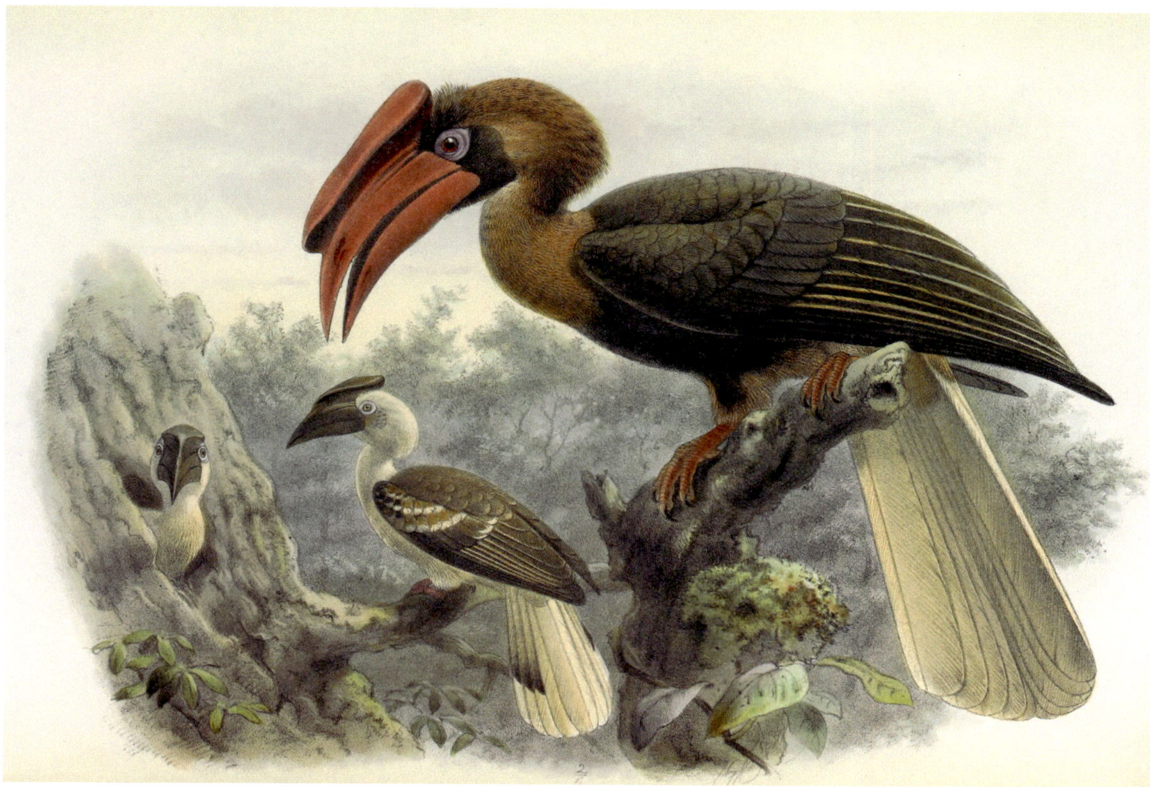

OPPOSITE:

SOUTHERN RUFOUS HORNBILLS
Buceros mindanensis

John Gerrard Keulemans, *A Monograph of the Bucerotidæ, or Family of the Hornbills*, 1877–82

Found in tropical and subtropical parts of Africa, Asia and the East Indies, the hornbill is characterized by its long, down-curved bill which is typically brightly coloured. Many hornbills also sport a remarkable decorative casque on the upper part of their mandible. Despite their fascinating appearance, knowledge about the hornbill remains relatively scarce and many birds in this family are noted for their elusive nature. In 1882 the American ornithologist Daniel Giraud Elliot completed the publication of a monograph on the family.

NORTHERN RUFOUS HORNBILL
Buceros hydrocorax

John Gerrard Keulemans, *A Monograph of the Bucerotidæ, or Family of the Hornbills*, 1877–82

The rufous hornbill is endemic to the Philippines and it is also simply known as the Philippine hornbill. The illustration here is by Keulemans and it shows a northern rufous hornbill with two of its young. Hornbills are monogamous breeders and will typically nest in the cavities of trees. The female lays her clutch in the nest and remains there for the majority of the nesting period. During this time, the male provides all of the food for the female as well as for its young.

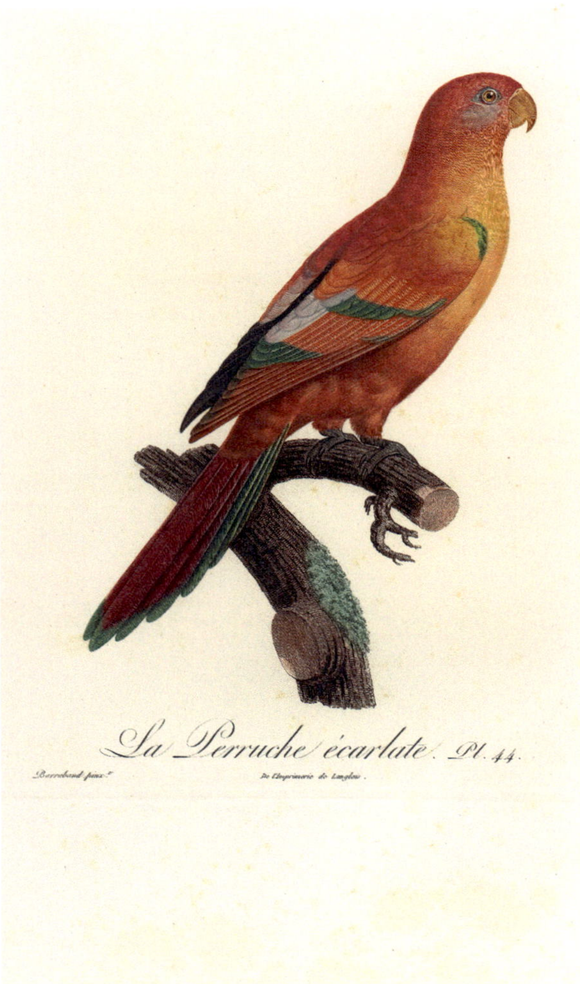

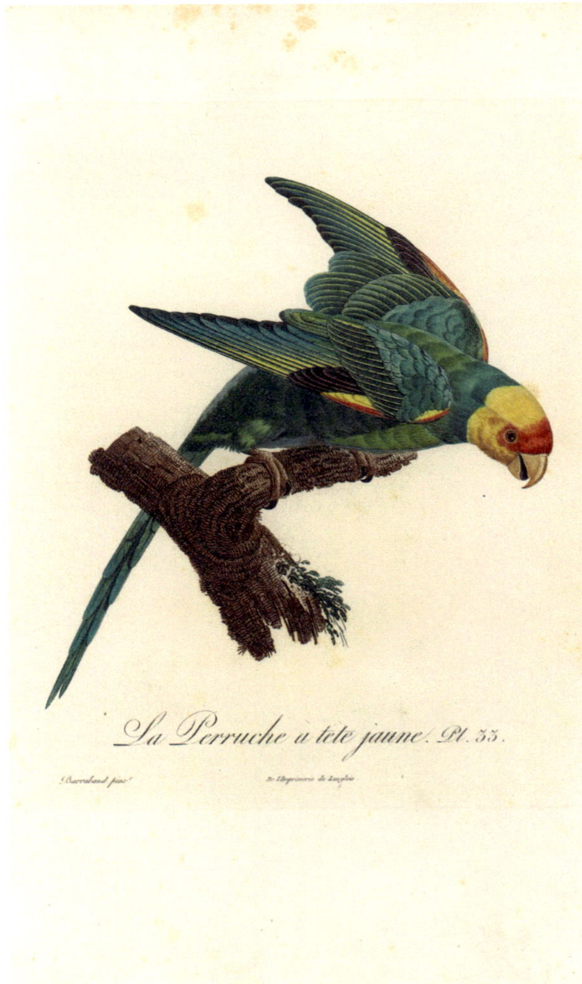

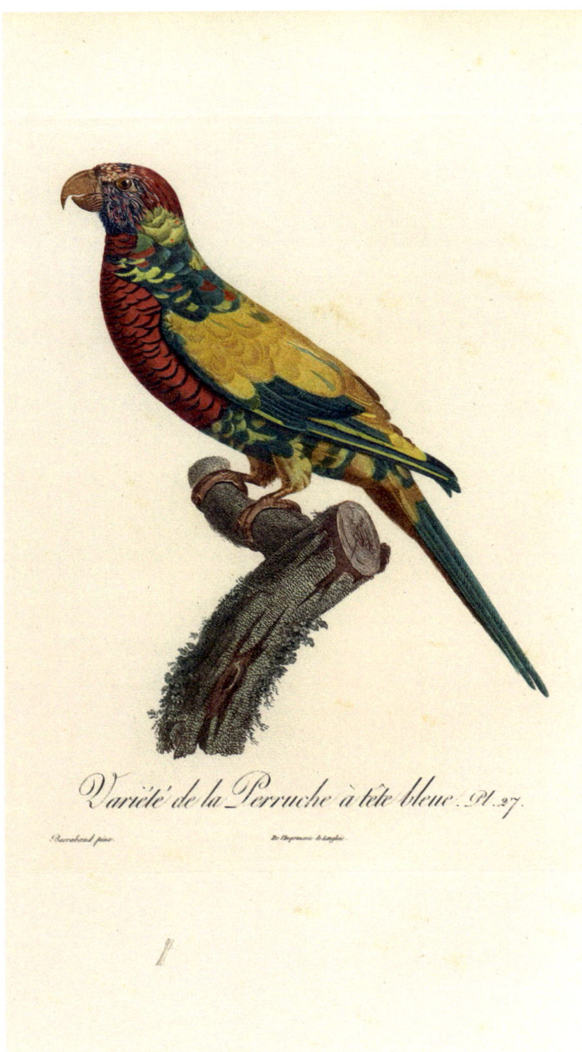

TOP LEFT:
PROBABLY RED LORY
Eos bornea

Jacques Barraband, *Histoire naturelle des perroquets*, 1801–5

BOTTOM LEFT:
COCONUT LORIKEET
Trichoglossus haematodus

Jacques Barraband, *Histoire naturelle des perroquets*, 1801–5

TOP RIGHT:
CAROLINA PARAKEETS
Conuropsis carolinensis

Jacques Barraband, *Histoire naturelle des perroquets*, 1801–5

OPPOSITE:
CAROLINA PARAKEET
Conuropsis carolinensis

John James Audubon, *The Birds of America*, 1827–38

The Carolina parakeet was the only native parrot species in the United States. It is seen here in a work by Audubon. Even he commented on their rapid decline of this species, noting in 1832 that 'where twenty-five years ago they were plentiful, scarcely any are now to be seen'. The last Carolina parakeet died in the early 1900s. At the time, there had been no scientific studies of this bird, so our only knowledge of it comes from museum specimens and accounts such as Audubon's.

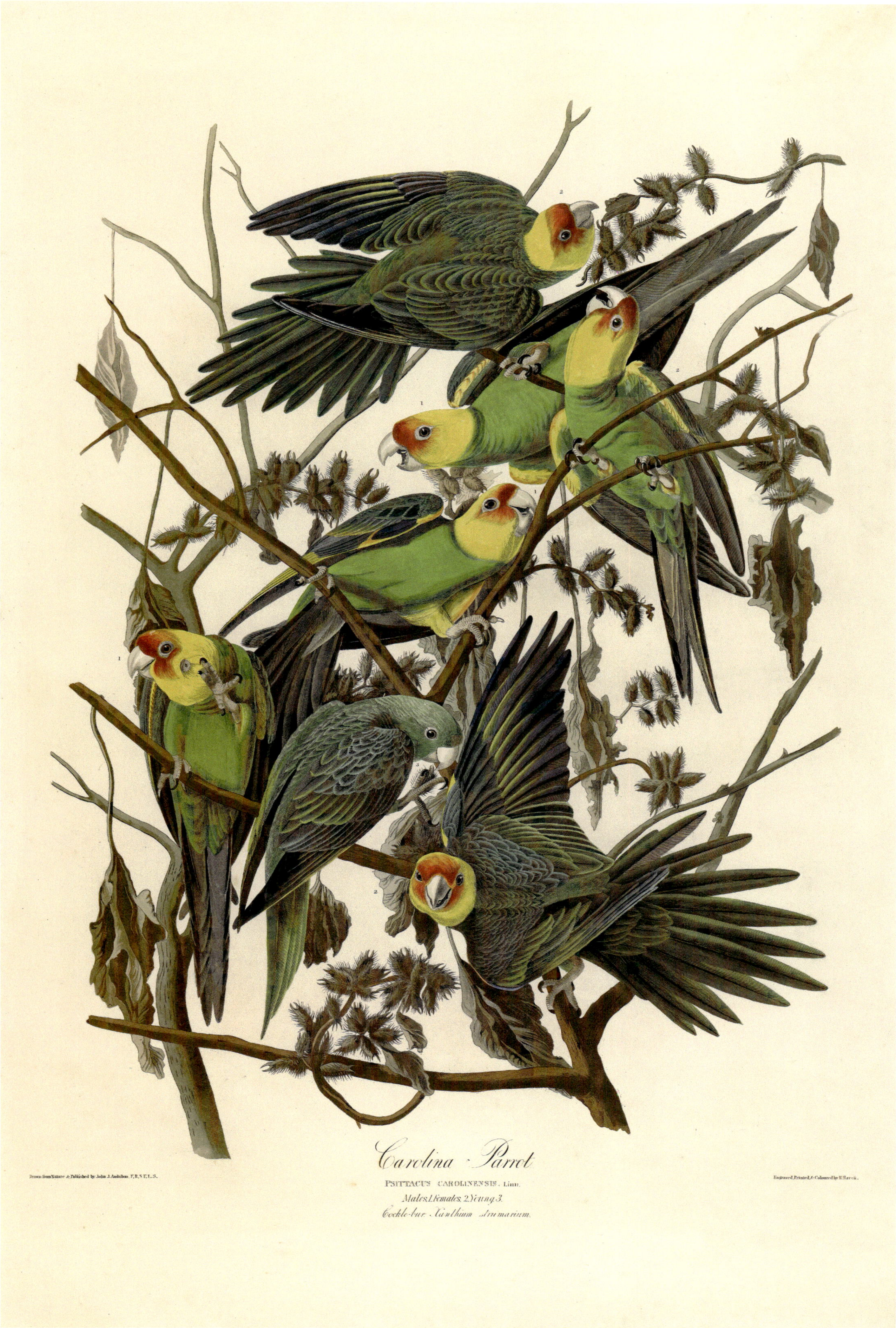
Carolina Parrot
PSITTACUS CAROLINENSIS. Linn.
Males 1. Females, 2. Young 3.
Cockle-bur. Xanthium strumarium.
Drawn from Nature & Published by John J. Audubon, F.R.S. F.L.S.
Engraved, Printed, & Coloured by R. Havell.

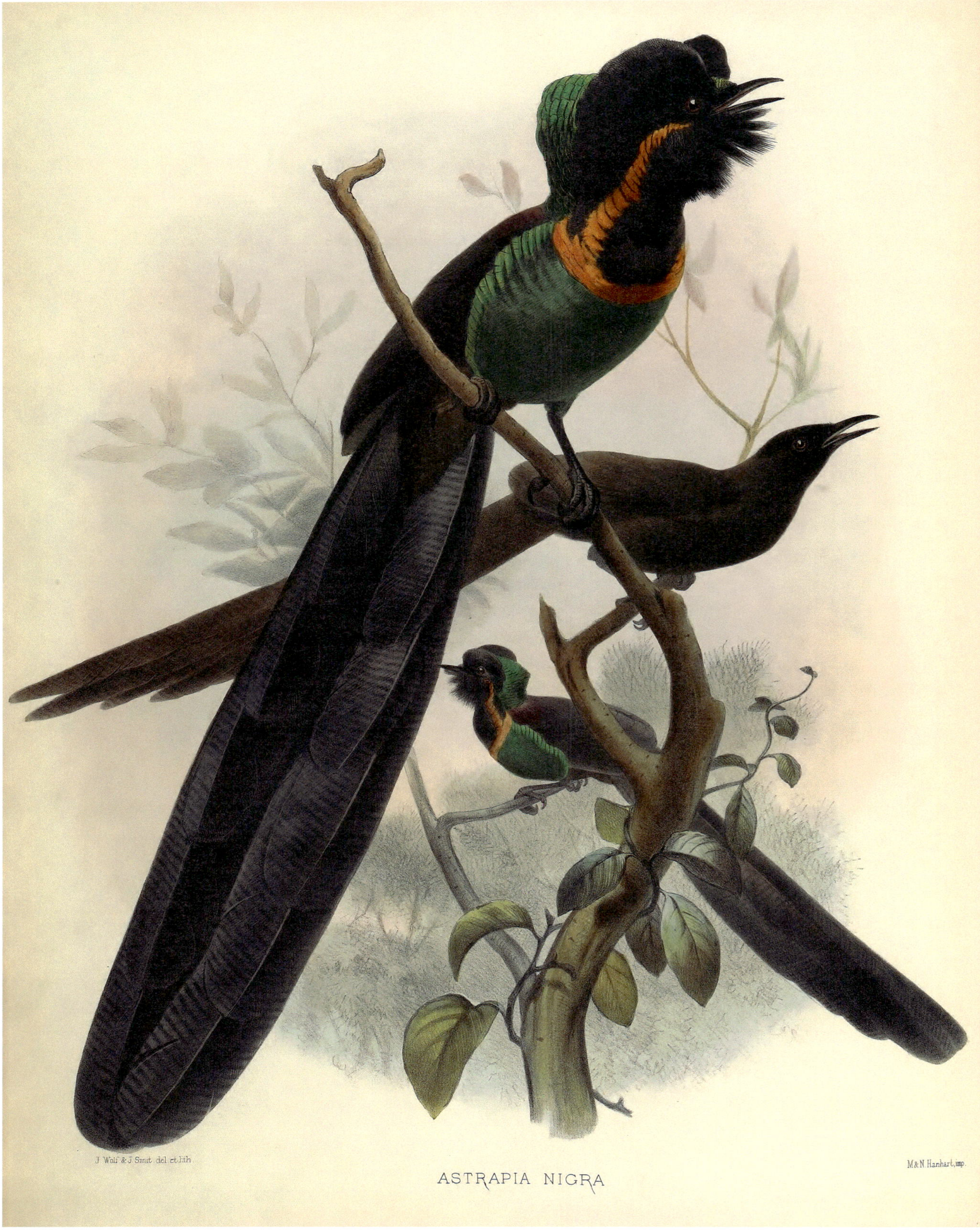

ARFAK ASTRAPIA
Astrapia nigra

Joseph Wolf, *A Monograph of the Paradiseidae*, 1873

The arfak astrapia is one of the more dramatic examples of sexual dimorphism. Found in the cloud forests of the Bird's Head Peninsula on the northwest portion of the island of New Guinea, the female is predominantly a drab blackish-brown, while the male boasts a far more impressive plumage, seen here in an illustration from Elliot's 1873 monograph of the birds-of-paradise. We can see the bird's greenish-yellow cape as well as its impressive tail, which can be almost two times the length of its body.

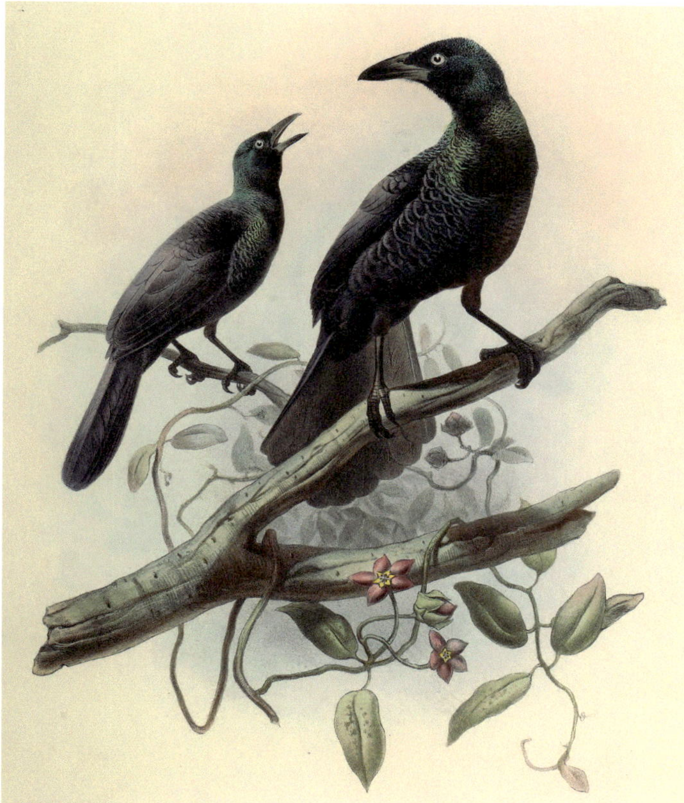

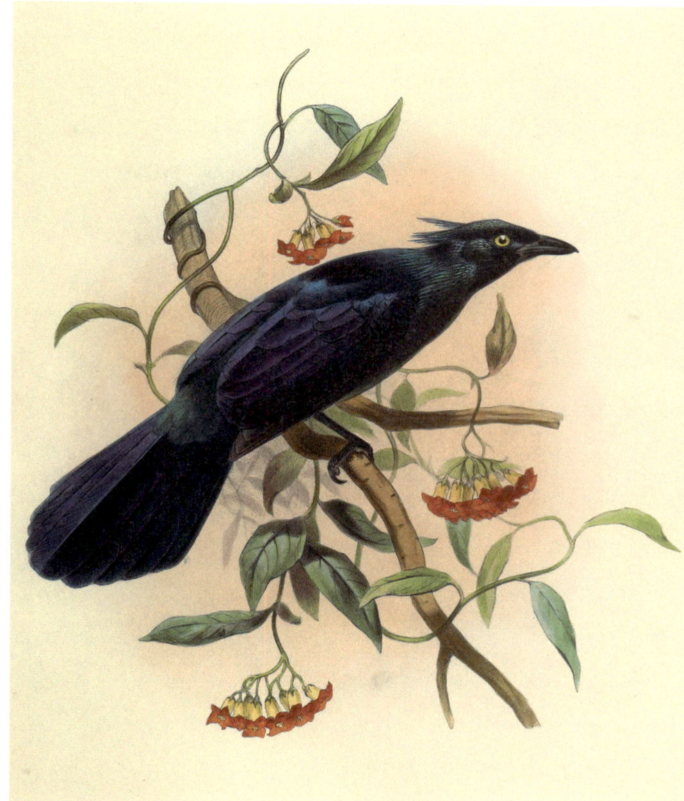

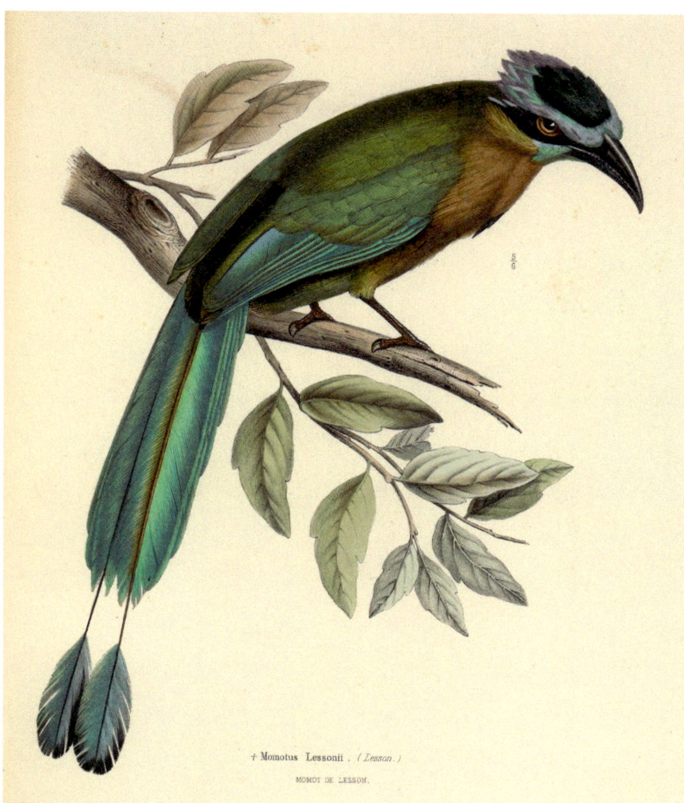

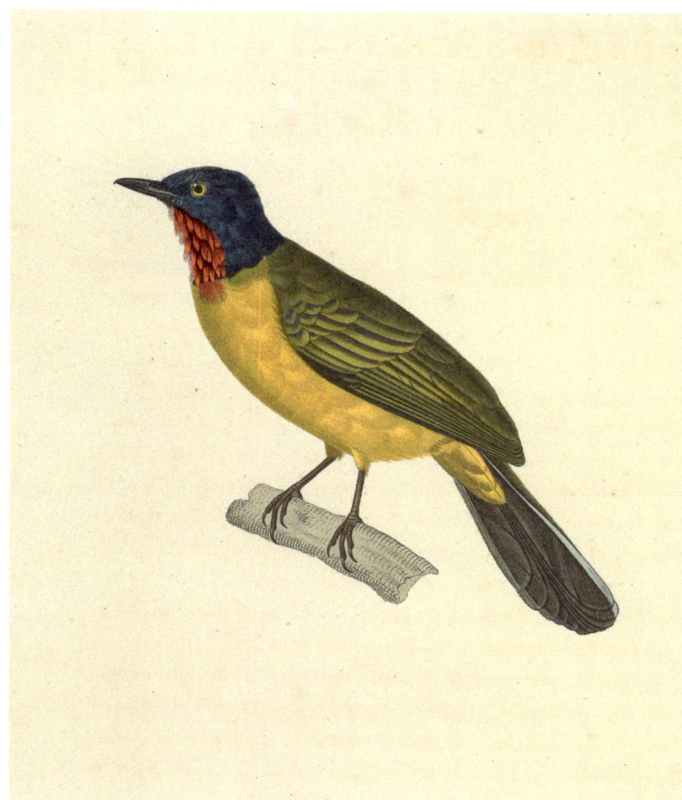

TOP LEFT:

CRINKLE-COLLARED MANUCODE
Manucodia chalybatus

Joseph Wolf, *A Monograph of the Paradiseidae*, 1873

BOTTOM LEFT:

LESSON'S MOTMOT
Momotus lessonii

Paul Louis Oudart, *Iconographie ornithologique*, 1845–9

TOP RIGHT:

TRUMPET MANUCODE
Phonygammus keraudrenii

Joseph Wolf, *A Monograph of the Paradiseidae*, 1873

BOTTOM RIGHT:

RUBY-THROATED BULBUL
Pycnonotus dispar

Jean-Gabriel Prêtre, *Nouveau recueil de planches coloriées d'oiseaux*, 1820–39

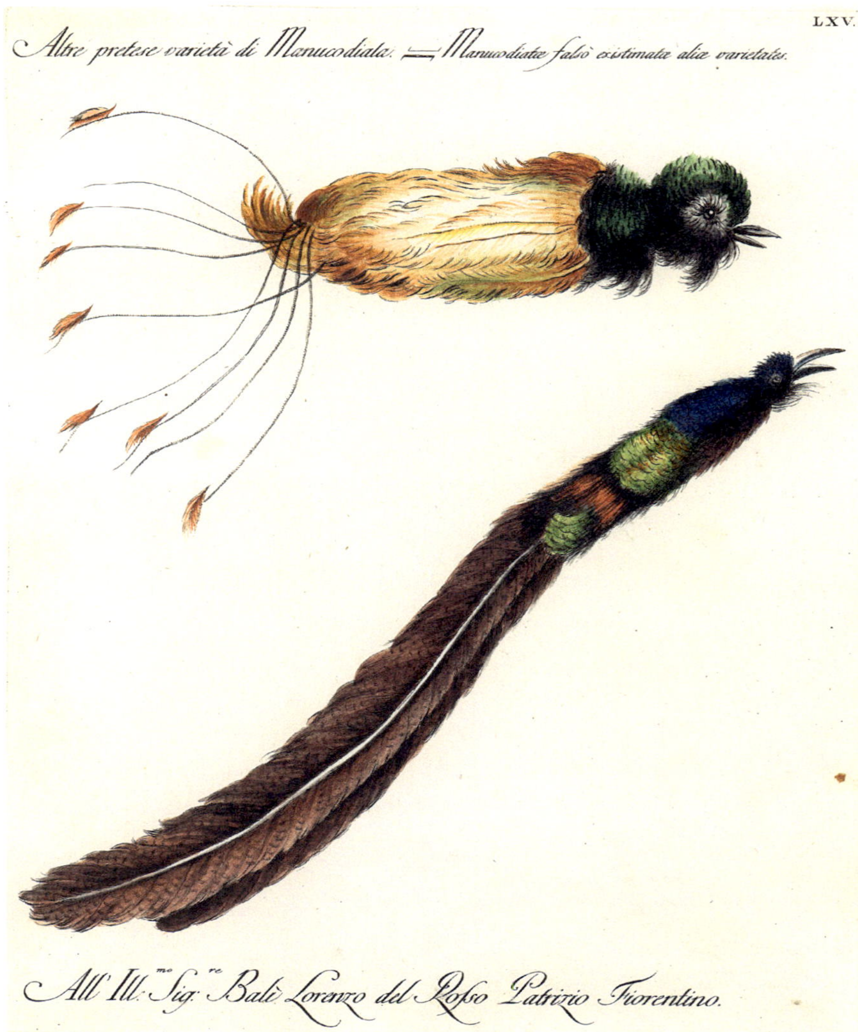

BIRD-OF-PARADISE

Lorenzo Lorenzi & Violante Vanni, *Ornithologia Methodice Digesta*, 1767–76

This curious rendering from the 1760s shows the tail feathers of two birds-of-paradise. Bizarrely, the illustrator has chosen to attach a head, beak and body to these tails but has excluded their wings. This type of work would have helped further the belief that these birds were, in fact, mythical creatures that had descended from heaven and could float through the air without wings.

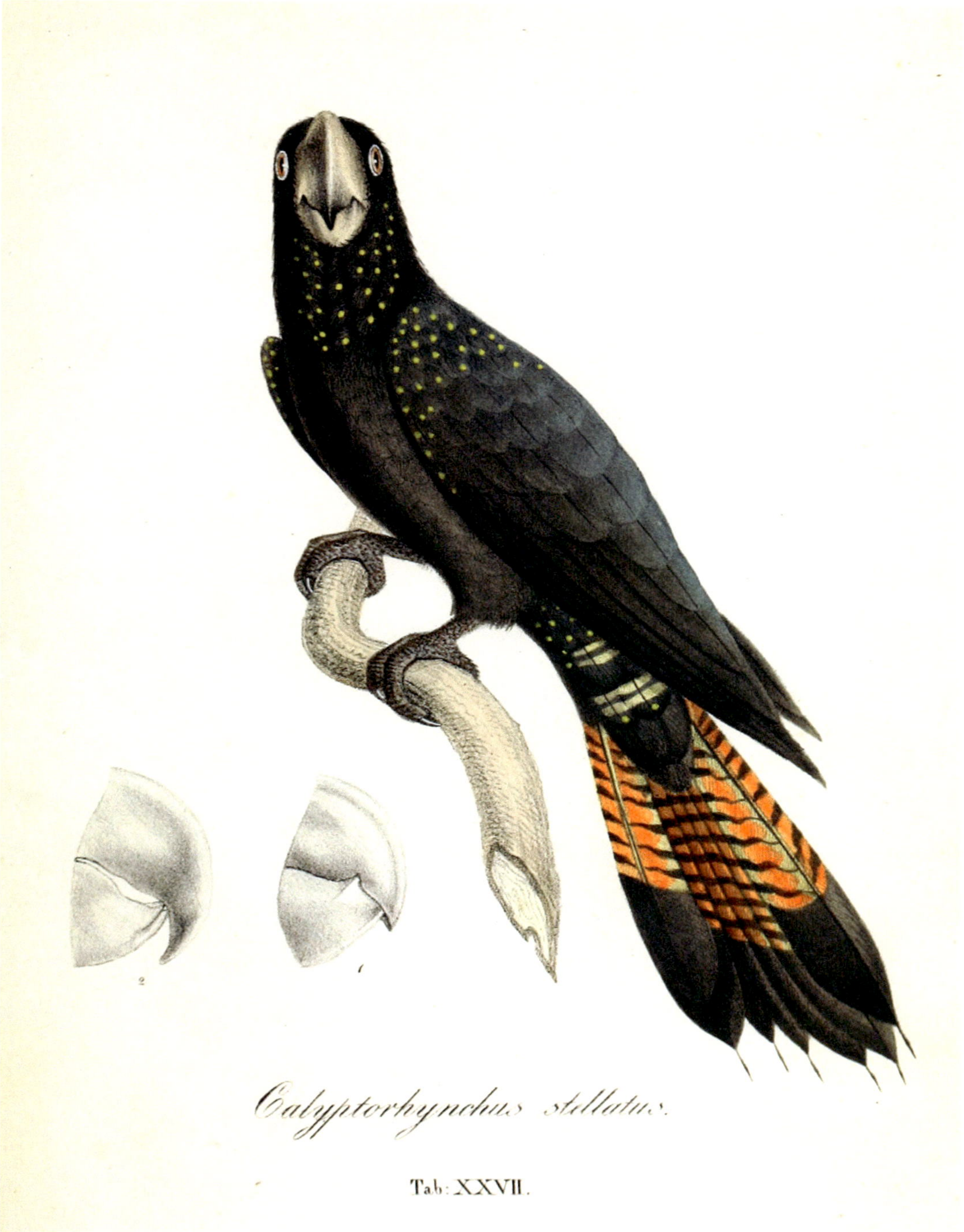

GLOSSY BLACK-COCKATOO
Calyptorhynchus lathami

Uncredited, *Monographia Psittacorum*, 1835

In the early 1800s, Johann Georg Wagler's *Monographia Psittacorum* was one of the world's most in-depth accounts of the parrot family. Born in Germany, Wagler travelled to Brazil as an assistant to the famed German biologist Johann Baptist von Spix. The specimens they collected went on to form the basis of the Bavarian State Collection of Zoology. Wagler worked extensively from this collection, and despite his connection to Brazil, this specific bird is native to parts of eastern Australia.

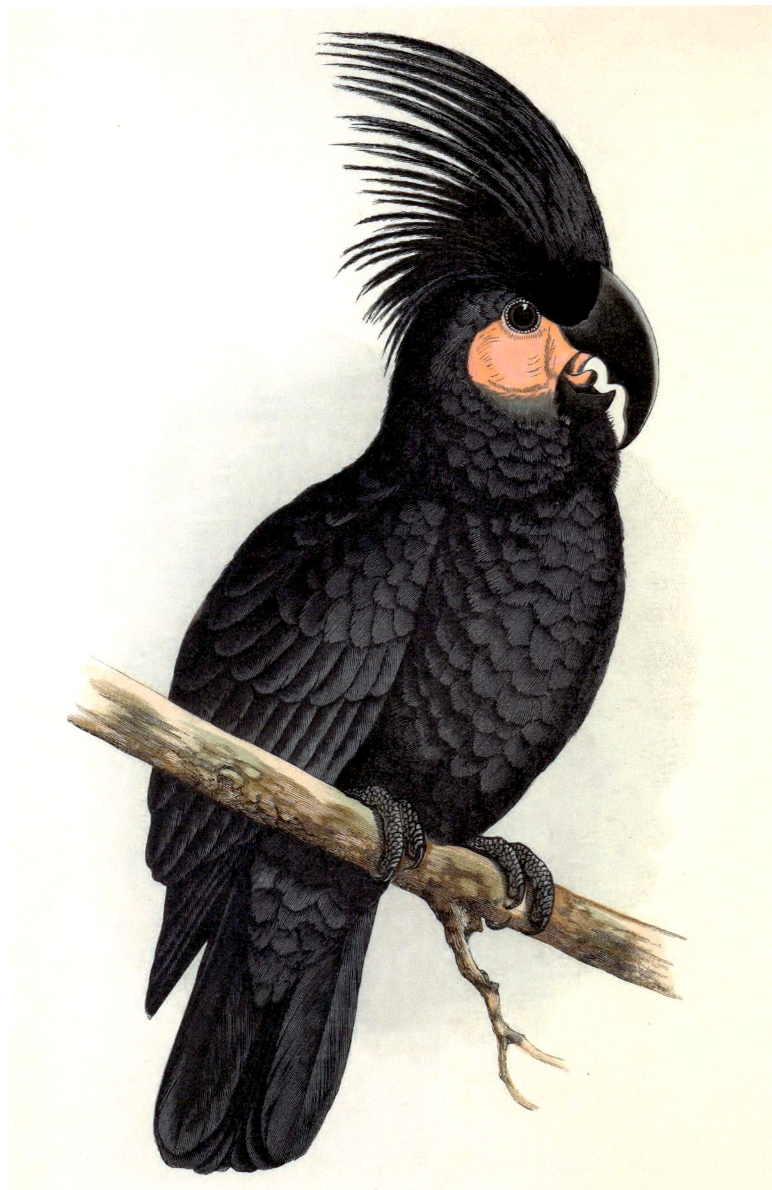

PALM COCKATOO
Probosciger aterrimus

Alexander Francis Lydon, *Parrots in Captivity*, 1884–8

Published in the 1880s, *Parrots in Captivity* was arguably the first guide on how to keep and raise parrots as pets. Containing notes on several species, it includes information on how to feed and rear each of these birds. Written by William Thomas Greene, it features illustrations by Alexander Francis Lydon. This – a black palm cockatoo – is the largest of the Australian cockatoos, and it remains a popular pet to this day, despite its size and assertive personality.

OPPOSITE:
RED-TAILED BLACK COCKATOO
Calyptorhynchus banksii

John Gould & Henry Constantine Richter, *The Birds of Australia*, 1840–8

This bird is sometimes called the banksian cockatoo, a name was given to commemorate English botanist Sir Joseph Banks, who accompanied James Cook on his famed voyage to the Australian coast. Banks was the voyage's official botanist and, on his return, he advocated for the colonization of Australia. Gould was quite pleased that this bird was given Bank's name, noting in his text that it brought him great pleasure that the name had been given to such a 'noble and ornamental bird'.

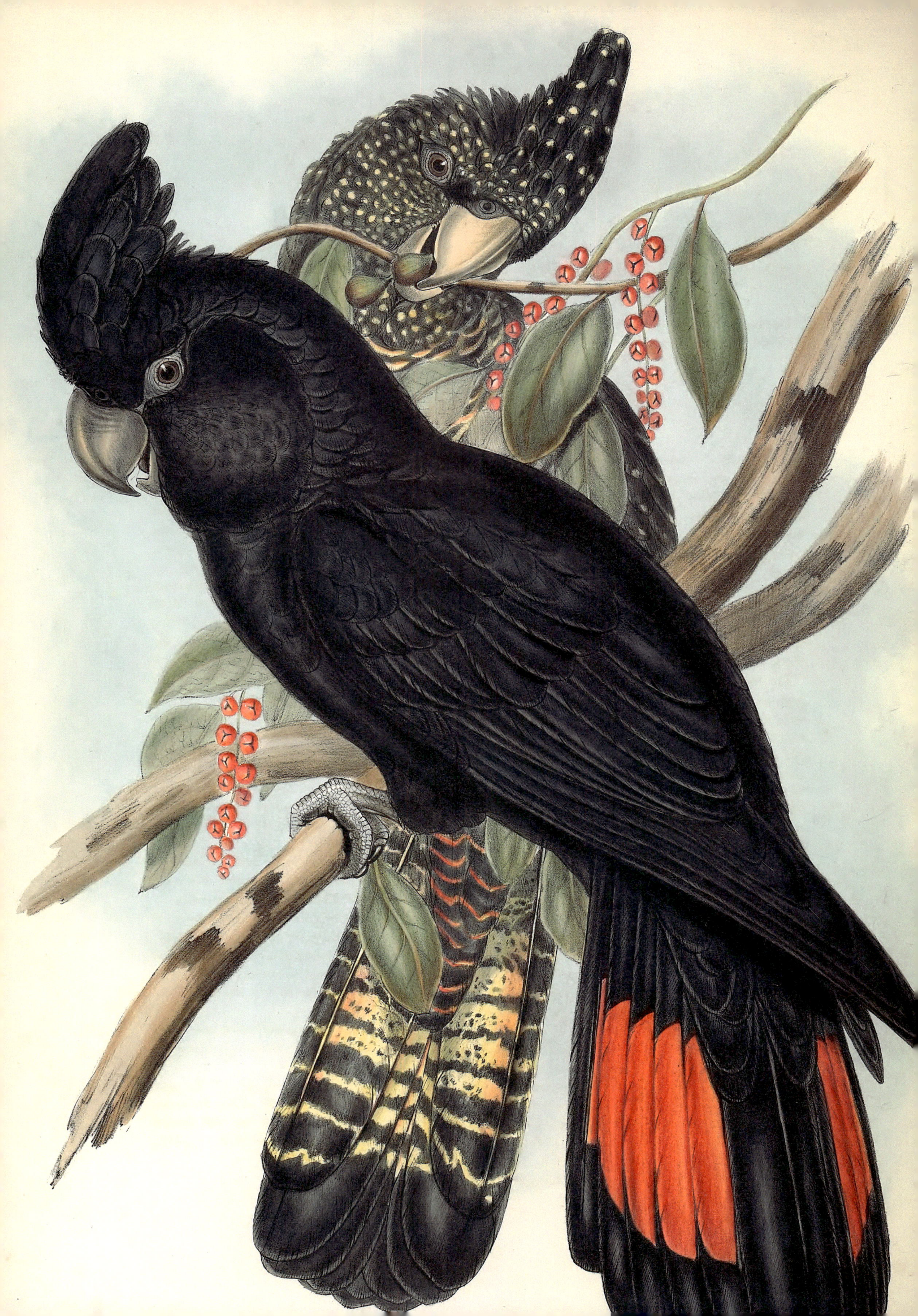

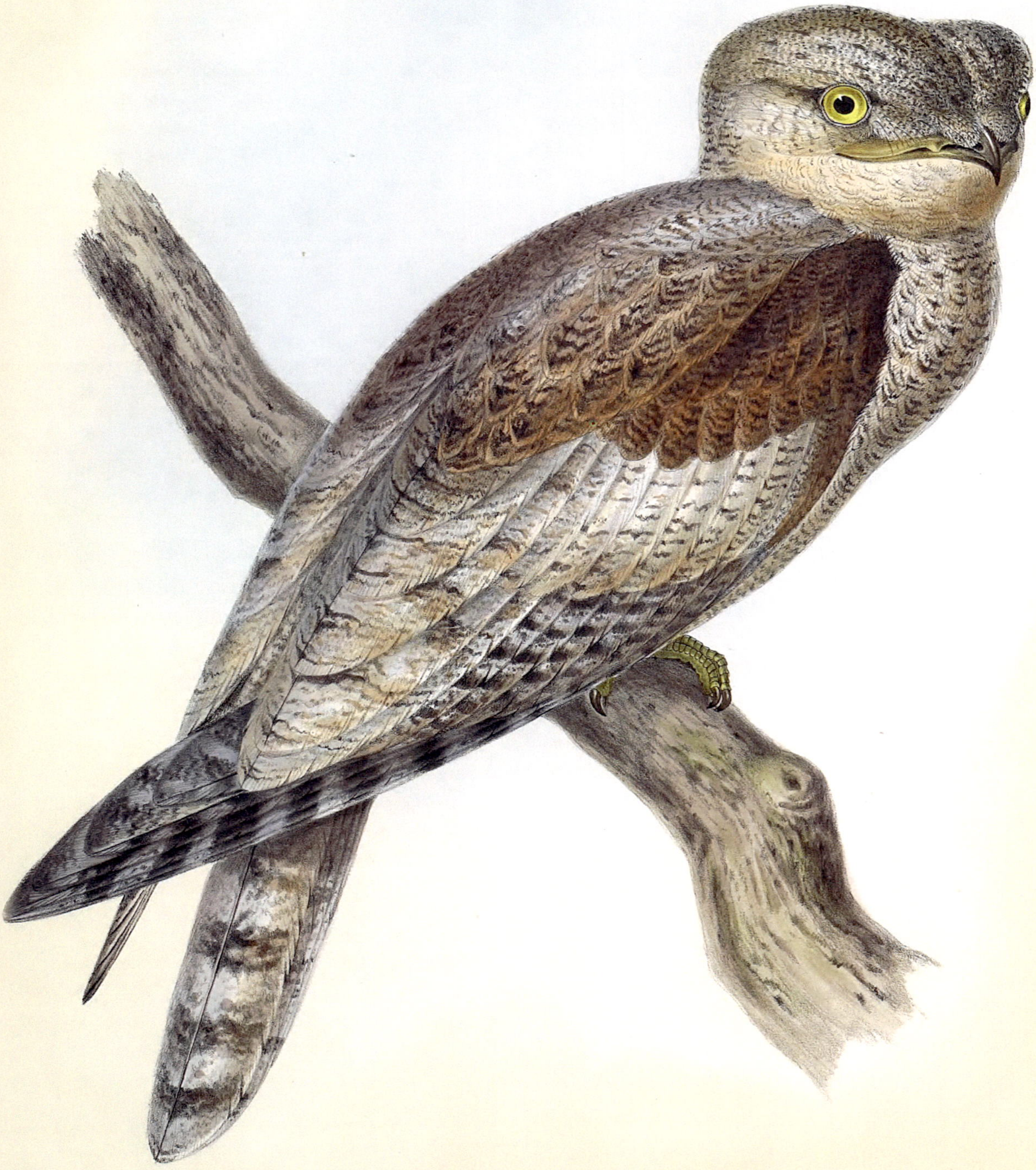

COMMON POTOO
Nyctibius griseus

David William Mitchell,
The Genera of Birds, 1844–9

David William Mitchell was an English zoologist and illustrator who also collected and sold skins and eggs. Employed by George Robert Gray to illustrate his *Genera of Birds*, the workload became too much for the artist, and he enlisted the assistance of Joseph Wolf on the project. The potoo is a nocturnal bird found in Central and South America. Its brown plumage forms excellent camouflage against bark, and in daylight its distinctive bulging eyes are often the only distinguishable feature that can be seen.

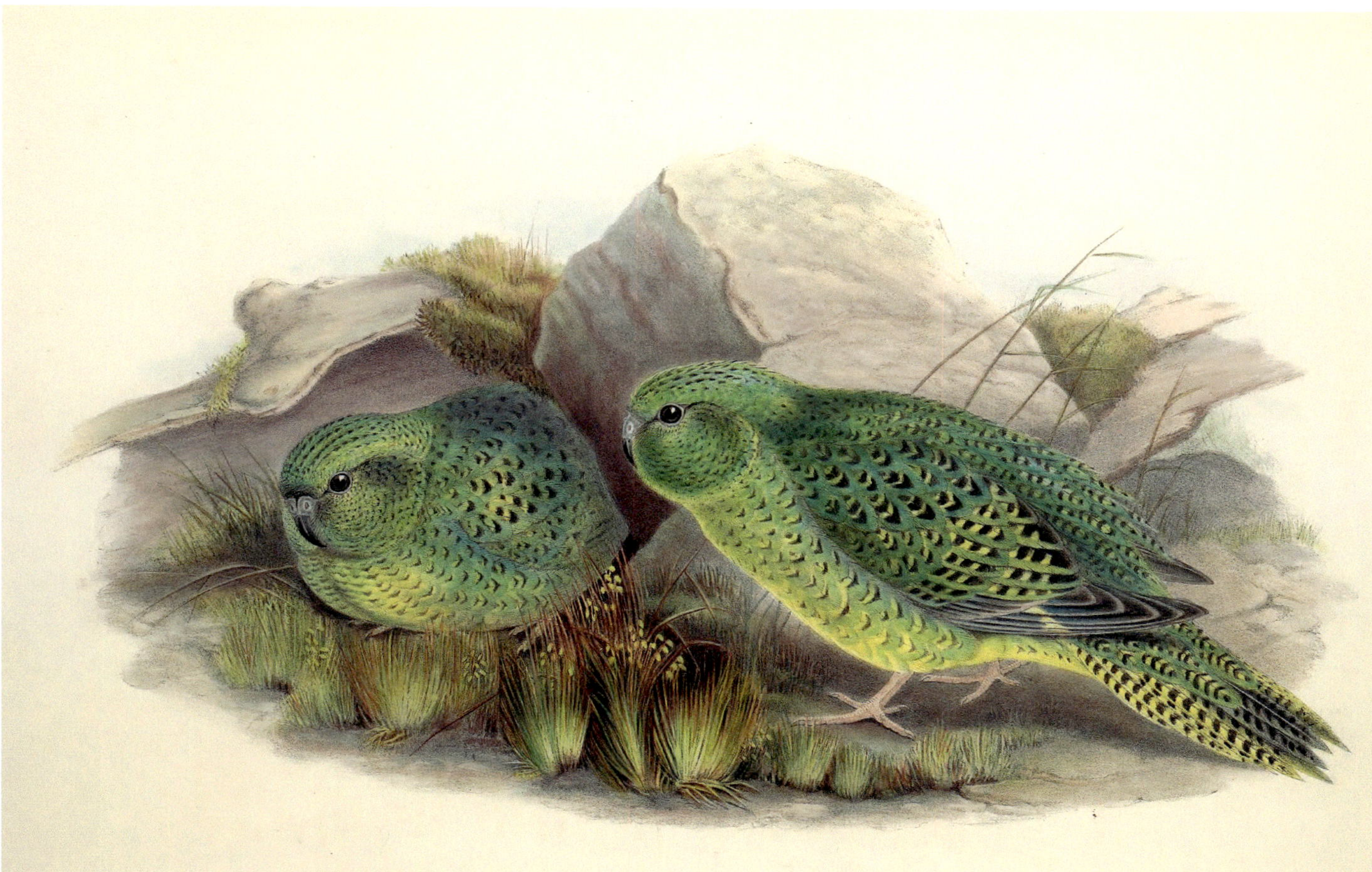

NIGHT PARROTS
Pezoporus occidentalis

John Gould & Henry Constantine Richter, *Supplement to the Birds of Australia*, 1851–69

One of the most elusive birds in the world, there were no confirmed sightings of this species between 1912 and 1979.

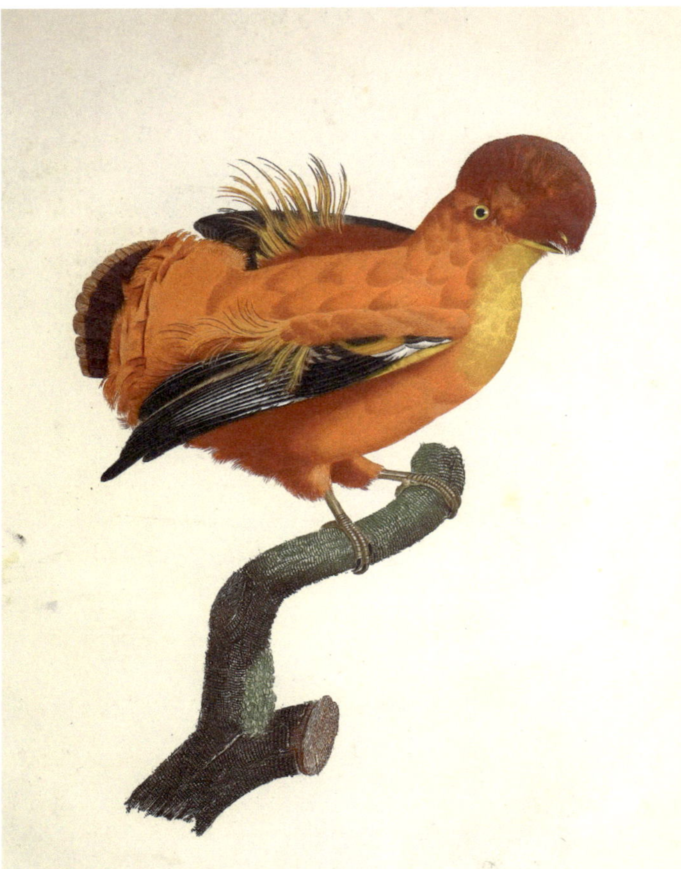

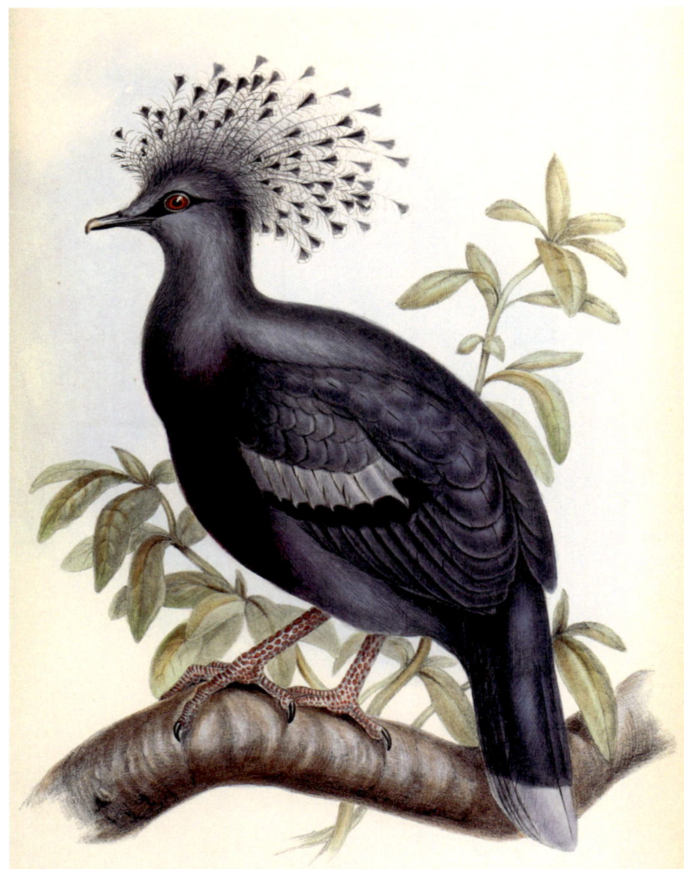

LEFT:

GUIANAN COCK-OF-THE-ROCK
Rupicola rupicola

Jacques Barraband, *Histoire naturelle des oiseaux de paradis et des rolliers*, 1806

Native to South America, the cock-of-the-rock is identified by its brilliant orange plumage and its distinctive half-moon crest.

RIGHT:

CROWNED PIGEON SP.
Goura sp.

David William Mitchell, *The Genera of Birds*, 1844–9

The crowned pigeon is the largest pigeon in the world and sports an impressively elegant crest of feathers on its head.

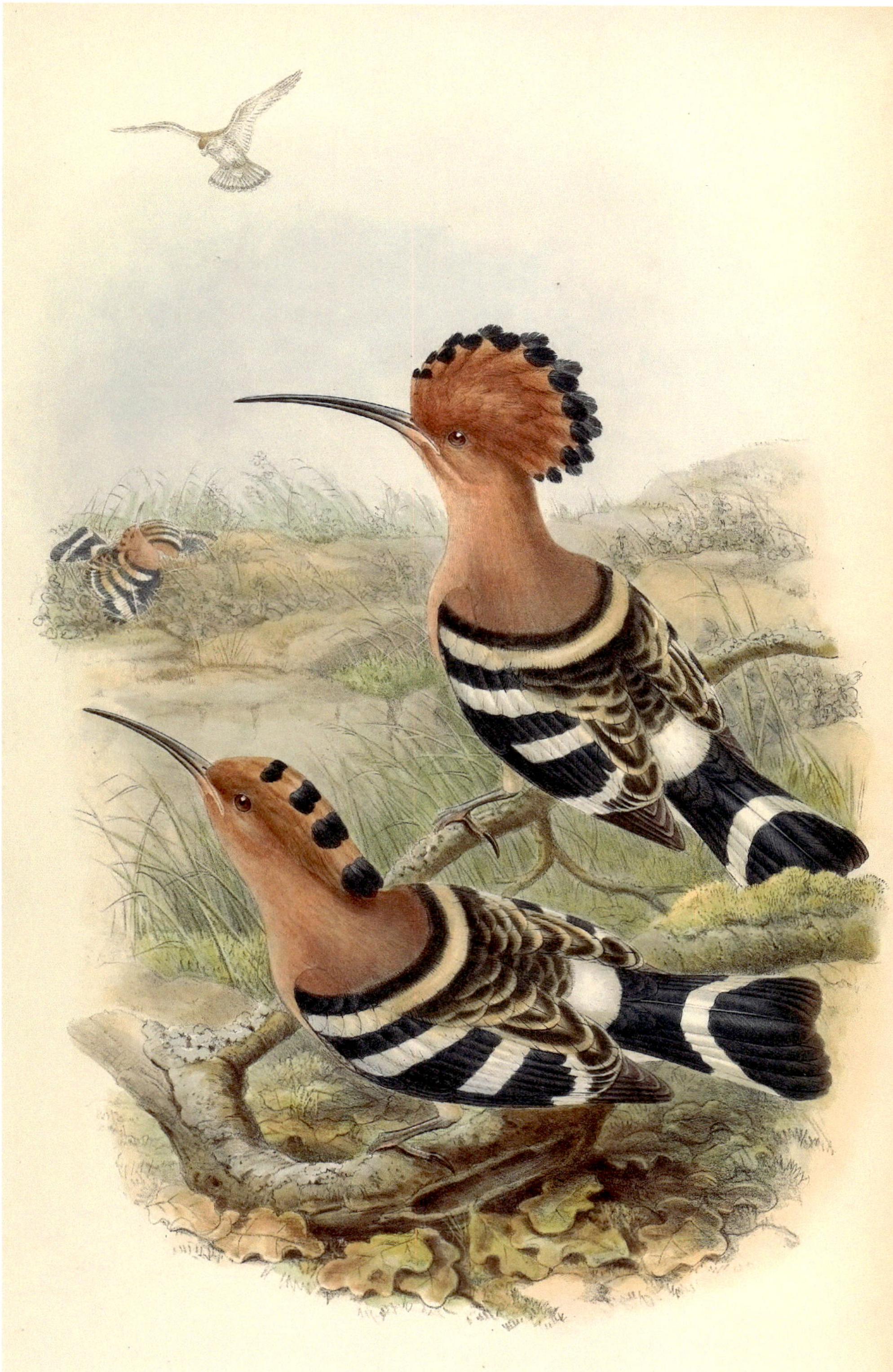

EURASIAN HOOPOE
Upupa epops

John Gould & William Matthew Hart, *The Birds of Asia*, 1850–83

Hoopoes are found throughout Eurasia and Africa and are identified by their impressive mohawk-style crest. This – a Eurasian hoopoe – is the most widespread species. Its name is onomatopoeic, a reference to its distinctive call. The text that accompanies this original illustration comments on how striking the bird's soft tones are; a melodious call that is 'accompanied by a movement of its handsome crest and an oscillation to and fro of its head at each note'.

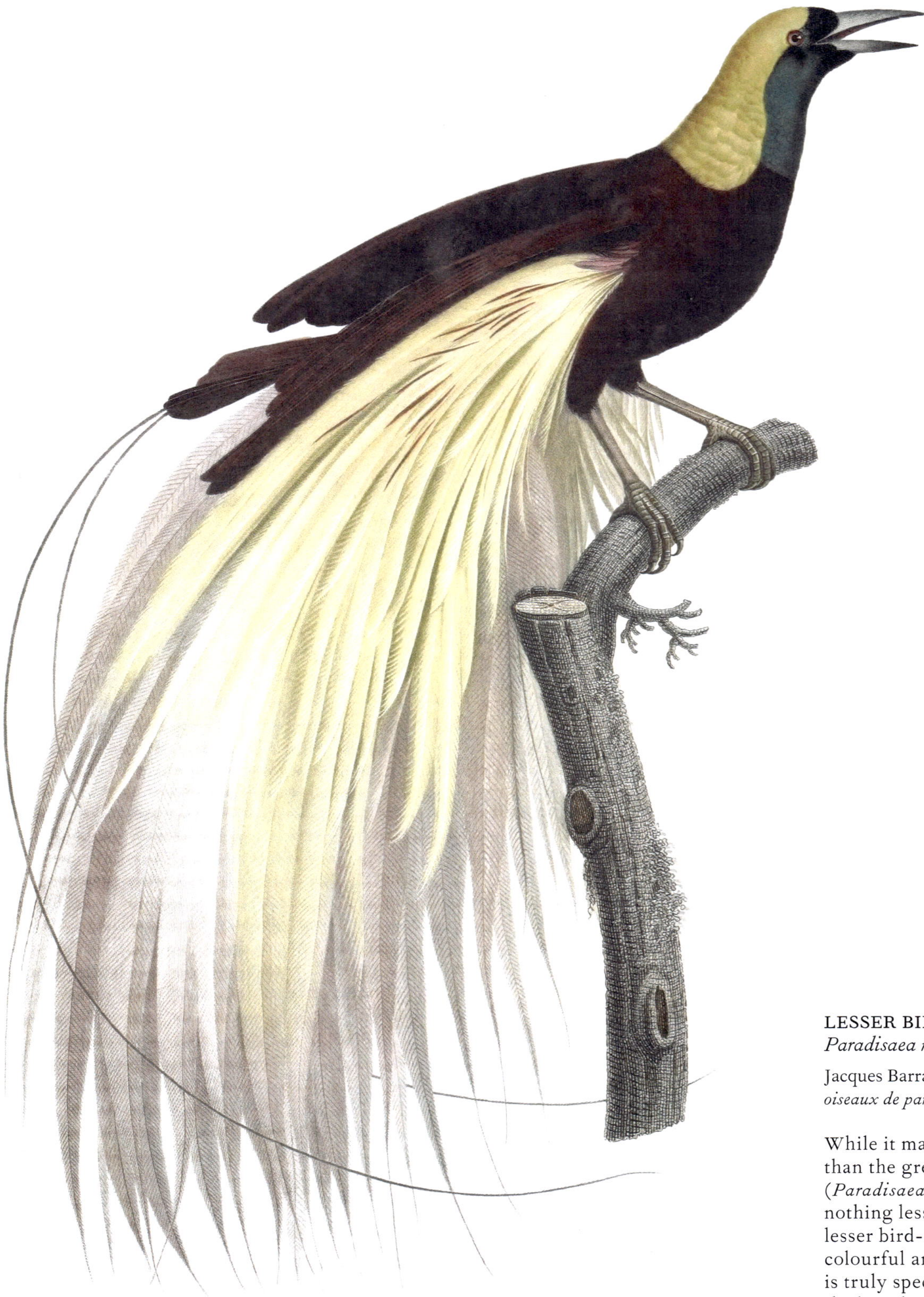

LESSER BIRD-OF-PARADISE
Paradisaea minor

Jacques Barraband, *Histoire naturelle des oiseaux de paradis et des rolliers*, 1806

While it may be smaller in size than the greater bird-of-paradise (*Paradisaea apoda*), there is nothing less spectacular about the lesser bird-of-paradise. The male's colourful and elaborate plumage is truly spectacular and, during the breeding season, they are keen to show off this impressive ornamentation through elaborate and ostentatious mating displays. Barraband's work effortlessly captures the impressive beauty of this bird's extravagant tail feathers.

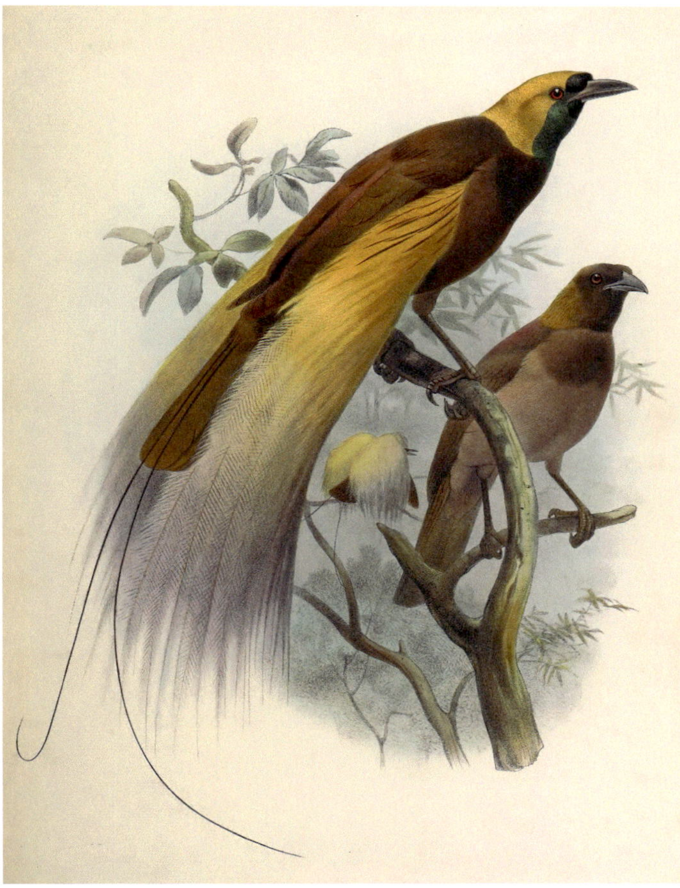

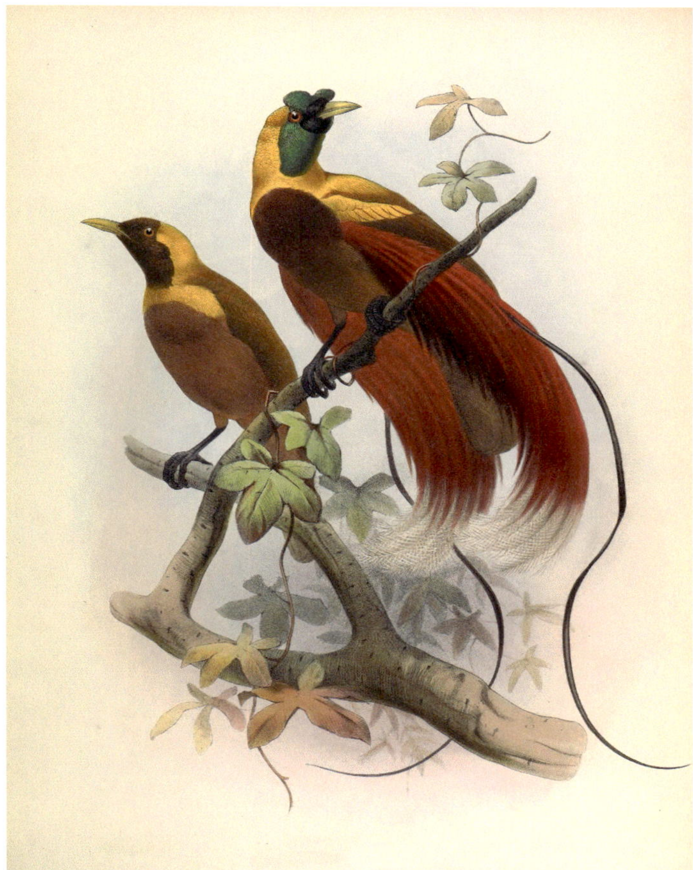

LEFT:

GREATER BIRD-OF-PARADISE
Paradisaea apoda

Joseph Wolf, *A Monograph of the Paradiseidae*, 1873

Elliot published his monograph of the birds-of-paradise in 1873. Illustrated by Joseph Wolf and lithographed by his friend Joseph Smit, it was one of a number of bird and mammal books that the pair worked on together. Together, they produced some of the finest plates of the late 1800s, with Wolf producing the drawings and Smith handling the lithography. Their work is noted for their scientific accuracy as well as their delicate and carefully considered backgrounds.

RIGHT:

RED BIRD-OF-PARADISE
Paradisaea rubra

Joseph Wolf, *A Monograph of the Paradiseidae*, 1873

The red bird-of-paradise is named for its impressive crimson train of tail feathers. These red plumes spring from the sides of its breast and require at least six years to attain fully. Like most birds-of-paradise, this species lives in New Guinea – specifically on the Waigeo and Batanta islands of Raja Ampat. Seen here in Elliot's monograph from 1873, this particular specimen would have originally been collected by British naturalist and explorer Alfred Russel Wallace during his time in New Guinea in the 1850s.

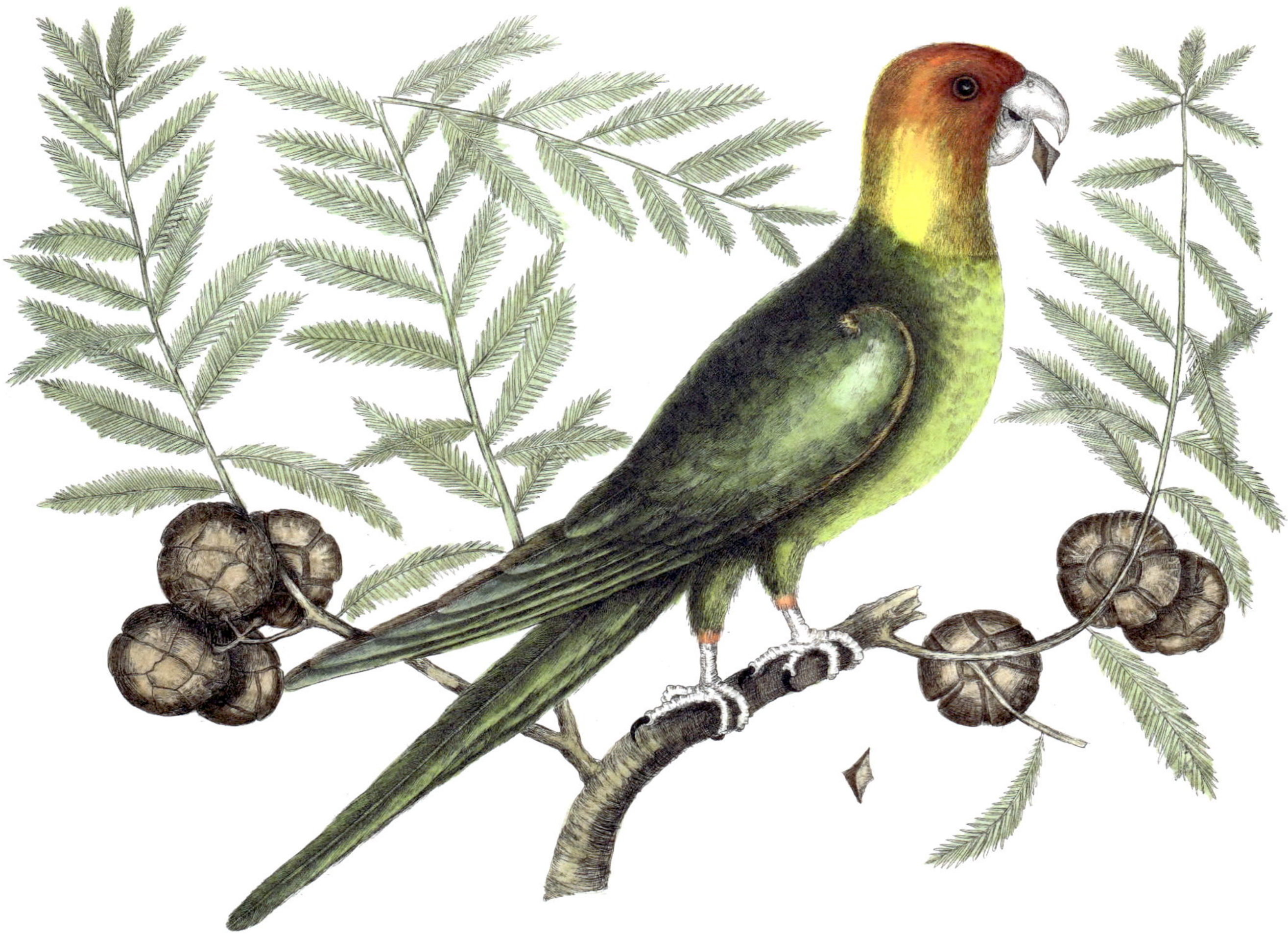

CAROLINA PARAKEET
Conuropsis carolinensis

Mark Catesby, *The Natural History of Carolina, Florida and the Bahama Islands*, 1729–47

Catesby's detailed etchings provide us with a rare glimpse at a now lost species. Unlike many later works, Catesby's illustrations here tend to place near-equal importance on the flora that surrounds his birds. His parakeet is seen feeding in a bald cypress (*Taxodium distichum*). While other birds struggle to open the tough fruit of these trees, the parakeet's tough bill was strong enough to do so. These trees lost an important means of seed dissemination with the extinction of the species.

OPPOSITE:
NEW ZEALAND KAKA
Nestor meridionalis

Charles Reuben Ryley, *Museum Leverianum*, 1792–6

Museum Leverianum contains several plates depicting species that had been newly imported from Australia, South America and the East Indies. The New Zealand kaka is endemic to New Zealand and takes its name from the Māori language, where the word *kākā* means 'parrot'. These birds are an important pollinator for many native plants in New Zealand, but sadly they are a threatened species.

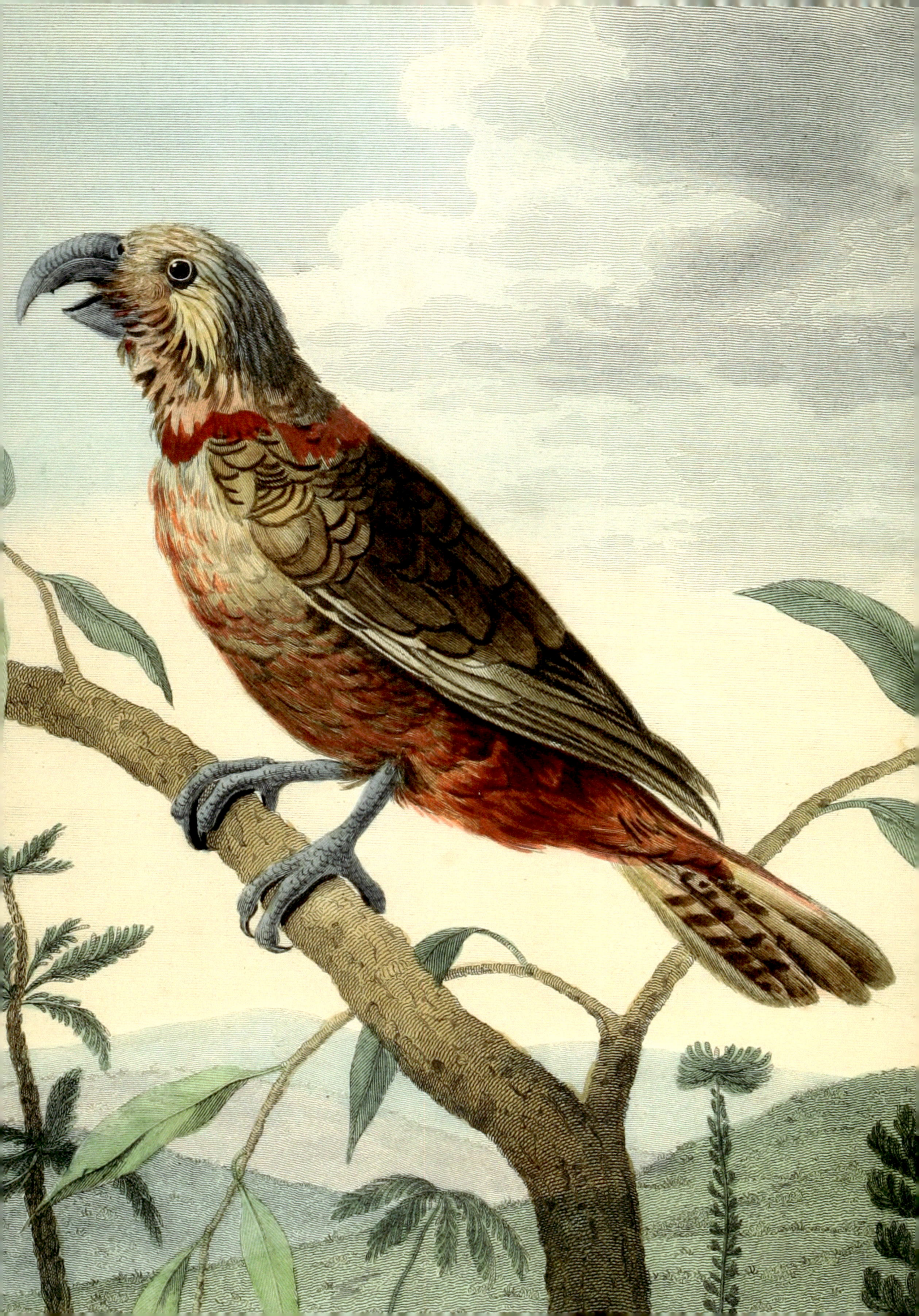

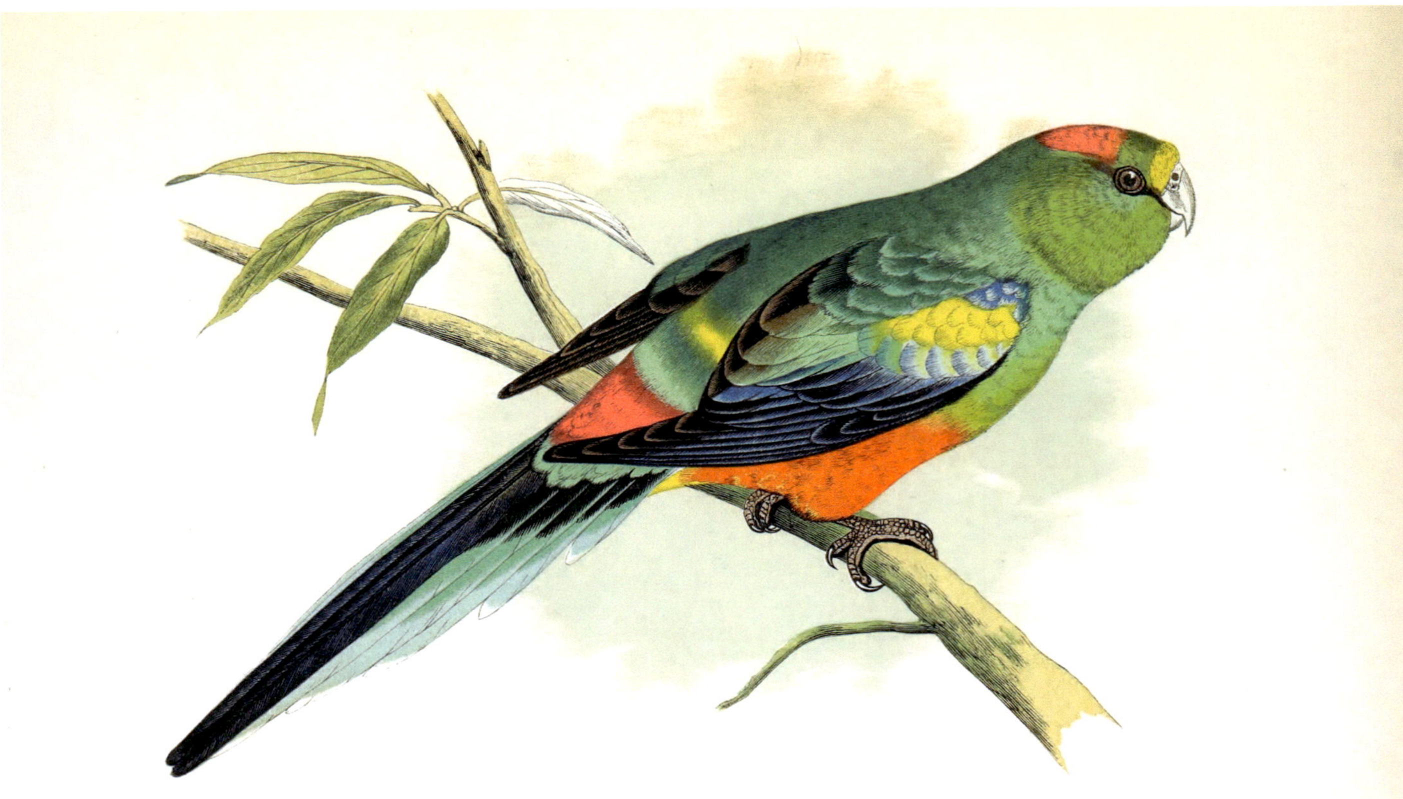

MULGA PARROT
Psephotellus varius

Alexander Francis Lydon,
Parrots in Captivity, 1884–8

Alexander Francis Lydon produced the illustrations for William Thomas Greene's *Parrots in Captivity*. Relatively uncelebrated today, Lydon was an English watercolour artist, illustrator and engraver who specialised in natural history and landscapes. He had a talent for drawing parrots and contributed to several books, as well as several of the era's most popular bird magazines.

OPPOSITE:
PRINCESS PARROTS
Polytelis alexandrae

John Gould & Henry Constantine Richter, *Supplement to the Birds of Australia*, 1851–69

Endemic to Australia, the scientific name of the princess parrot is *Polytelis alexandrae*. This name was given in honour of Princess Alexandra of Denmark. Born in 1844, the princess married Edward VII, then Prince of Wales and, in 1901, she became Queen consort of the United Kingdom. As the title of royals changed, so too did the title of this bird, and accounts sometimes list it as Princess Alexandra's parrot, the Princess of Wales parakeet and Queen Alexandra's parakeet. Others simply call it the rose-throated parakeet in reference to its distinctive pink throat.

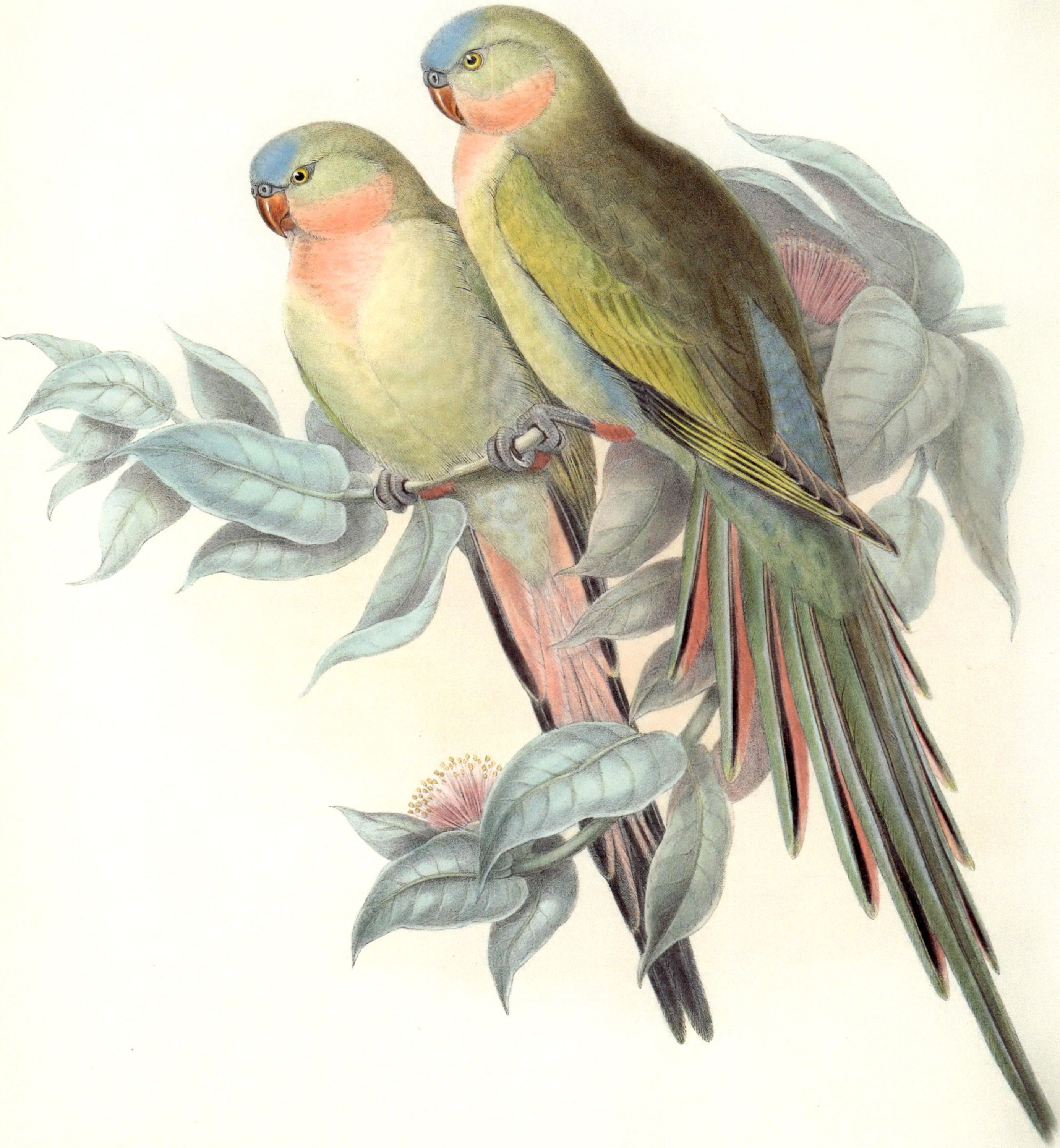

POLYTELIS ALEXANDRÆ, *Gould.*

J. Gould and H.C. Richter del. et lith. *Walter, Imp.*

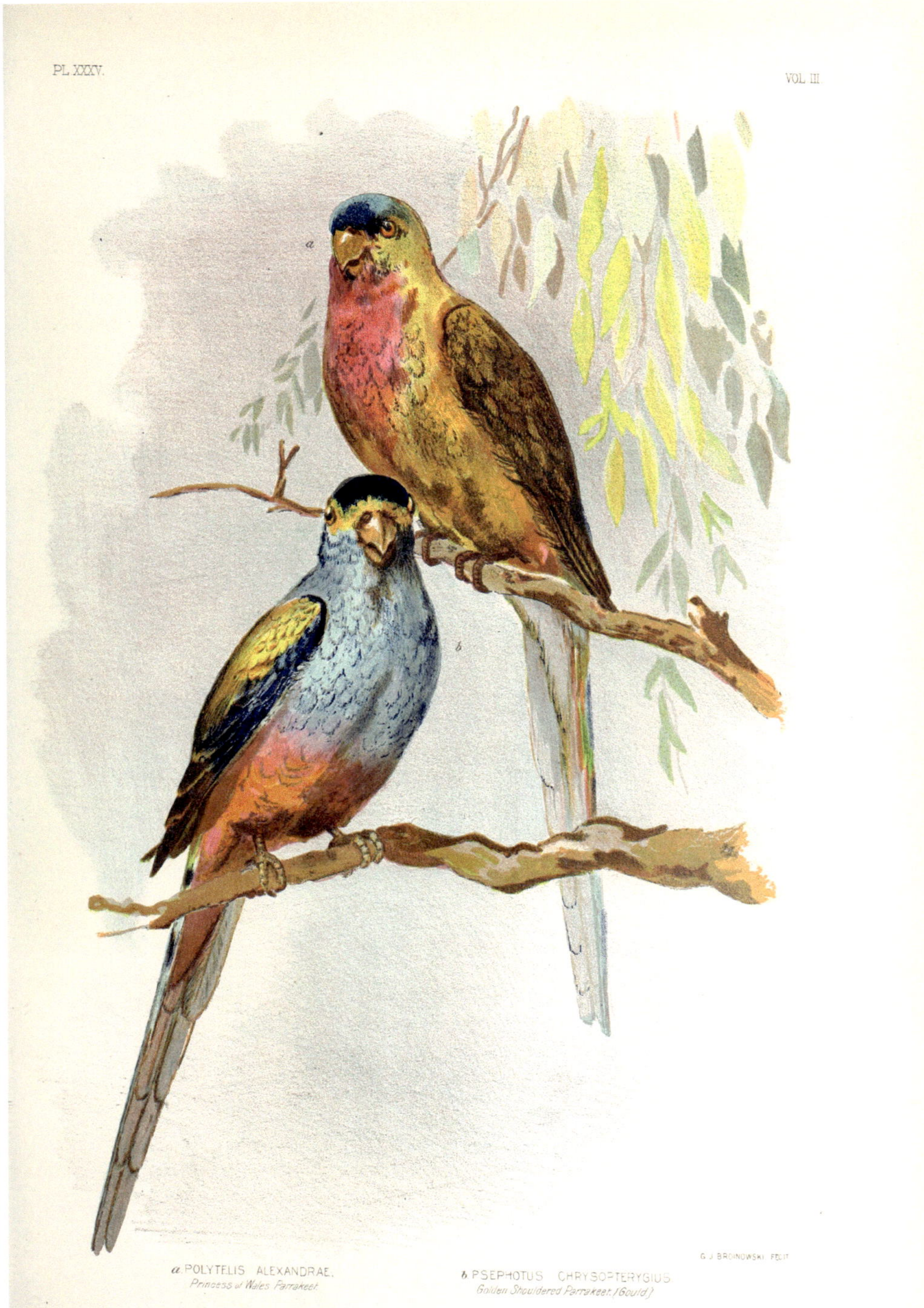

GOLDEN-SHOULDERED PARROT (LEFT) & PRINCESS PARROT (RIGHT)
Psephotellus chrysopterygius & Polytelis alexandrae

Gracius Broinowski,
The Birds of Australia, 1890–1

Gracius Broinowski was a Polish-born ornithologist and illustrator who travelled to Australia in the 1850s. Working initially as an artist and lecturer, he started to publish ornithological books in his forties. Of these, his most celebrated is his *Birds of Australia*. Intended to cover all the birds of the Australian continent, the final work contained a staggering 700 species. It took many years to complete, with Broinowski travelling the country and depositing his paintings in the art unions of the towns that he passed through.

OPPOSITE:
LAUGHING KOOKABURRA
Dacelo novaeguineae

Neville H.P. Cayley, watercolour, 1892

Found in Australia, the laughing kookaburra is the largest member of the kingfisher family. This unassuming bird is a force to be reckoned with. Typically hunting a variety of small animals such as lizards, mice and insects, the bird has earned an impressive reputation for its killing of snakes. It does this by either snatching the snake from behind its head and dropping it from a height, or by beating it against a hard surface.

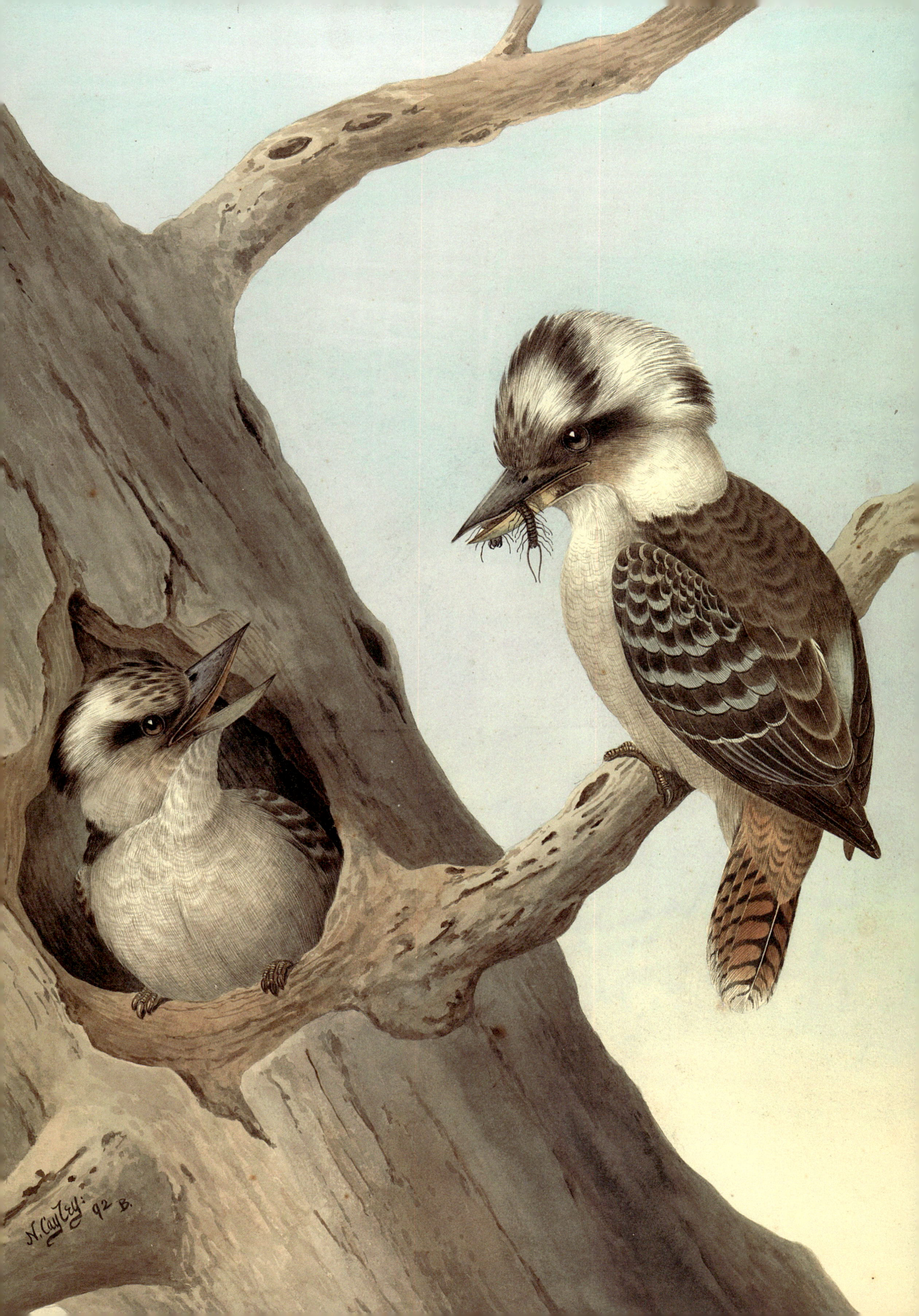
N. Cayley: 92 B.

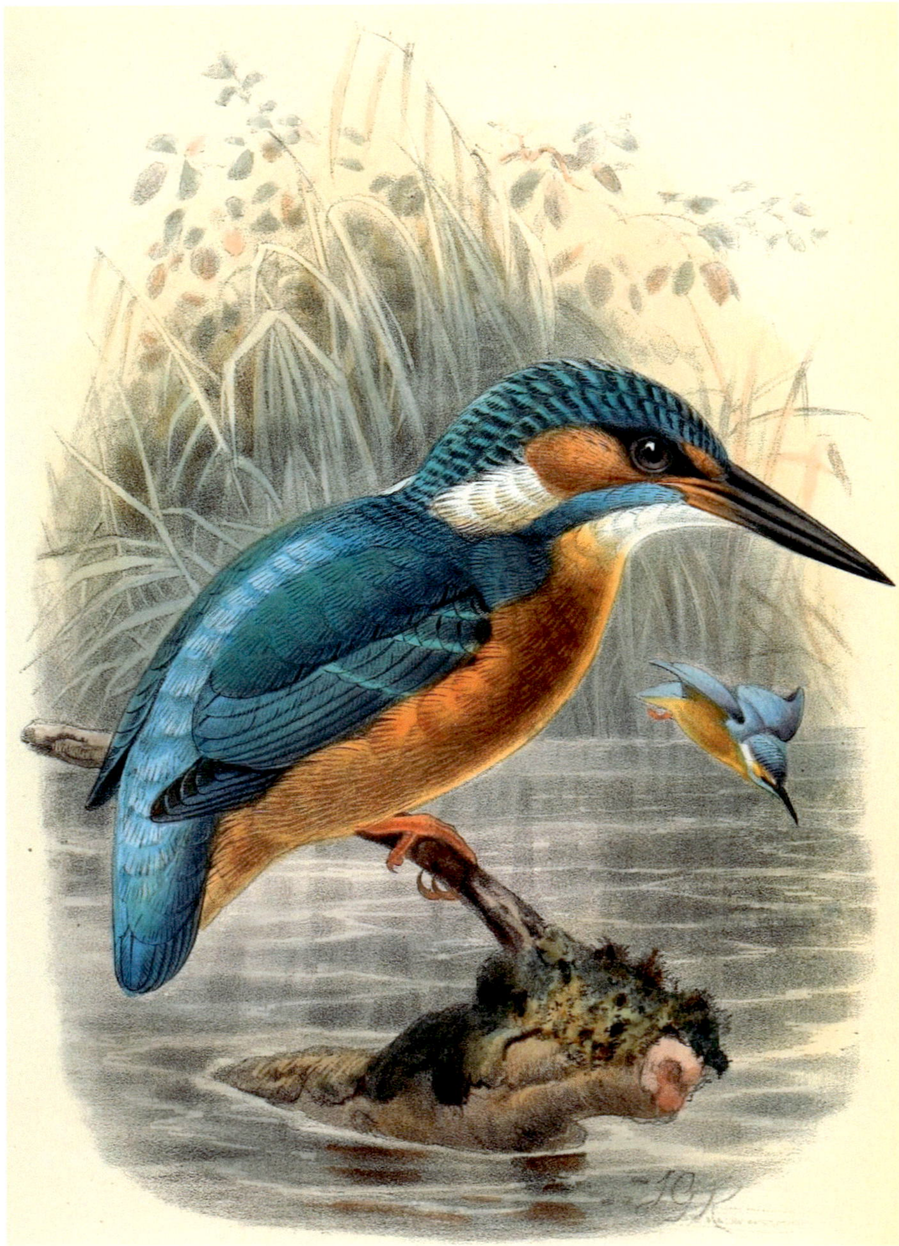

COMMON KINGFISHER
Alcedo atthis

John Gerrard Keulemans, *Coloured Figures of the Birds of the British Islands*, 1885–97

Found at lakes, canals and rivers, the common kingfisher is an impressive bird that has a shrill whistle and brilliant electric-blue colouring. Its distinctive dagger-shaped bill is perfect for hunting fish. In total, the kingfisher family contains 129 species, but not all of these are associated with fishing.

OPPOSITE, FROM TOP:
SACRED KINGFISHER
Todiramphus sanctus

COLLARED KINGFISHER
Todiramphus chloris

GUAM KINGFISHER
Todiramphus cinnamominus

John Gould, *A Synopsis of the Birds of Australia and the Adjacent Islands*, 1837–8

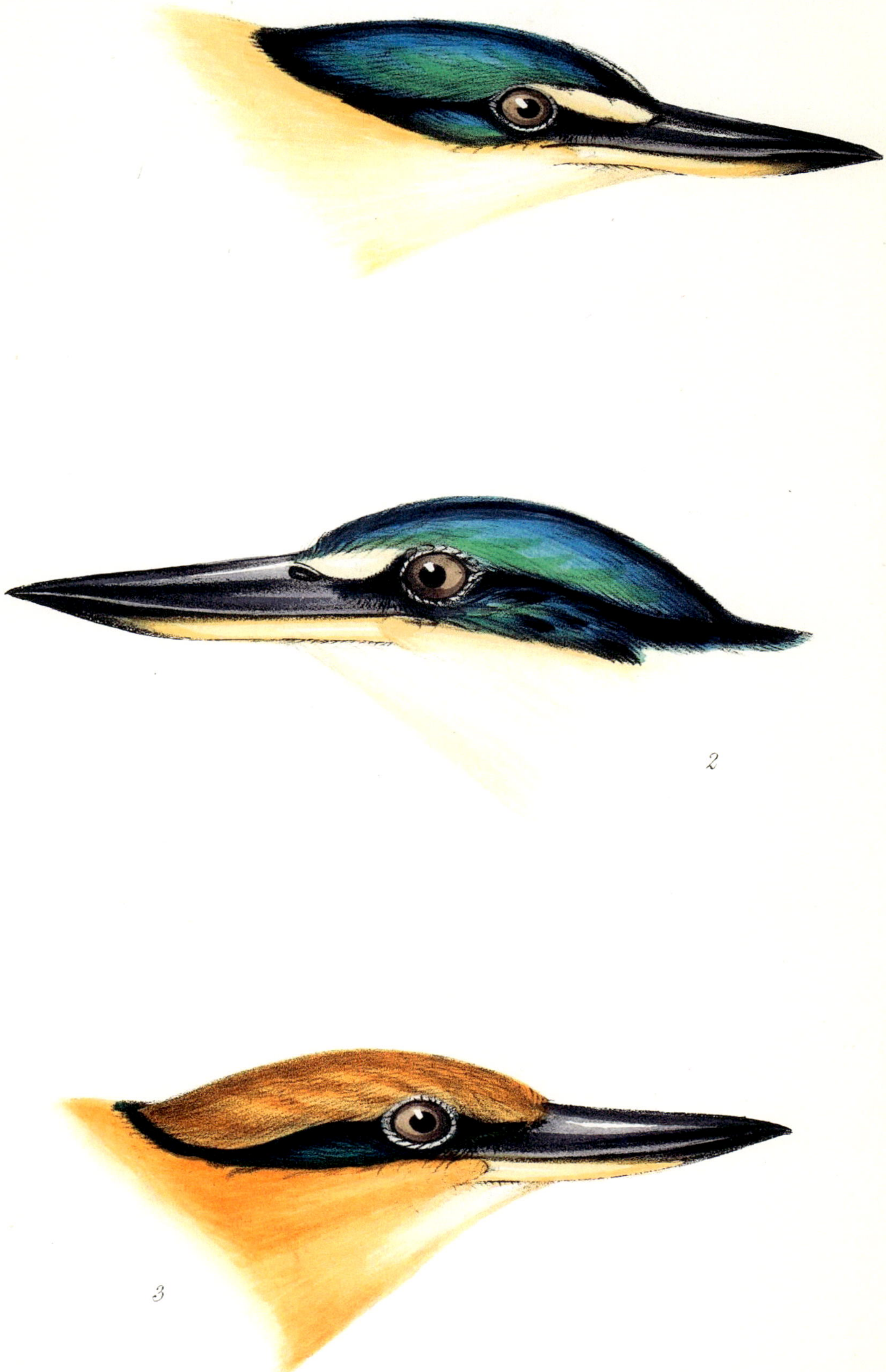

1 *Halcyon sanctus*
2 ——— *collaris*
3 ——— *cinnamominus.*

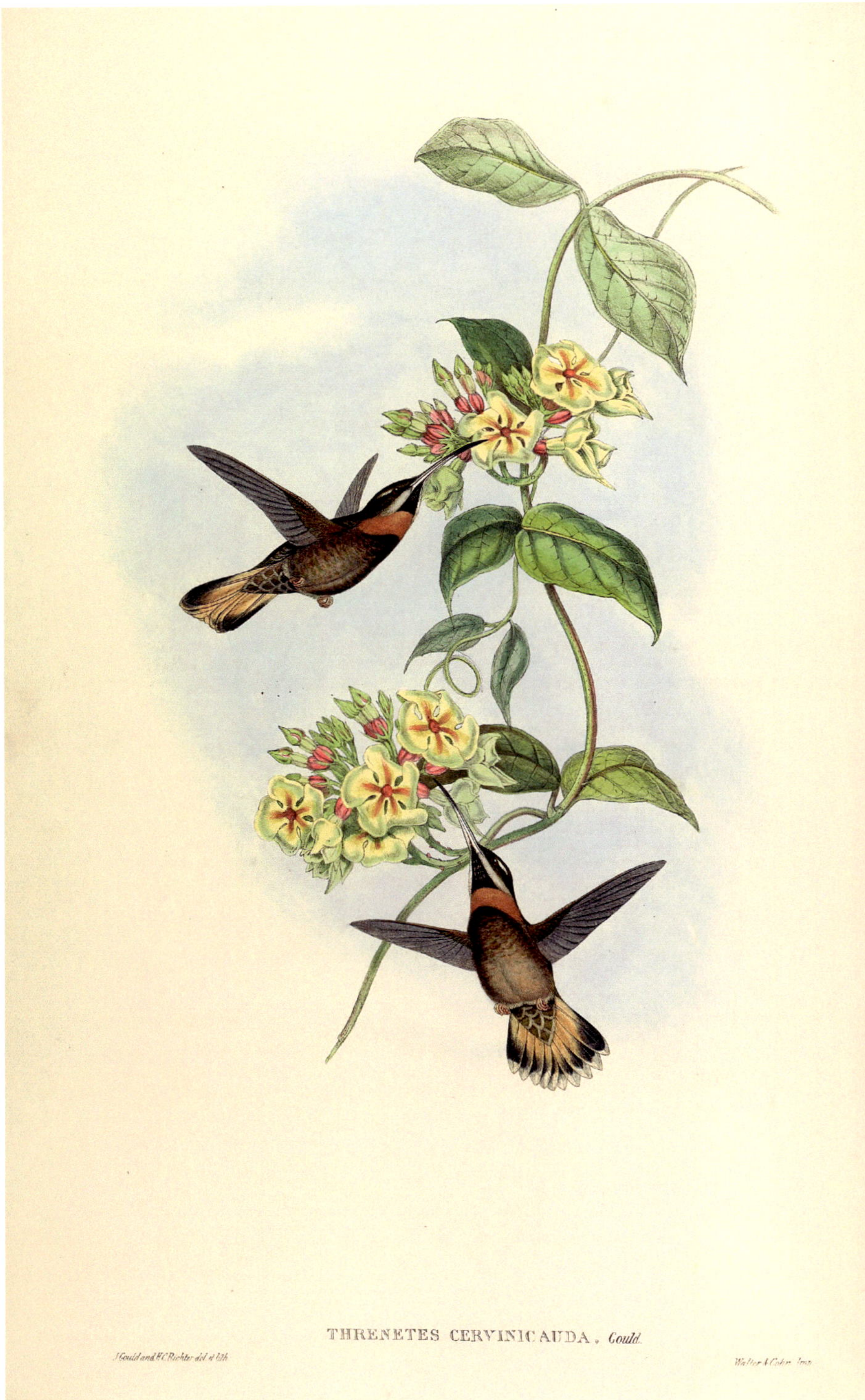

OPPOSITE, CLOCKWISE FROM TOP LEFT:
RUBY-THROATED HUMMINGBIRD
Archilochus colubris

HORNED SUNGEM
Heliactin bilophus

CRIMSON TOPAZ
Topaza pella

TUFTED COQUETTE
Lophornis ornatus

RED-TAILED COMET
Sappho sparganurus

SWORD-BILLED HUMMINGBIRD
Ensifera ensifera

WHITE-BOOTED RACKET-TAIL (MALE)
Ocreatus underwoodii

BUFF-TAILED SICKLEBILL
Eutoxeres condamini

DOT-EARED COQUETTE
Lophornis gouldii

WHITE-VENTED VIOLETEAR
Colibri serrirostris

HOODED VISORBEARER
Augastes lumachella

JUAN FERNÁNDEZ FIRECROWN
Sephanoides fernandensis

Hummingbirds (Trochilidae), Ernst Haeckel, *Kunstformen der Natur*, 1899

PALE-TAILED BARBTHROATS
Threnetes leucurus

John Gould & Henry Constantine Richter, *A Monograph of the Trochilidae*, 1849–61

Hummingbirds take their name from the high-frequency sound that their wings produce as they fly. Strikingly colourful with their iridescent feathers, Audubon described one species as 'a glittering fragment of a rainbow'. While some of the smallest birds in the world are hummingbirds, they belong to one of the largest bird families – with well over 300 species in total. The illustration above comes from Gould's five-volume monograph that contains a staggering 418 hand-coloured plates, each illustrating the variety and beauty of this wonderful bird.

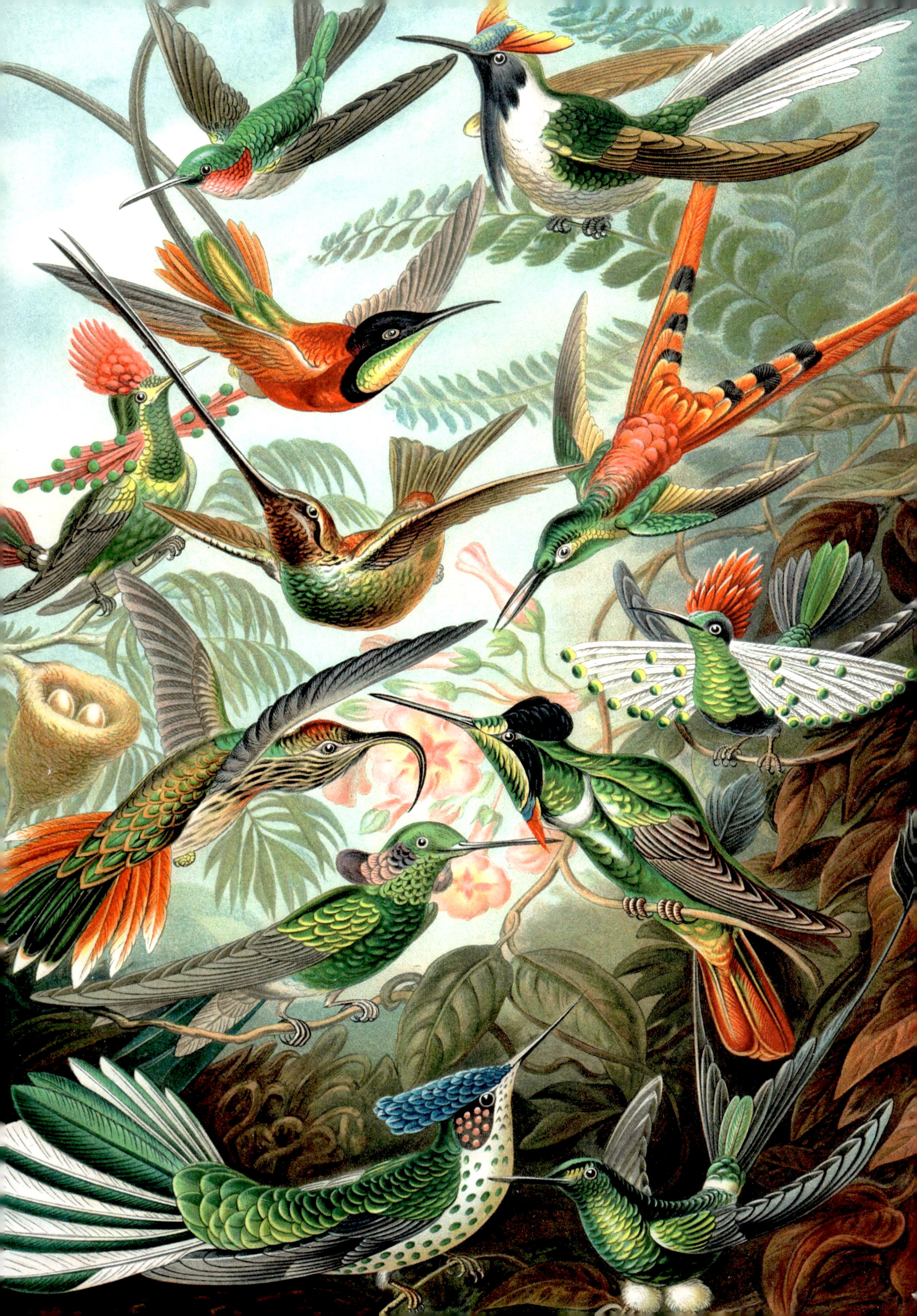

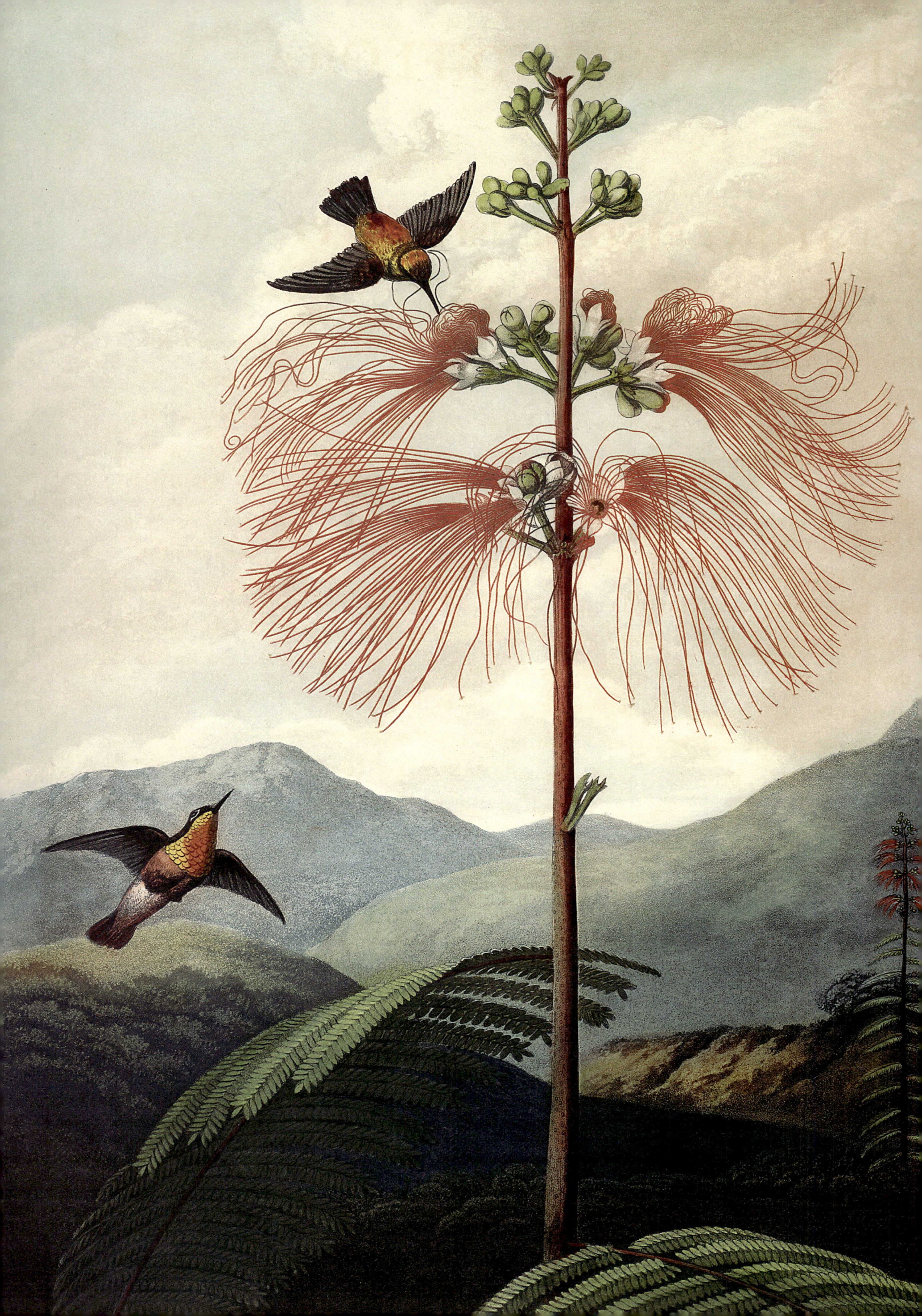

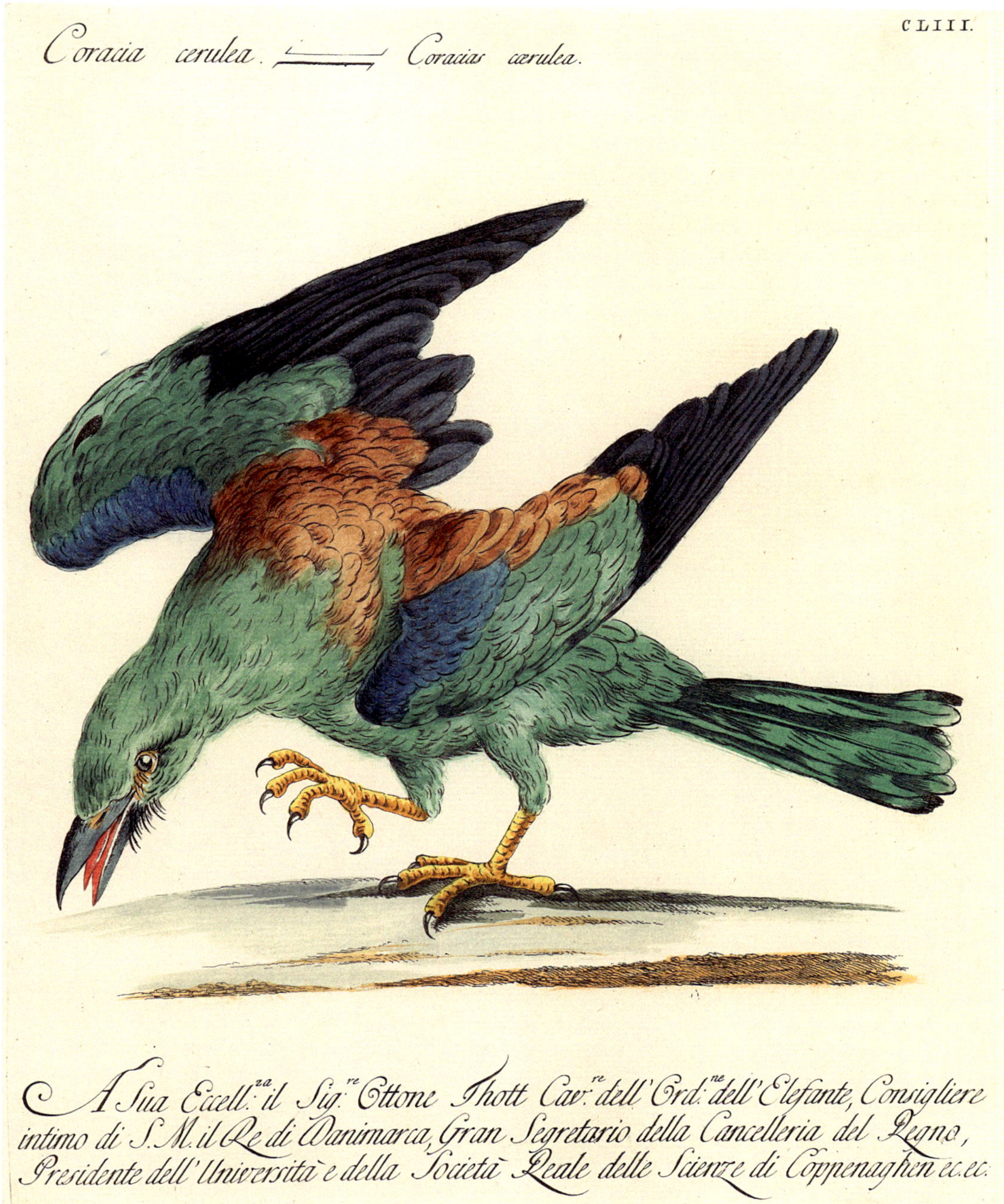

OPPOSITE:
POSSIBLY SHINING SUNBEAM
Aglaeactis cupripennis

Large Flowering Sensitive Plant, Philip Reinagle, New Illustration of the Sexual System of Carolus von Linnaeus, 1799–1810

EUROPEAN ROLLER
Coracias garrulus

Lorenzo Lorenzi & Violante Vanni, *Ornithologia methodice digesta*, 1767–76

Personality and a sense of playfulness can be seen in this early ornithological illustration produced for *Ornithologia Methodice Digesta*. Since it depicts a European roller, the bird would have presumably been drawn from life, with artists Lorenzo Lorenzi and Violante Vanni working from the collection of Giovanni Gerini. Their depiction captures some of the fantastically bright-coloured plumage of the bird – yet in reality this roller tends to be a more brilliant shade of blue.

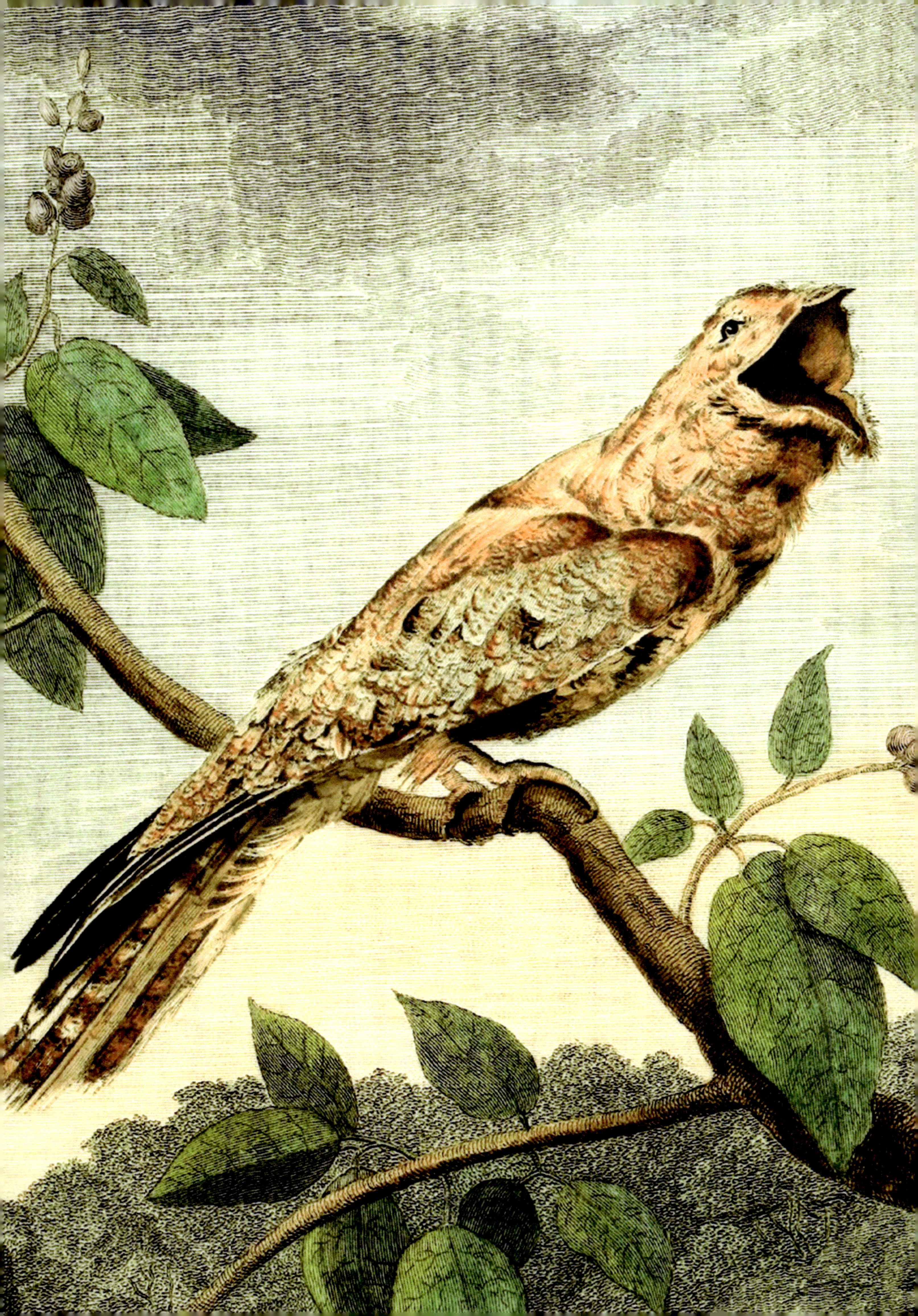

OPPOSITE:

EUROPEAN NIGHTJAR
Caprimulgus europaeus

Charles Reuben Ryley, *Museum Leverianum*, 1792–6

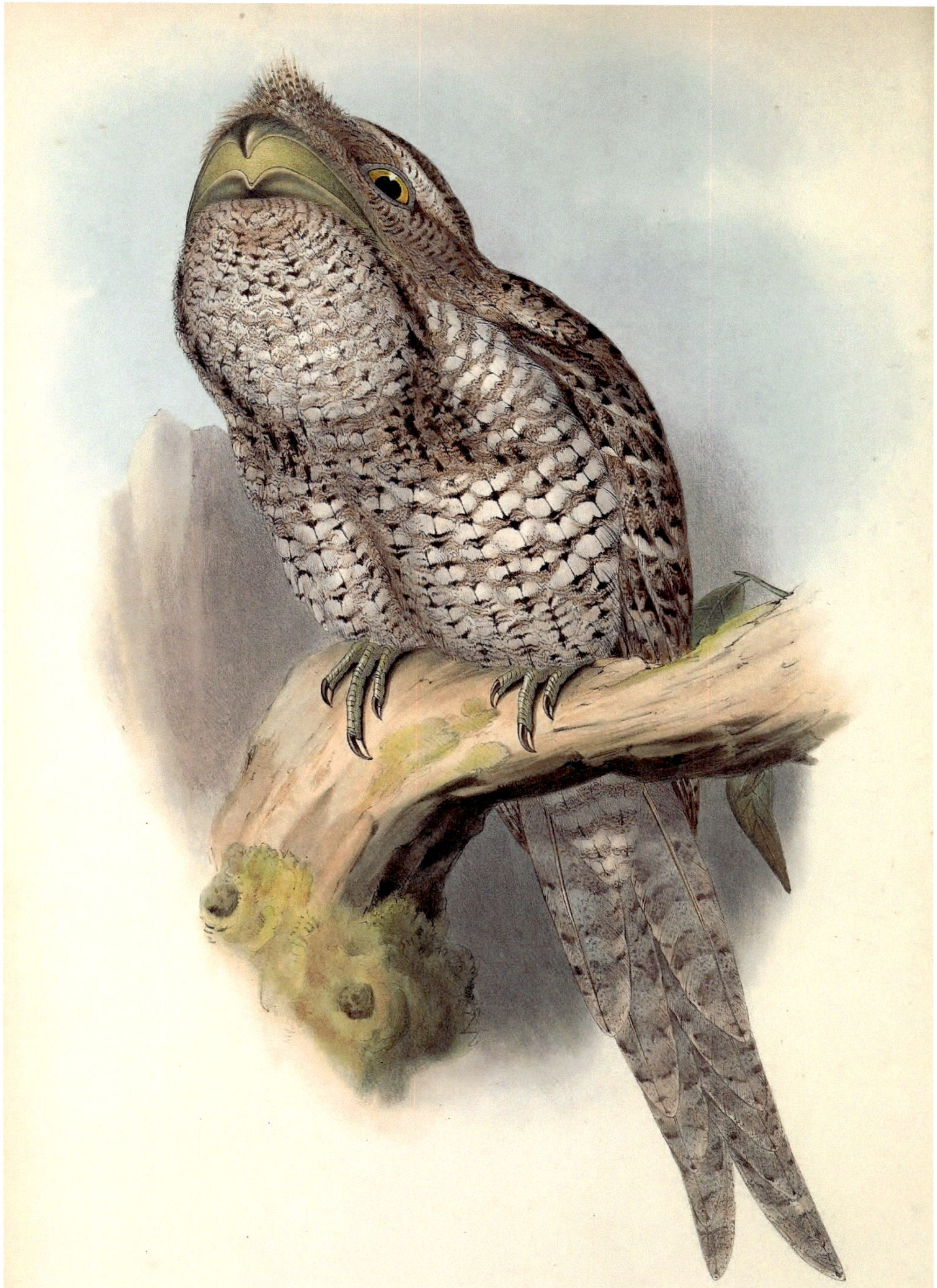

PAPUAN FROGMOUTH
Podargus papuensis

John Gould & Henry Constantine Richter, *Supplement to the Birds of Australia*, 1851–69

Both the nightjar and the frogmouth belong to an order of birds known as Caprimulgiformes. Birds in this order are typically nocturnal and feed on insects. The name of the order is derived from the Latin word for 'goat-sucker' – a reference to the feeding habits of the European nightjar. These birds have small bills but large mouths that they open wide to catch insects. Often found in fields with goats, their unusual mouths led to a myth that they could suck milk from the teats of goats.

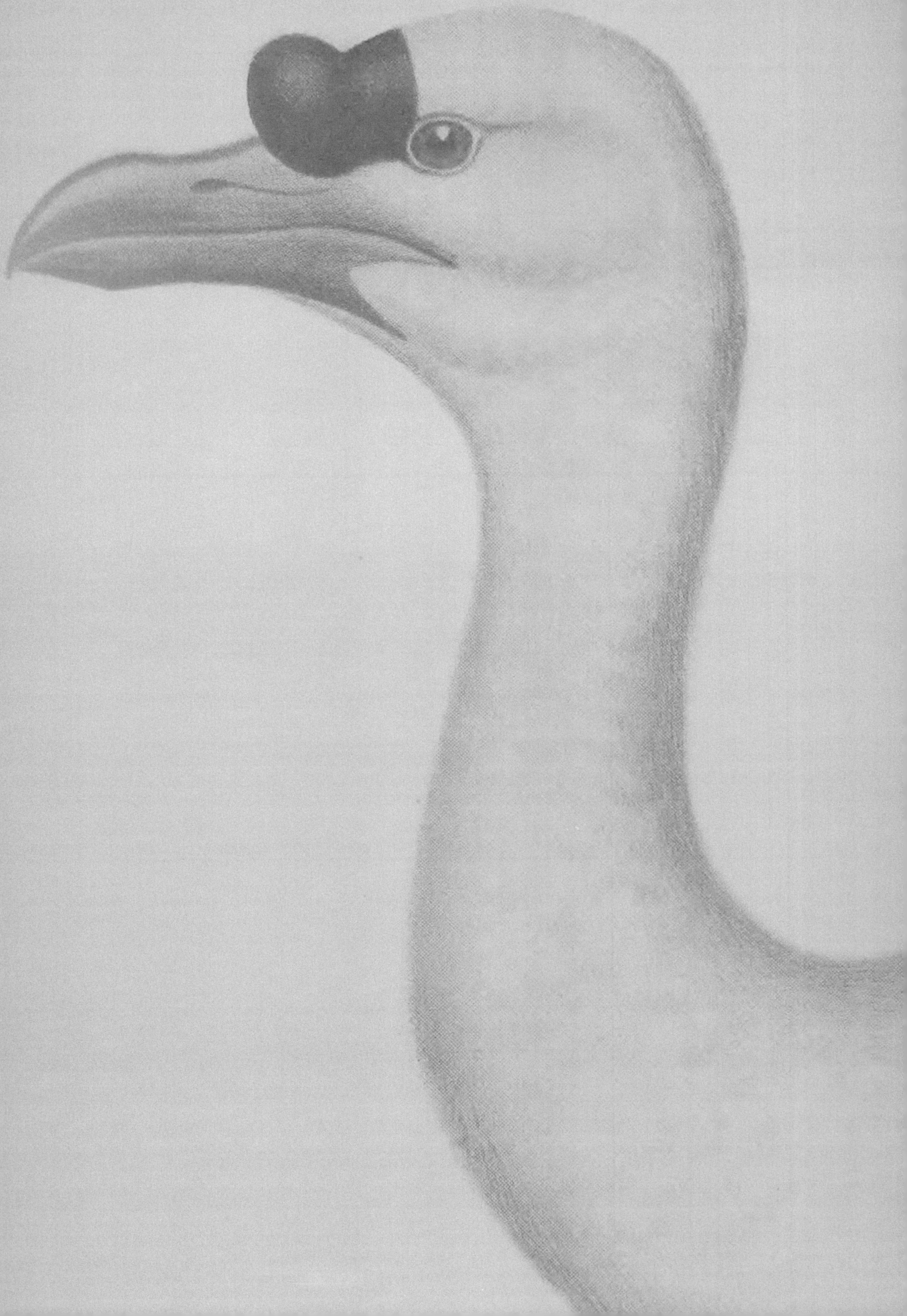

Flightless Birds

Nearly all birds have wings, which originally evolved from forelimbs and gave birds the ability to fly. Over time, further evolution led to some birds losing this ability. In environments where food was plentiful and predators were scarce, flight became unnecessary for survival. Yet, as this chapter will show, this evolution made these birds highly vulnerable and led to many becoming extinct.

Throughout this book, we have seen many examples of birds that have adapted and evolved to cope with their environments. Some have developed long legs to help them forage for food, others have webbed feet to move through water and some have large wingspans that allow them to spend the majority of their life at sea. Yet there are some birds that have adapted in an unexpected way: they have let go of the need to fly. Such birds have evolved in the absence of ground predators.

An example of this is seen in the family commonly known as cormorants (*Phalacrocoracidae*). Typically, these birds are found flying along coastlines throughout the world. They are skilled fishers who are able to dive underwater and catch prey with their hooked bills. Their ability to fly has allowed them to quickly escape threats and to bring food back to their colonies.

On the Galapagos Islands, however, these birds found a habitat that, up until recently, had no natural predators. Its food-rich shorelines meant that they did not need to fly and so their wings became stunted. This allowed them to dive effortlessly under the sea and propel themselves smoothly through the water. When Charles Darwin first encountered these birds on his famous 1830s voyage to the islands, they captured his imagination. He was quick to note that their ability to swim and dive was far more impressive than that of their flying cousins.

The Galapagos cormorant (*Phalacrocorax harrisi*) is only one of several endemic species that piqued the young naturalist's curiosity while on the Islands. The remoteness of the Galapagos made them the perfect habitat for such unique species to develop and thrive. Islands tend to provide the type of isolation needed to foster the development of distinct and unique creatures and, when it comes to flightless birds, one set of islands stands out in particular. Those are the islands of New Zealand – a country that has more species of flightless bird than any other place in the world.

New Zealand was an archipelago where birds truly ruled. Before human settlement, the islands had no mammals apart from a few bats and marine species. Isolated from these predators for roughly twenty million years, the bird population faced few threats and so they proliferated and diversified. Birds were safe to feed freely on forest floors, an ecosystem typically dominated by mammals. Food was plentiful and the climate was mild. Nests no longer needed to be in trees. The need to fly became redundant and the birds grew in size. Some of these birds fed in forests and grasslands, not unlike cattle, deer, goats and sheep. Without the threat of mammals, flightless birds such as the weka (*Gallirallus australis*), kākāpō (*Strigops habroptila*) and South Island takahe (*Porphyrio hochstetteri*) evolved.

When humans arrived in around the thirteenth century they wrought havoc. Over-hunting and deforestation greatly impacted upon the bird population. Humans also brought with them other animals such as ferrets, stoats, dogs, cats and rats. These posed new threats to the native bird population while also bringing increased competition for food in the forests and wetlands.

Throughout recent human history, in many parts of the world, flightless birds have been defenceless in the face of these threats. Many have become extinct and a lot more are now endangered. The number of endangered bird species has risen at an exorbitant rate and bird populations have been in steep decline. From the destruction of habitats to the effects of climate change, the activities of man have caused great destruction and devastation throughout our natural world. Despite the grim reality of this situation, conservationists have been working hard to maintain bird species and there have been a number of success stories. At the heart of successful bird conservation lies the need to understand the world of birds, an understanding that has its origins in the efforts of the illustrators and naturalists in this book.

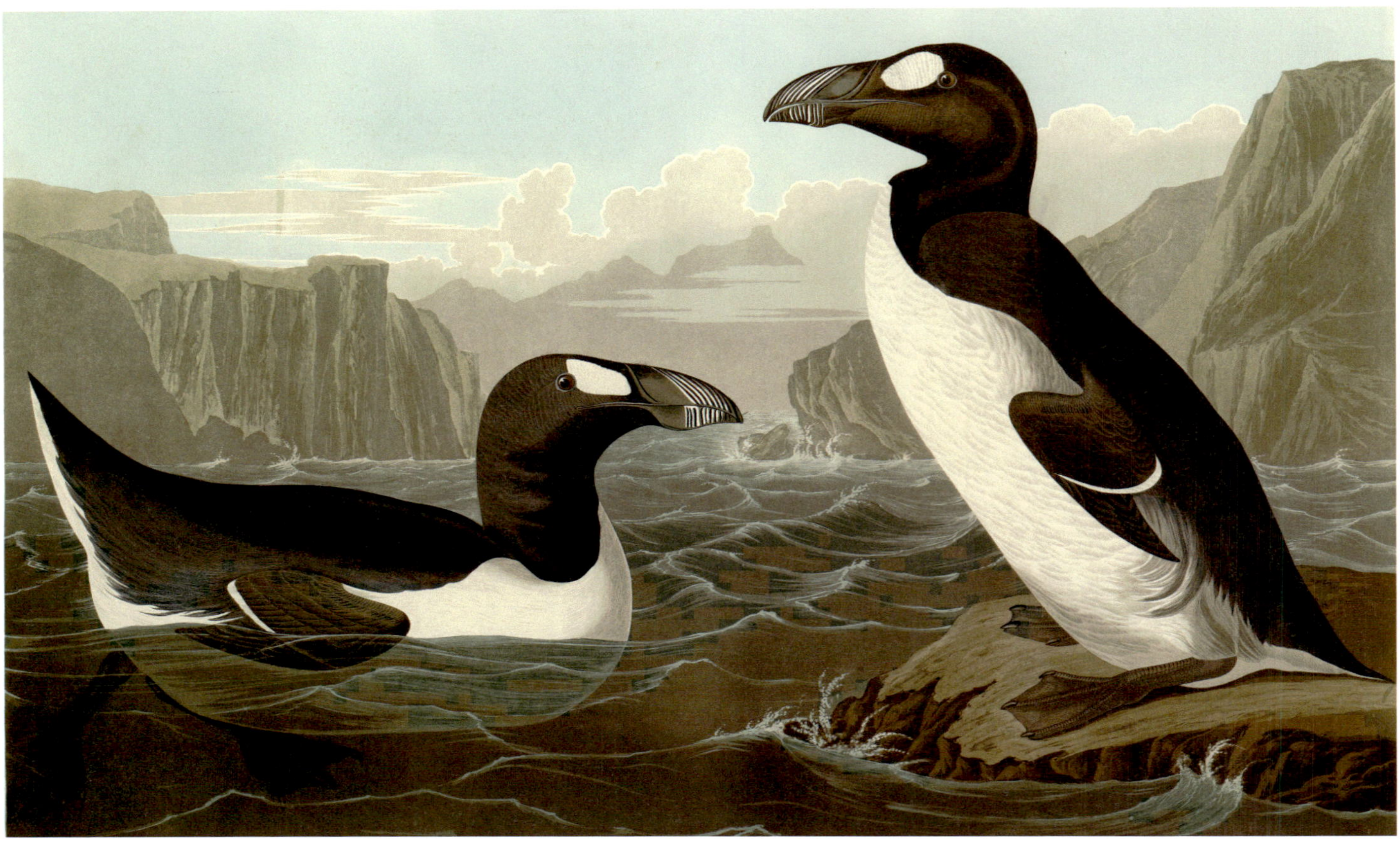

GREAT AUKS
Pinguinus impennis

John James Audubon,
The Birds of America, 1827–38

The great auk was once a populous species of bird that was distributed widely across the North Atlantic. They would breed in colonies dotted across the ocean and spend the remainder of their time on the water. These birds were roughly the same size as a goose and their inability to fly made them easy prey for passing sailors who were eager for fresh meat and eggs. This practice led to the extinction of the species at the end of the nineteenth century, and today only a small number of these birds can be found preserved in museums around the world.

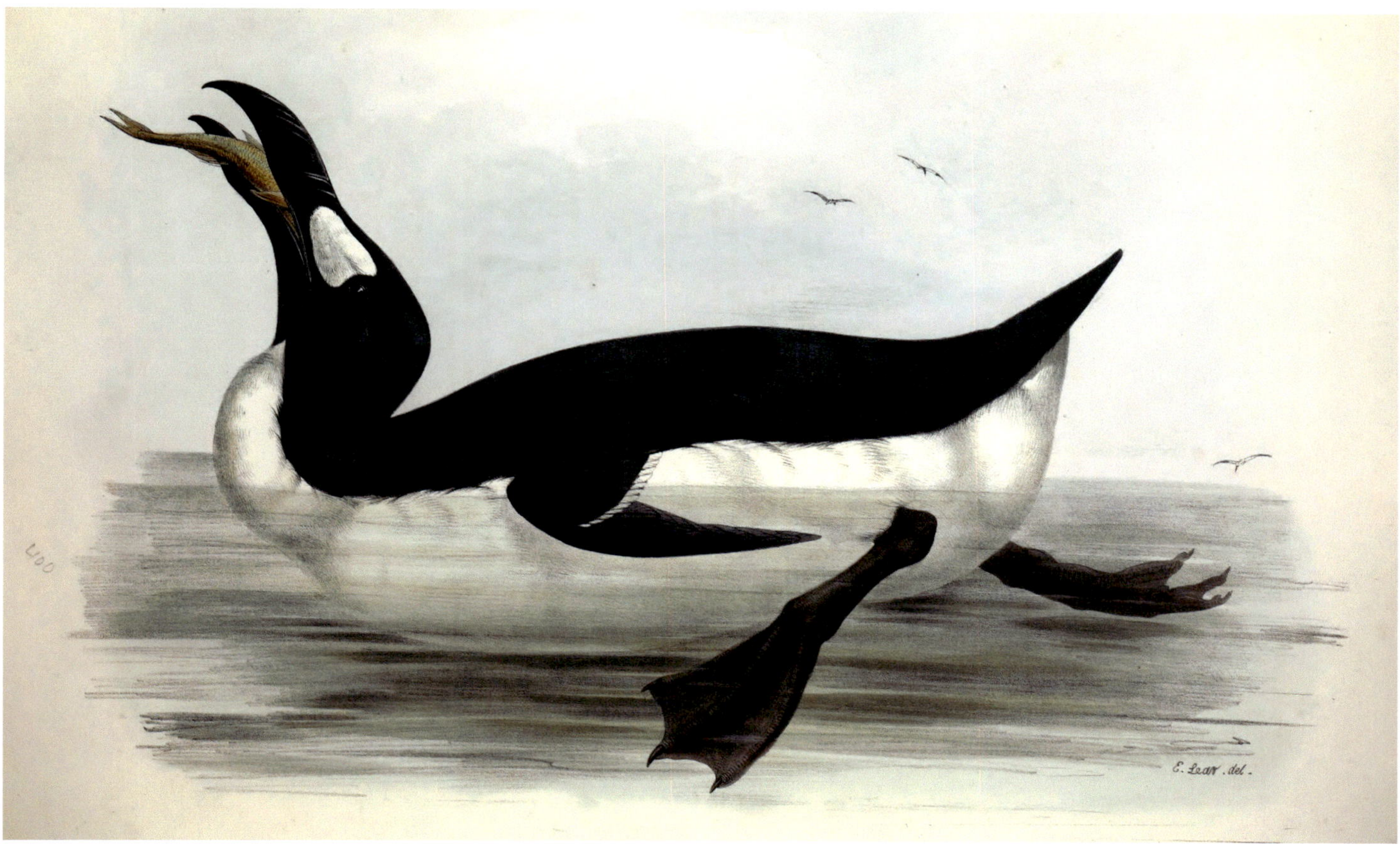

GREAT AUK
Pinguinus impennis

Edward Lear, *The Birds of Europe*, 1832–7

The last two confirmed specimens of the great auk were killed just seven years after the publication of this image.

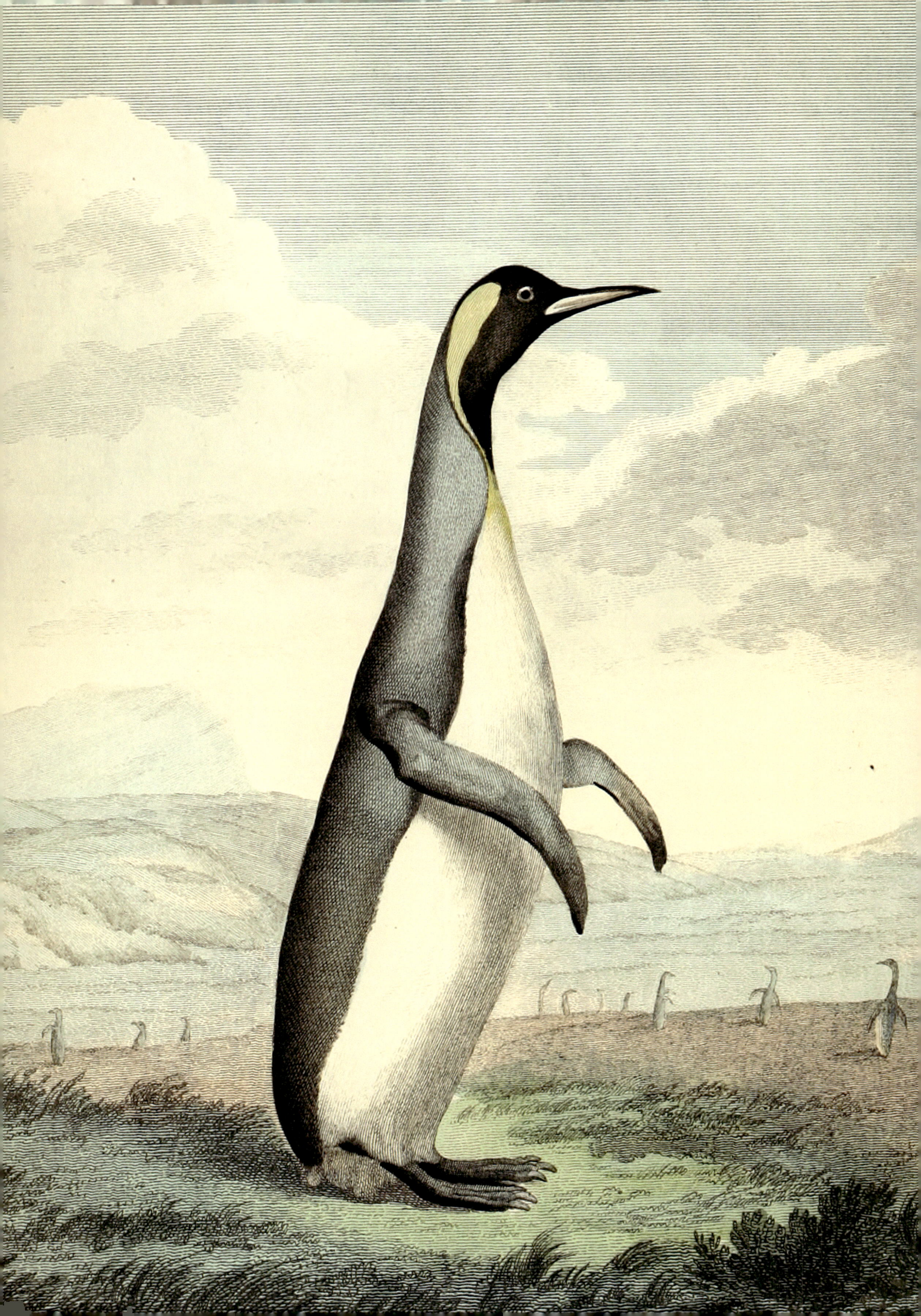

OPPOSITE:
KING PENGUIN
Aptenodytes patagonicus

Charles Reuben Ryley, *Museum Leverianum*, 1792–6

Ryley's illustration is accompanied by text that states that this 'curious species is a native of the southern hemisphere, and is principally found about [the] Falkland Islands'. The king penguin is one of five species of the bird that breed on the islands, but its population was nearly wiped out in the 1920s by whalers who would eat the bird and burn its blubber for fuel.

TOP:
LITTLE PENGUINS
Eudyptula minor

John Gould & Henry Constantine Richter, *The Birds of Australia*, 1840–8

The little penguin is the smallest species of penguin in the world.

BOTTOM:
EASTERN ROCKHOPPER PENGUINS
Eudyptes chrysocome filholi

John Gould & Henry Constantine Richter, *The Birds of Australia*, 1840–8

Rockhoppers take their name from their ability to hop from rock to rock.

OVERLEAF:
KING PENGUINS
Aptenodytes patagonicus

Wilhelm Kuhnert, *The Royal Natural History* (Vol. 4, Section 8), 1893–6

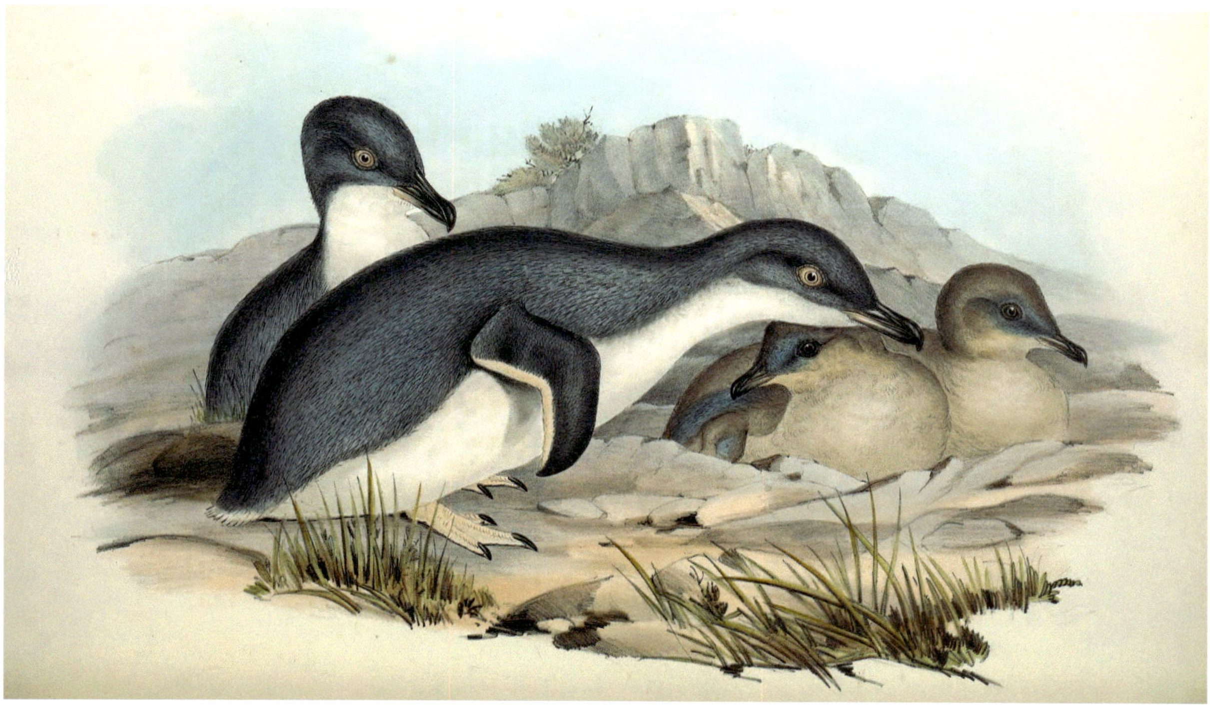

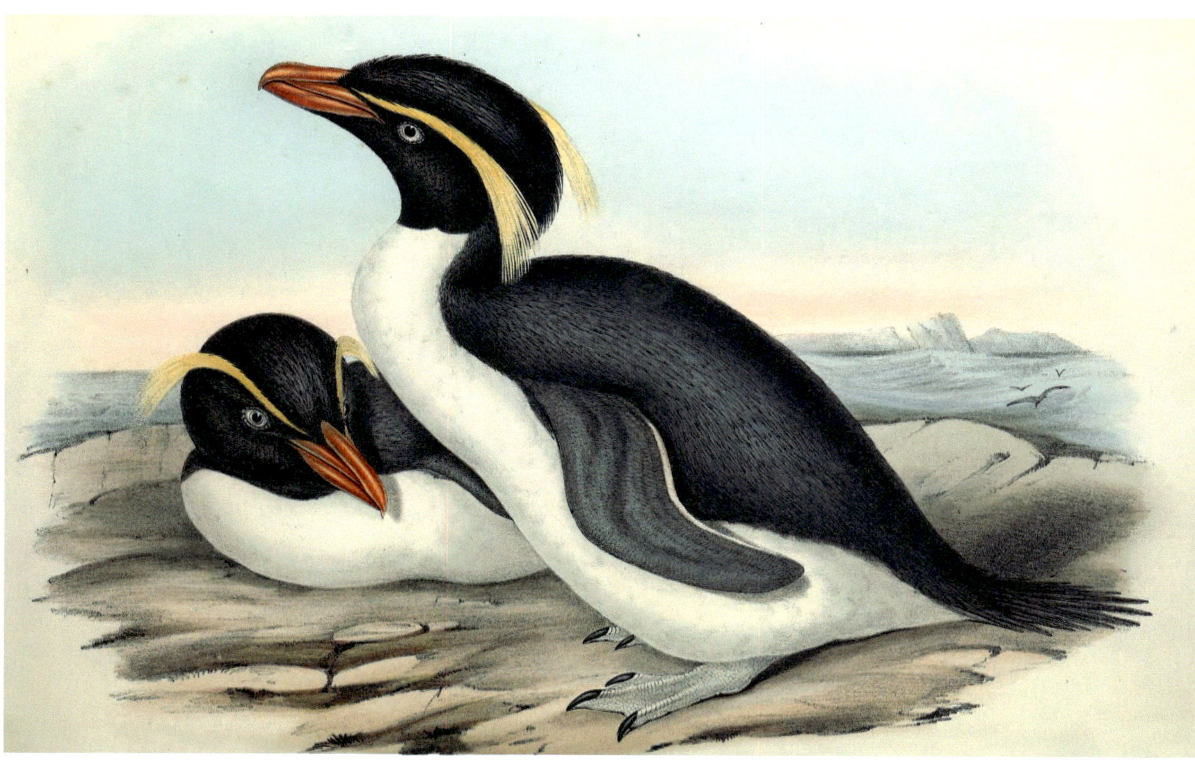

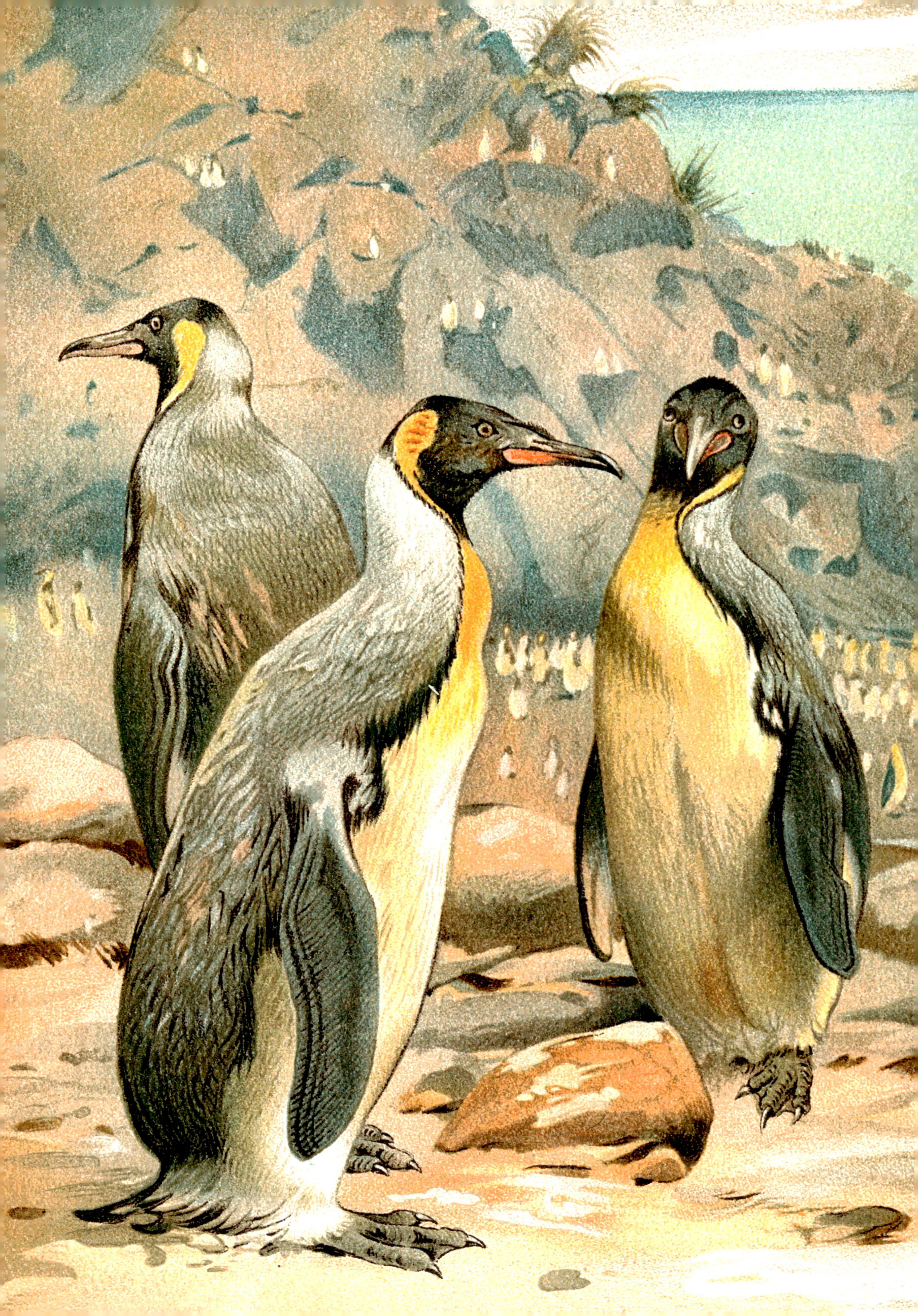

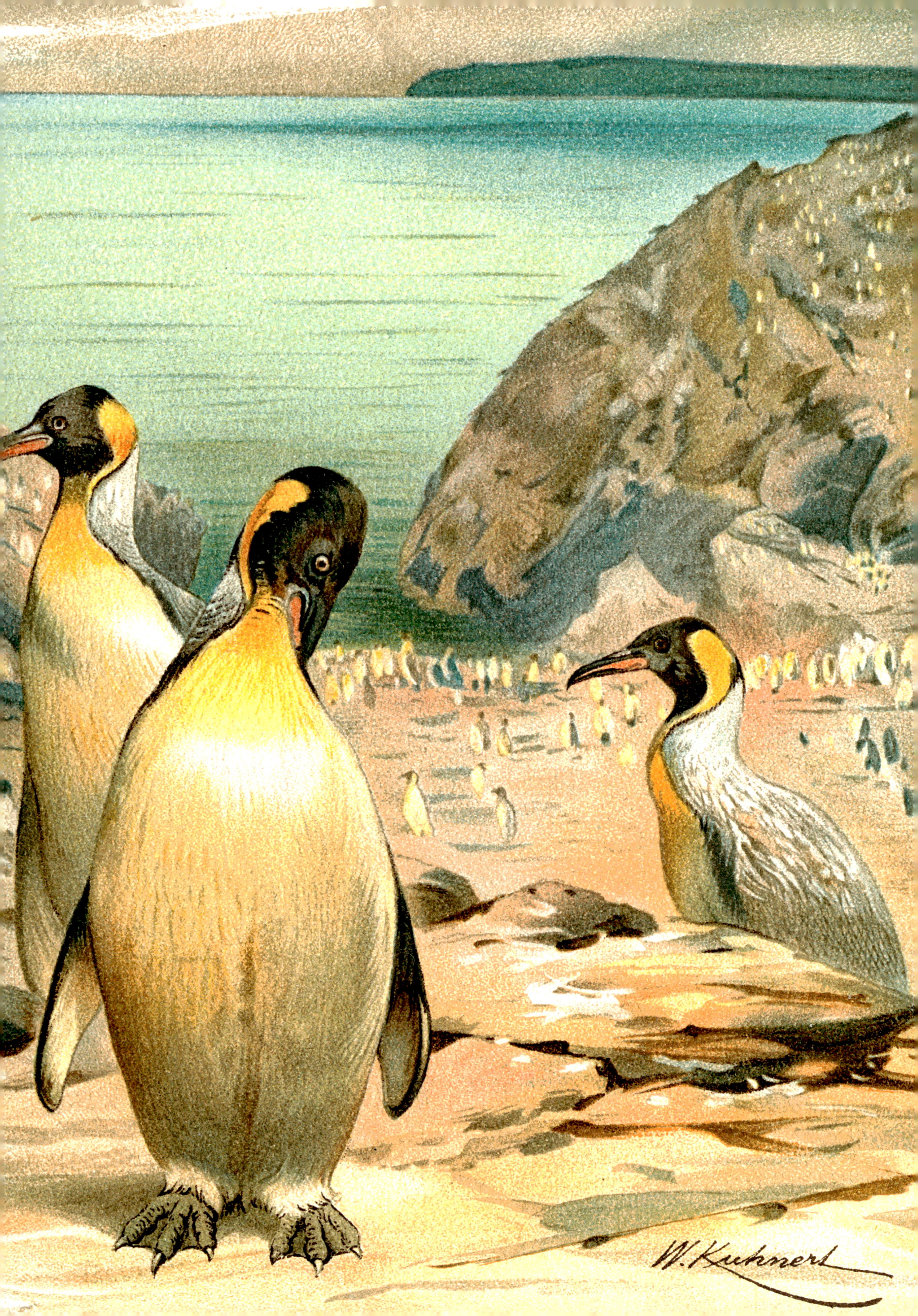
W. Kuhnert

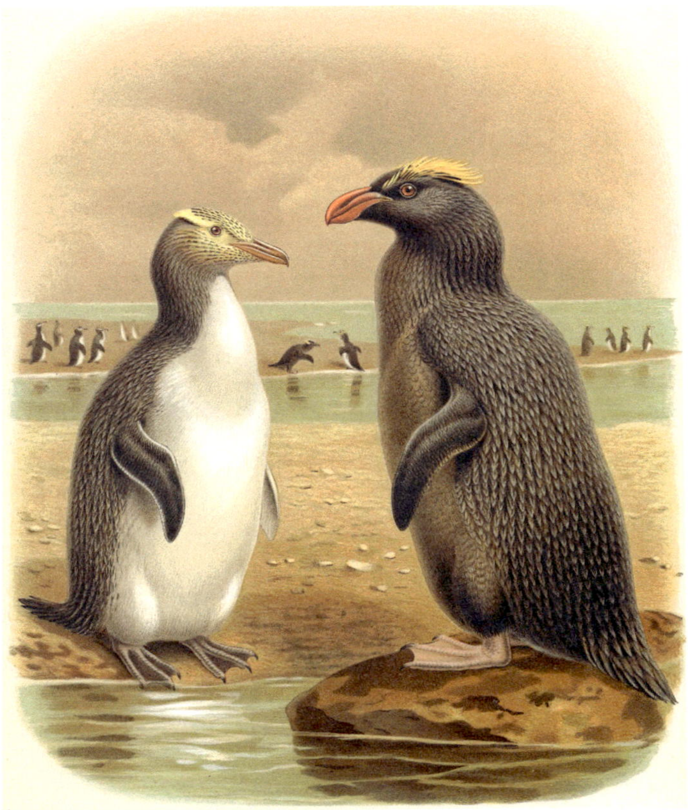

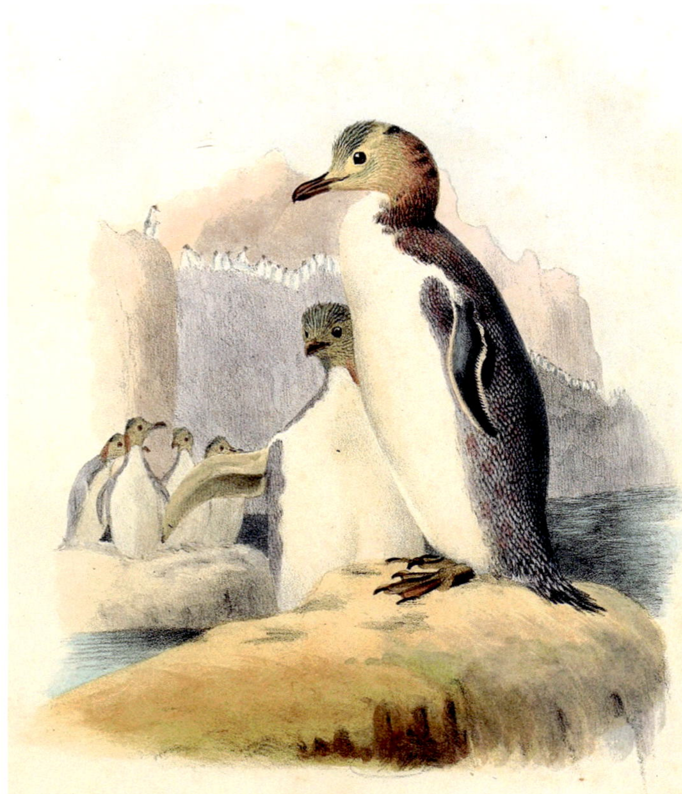

LEFT:

YELLOW-EYED PENGUIN & ERECT-CRESTED PENGUIN
Megadyptes antipodes & Eudyptes sclateri

John Gerrard Keulemans, *A History of the Birds of New Zealand*, (1872–3, this edition: 1887–8)

The yellow-eyed penguin and the erect-crested penguin are endemic to New Zealand. Seen here together, both would have at one stage bred on the Southern Island's Otago Peninsula, but today the erect-crested penguin only breeds on the remote islands of Bounty and Antipodes.

RIGHT:

YELLOW-EYED PENGUIN
Megadyptes antipodes

Joseph Wolf, *The Zoology of the Voyage of the H.M.S. Erebus & Terror*, 1844–75

The yellow-eyed penguin is also known by the Māori name *hoiho*. This means 'noise shouter' and is a reference to the bird's distinctive bray-like call. It is now the largest living penguin to breed on the mainland of New Zealand.

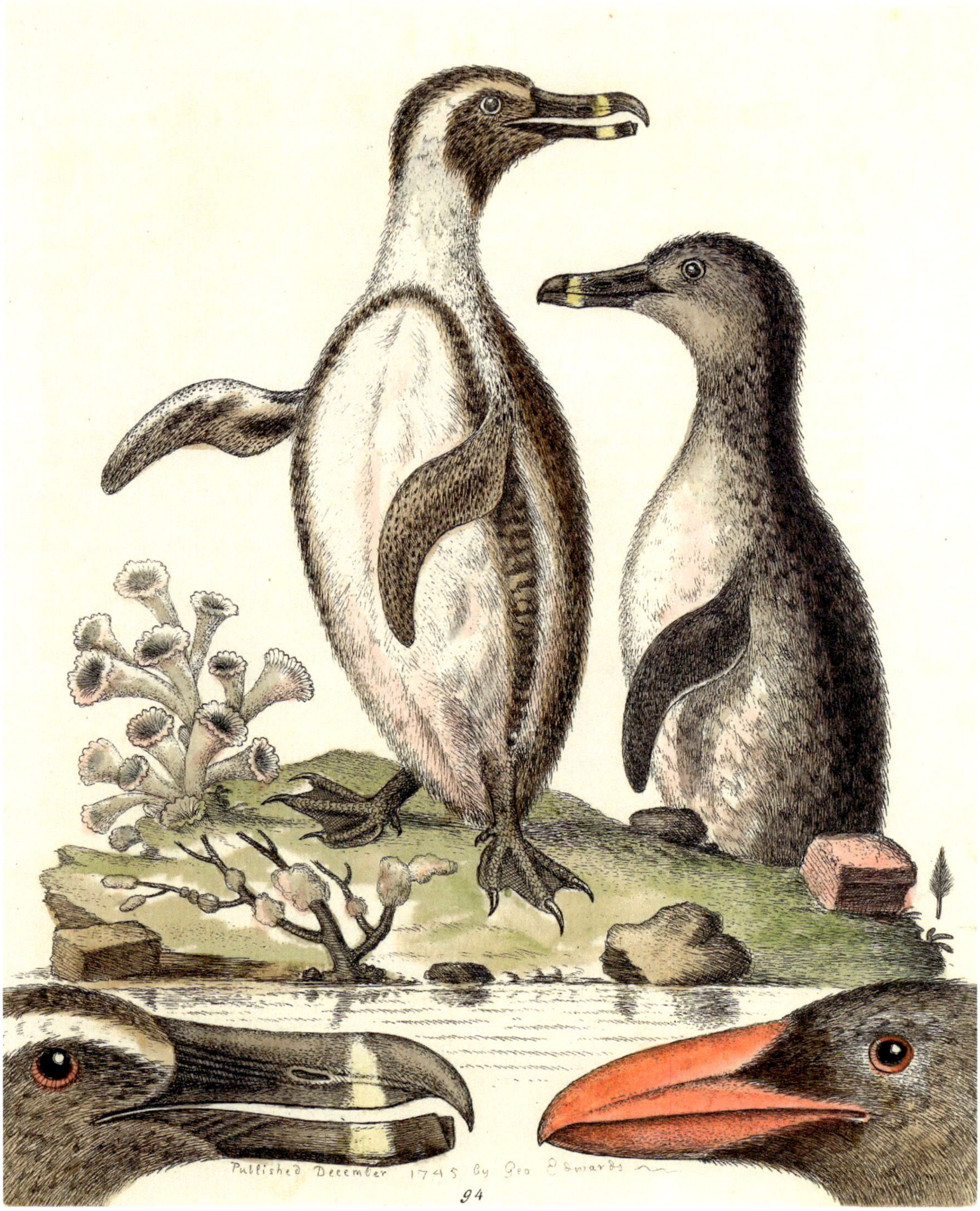

AFRICAN PENGUIN
Spheniscus demersus

George Edwards, *A Natural History of Uncommon Birds*, 1743–51

Edwards featured two penguins in his *Natural History of Uncommon Birds*. This – the second – is labelled the 'black-footed penguin' to distinguish it from 'the penguin' which Edwards had described earlier as having feet 'of a dirty red colour'. Edwards had no knowledge of where these birds had come from. The first, he notes, was lent to him by a Mr Cowell and this specimen had come from a Mr Holms. Both had come from East-India ships and so Edwards had rightly assumed that this bird had come from the Cape of Good Hope.

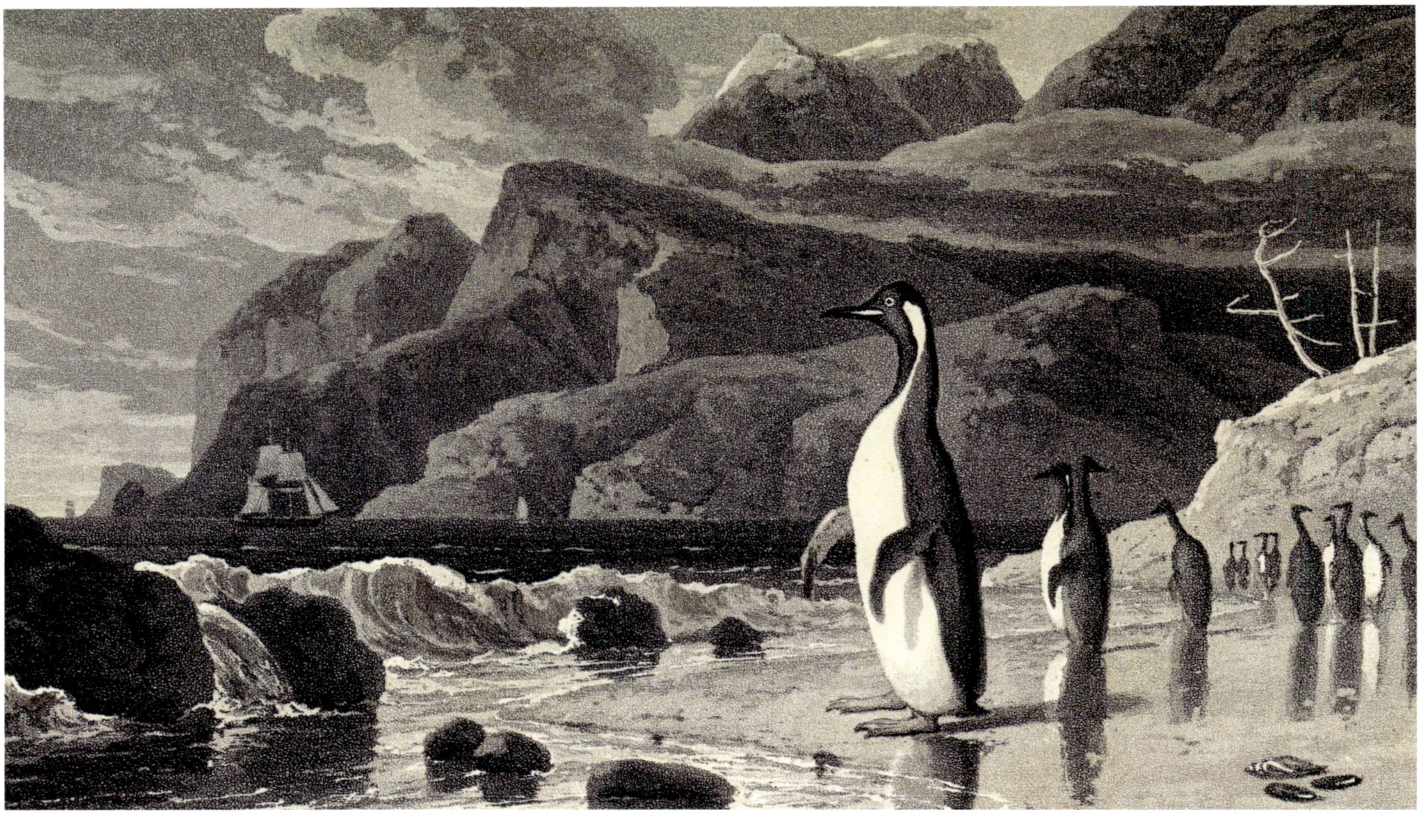

PENGUINS
Sphenisciformes

William Daniell, *Zoography*, 1807

This black-and-white aquatint from William Woods's *Zoography* shows a waddle of stylized penguins against the dramatic backdrop of a rugged coastline. Created by celebrated English landscape and marine painter William Daniell, the illustrations in this work highlight a large variety of wonders in nature, and Daniell's work tends to emphasize the landscape in which each of his subjects sit.

OPPOSITE:
FLIGHTLESS CORMORANTS
Phalacrocorax harrisi

John Gerrard Keulemans,
Novitates Zoologicae, vol. 9, 1902

Also known as the Galapagos cormorant, this species is endemic to the Galapagos Islands. It is unique in that it is the only known cormorant that has lost its ability to fly. This evolution occurred because the islands did not present any natural enemies for this bird. They are very efficient swimmers and, while their legs may be clumsy on land, they work incredibly well in water. Their stubby wings help them to become an impressive hydrodynamic shape while underwater; on land they often use their wings as a parasol to protect their chicks from the sun.

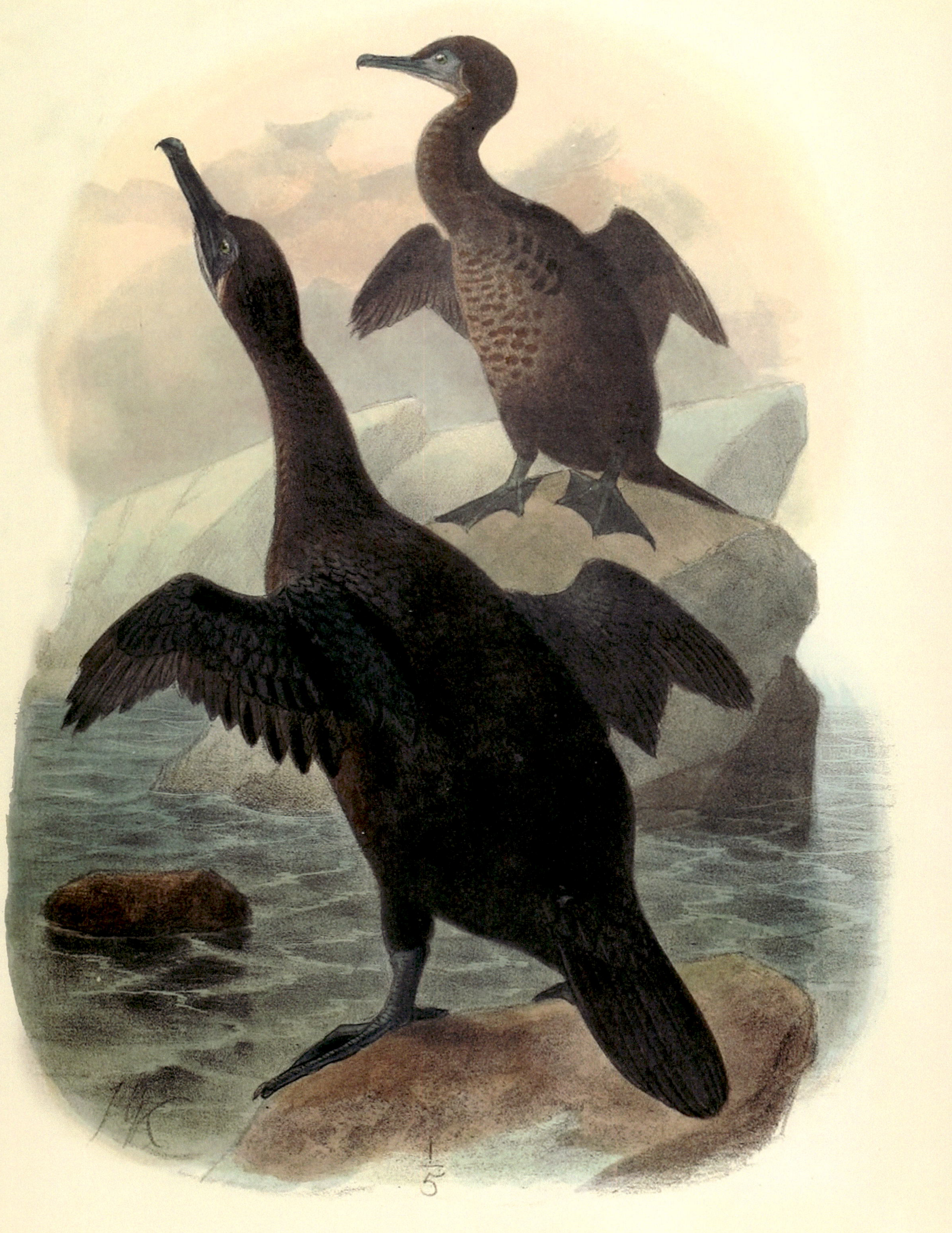

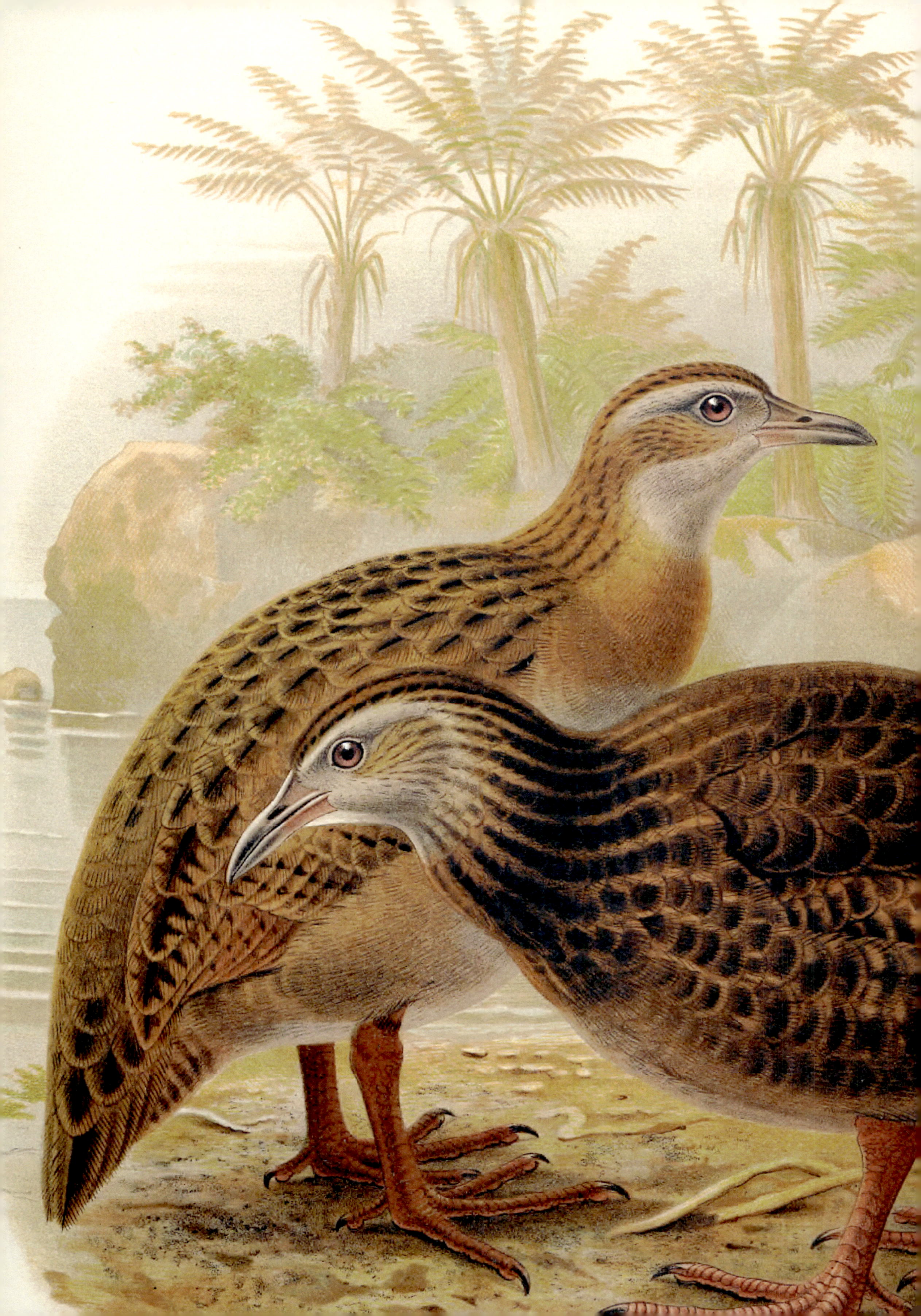

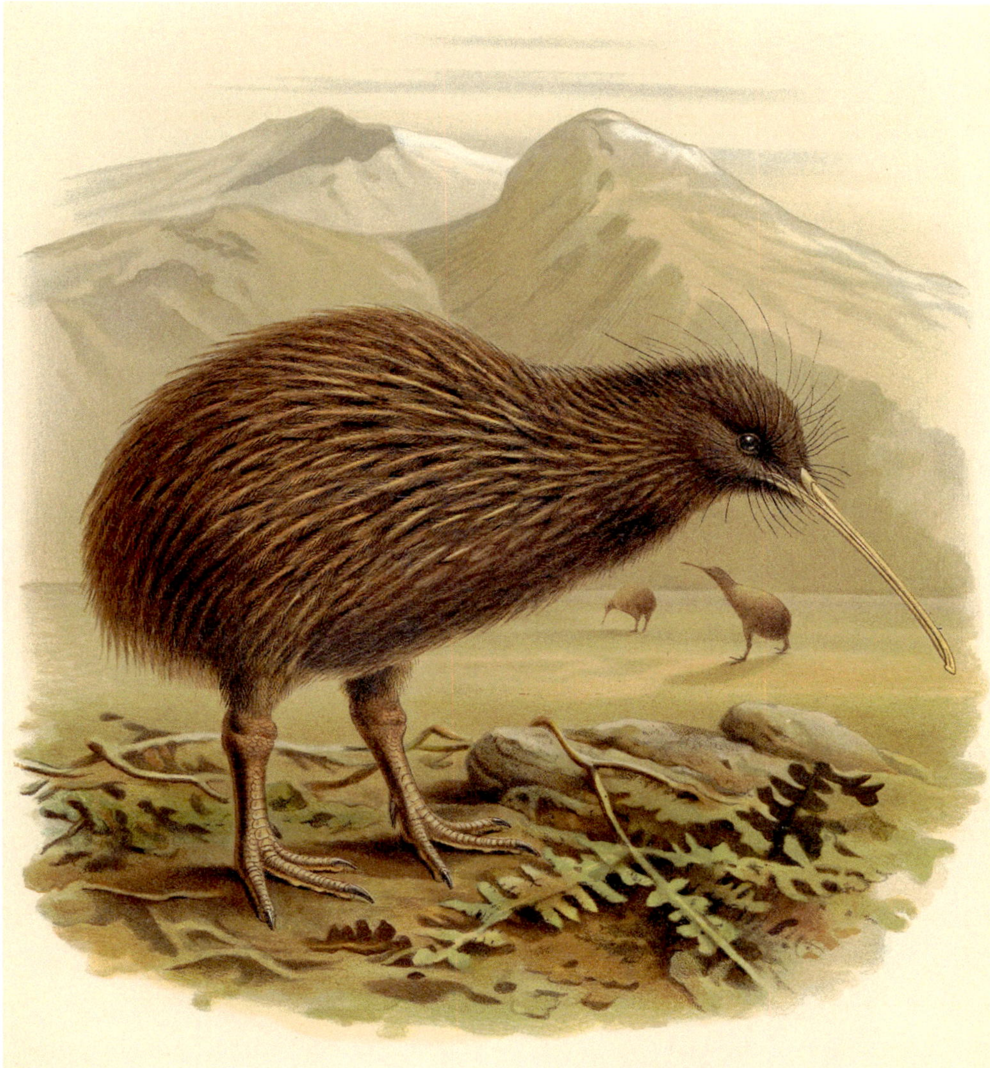

OPPOSITE:

WEKA (NORTH ISLAND AND SOUTH ISLAND TYPE)
Gallirallus australis greyi & *Gallirallus australis australis*

John Gerrard Keulemans, *A History of the Birds of New Zealand*, (1872–3, this edition: 1887–8)

Sometimes called the Māori hen, the weka is a flightless member of the rail family and is endemic to New Zealand. This plate by Keulemans shows the darker South Island form (*australis*) in the foreground and the lighter North Island weka (*greyi*) behind. The bird is likely to have derived from a flighted ancestor; the banded rail is considered to be their nearest flying relative. They are incredibly curious and charismatic birds and are often attracted to human activity.

NORTH ISLAND BROWN KIWI
Apteryx mantelli

John Gerrard Keulemans, *A History of the Birds of New Zealand*, (1872–3, this edition: 1887–8)

The kiwi is so beloved in New Zealand that the country and its people are colloquially known as Kiwis. The origin of this national symbol is probably rooted in a 1904 cartoon which depicted the country as a plucky kiwi that morphed into a powerful moa as the All Blacks defeated Britain in the first rugby test between the motherland and the then newly self-governing colony. There are five recognized species of kiwi and all have been affected by historic deforestation. This, the North Island brown kiwi, is the most common species.

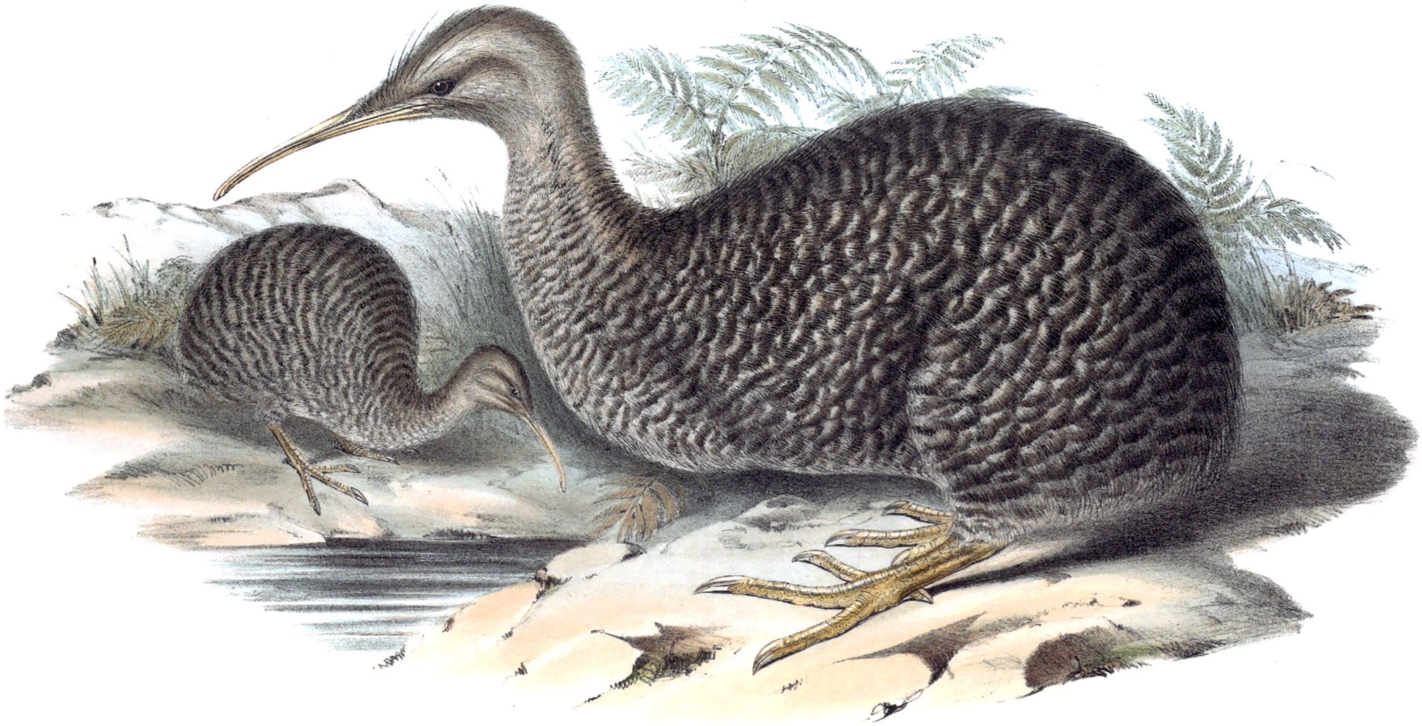

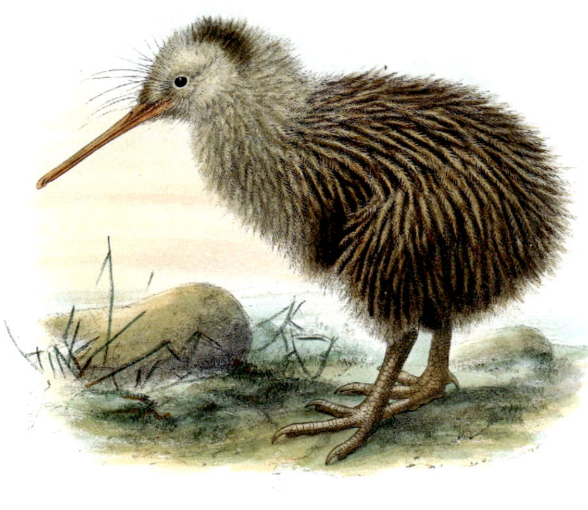

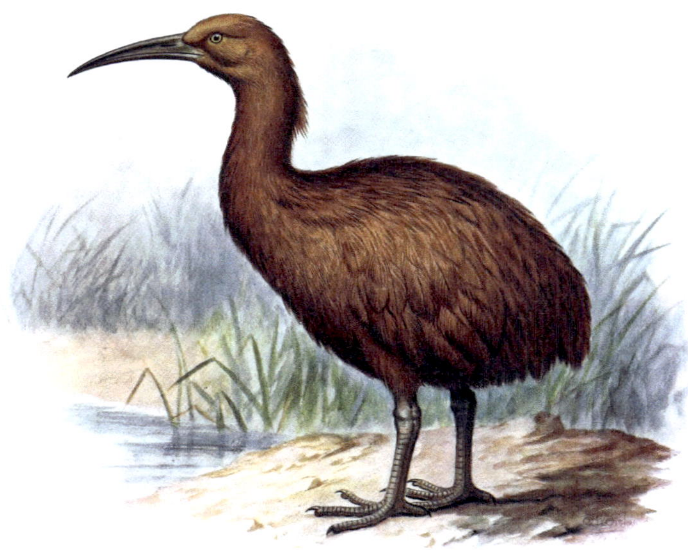

TOP:
LITTLE SPOTTED KIWI
Apteryx owenii

John Gould & Henry Constantine Richter, *The Birds of Australia*, 1840–8

Among the kiwi's other distinctive characteristics is its lack of tail feathers. Its feathers are coarse and shaggy and give the appearance of fur; its wings appear invisible against its body. It also is known for laying the biggest eggs in proportion to its body size.

BOTTOM LEFT:
SOUTHERN BROWN KIWI
Apteryx australis

John Gerrard Keulemans, *Ornithological Miscellany*, 1876–8

Kiwis eat a broad diet that includes seeds, grubs, fruit and small invertebrates. It finds its prey using its highly developed sense of smell. A strong sense of smell is uncommon in birds and the kiwi is, in fact, the only bird to have nostrils located at the end of its beaks.

BOTTOM RIGHT:
RED RAIL
Aphanapteryx bonasia

Frederick William Frohawk, *Extinct Birds*, 1907

The red rail was a flightless bird endemic to the island of Mauritius and is thought to have gone extinct around 1700. Descriptions and illustrations of the bird were our only link to it until subfossils of the bird were discovered in the 1860s.

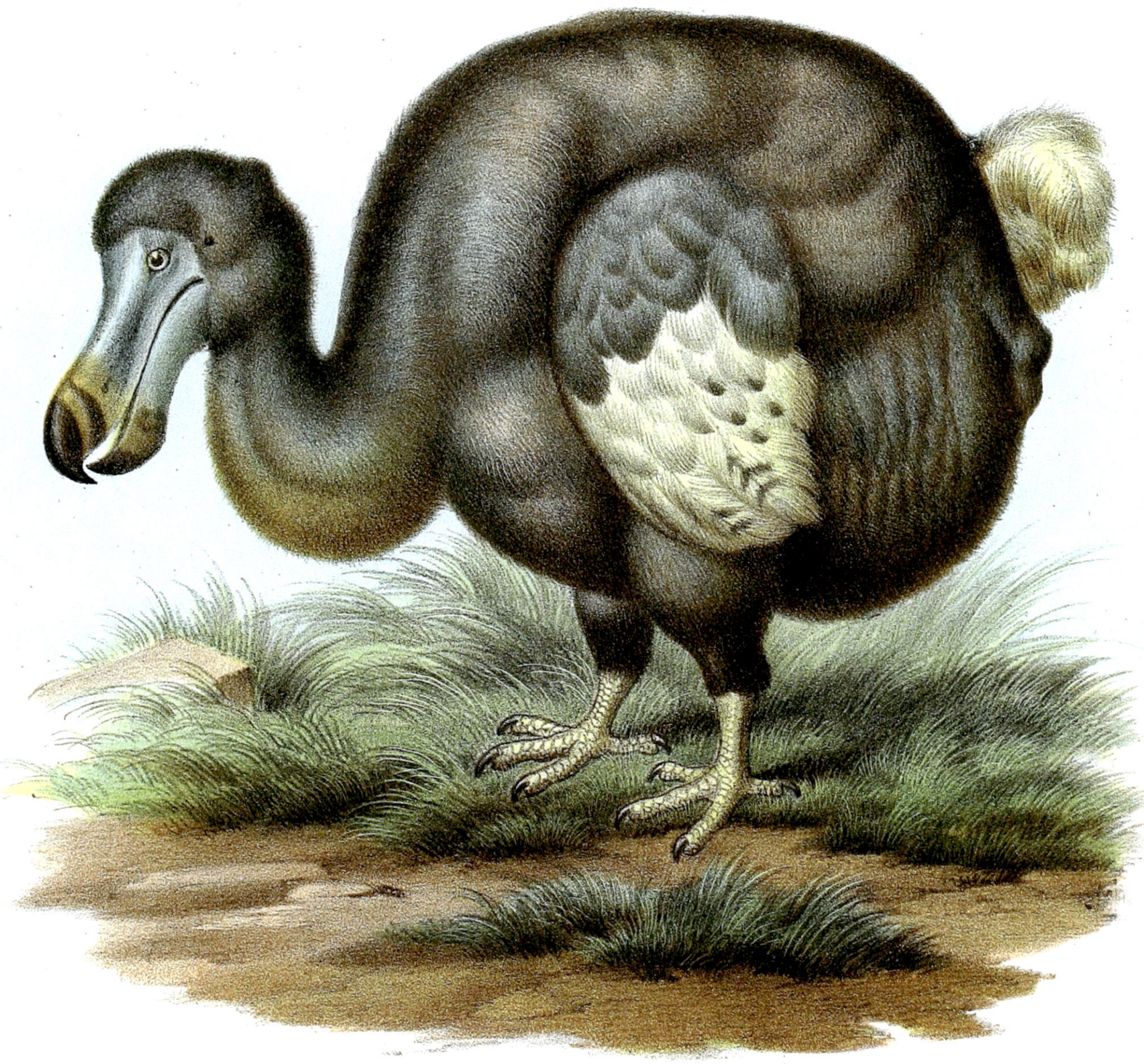

DODO
Raphus cucullatus
Uncredited, *Bilder-atlas*, 1860–4

The dodo is an extinct species that was endemic to the island of Mauritius and died out in the seventeenth century. Frequently cited as one of the most well-known examples of human-induced extinction, the dodo's first encounter with humans probably occurred when Dutch sailors arrived on the island in 1598. Some accounts say that these tame birds were hunted to extinction, but it is far more likely that it was rats and other animals aboard the ship that were its true predators, and brought an end to this endearing bird.

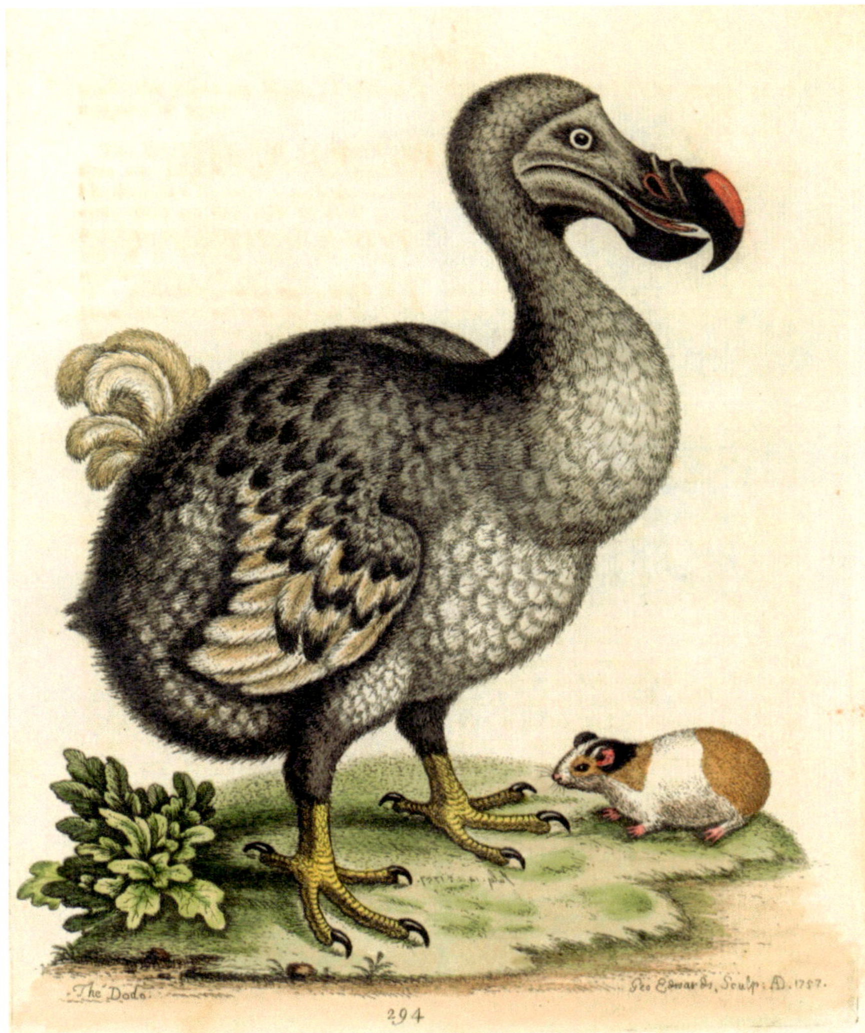

DODO
Raphus cucullatus

George Edwards, *Gleanings of Natural History*, 1758–64

In 1757 Edwards created this illustration of a dodo by copying an oil painting that had been produced by the Dutch painter Roelandt Savery. This original work had been drawn in 1626 from a living bird in Holland that had come from Mauritius. The painting had originally been owned by Hans Sloane and was passed down to Edwards following the collector's death in 1753. Now called *Edwards's Dodo*, it has become the standard image of a dodo and is housed in the Natural History Museum, London.

OPPOSITE, TOP:
WHITE DODO
Raphus cucullatus

Frederick William Frohawk, *Extinct Birds*, 1907

This restoration of a supposed 'white dodo' by Frederick William Frohawk is now considered to be erroneous. Published in 1907, discussion of this species arose from a number of travellers' tales that spoke of a large whitish bird on the island of Réunion. Several paintings and drawings of white dodos were also produced in the mid-seventeenth century.

OPPOSITE, BOTTOM:
DODO
Raphus cucullatus

Vincent Brooks, *The Dodo and its Kindred*, 1848

In 1848 the naturalists Hugh Edwin Strickland and Alexander Gordon Melville published a monograph on the dodo that aimed to separate reality from myth. Titled *The Dodo and its Kindred*, the book paid tribute to Roelandt Savery's painting in a frontispiece illustration that saw Savery's name carved in stone.

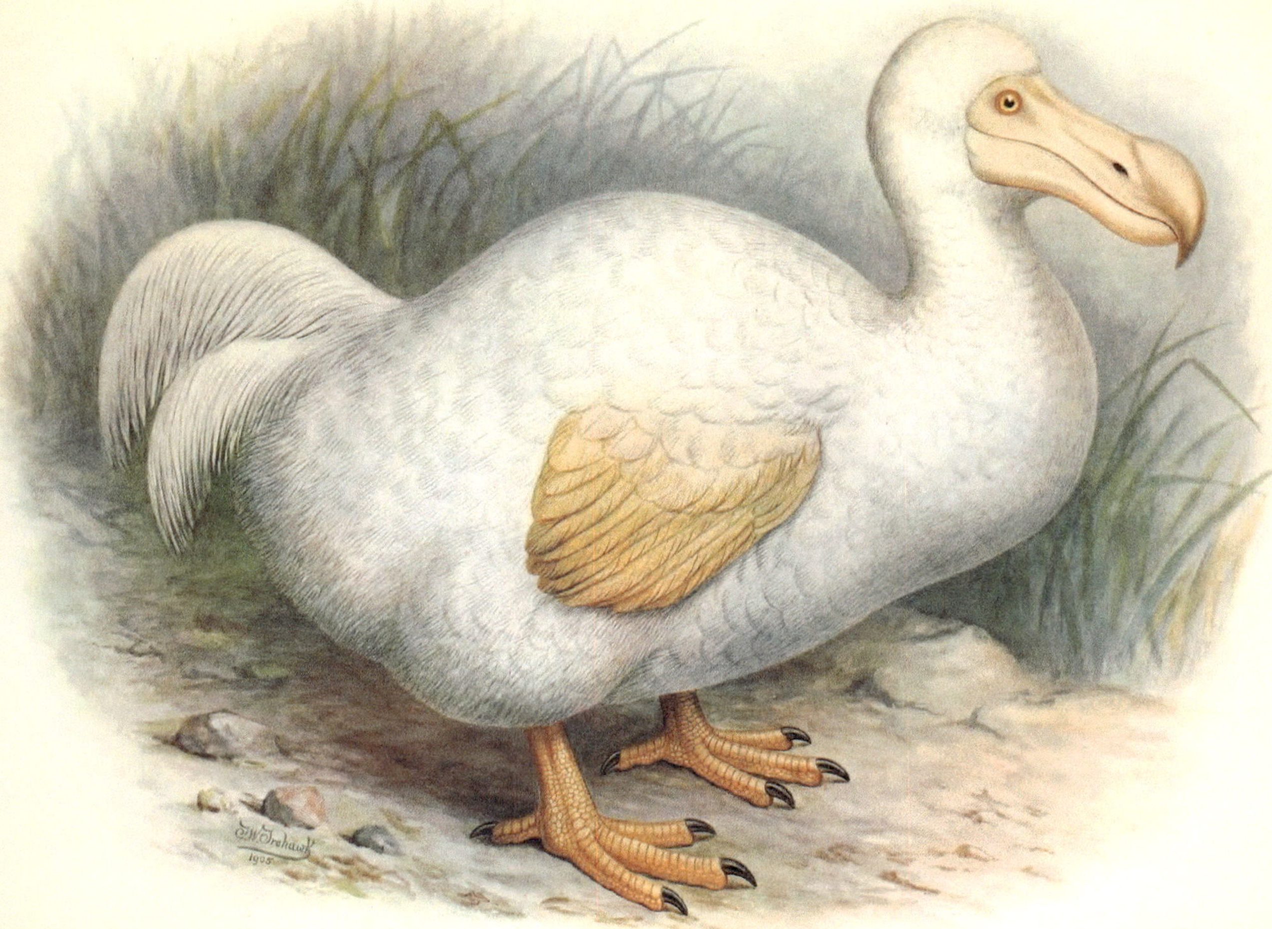

DIDUS SOLITARIUS
(ONE-THIRD NATURAL SIZE—*from a Dutch picture taken from living bird in Amsterdam, beak and wing restored*)

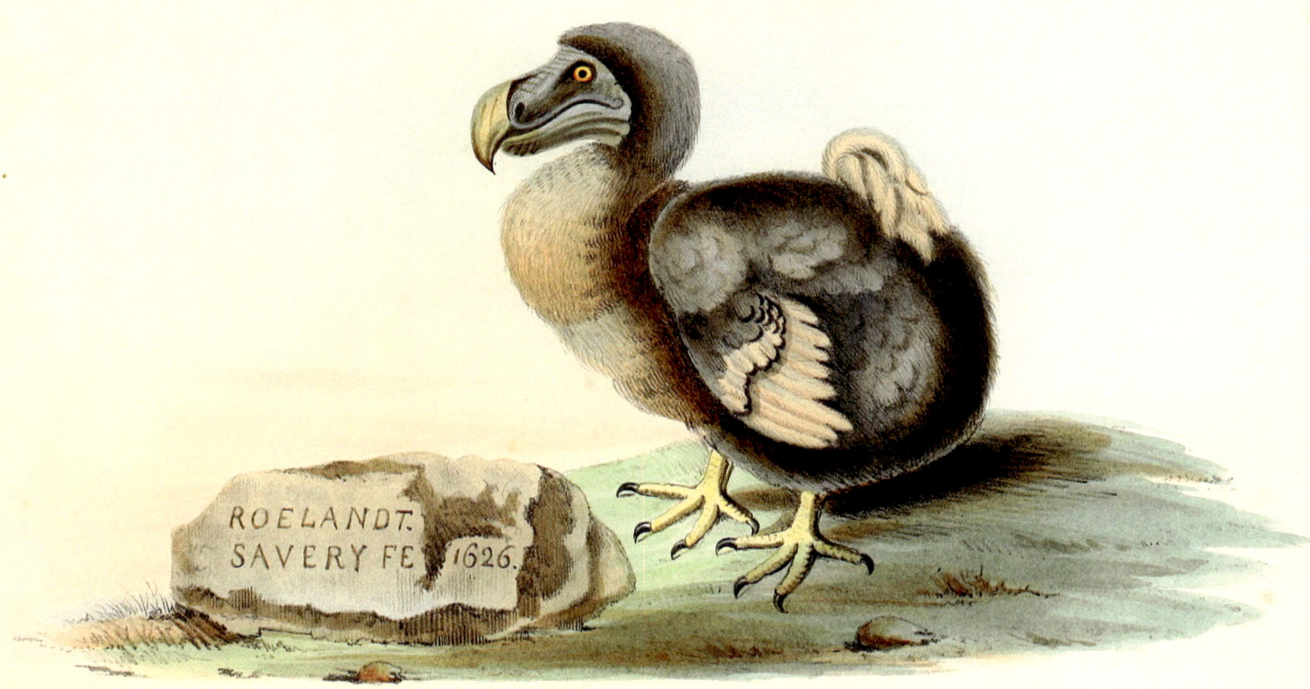

Fac-simile of Savery's picture of the DODO, in the Royal Gallery at Berlin.

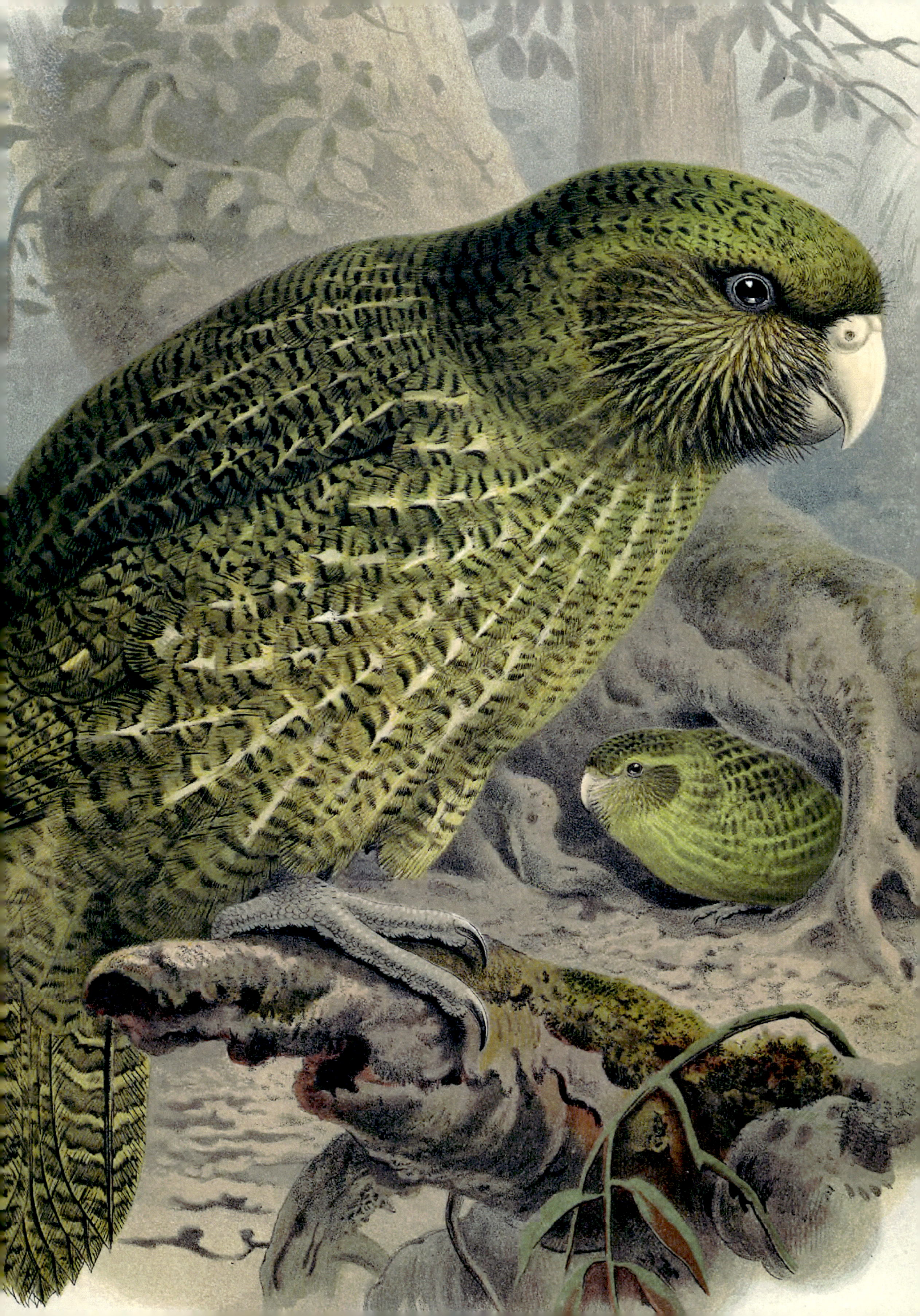

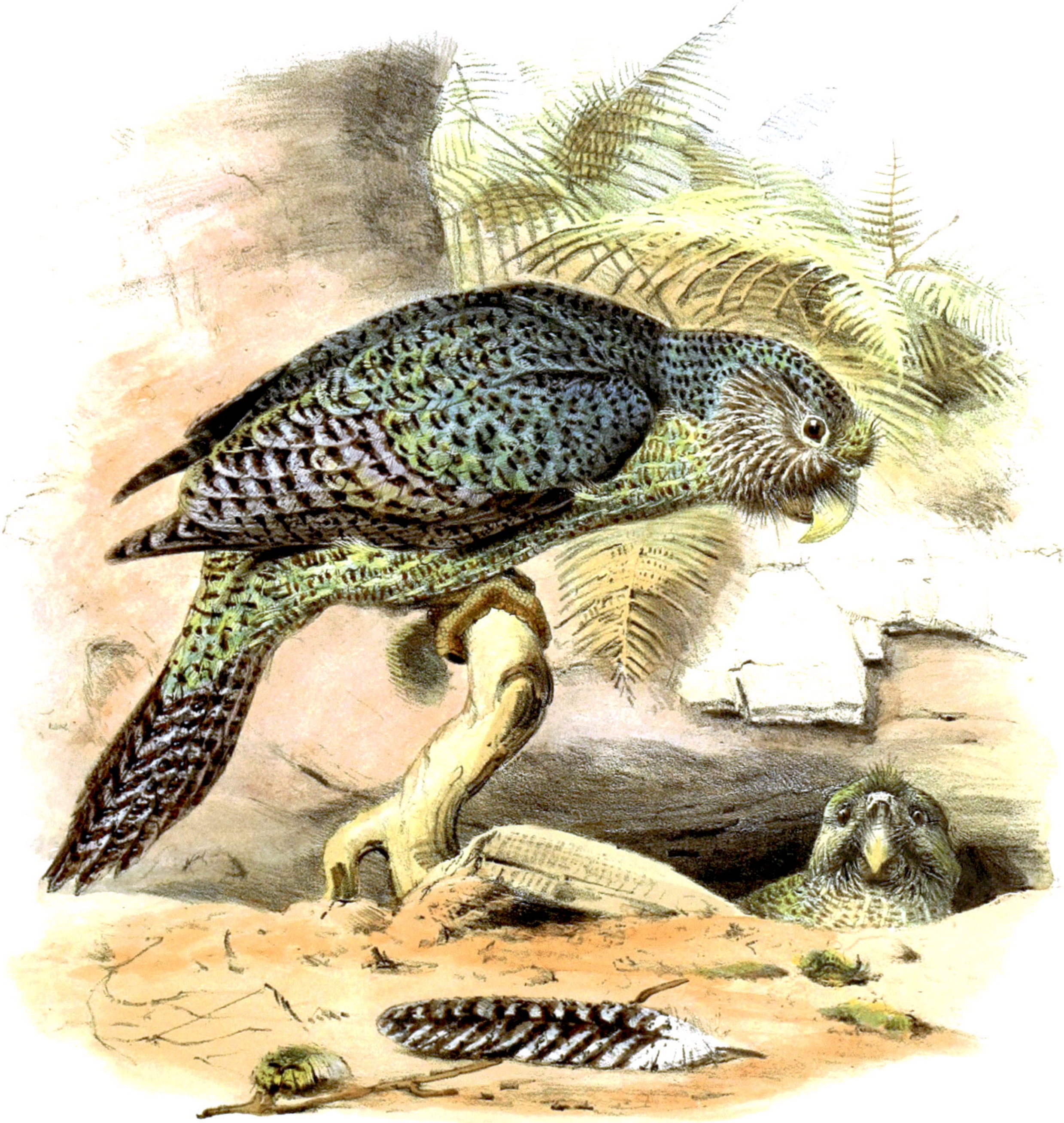

OPPOSITE:

KAKAPO
Strigops habroptila

John Gerrard Keulemans,
A History of the Birds of New Zealand,
(1872–3, this edition: 1887–8)

KAKAPO
Strigops habroptila

Joseph Wolf, *The Zoology of the Voyage of the H.M.S. Erebus & Terror*, 1844–75

Kakapo are large flightless parrots endemic to New Zealand. They are the largest parrots in the world and are also one of the world's longest living birds. In pre-Polynesian times, they were one of the most common birds on all three of the country's main islands, but when the new settlers arrived they brought cats, rats and other animals that devastated the population. Today, they remain critically endangered and a kakapo recovery programme has been in operation since 1989.

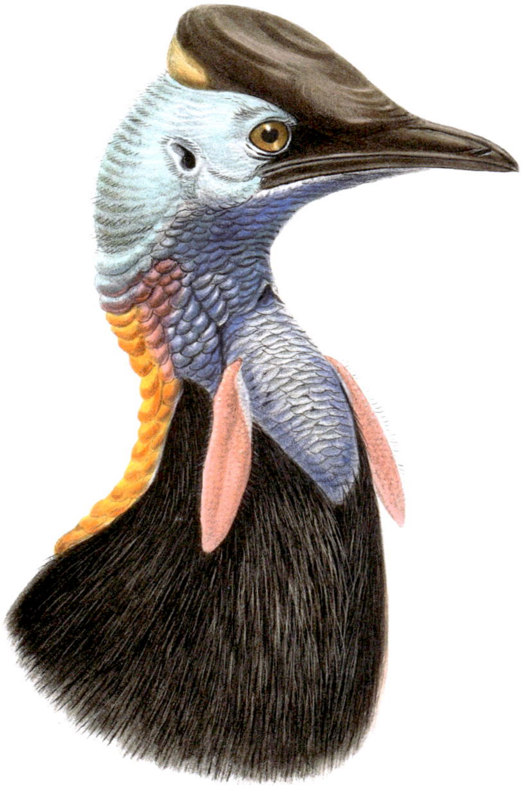

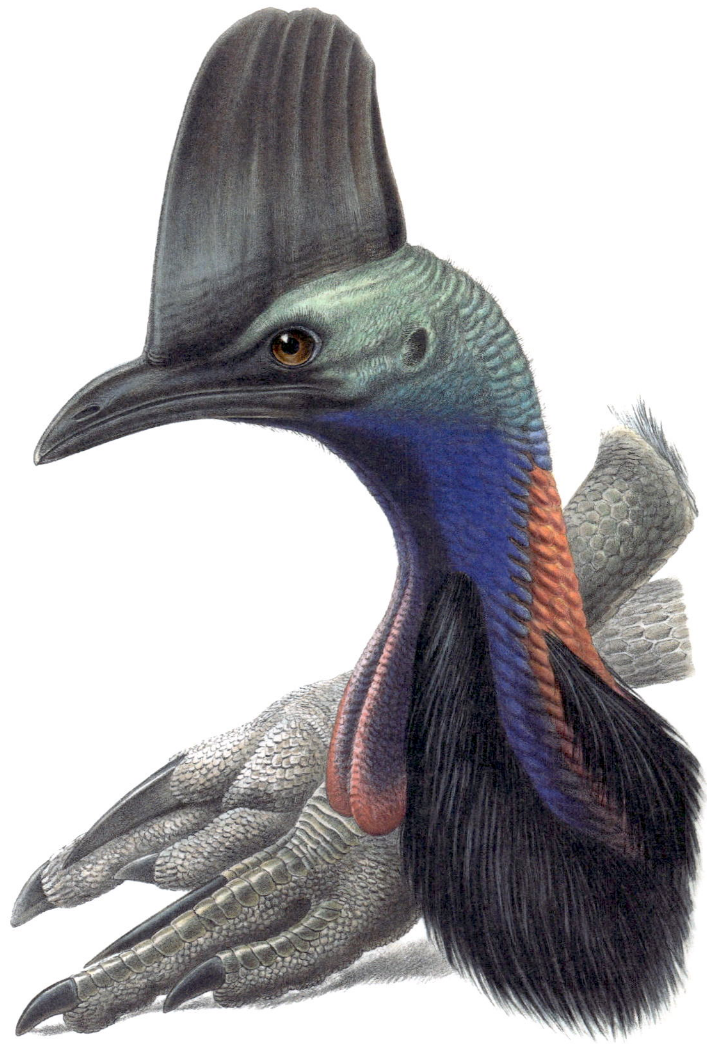

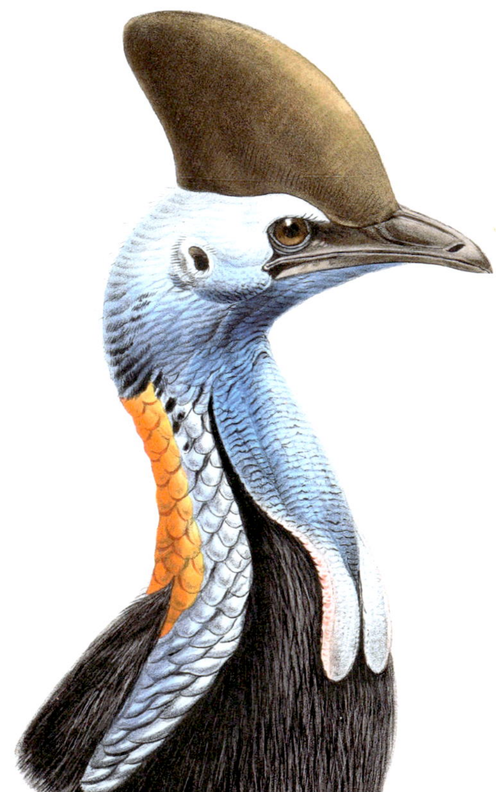

TOP LEFT:

SOUTHERN CASSOWARY
Casuarius casuarius

John Gerrard Keulemans, *A Monograph of the Genus Casuarius*, 1900

BOTTOM LEFT:

SOUTHERN CASSOWARY
Casuarius casuarius

John Gerrard Keulemans, *A Monograph of the Genus Casuarius*, 1900

RIGHT:

SOUTHERN CASSOWARY
Casuarius casuarius

John Gould & Henry Constantine Richter, *Supplement to the Birds of Australia*, 1851–69

Cassowaries are large flightless birds that are closely related to the emu. Their plumage is jet black, yet they have vivid skin on their neck and face as well as an impressive casque atop their head. Their look can vary depending on their species and location. This can be seen in the three illustrations on this page. The two images to the left were published in a monograph on the species by British zoologist Walter Rothschild. Rothschild was an enthusiastic collector of cassowaries and amassed one of the largest collections of them in the world.

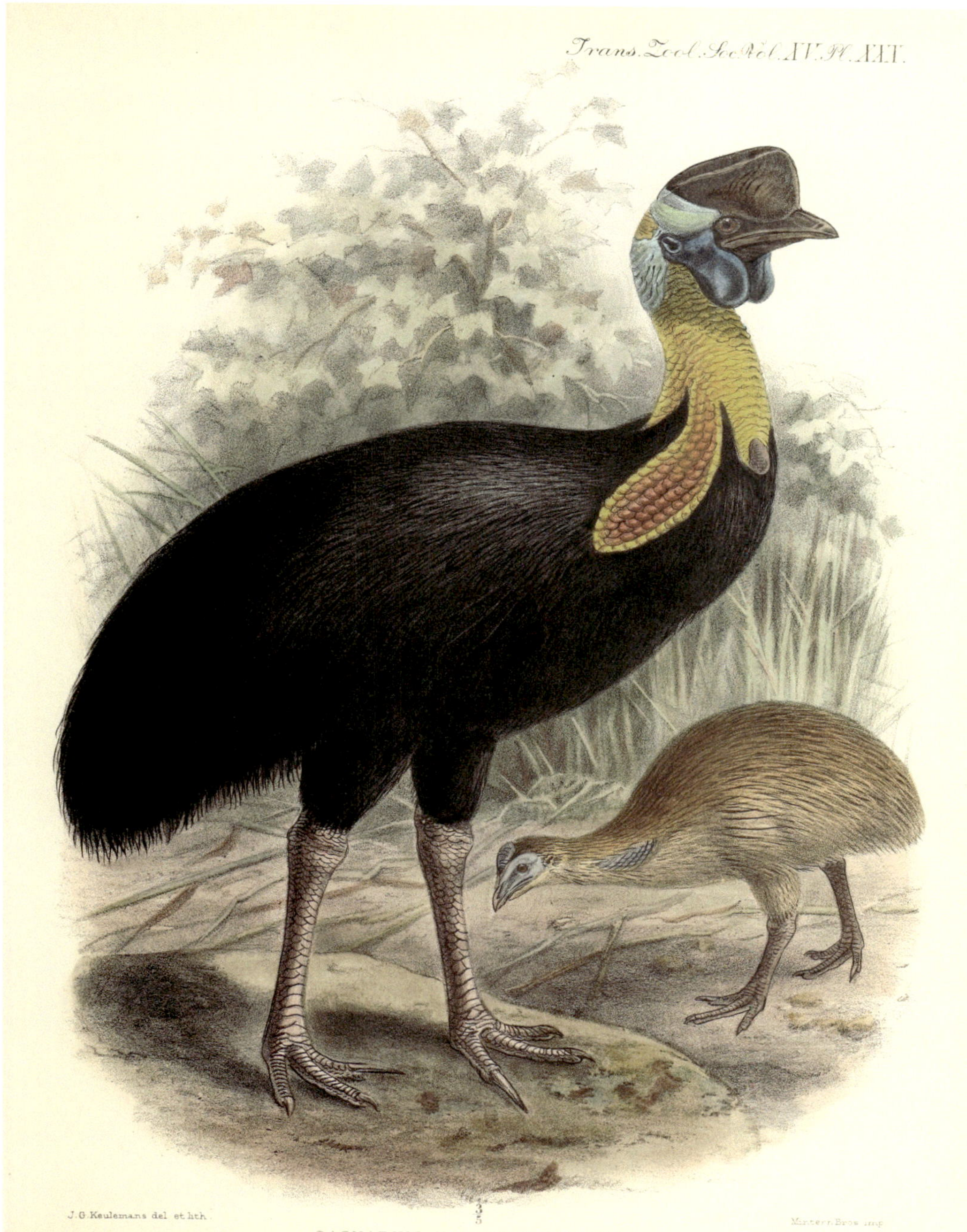

NORTHERN CASSOWARY
Casuarius unappendiculatus

John Gerrard Keulemans, *A Monograph of the Genus Casuarius*, 1900

Cassowaries are considered to be foul-tempered birds, but in reality, they are simply highly territorial. This northern cassowary is typically found in the lowland swamps and rainforests of Northern New Guinea, as well as on the Yapen, Salawati and Batanta islands. Its wings are small and almost non-existent yet these large birds would once have been strong and powerful flyers. Keulemans' illustration shows both an adult and juvenile; the latter's plumage is brown and its casque is much smaller.

John & Elizabeth Gould

1804–81 / 1804–41

Fuelled by a boundless ambition to be the world's premier authority on birds, John Gould produced some of the most impressive ornithological works of all time. Working with a large team of artists that included his wife Elizabeth, he produced more than forty volumes of work and more than 3,000 coloured plates.

The most celebrated British ornithologist of the nineteenth century, John Gould, was the author and publisher of some of the most impressive and financially successful works on birds ever created. A man of immense ambition, he developed a reputation for being Britain's defining authority on the subject. Despite his work's unrivalled financial success, Gould himself had come from modest beginnings. The son of a gardener, he originally followed in his father's footsteps, working initially as a naturalist in the Royal Gardens of Windsor. It was his ability in taxidermy that led him to a life in science. He taught himself the skill at a young age, and it eventually led to him becoming the first curator and preserver at the museum of the Zoological Society of London.

Eager to be accepted into the world of science and stature, this role presented the young Gould with a plethora of contacts and connections. When a collection of bird specimens arrived from the Himalayas, Gould saw this as an opportunity for career advancement. Many of these birds had not been described before, and while his skills as an artist were moderate at best, his true strength lay in his business acumen. He had witnessed the potential of lithographic printing through the work of Edward Lear and, utilizing both Lear's talents and that of his wife Elizabeth, he produced *A Century of Birds from the Himalaya Mountains* (1830–2).

The work was self-financed but Gould did well out of it. As the years progressed, his output turned from a small business into a full-on enterprise. Over the course of his career, he produced more than forty volumes of work, with more than 3,000 coloured plates. Yet to frame this work solely as the achievement of Gould would be unfair and unjust. This was the work of a whole team of people, with projects involving taxidermists, draughtsman, lithographers, colourists and printers. Often these names went unacknowledged, and Gould kept the inner workings of his practice well guarded.

The work of Elizabeth Gould deserves just as much attention and praise as that of her husband. Born Elizabeth Coxen in 1804, her skills in watercolour and lithography played an integral role in the success of her husband's work. For *Birds of Australia*, the Goulds ventured with the oldest of their four surviving children to the colonies of the New World. There, they spent two years exploring the occupied territories of Australia and Tasmania. John would travel extensively and collect many specimens, while Elizabeth would work from a house in Hobart producing hundreds of drawings based on the skins that John sent back to her. When they returned to England in 1840, they did so with countless drawings, specimens, notes and records. They even had another son while living there. The following year Elizabeth gave birth to yet another child but sadly, in the year that followed, she passed away. She was just thirty-seven years of age.

Despite this tragedy, Gould persevered. Elizabeth had completed some eighty-four plates for the project before her death, and the majority of the remaining work was then completed by English illustrator Henry Constantine Richter (1821–1902) who based most of his illustration on Elizabeth's original drawings. Published between 1840 and 1848, *The Birds of Australia* was a landmark publication and the first comprehensive illustrated account of Australian birds. Among its seven volumes and 600 plates, it includes 328 species that were new to science and which Gould had named and described. The expedition also led Gould to publish *A Monograph of the Macropodidae, or Family of Kangaroos* (1841–2) and the three-volume work *The Mammals of Australia* (1845–63).

While these works satisfied the colonial curiosities of its readers, another large reason for their success had to do with Gould's keen ability to define – and then stick to – a successful formula. His work put shape to a certain type of formalized aesthetic within ornithology. He set about, and then refined, the composition and design of plates. His images tended to show both the male and female of the species to highlight sexual dimorphism. He placed a stronger emphasis on certain elements of the background to denote a bird's habitat. It was an approach that worked incredibly successfully and these elements were then adopted by others as the norm for many years after. Much like Audubon, Gould had an ability to marry two seemingly opposing traits: the modernity of science with the romanticism of art. Unlike Audubon's dramatic work, however, Gould's compositions have a delicate beauty, which seemed to chime well with the British sensibilities of his time.

Having defined this formula for his work, and having the team and resources in place, Gould continued to produce commercially successful work well into old age. Having already published his *Birds of Europe* with Elizabeth between 1832 and 1837, he worked with his team to make follow-on projects that included *The Birds of Asia* (1850–83), *The Birds of New Guinea and the Adjacent Papuan Islands* (1875–88) and his hugely popular *Birds of Great Britain* (1862–73).

John Gould passed away in 1881 at the age of seventy six. His life and work had been fuelled by a boundless ambition to be the world's premier authority on birds. Although today many remember him for the beauty of his folios, his work in science helped to identify hundreds of species and led to a greater appreciation of the natural world within Victorian society.

EMU
Dromaius novaehollandiae

John Gould & Elizabeth Gould, *The Birds of Australia*, 1840–8

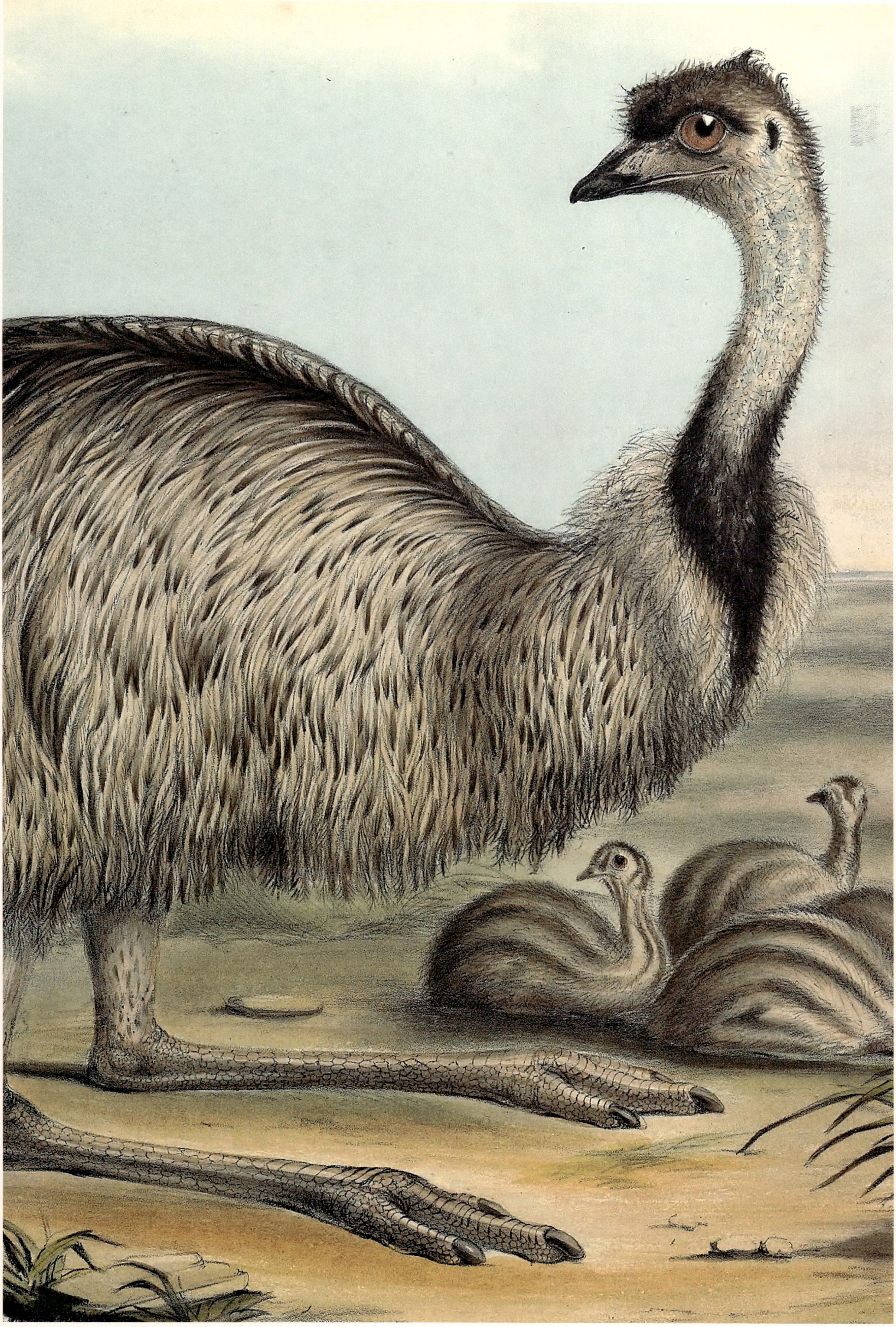

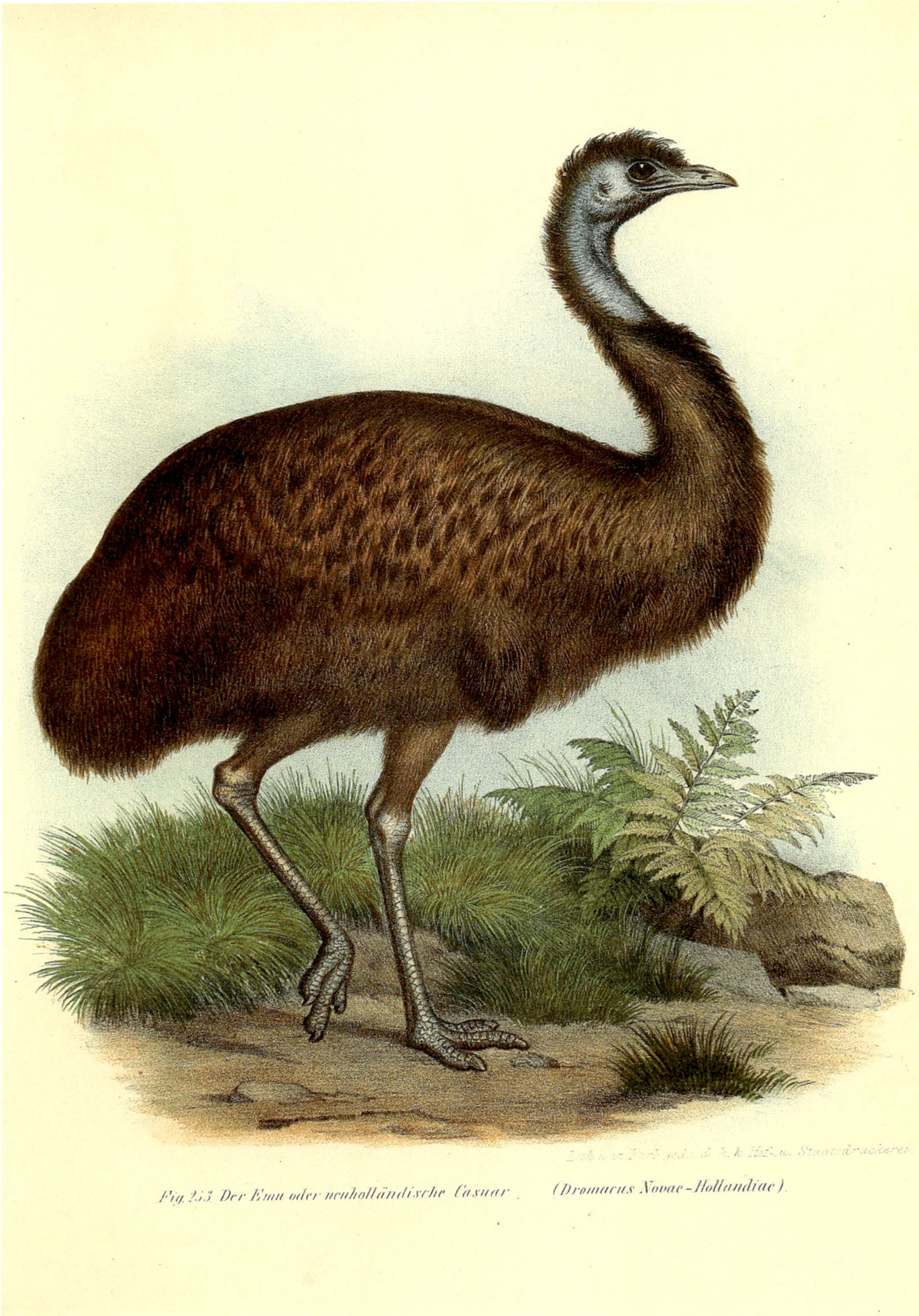

EMU
Dromaius novaehollandiae
Uncredited, *Bilder-atlas*, 1860–4

Emus are the second tallest birds in the world and are one of a number of birds that form the extensive group of birds known as ratites. These are mostly large, flightless birds that have a flat ratite breastbone that makes flight impossible. Other ratites include African ostriches, South American rheas and New Zealand kiwis. The emu is endemic to Australia and it is the country's largest native bird. Today, it can be found in many other countries, where it is typically farmed for its meat, eggs, feathers, oil and even leather.

COMMON OSTRICH
Struthio camelus

Peter Brown, *New Illustrations of Zoology*, 1776

The ostrich has the largest eyes of any bird in the world, and this is shown here in a rather over-pronounced manner by illustrator Peter Brown. Measuring fifty millimetres in diameter, their eyes help them to see predators from a great distance. They are also one of a very small number of birds that have upper eyelashes. This protects their sight from the sand. Brown's illustration lists the origin of this specimen as the Cape of Good Hope, and states that it comes from the collection of Welsh naturalist Thomas Pennant.

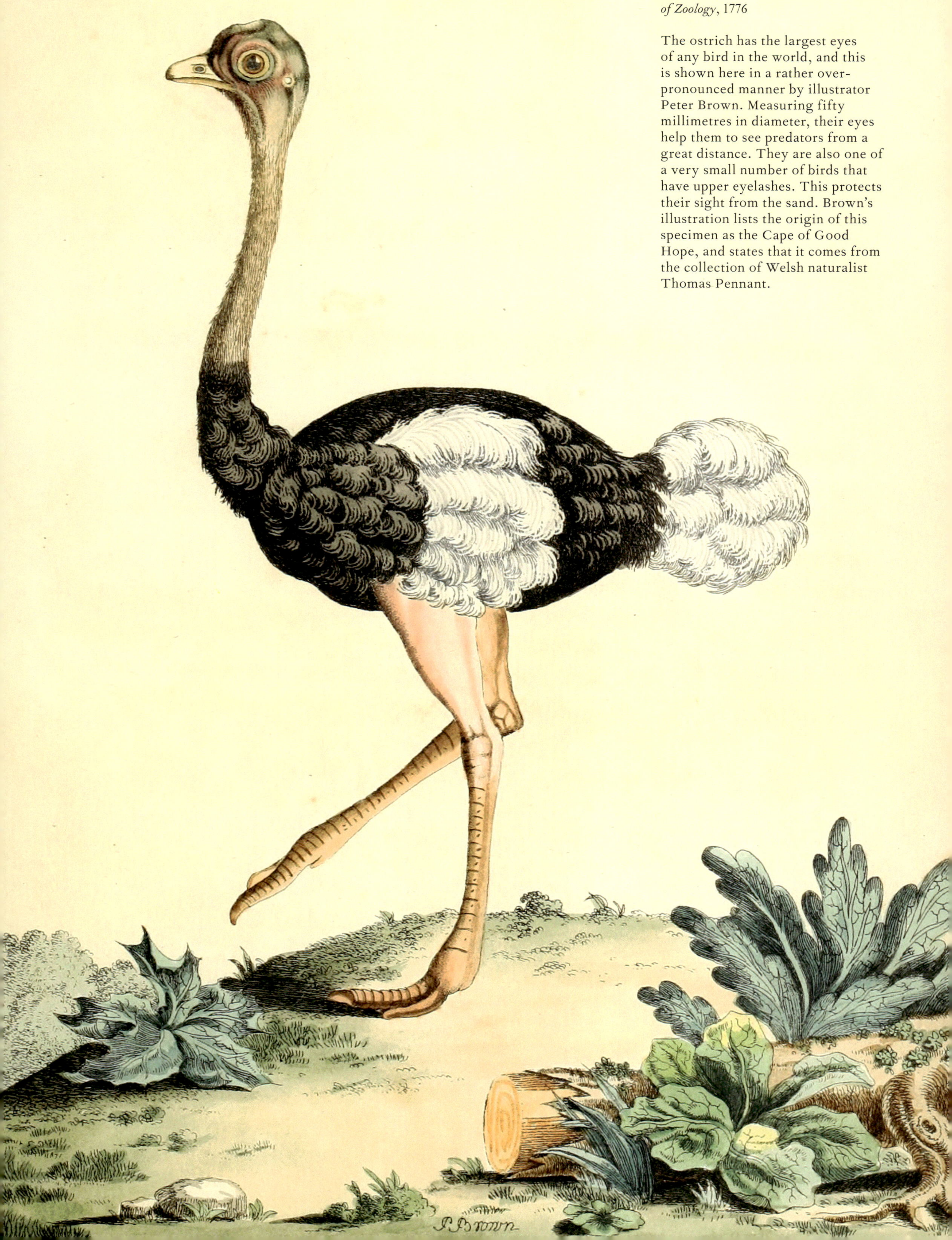

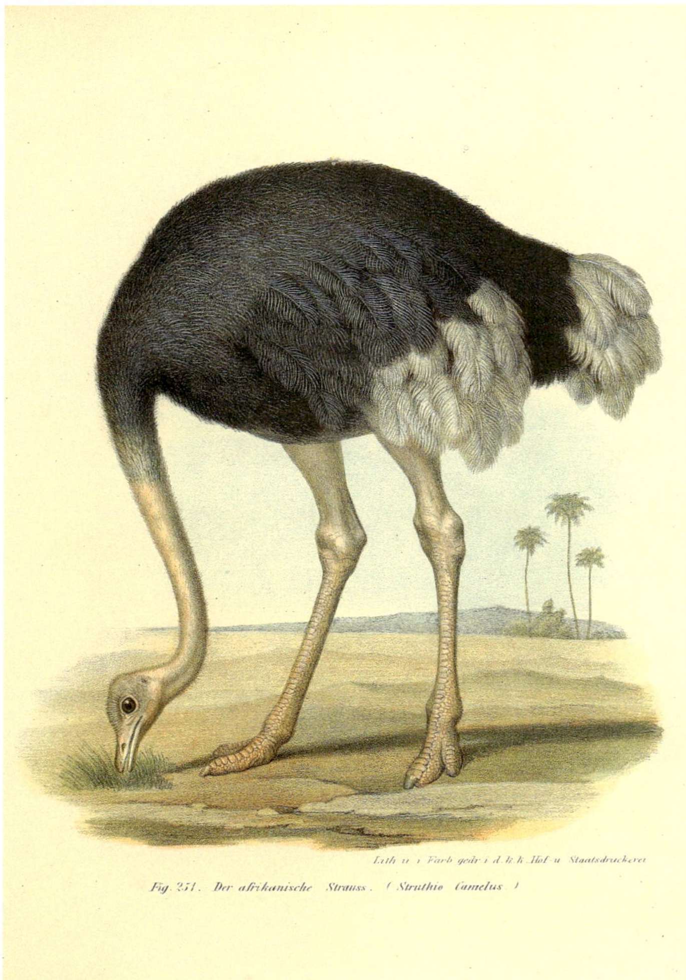

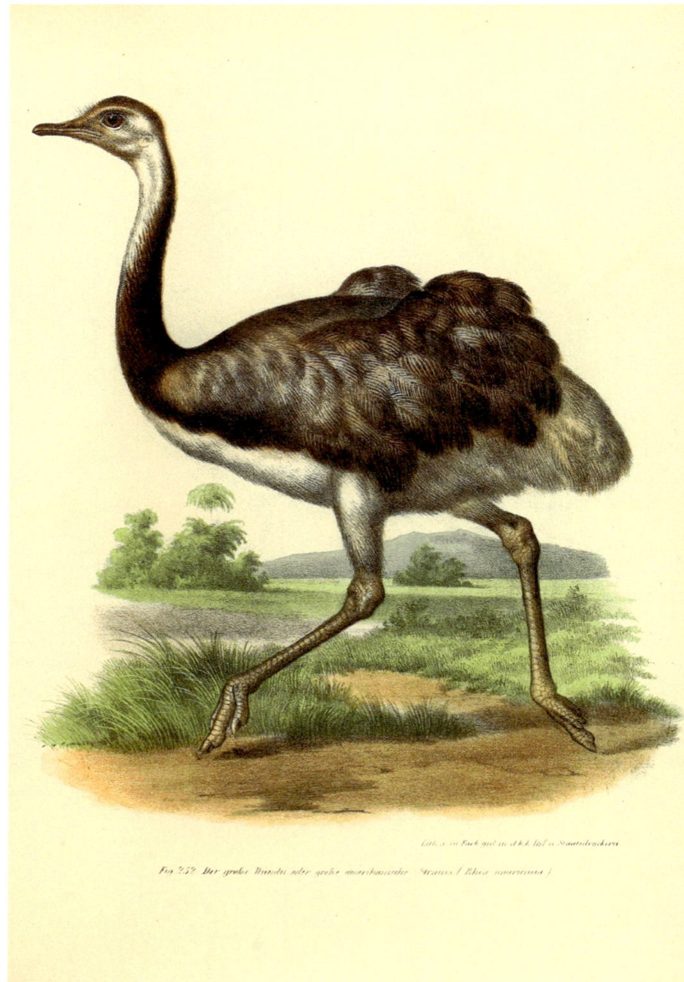

LEFT:

COMMON OSTRICH
Struthio camelus

Uncredited, *Bilder-atlas*, 1860–4

The ostrich is by far the world's largest bird and – with a top speed of seventy kilometres an hour – it is also the fastest bird on land. Unlike most other birds, which have four toes, the ostrich has just two which is likely an adaptation it developed to run at such speeds. This plate shows an ostrich feeding on grass. Typically their diet consists of seeds, leaves and roots but they will also eat insects, rodents and reptiles. They lack teeth for chewing, so they will also swallow stones which they use to grind their food in their gizzard.

RIGHT:

GREATER RHEA
Rhea americana

Uncredited, *Bilder-atlas*, 1860–4

This rhea appears in *Bilder-atlas*. Published between 1860 and 1864 by the Austrian zoologist Leopold Joseph Fitzinger, the work has very little text but contains some 347 colour lithographed illustrations. The name of the illustrator, or illustrators, responsible for these works is left unattributed.

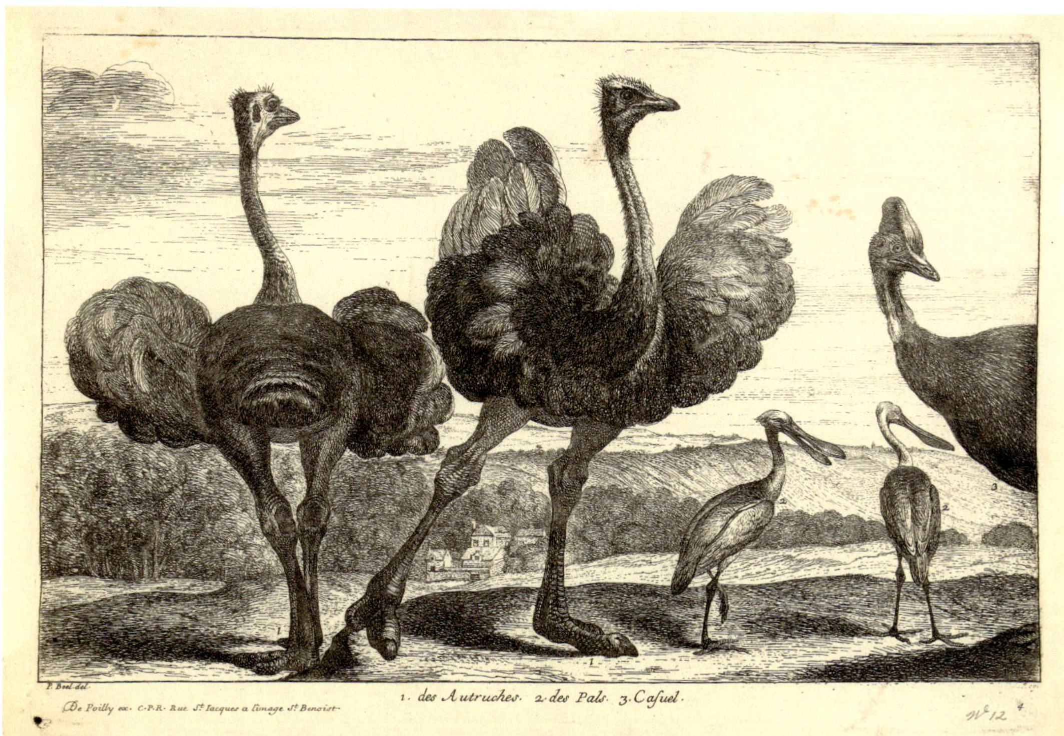

OSTRICHES, CASSOWARY
& SPOONBILLS
Struthio, *Casuarius* & *Platalea*

Struisvogels, kasuaris en lepelaar,
Peeter Boel, etching, *ca*. 1670–4

Flemish artist Peeter Boel is known for his lavish still life paintings of animals, birds, fruit and flowers. This, an etching of ostriches, a cassowary and spoonbills, is a rare example of his work in print. In the 1660s he moved to Paris and eventually became court painter to Louis XIV. In this role he often drew the animals he observed at the Versailles Ménagerie. At the time, drawing live animals was uncommon and Boel's approach is said to have helped revolutionize animal painting.

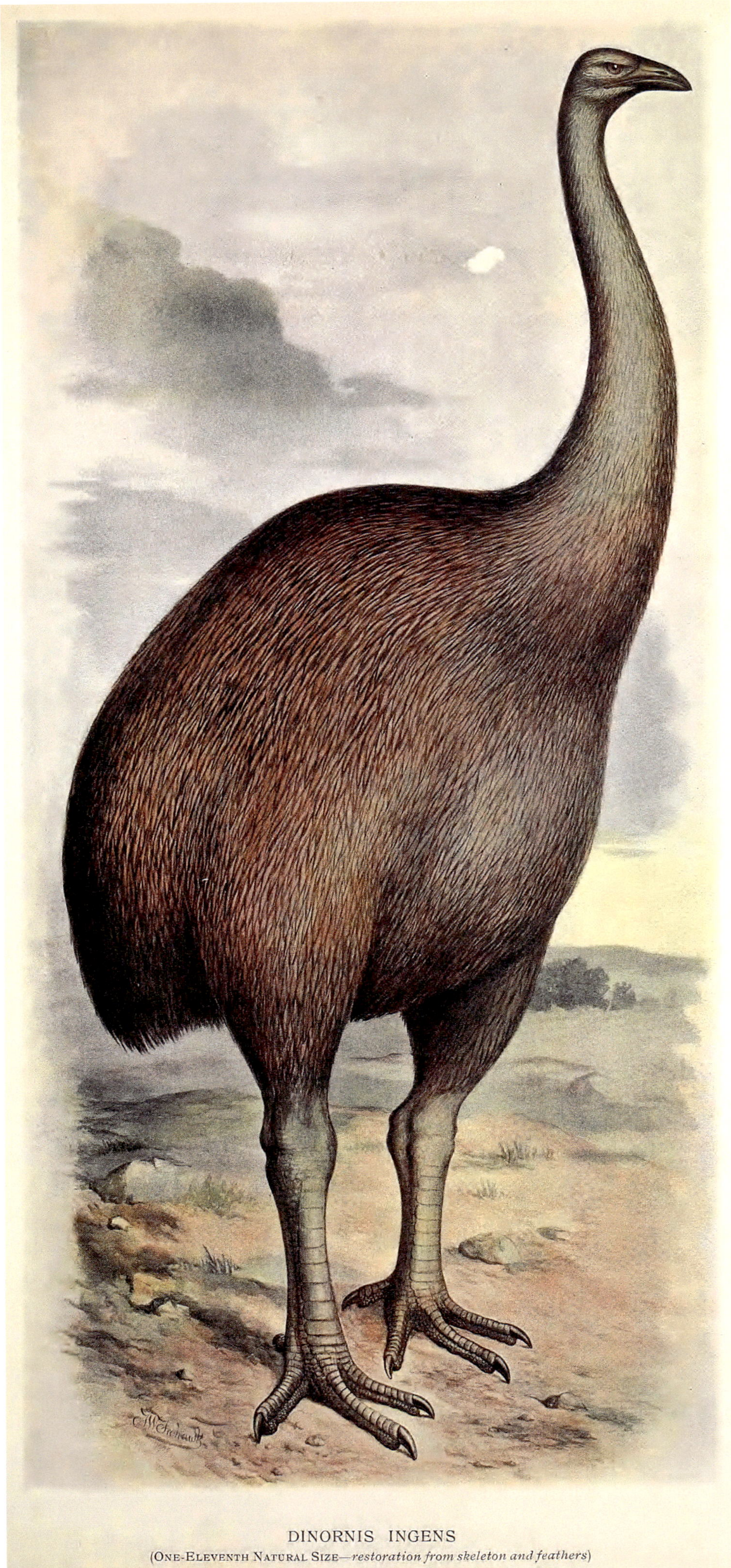

NORTH ISLAND GIANT MOA
Dinornis novaezealandiae

Frederick William Frohawk,
Extinct Birds, 1907

The moa were the tallest birds to ever exist. For millions of years, these nine species of large flightless birds lived in New Zealand where they had encountered few predatorial risks. The arrival of Polynesian settlers quickly led to the extinction of this dramatic bird. Unlike other birds, who suffered at the hands of habitat reduction and invasive species, the moas' demise was predominantly caused by human hunters. Frederick William Frohawk's restoration shows the North Island giant moa, who – at a height of up to three metres – was the second tallest of the nine moa species.

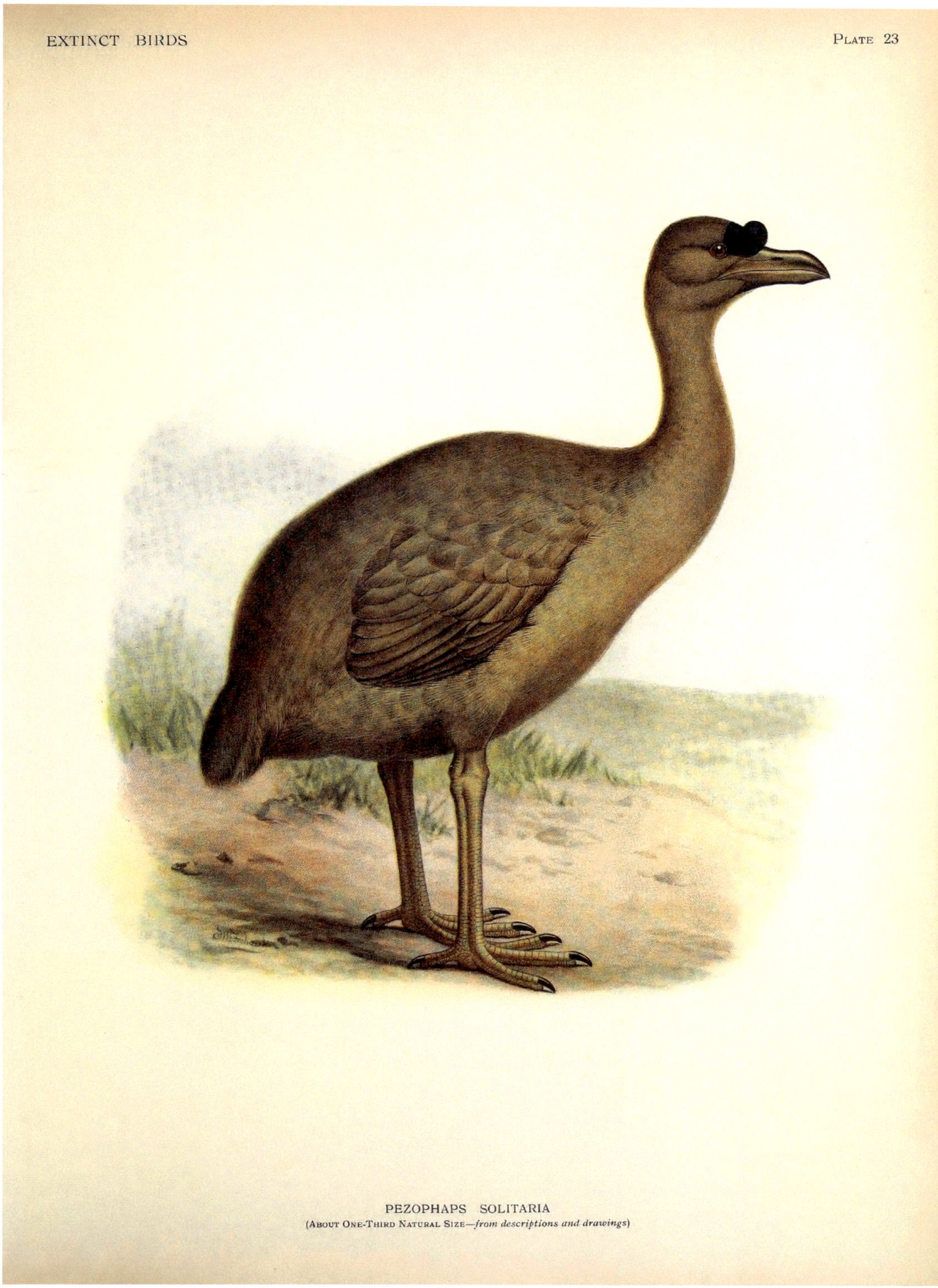

RODRIGUES SOLITAIRE
Pezophaps solitaria

Frederick William Frohawk, *Extinct Birds*, 1907

A close relative of the dodo, the Rodrigues solitaire was a large flightless bird that became extinct soon after European explorers settled in its habitat. Endemic to the small island of Rodrigues, east of Mauritius, it belongs to the Columbiformes order of birds, which also contains pigeons and doves. The first mention of this bird occurs in the writings of François Leguat. A French explorer and naturalist, Leguat was marooned on the then uninhabited island of Rodrigues from 1691 until 1693.

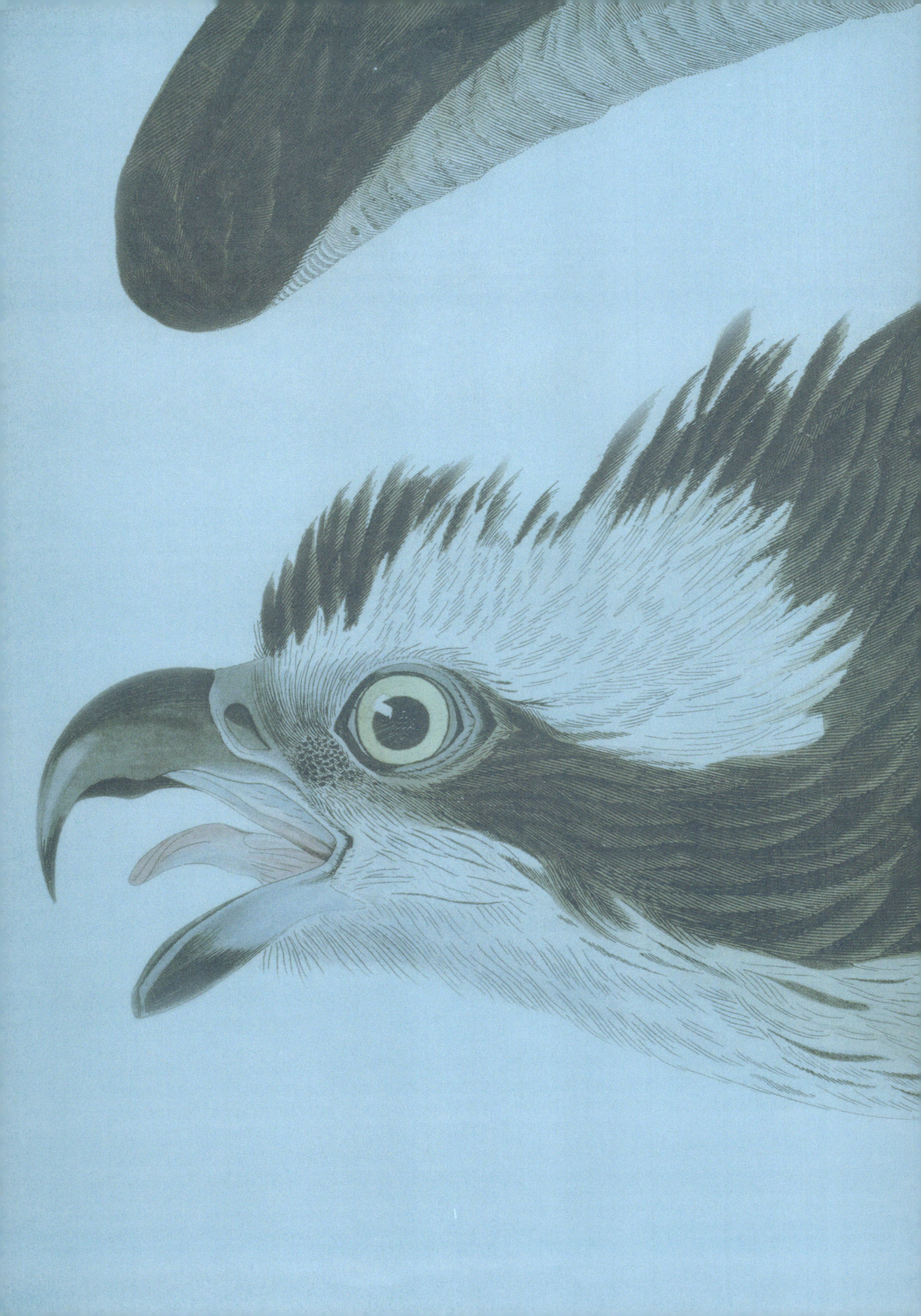

Birds of Prey

With swift flight, acute hearing and keen eyesight, birds of prey are expert hunters. They are sometimes called raptors, a name derived from the Latin word raptare, which means 'to seize violently and drag away'. Despite their violent nature, these birds are typically associated with power and majesty. Illustrators took great pleasure in capturing them in action and aimed to emphasize their magnificence and grandeur.

Few birds are as impressive or as captivating as those that are classed as birds of prey. Often called raptors, this group of birds include some of nature's finest hunters, including families such as eagles, falcons, harriers, hawks, ospreys and owls. These birds are distinguished by their hooked bills, their sharp talons and their keen eyesight. While a number of birds of prey are scavengers, eating meat from dead animals, most catch live food, either in flight or on the ground, often using ambush tactics.

As skilled hunters, these birds have each developed traits that allow them to catch prey with grace and accuracy. Falcons, for example, are swift flyers who only need to kill once a day to feed. These birds are found widely throughout the world and for centuries they have been kept and trained by humans to pursue and attack wildfowl and game. The tradition of falconry extends past falcons and will also sometimes include other raptors such as hawks and eagles. Their name comes from the Latin word *falco*, or *falx*, meaning 'sickle' – a reference to their long, sharp claws. Their wings are distinctively pointed, making them incredibly aerodynamic.

Owls have adapted too. They have large front-facing eyes, which allows them to judge distances more accurately and also makes them experts at detecting movement. Owls hunt both at dusk and at night, and their large eyes allow them to capture more light than most other birds. They also have acute hearing which lets them listen out for prey. For this reason, owls are incredibly stealthful, flying almost silently through the night air as they hunt out prey.

Vultures are scavengers, and hunt only for dead animals. They have sharp bills and keen eyesight, but their talons are relatively weak. Their vision is particularly good, allowing them to soar high in the sky as they search for the remains of a dead animal to feed on. These birds rarely attack healthy animals, but they do occasionally kill those that are wounded or sick. Vultures are exceptional scavengers: their stomachs contain highly acidic gastric juices which are strong enough to kill most types of bacteria. For this reason, they can consume meat that would harm most other birds. Once a vulture has found a carcass, other vultures will soon follow and the group will devour an entire animal in record time. As scavengers, these birds play a fundamental role in our ecology, providing natural environmental sanitation. Yet despite this, vultures typically have a poor reputation and their association with death has made them figures of fear and repulsion.

Often considered to be King of the Birds, the eagle ranks among the biggest and most powerful birds of prey. Associated with power and majesty, its attributes have been borrowed by kings, commanders and emperors all around the globe. In religion, they also have taken on respected and symbolic roles: Zeus' personal messenger and animal companion was the eagle Aetos Dios. Native Americans regard the bird as sacred and a messenger of the Great Spirit. In Christianity, the bird has been a symbol of hope and strength, often representing the resurrection of Christ and acting as a symbol of St. John.

A number of the illustrations in this chapter capture the impressive nature of these birds by emphasizing their skills as hunters. Eagles are seen catching fish, while owls and condors are seen feeding on rabbits and lambs. For illustrators, birds of prey were impressive and dramatic birds and, as these illustrations will show, they took great pleasure in attempting to capture them in all their magnificence.

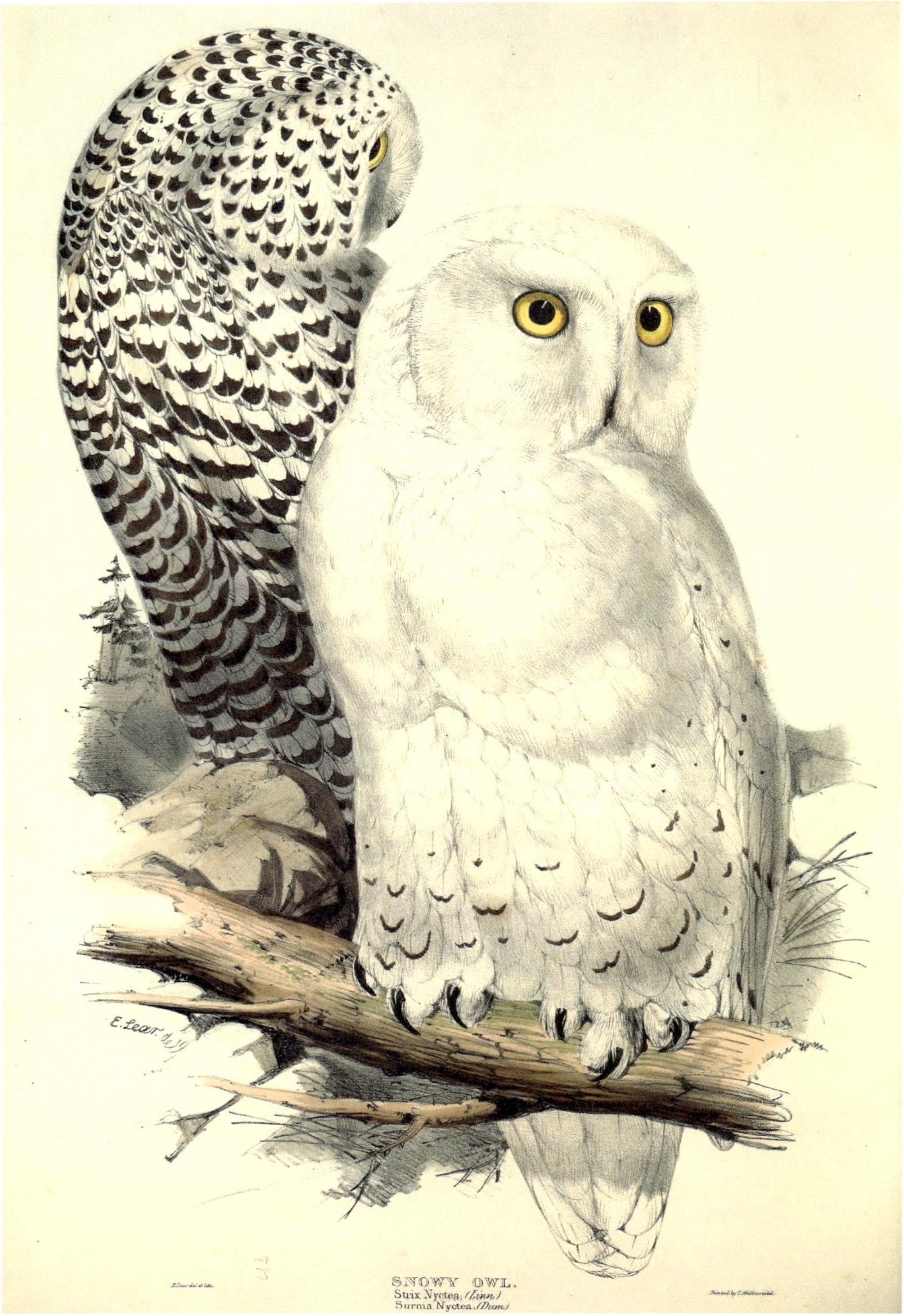

SNOWY OWLS
Bubo scandiacus

Edward Lear, *The Birds of Europe*, 1832–7

A bird of the Arctic, the snowy owl is found in the northern reaches of North America, Europe and Asia. The male's distinctive white plumage (right) provides camouflage in the snow, while the female (left) has a more mottled appearance that helps it to camouflage while it nests on the ground. This is a work by Edward Lear, who produced several owl illustrations for Gould. He later gained popularity for his literary nonsense in poetry and prose – his most famous work being *The Owl and the Pussycat* (1871).

AMERICAN BARN OWLS
Tyto furcata

John James Audubon,
The Birds of America, 1827–38

This vibrant illustration by Audubon demonstrates the artist's unique talent for creating dynamic and exciting compositions. Audubon's work not only captured the beauty of the natural world, but also captured its drama. His process for creating such work involved shooting his subjects and then pinning them to wooden frames. This allowed him to produce striking compositions that were brilliantly theatrical while also being scientifically accurate.

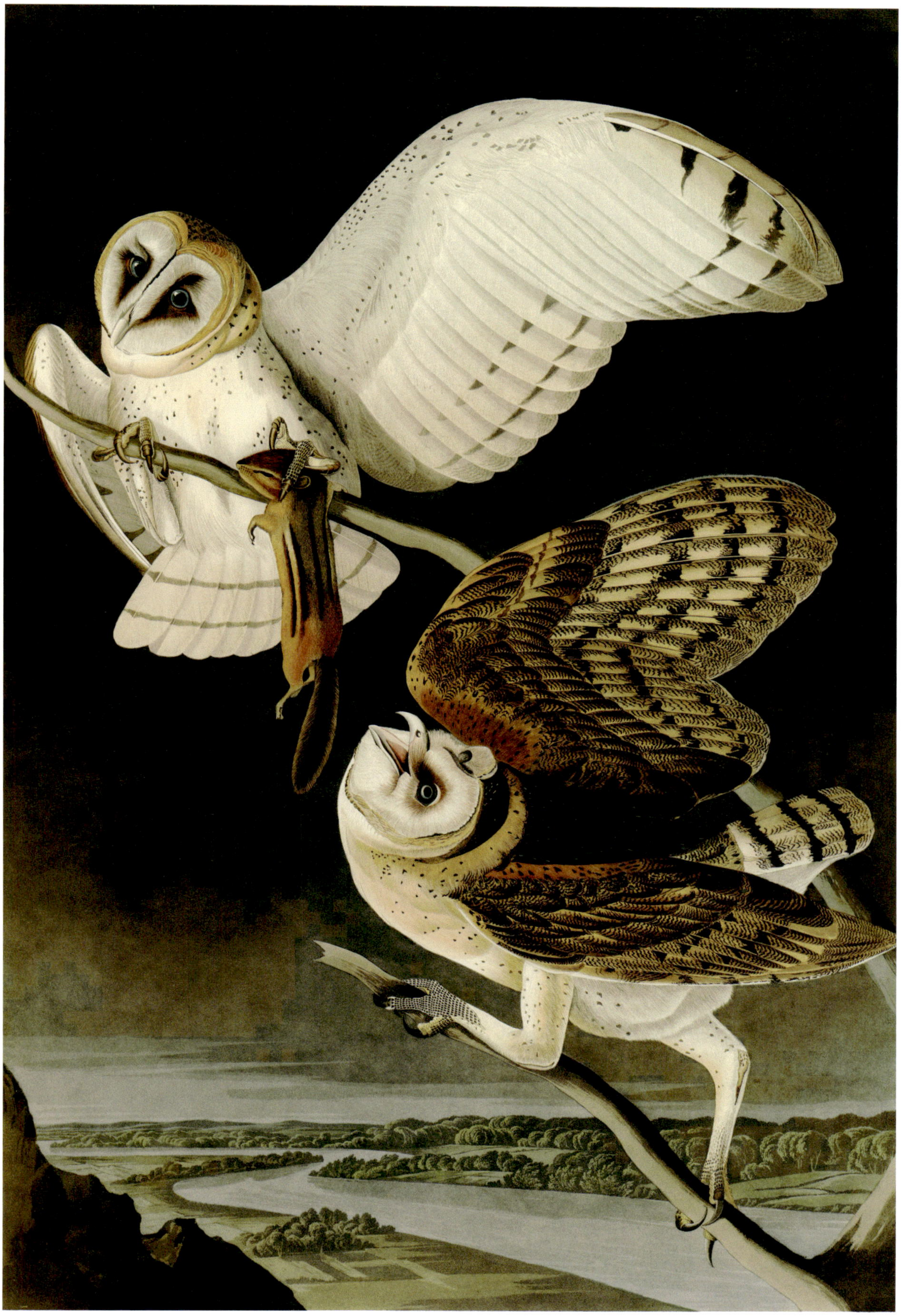

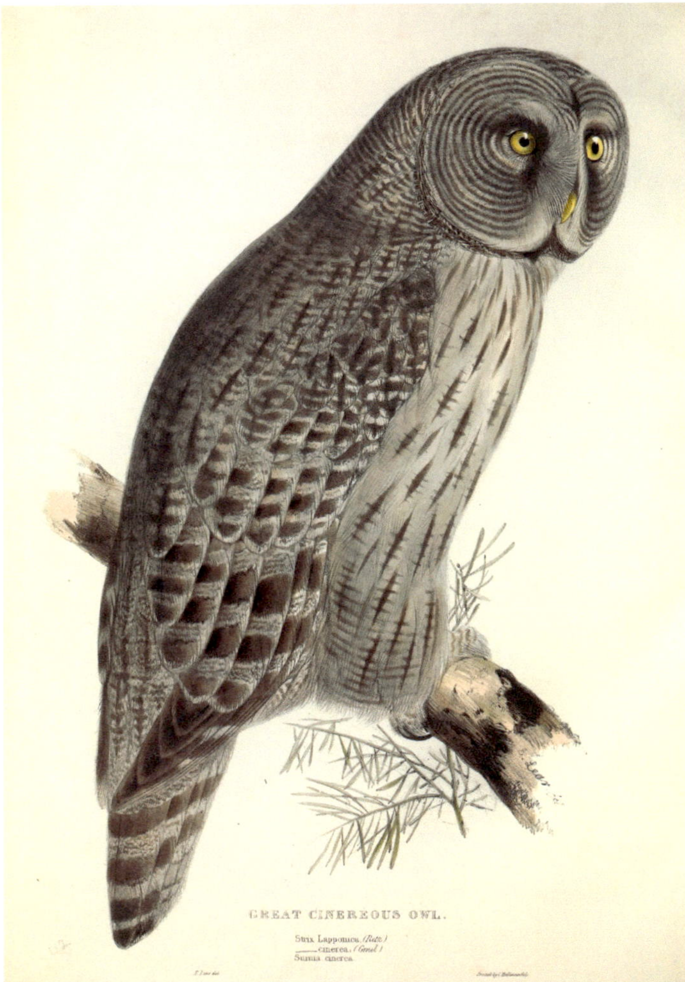

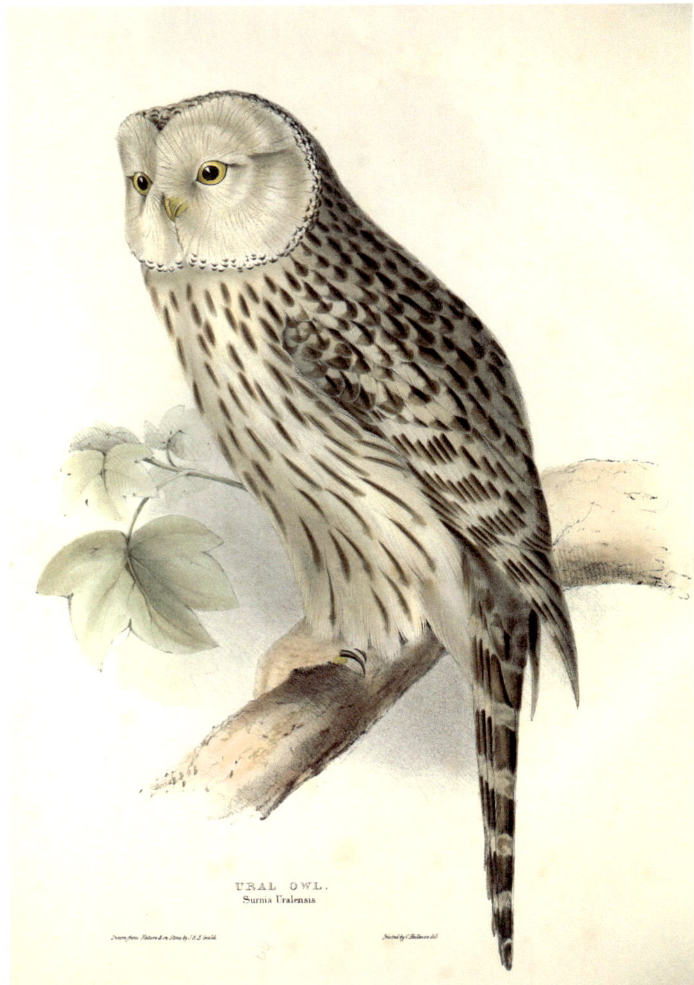

LEFT:

GREAT GREY OWL
Strix nebulosa

Edward Lear, *The Birds of Europe*, 1832–7

The great grey owl hunts in the Arctic, where its prey is almost invisible. It feeds on small rodents that are often buried deep under the snow, yet its hearing is so accurate that it can hunt out its prey through the use of sound alone. Its large satellite-dish-shaped face, known as its 'ruff', is specifically designed to focus sounds and, at certain frequencies, its hearing can be ten times more sensitive than our own.

RIGHT:

URAL OWL
Strix uralensis

John Gould & Elizabeth Gould, *The Birds of Europe*, 1832–7

The Ural owl takes its name from the Ural mountains, where specimens of the bird had originally been collected. Despite its name, this bird actually has an extremely large range that spans as far west as Scandinavia and as far east as Japan. When Gould documented it in the 1830s, he believed his specimen to be the only one in England. In fact, the Ural owl is typically quite a common bird and the absence of specimens reflects how their habitat had remained relatively unexplored during Gould's time.

OPPOSITE:

AUSTRALIAN MASKED-OWL & TASMANIAN MASKED-OWL
Tyto novaehollandiae & Tyto castanops

Elizabeth Gould, *A Synopsis of the Birds of Australia and the Adjacent Islands*, 1837–8

Elizabeth Gould's illustration shows the heads of two barn owls. The top is a species known as the Australian masked-owl, while the lower, the Tasmanian masked-owl, was once considered a subspecies but is now treated as a full species. This plate was published in John Gould's *A Synopsis of the Birds of Australia and the Adjacent Islands* in 1837 and is the first account of this particular species. This synopsis introduced many first descriptions of Australian birds and is unique for Gould as it presents only head studies of its birds.

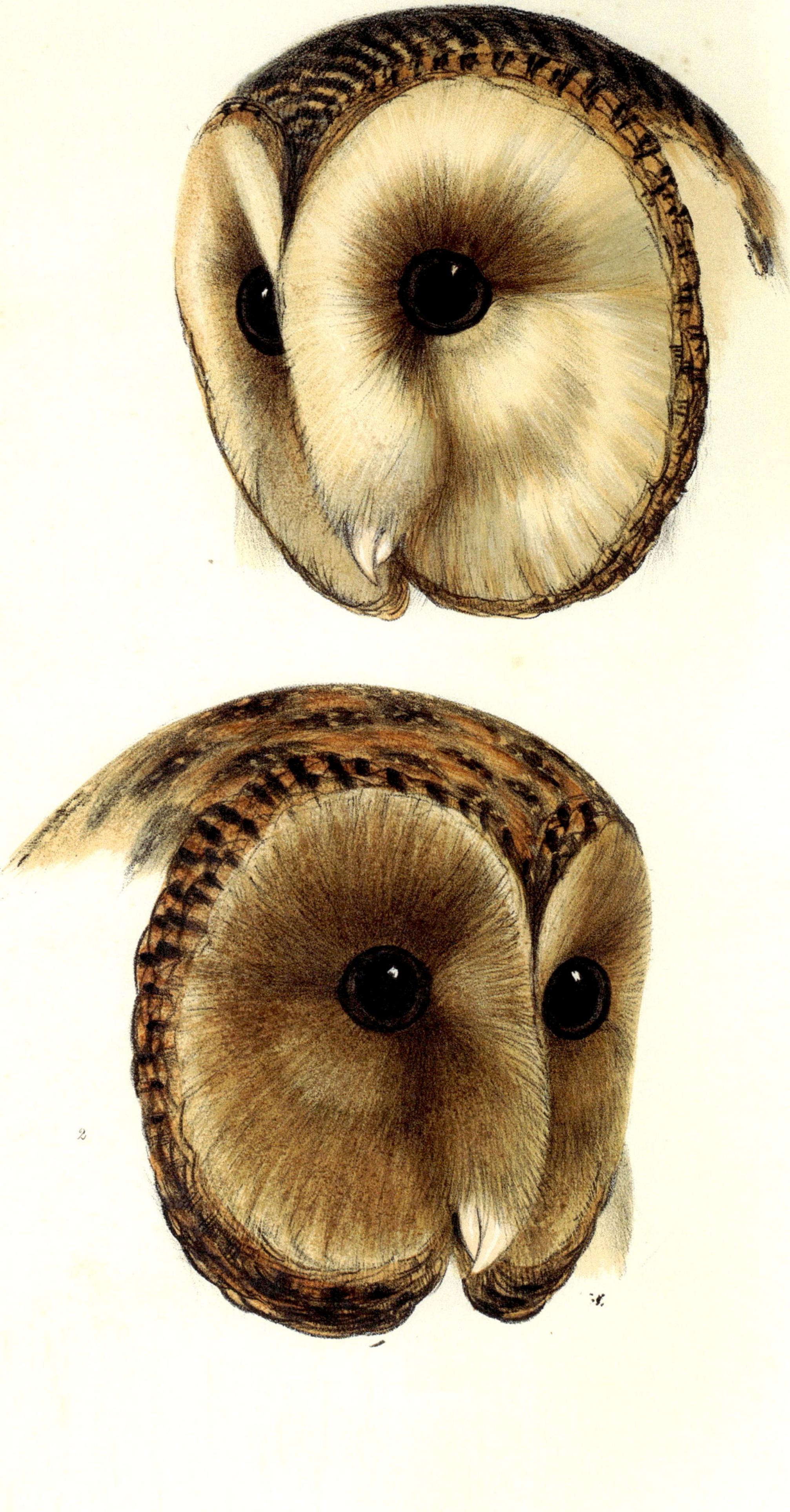

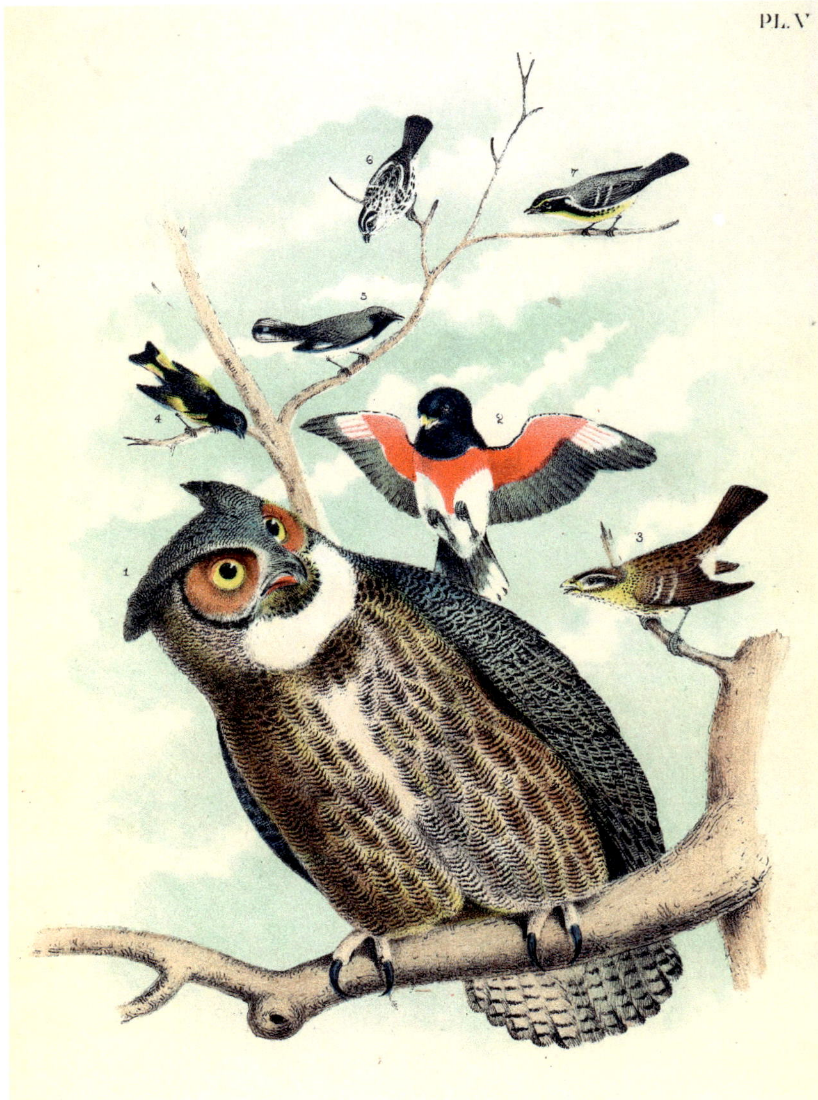

1. GREAT HORNED OWL
   Bubo virginianus
2. & 3. ROSE-BREASTED GROSBEAK
   Pheucticus ludovicianus
4. AMERICAN REDSTART
   Setophaga ruticilla
5. BLACK-THROATED BLUE WARBLER
   Setophaga caerulescens
6. BLACK-AND-WHITE WARBLER
   Mniotilta varia
7. YELLOW-THROATED WARBLER
   Setophaga dominica

Jacob H. Studer, *The Birds of North America*, 1903

Jacob H. Studer was a popular American ornithologist at the turn of the twentieth century. He was also a painter, printer and lithographer.

OPPOSITE, CLOCKWISE FROM TOP LEFT:

1. BARRED OWL
   Strix varia
2. NORTHERN HAWK-OWL
   Surnia ulula
3. AMERICAN BARN-OWL
   Tyto furcata
4. BURROWING OWL
   Athene cunicularia
5. NORTHERN SAW-WHET OWL
   Aegolius acadicus

Alexander Wilson, *Illustrations of the American Ornithology of Alexander Wilson and Charles Lucian Bonaparte*, 1835

Alexander Wilson's *American Ornithology* spanned nine volumes and contained 268 species of bird. As a means to cut costs, his compositions tended to include several species together.

XIII

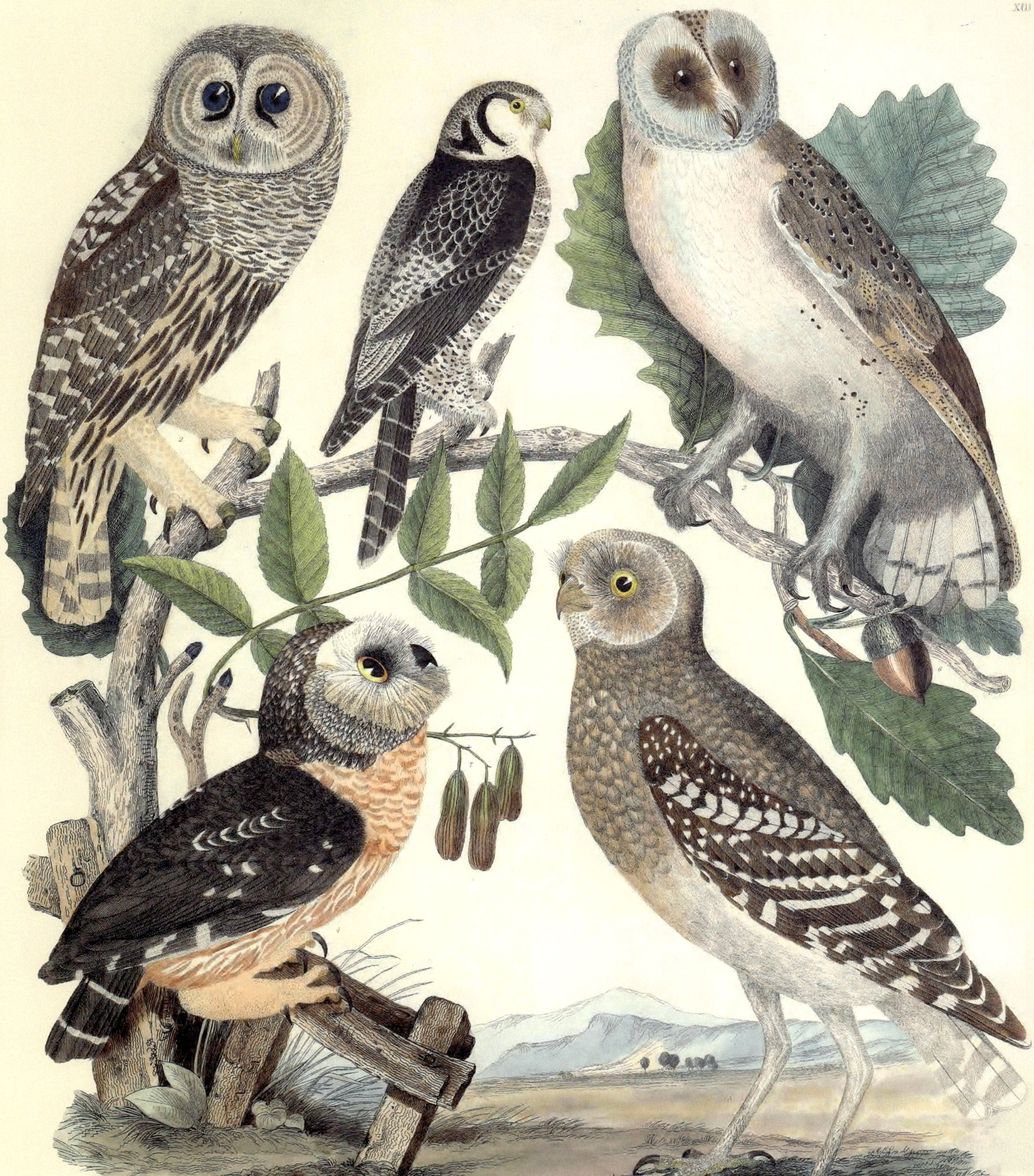

STRIX.

OWLS.

1 Hawk Owl. *Strix Hudsonia.*
2 Barred Dº — *Nebulosa.*
3 Barn Dº — *Flammea.*
4 Little Dº — *Passerina.*
5 Burrowing Dº — *Cunicularia.*
6 Swamp White Oak.
7 Black Ash.

Engraved by R. Scott

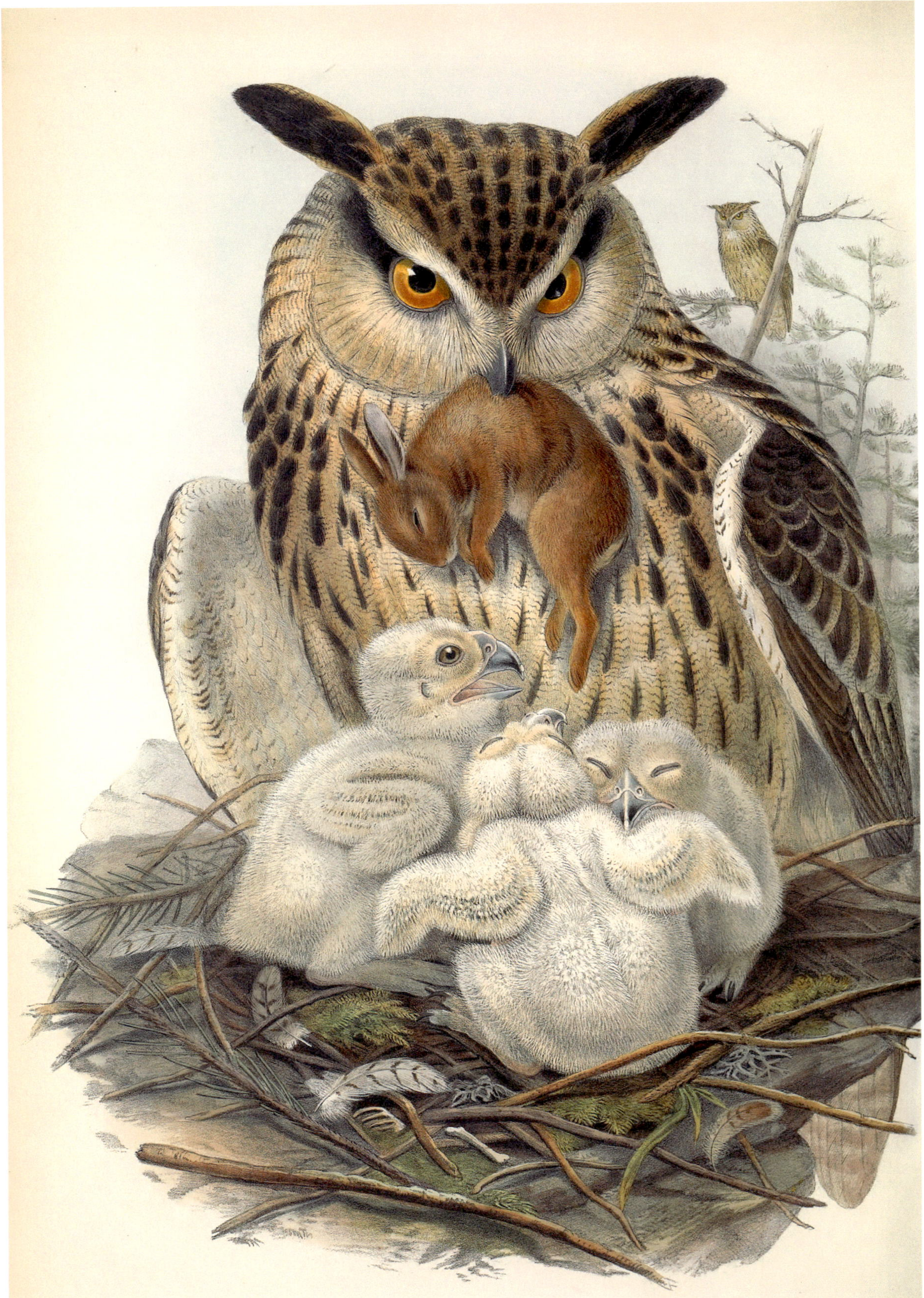

EURASIAN EAGLE-OWL
Bubo bubo

Joseph Wolf & Henry Constantine Richter, *The Birds of Great Britain*, 1862–73

This work from Gould's *The Birds of Great Britain* is distinctive in that it presents an image that emphasizes the family life of a bird. While Gould would often show both sexes to highlight sexual dimorphism, this is a rare example where the focus is placed on the nesting and feeding habits of the bird. Here, we see an eagle-owl feeding its young a small hare. These owls are one of the largest owl species in the world and they feed predominantly on small mammals such as rats, mice, voles, rabbits and hares.

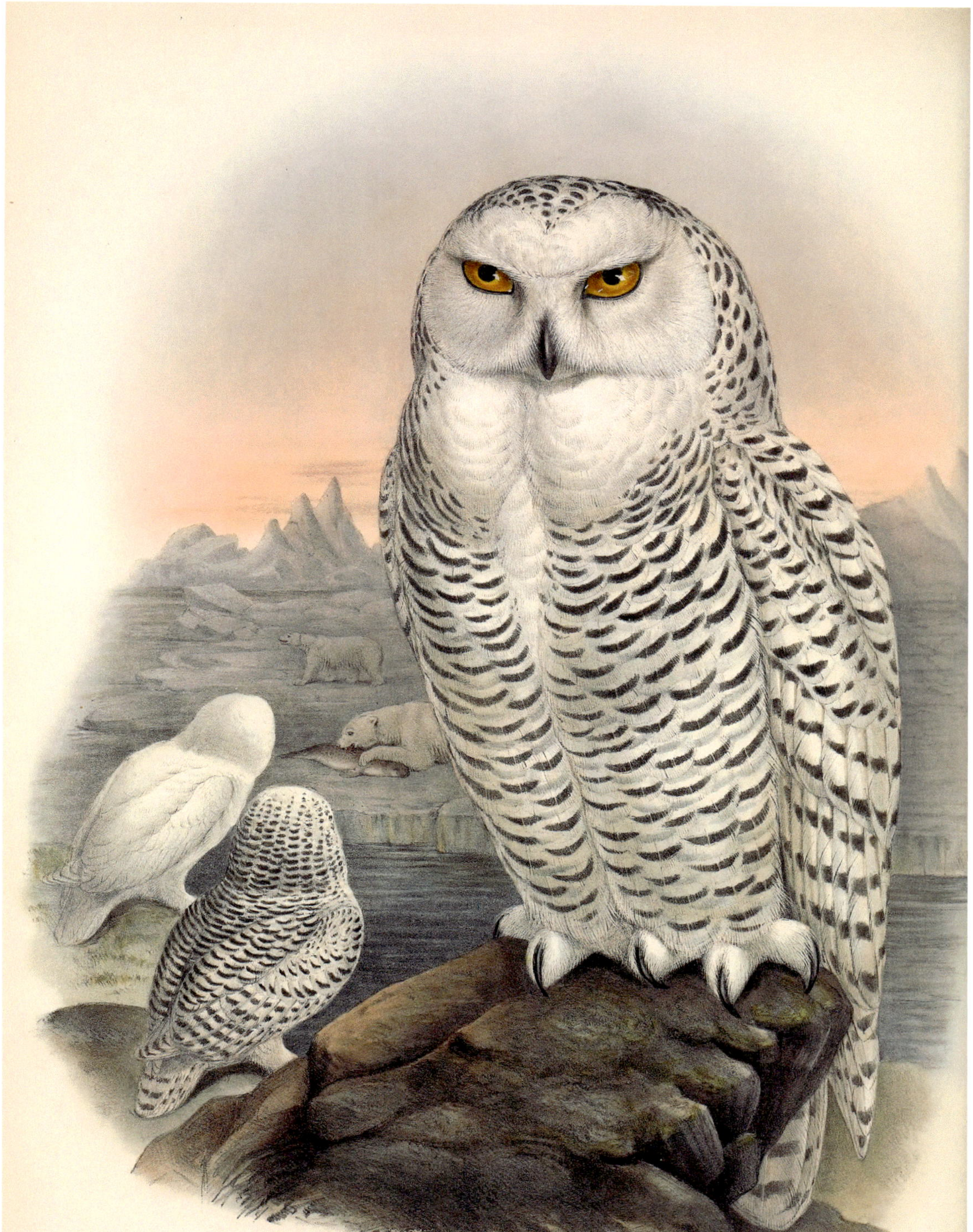

SNOWY OWL
Bubo scandiacus

Joseph Wolf & Henry Constantine Richter, *The Birds of Great Britain*, 1862–73

The snowy owl occasionally – if somewhat rarely – travels as far south as Britain, yet it appears here in Gould's *The Birds of Great Britain*. Despite the book's title, Wolf and Richter do not shy away from depicting it in its Arctic habitat and show a female owl against a backdrop of glaciers that even includes a polar bear feeding on a seal. Snowy owls are hunters who feed primarily on lemmings and other small rodents.

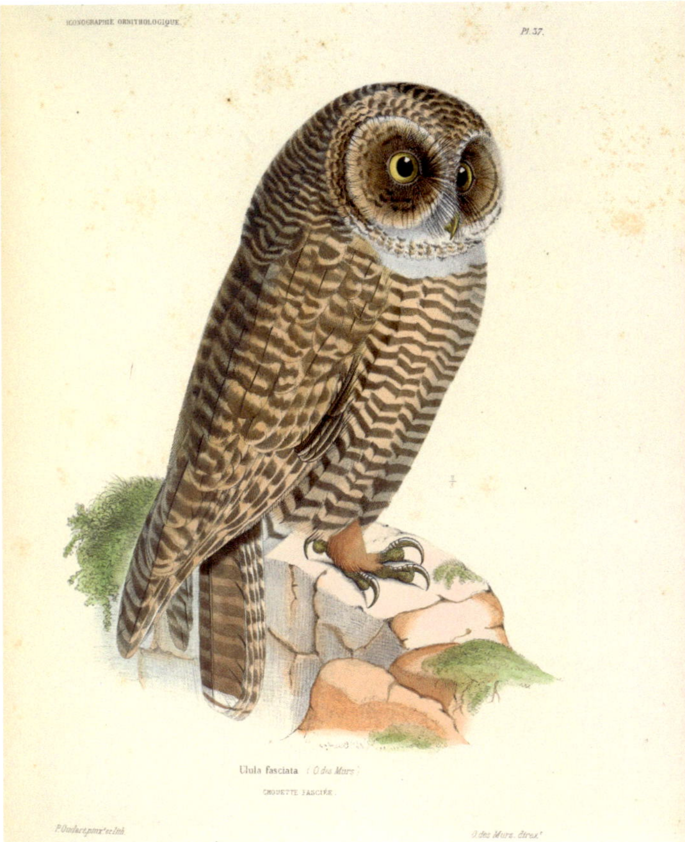

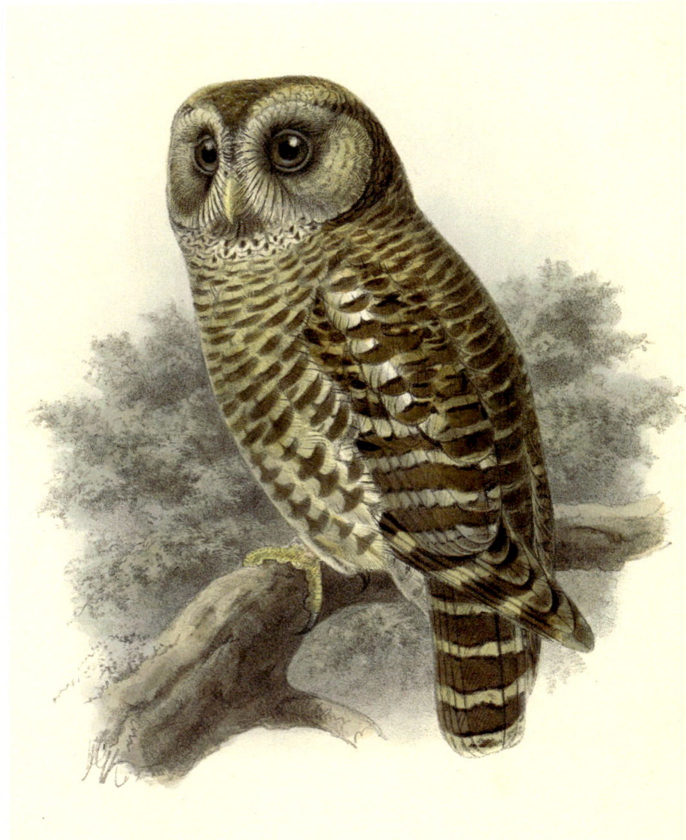

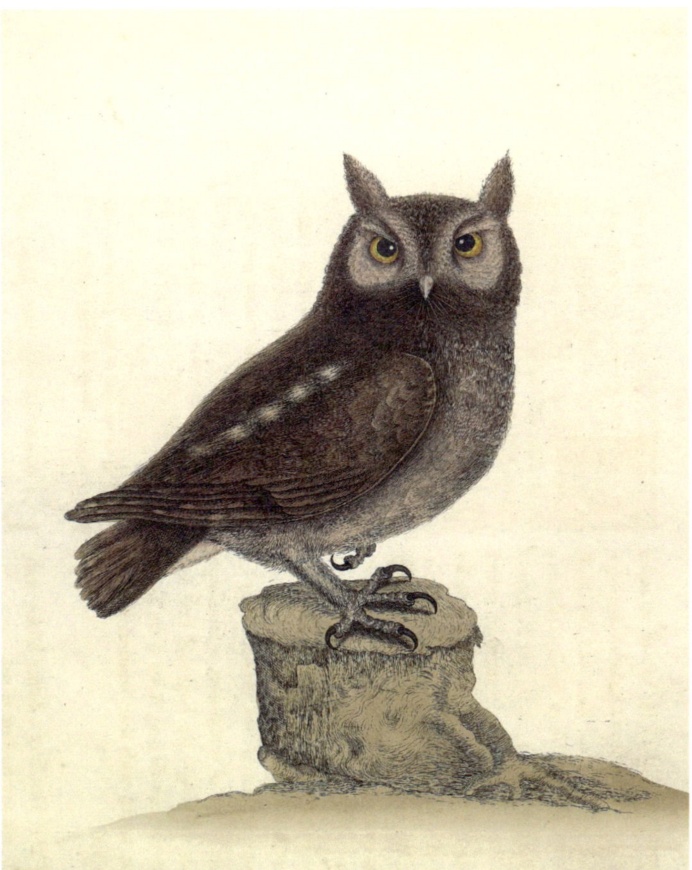

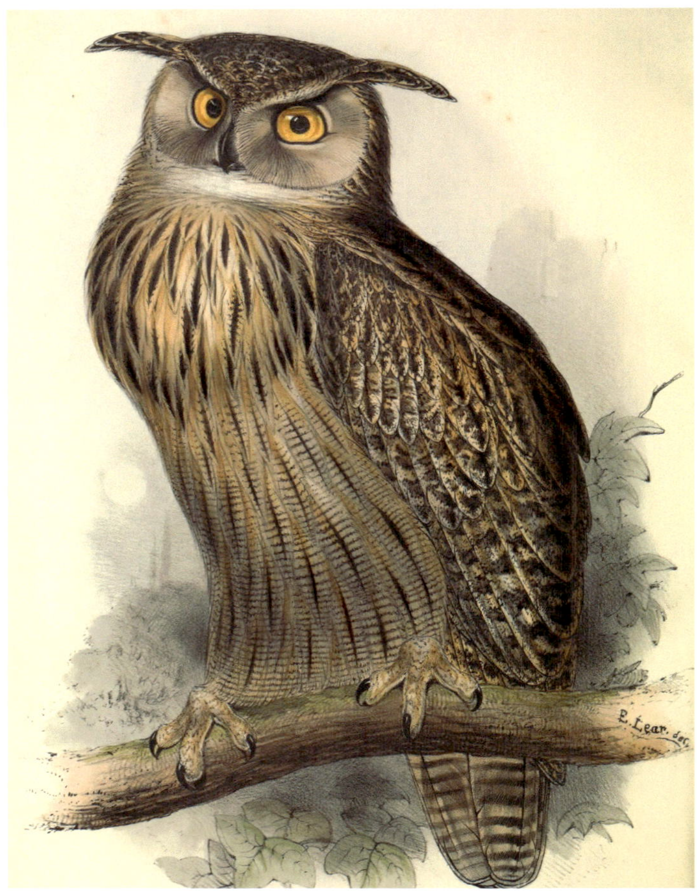

TOP LEFT:
OWL
Strix

Paul Louis Oudart, *Iconographie ornithologique*, 1845–9

BOTTOM LEFT:
EASTERN SCREECH OWL
Megascops asio

Mark Catesby, *The Natural History of Carolina, Florida and the Bahama Islands*, 1729–47

TOP RIGHT:
FULVOUS OWL
Strix fulvescens

John Gerrard Keulemans, *Biologia Centrali-Americana*, 1879–1915

BOTTOM RIGHT:
EURASIAN EAGLE-OWL
Bubo bubo

Edward Lear, *The Birds of Europe*, 1832–7

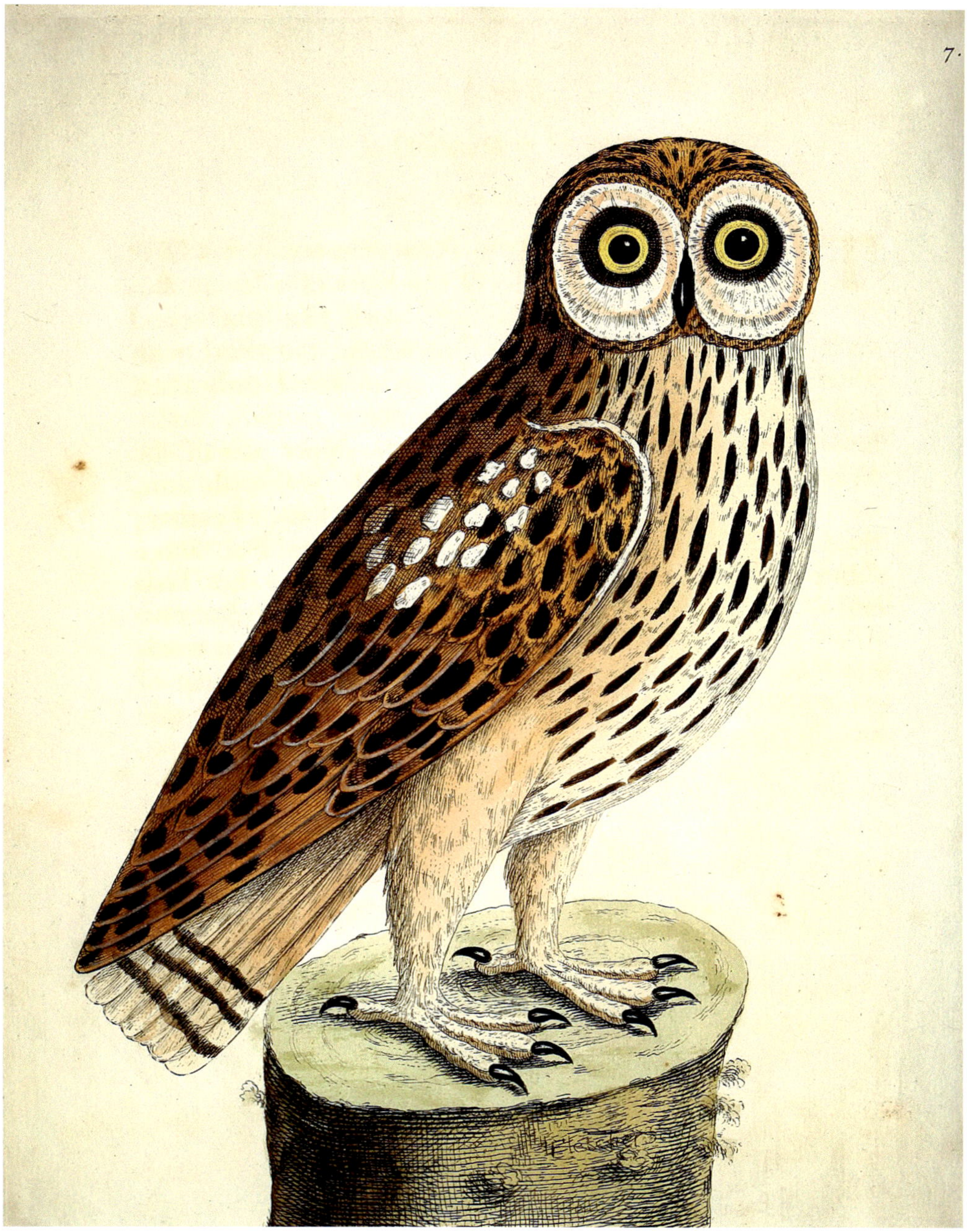

PROBABLY SHORT-EARED OWL
Asio flammeus

Eleazar Albin & Elizabeth Albin, *A Natural History of Birds*, 1731–8

Eleazar Albin identifies this bird as the great brown owl and notes that it 'differs from our *brown Owl* borth (*sic*) in the Bigness and Colour of its Feathers'. Today the brown owl is commonly called the tawny owl (*Strix aluco*), and we can assume that this bird is a short-eared owl (*Asio flammeus*). Albin mentions that this specimen had been borrowed from the collection of Sir Thomas Lowther. Lowther was an English landowner, and several birds from his collection were published in Albin's *A Natural History of Birds*.

NORTHERN SAW-WHET OWLS
Aegolius acadicus

John James Audubon,
The Birds of America, 1827–38

Audubon described this bird as being both beautiful and lively, and this is reflected perfectly in his illustration. While he identified the bird as a little owl (*Athene noctua*), it is more likely to be a northern saw-whet owl as the little owl is not native to North America. These owls feed predominantly on small rodents such as deer mice, voles and shrews. They hunt almost exclusively at night, when they wait silently on low perches and swoop down to catch their prey.

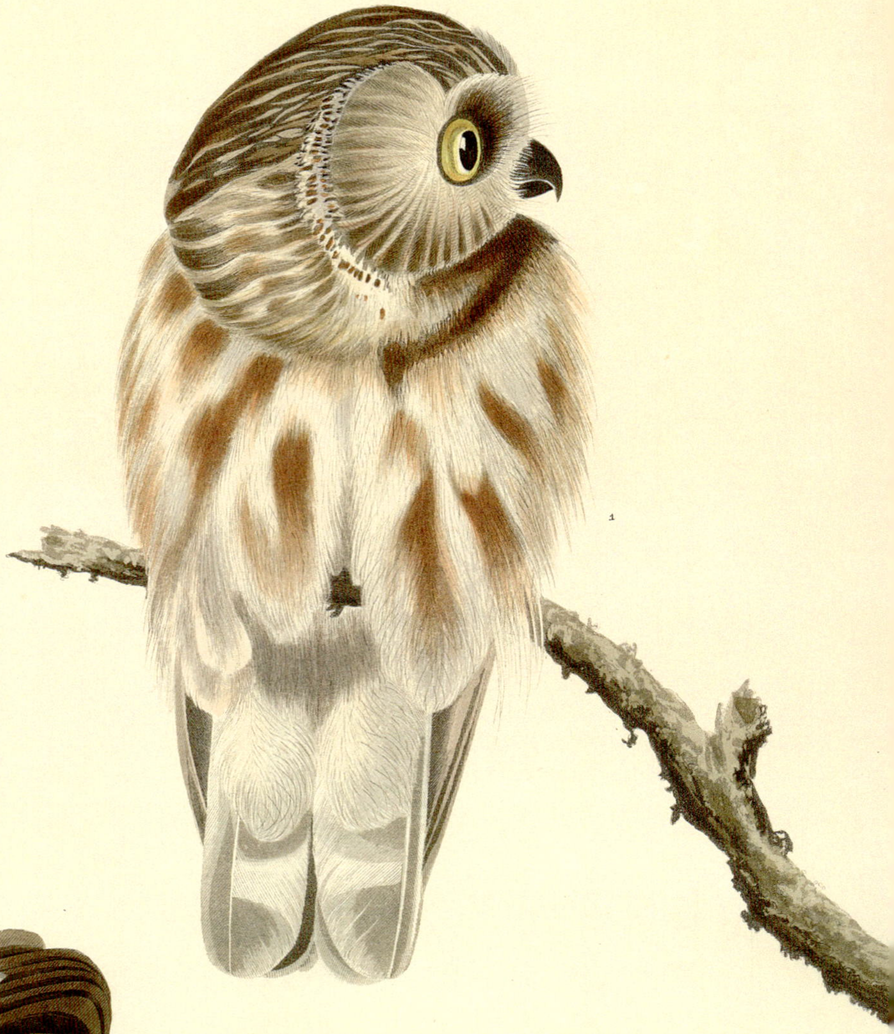

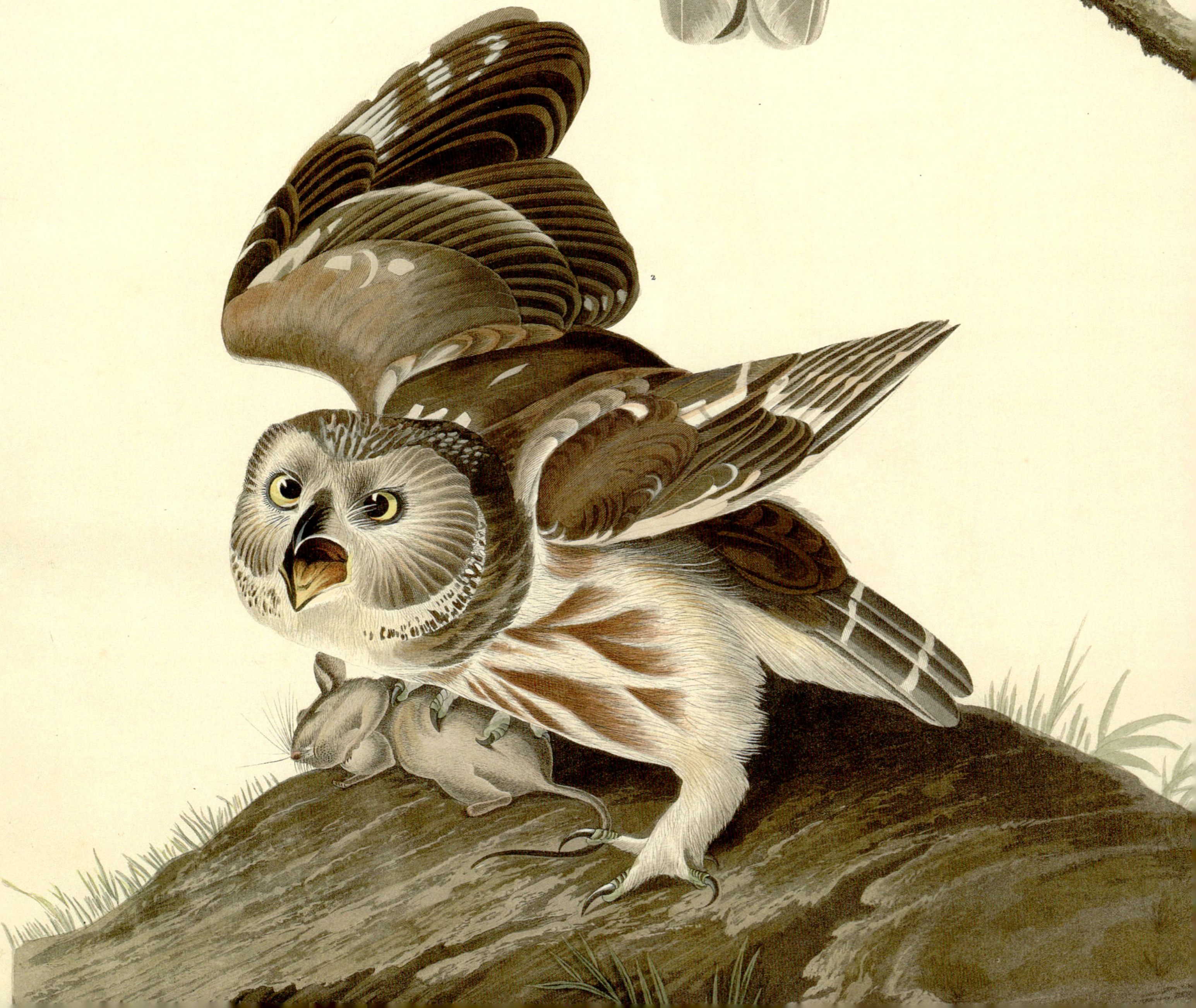

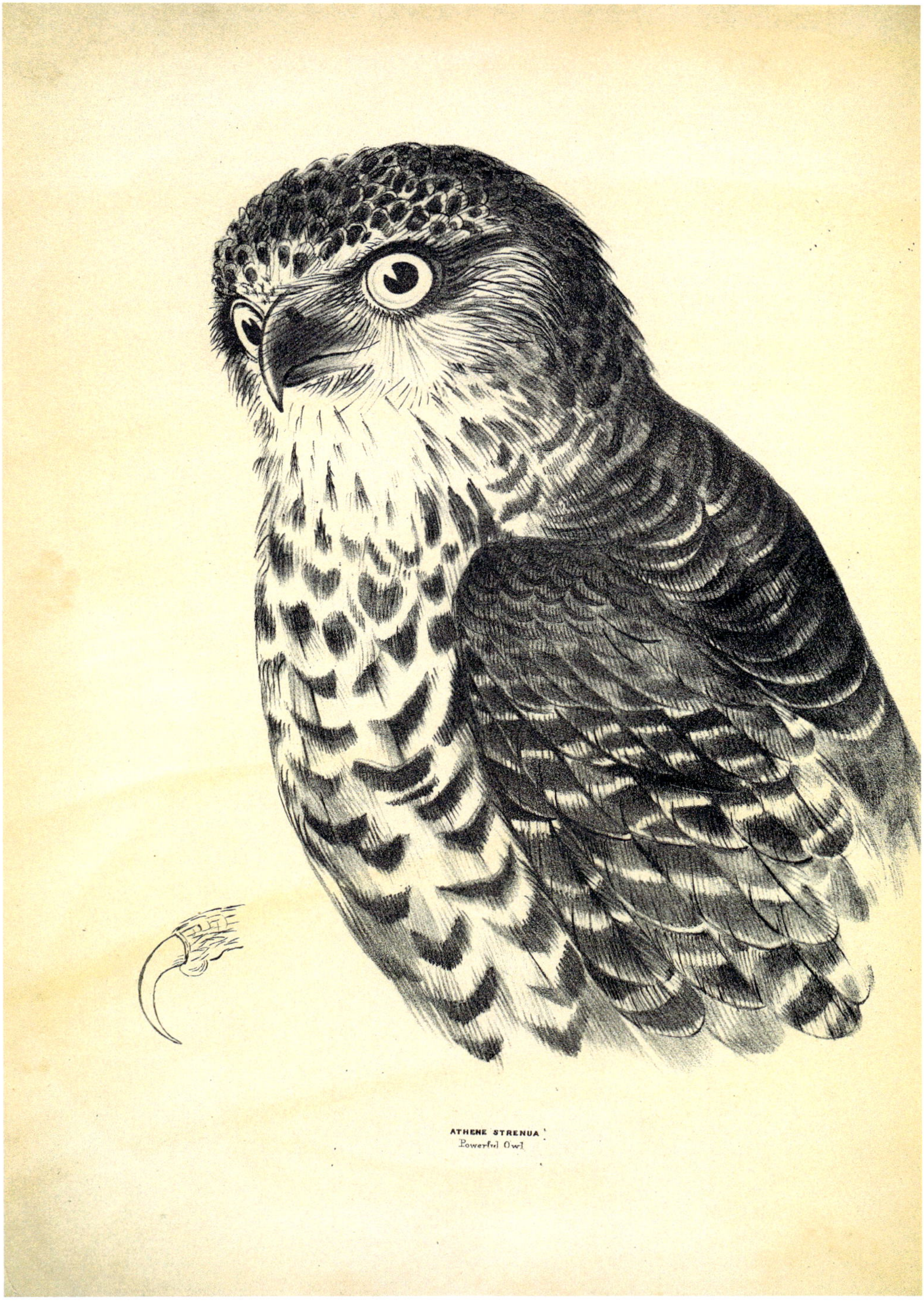

POWERFUL OWL
Ninox strenua

Silvester Diggles, Companion to Gould's Handbook, 1877

The powerful owl can reach up to 60 centimetres in length and is the largest owl to be found on the Australian continent. This depiction is by Silvester Diggles. An immigrant to Australia, Diggles was a musician, artist and amateur ornithologist. Between 1865 and 1870 he attempted to publish his ambitious *Ornithology of Australia*. A lack of funds meant that the project was uncompleted, but in 1877 he managed to republish the majority of his prints under the new title *Companion to Gould's Handbook*.

COMMON BUZZARD
Buteo buteo

Edward Lear, *The Birds of Europe*, 1832–7

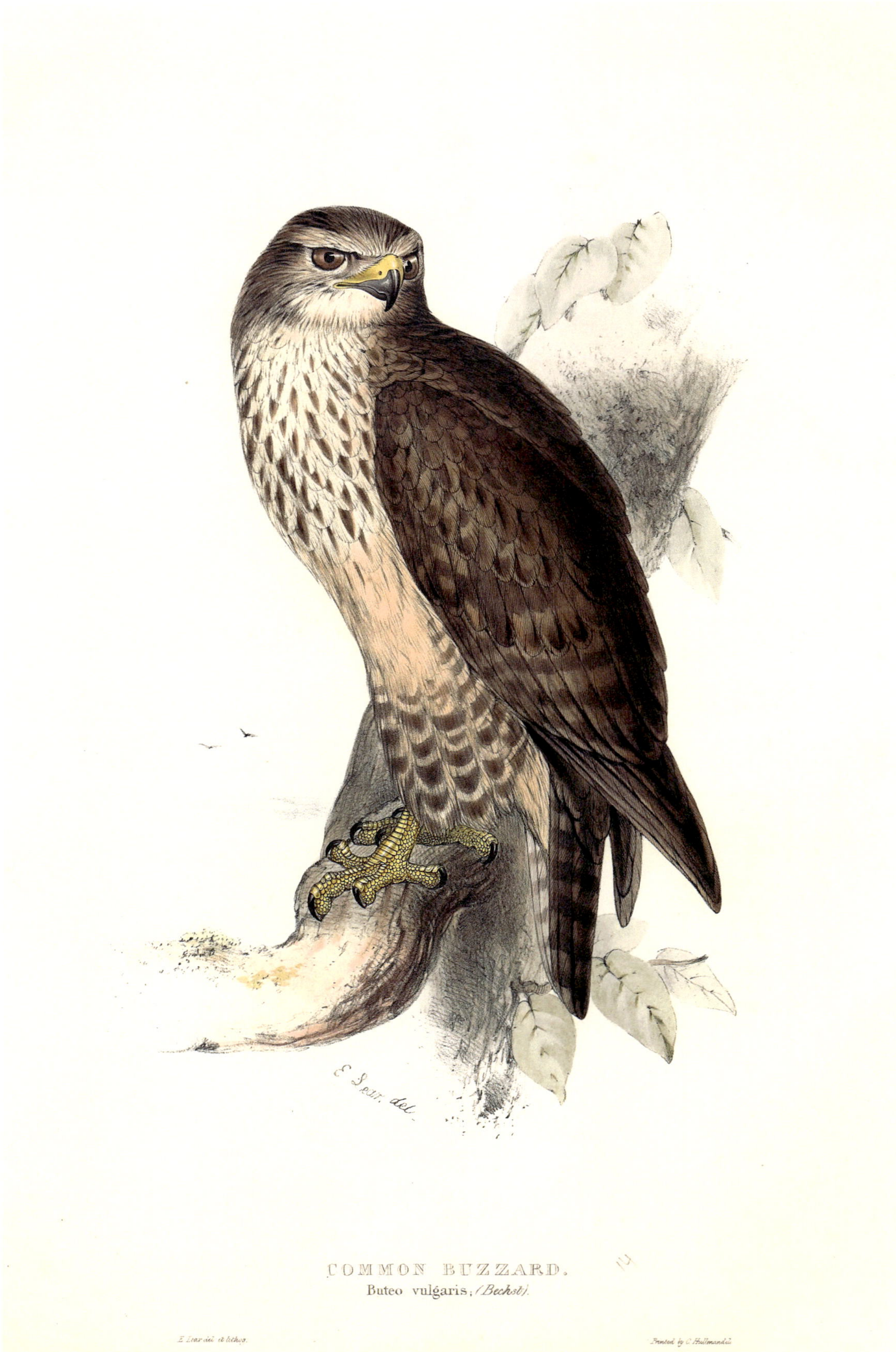

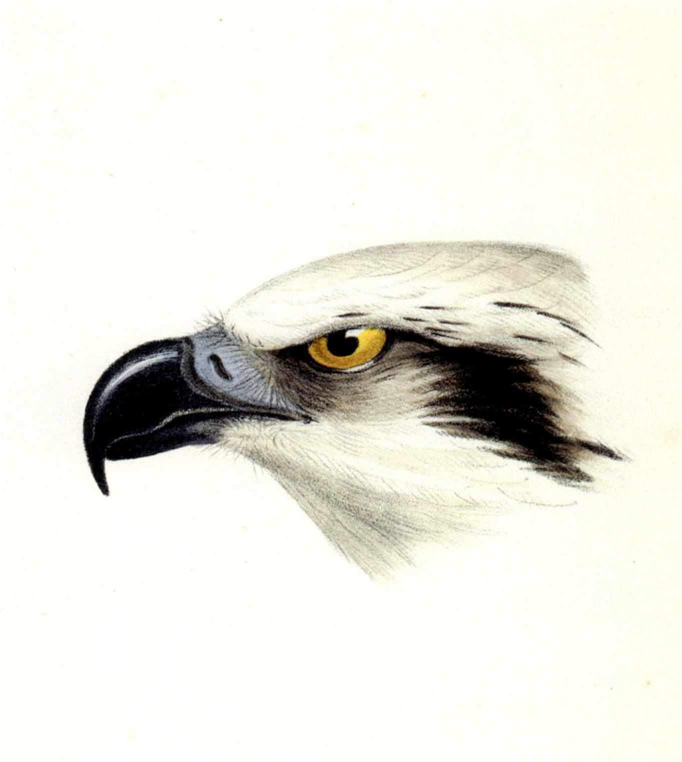

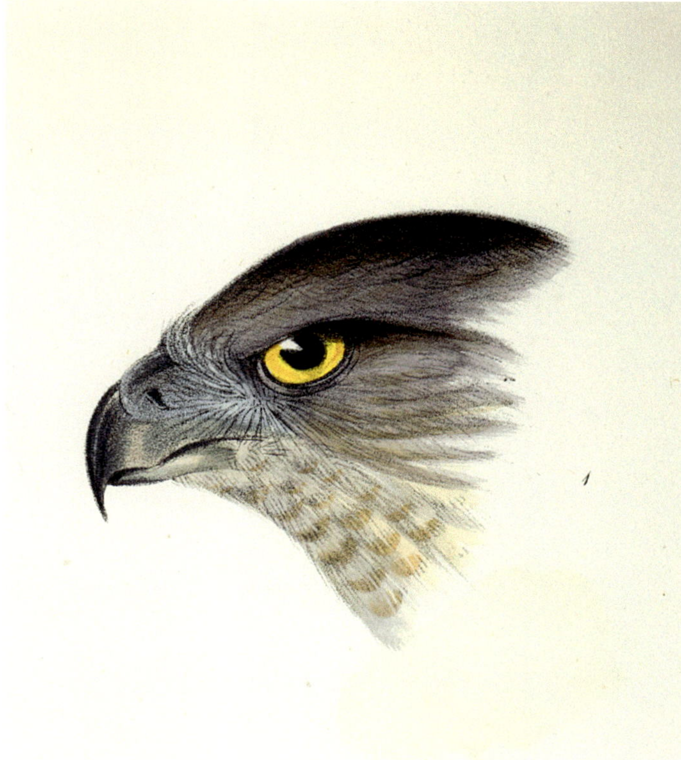

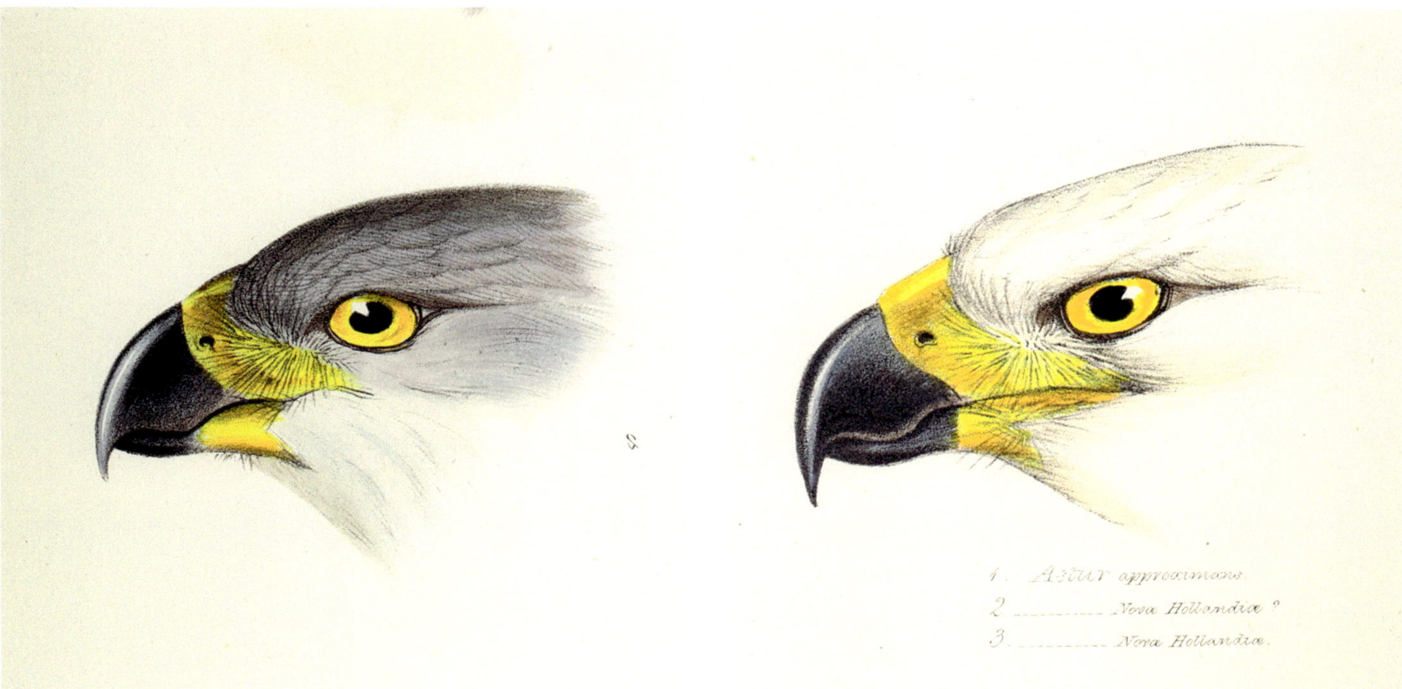

TOP LEFT:
EASTERN OSPREY
Pandion cristatus

BOTTOM LEFT:
GREY GOSHAWK
Accipiter novaehollandiae

TOP RIGHT:
SWAMP HARRIER
Circus approximans

BOTTOM RIGHT:
GREY GOSHAWK
Accipiter novaehollandiae

Elizabeth Gould, *A Synopsis of the Birds of Australia and the Adjacent Islands*, 1837–8

These works by Elizabeth Gould appear in *A Synopsis of the Birds of Australia and the Adjacent Islands*. Published between 1837 and 1838, the work came two years before the first volume of Gould's celebrated *Birds of Australia*. Produced in England before the couple ventured to Australia, this small work was designed to test the commercial viability of his ambitious venture as well as to highlight the scientific merit of his work.

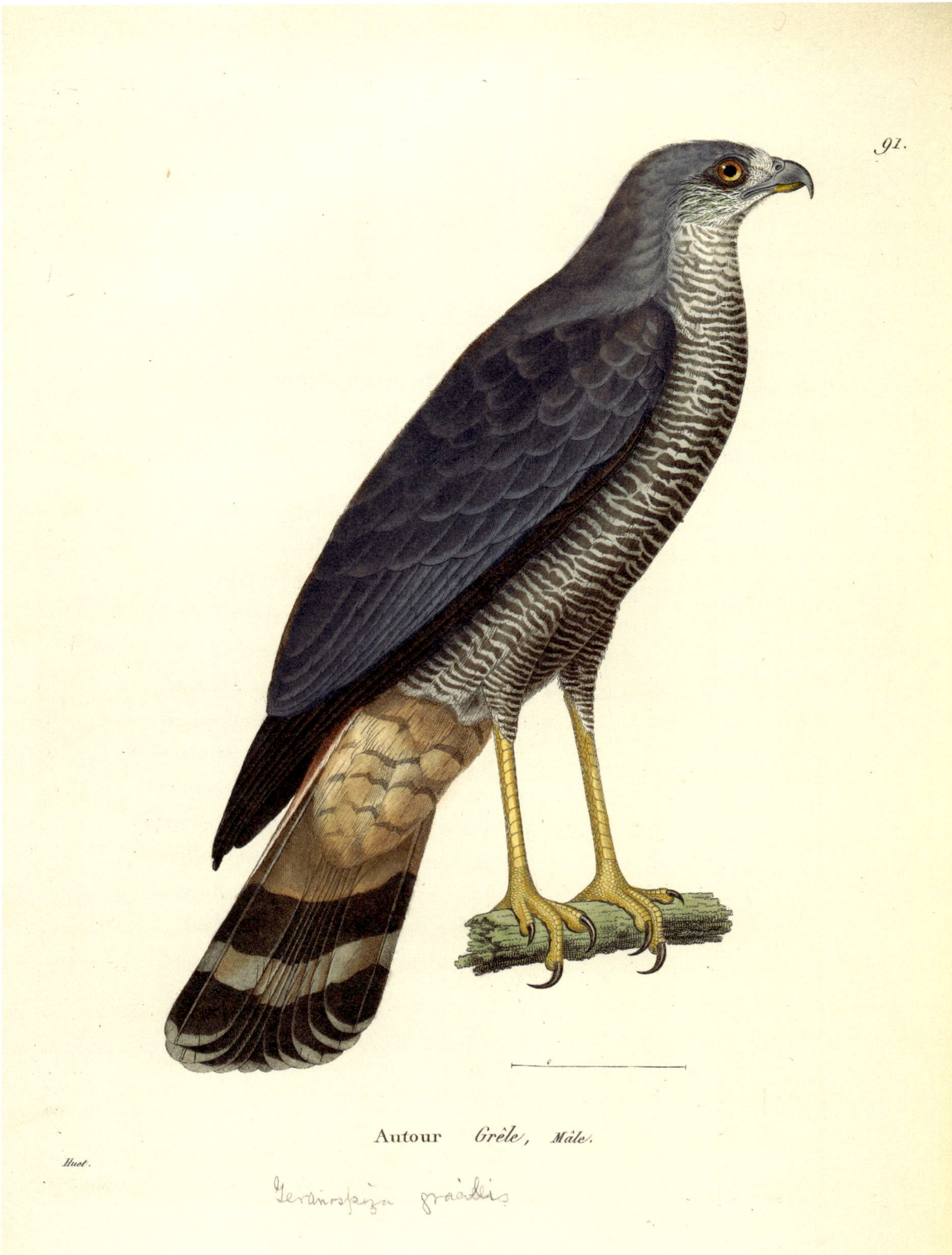

CRANE HAWK
Geranospiza caerulescens
Nicolas Huet II, *Nouveau recueil de planches coloriées d'oiseaux*, 1820–39

Crane hawks are found throughout Central and South America and can be found perching in lowland forests and near water. Their legs are long and double jointed, allowing them to reach deep into tree cavities to extract their prey. These birds typically feed on small creatures such as rodents, bats, lizards and smaller birds. This particular crane hawk is identified as being from the southern part of South America due to its yellow eyes; birds from Central America and northern South America have red eyes.

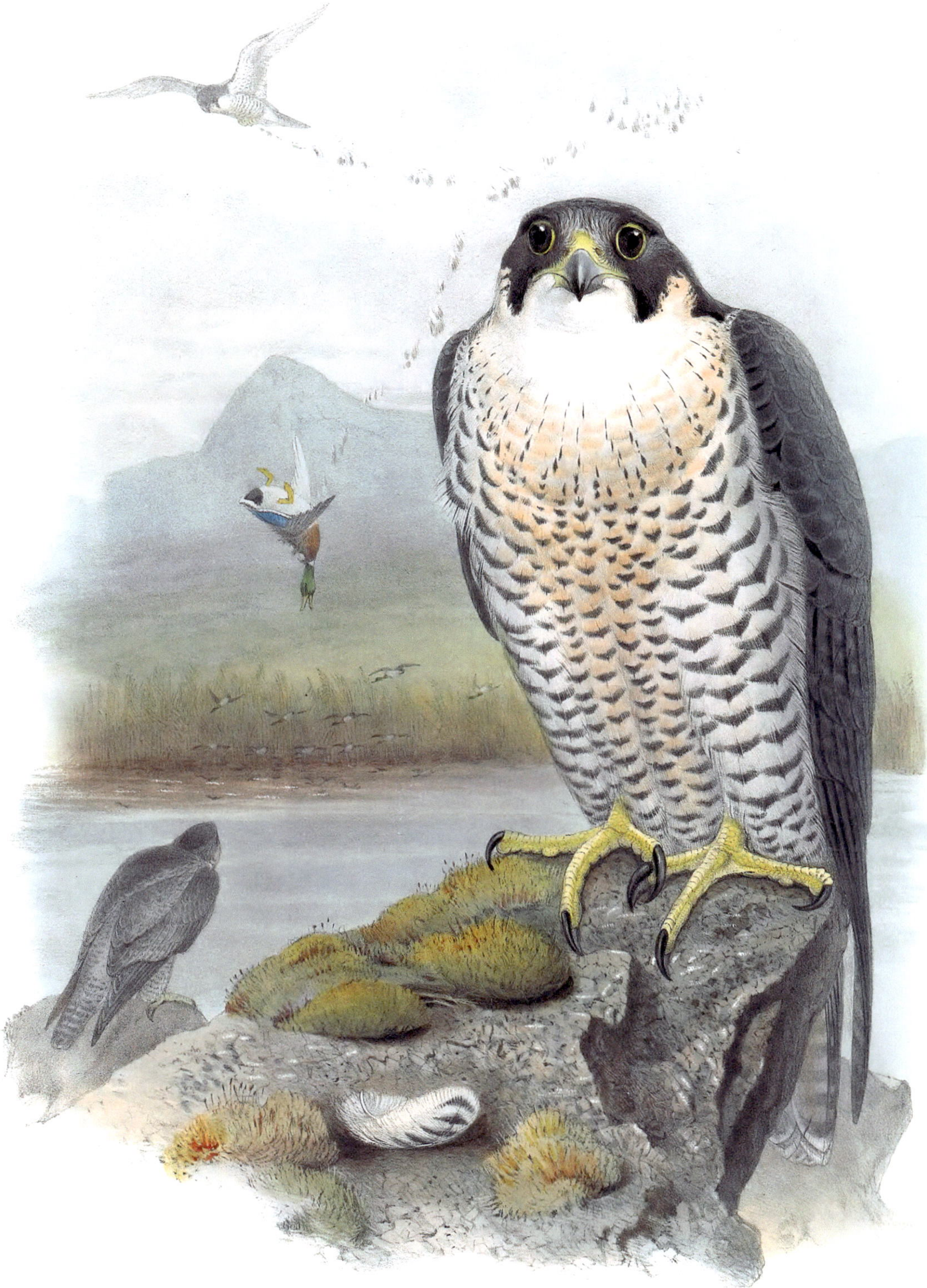

PEREGRINE FALCON
Falco peregrinus

Joseph Wolf & Henry Constantine Richter, *The Birds of Great Britain*, 1862–73

A mallard duck is seen falling from the sky after a mid-air attack by a peregrine falcon. Falcons feed almost exclusively on other birds and tend to hunt medium-sized prey such as doves, pigeons, waterfowl and songbirds. This species is thought to be one of the fastest animals on the planet, reaching speeds of up to 320 kilometres per hour while hunting. Indeed, the species is such a skilled hunter that it is commonly used in falconry due to its ability to be trained, and also its ability to be easily bred in captivity.

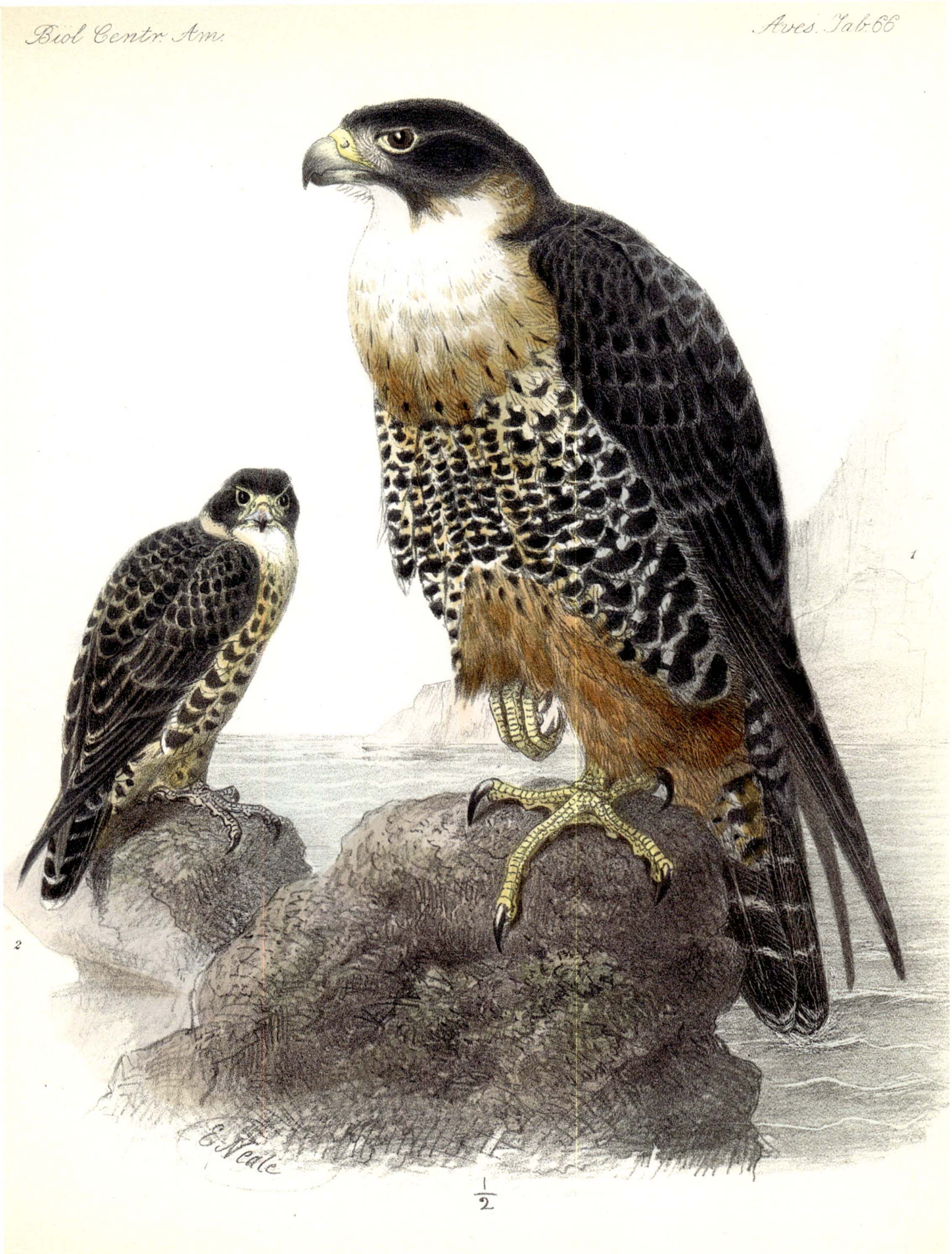

ORANGE-BREASTED FALCON
Falco deiroleucus

Edward Neale, *Biologia Centrali-Americana*, 1879–1915

Birds of prey are made up of several unique bird families. The falcon (above) belongs to the family *Falconidae*; a grouping that also includes caracaras. The merlin (opposite) also belongs to this family. Despite their small size, merlins can be fierce and powerful birds.

OPPOSITE:
MERLIN
Falco columbarius

Joseph Wolf & Henry Constantine Richter, *The Birds of Great Britain*, 1862–73

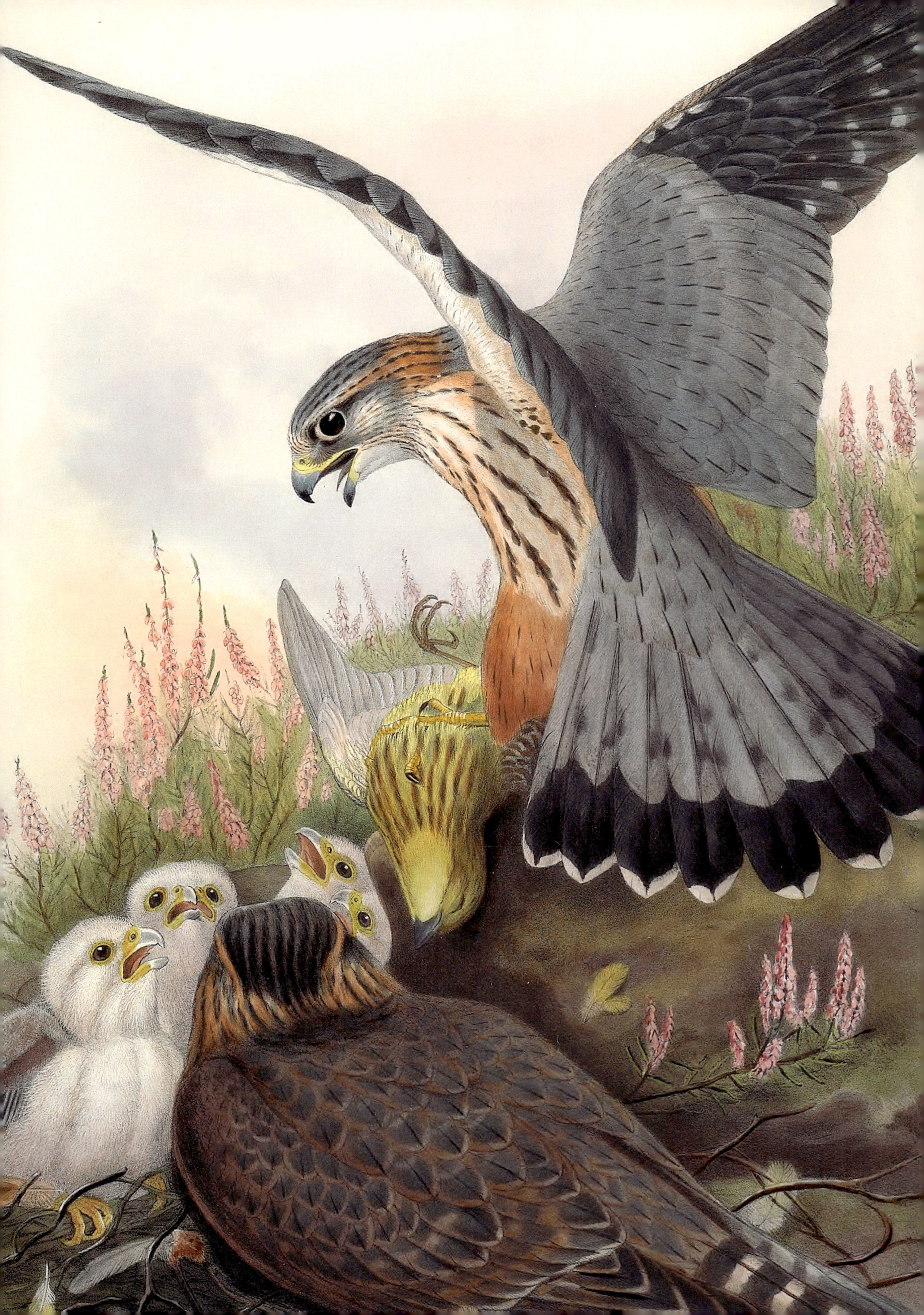

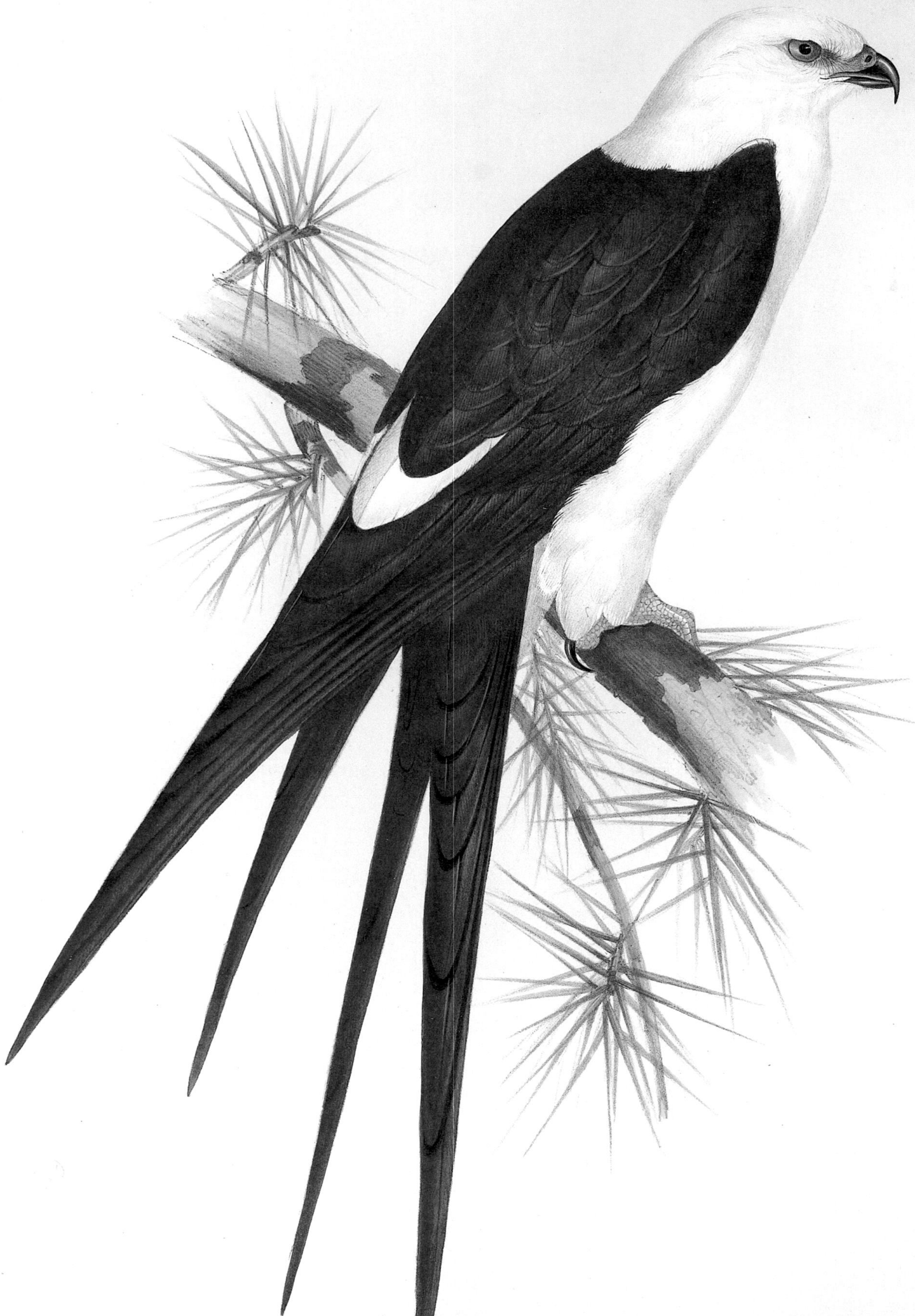

SWALLOW TAILED KITE.
Nauclerus furcatus. (Vig.)

Drawn from Nature & on Stone by J & E Gould.   Printed by C. Hullmandel.

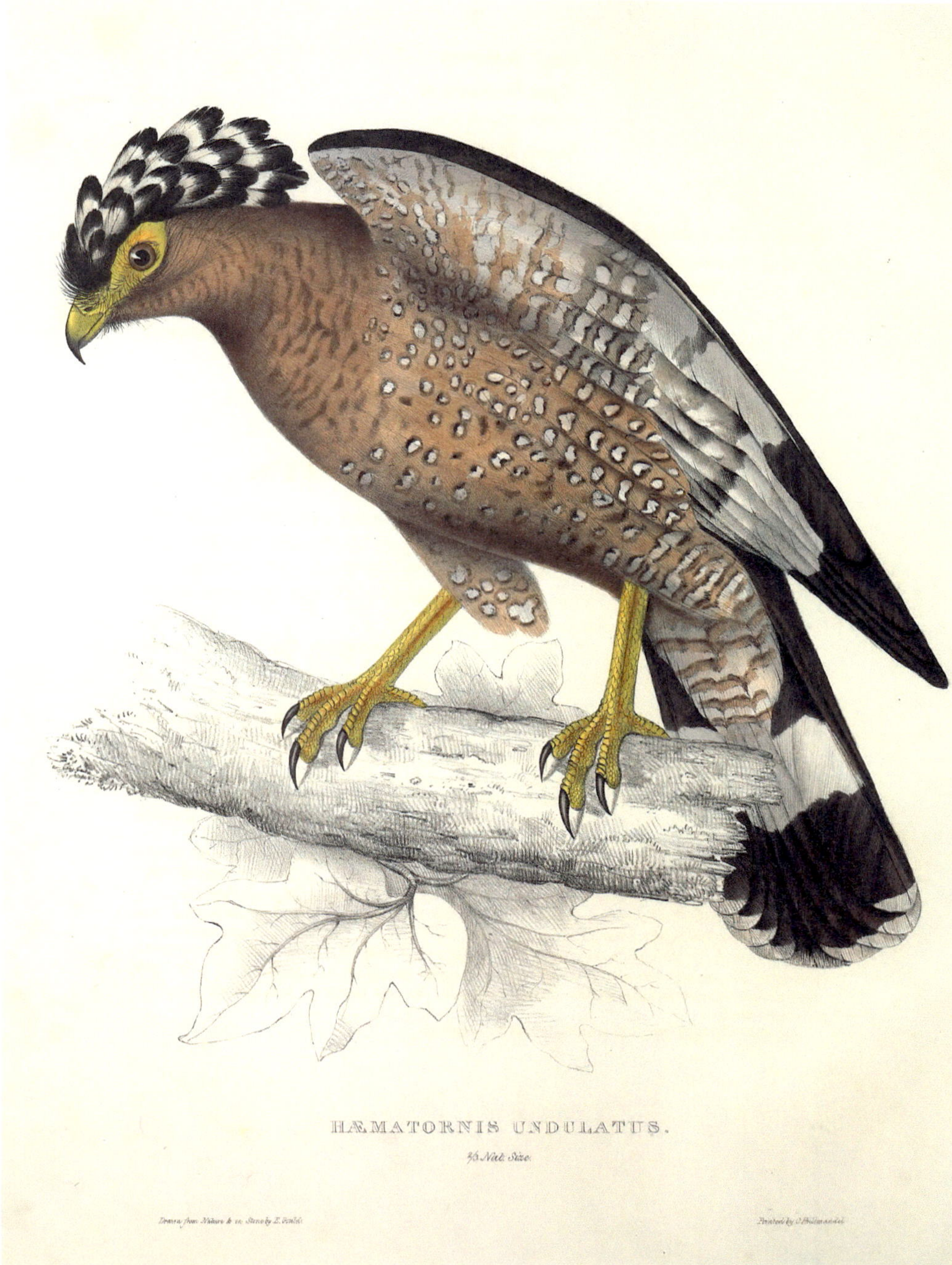

OPPOSITE:

SWALLOW-TAILED KITE
Elanoides forficatus

John Gould & Elizabeth Gould,
The Birds of Europe, 1832–7

This swallow-tailed kite appears in Gould's *The Birds of Europe*. This work was a monumental undertaking that spanned five volumes and was published between 1832 and 1837. The project was partly undertaken to redress an imbalance between the study of local and foreign ornithology.

Gould wrote in the introduction to the book: 'The Birds of Europe, in which we are, or ought to be, most interested, have not received that degree of attention which they naturally demand'. Ironically, the swallow-tailed kite is an American bird and has never been seen in Europe.

CRESTED SERPENT EAGLE
Spilornis cheela

Elizabeth Gould, *A Century of Birds from the Himalaya Mountains*, 1830–2

Joseph Wolf

1820–99

Combining an artistic sensibility with an unrivalled degree of scientific accuracy, Joseph Wolf established the genre of wildlife art, contributing illustrations to numerous books as well as creating stunning oil and watercolour paintings that often portrayed gamebirds and birds of prey.

Arguably one of the world's finest animal artists, Joseph Wolf is known today primarily for his illustrations – yet his work is a link between natural history illustration and wildlife art. Noted for his realism, his success lay in his ability to combine artistic sensibility with an unrivalled degree of scientific accuracy. During his life, he contributed bird and animal illustrations to numerous works and also produced stunning oil and watercolour paintings that typically portrayed gamebirds and birds of prey.

Born in the small German village of Mörz, Wolf was the first son of a farmer. He spent much of his youth exploring the countryside and sketching the wildlife. At sixteen, he travelled 24 kilometres northeast of his home to the city of Koblenz, where he was apprenticed to Gebrüder Becker and was trained in the art of lithography. Not only did the training give him the valuable skill of copying his drawings to limestone, but it also helped him to refine his drawing skills. Soon he was receiving commissions to illustrate scientific books for a number of naturalists.

After three years in Koblenz, he returned to the family farm. His apprenticeship had ended, so he spent some time creating miniature watercolours of birds and looking for employment. On a trip to Frankfurt, he had a chance encounter with German ornithologist and explorer Eduard Rüppell (1794–1884). Impressed by his work, Rüppell not only commissioned him to create drawings for his *Systematische Uebersicht der Vögel Nord-Ost-Afrikas* (1845), but he also introduced him to Johann Jakob Kaup (1803–73), who was then the director of the Grand Ducal natural history collection in Darmstadt. Equally impressed, Kaup took Wolf's drawings to a conference in Leyden where the work caught the attention of Hermann Schlegel (1804–84) at the city's Natural History Museum. He commissioned Wolf to produce some life-sized paintings of birds of prey for *Traité de fauconnerie* (1844–53). The work was highly acclaimed and Wolf's expressive falcon illustrations marked him out as a defining figure in the world of European ornithological illustration.

Despite the success of his illustration work, Wolf was keen to pursue a career as a painter. In 1847, at the age of twenty-seven, he enrolled at the Antwerp Academy to gain a formal education in Dutch oil painting techniques. The following year, Europe found itself awash with revolution and political upheaval. Keen to avoid the political unrest on the continent, he accepted an invitation from David William Mitchell (1813–59), secretary of the Zoological Society of London, to travel to England. There he assisted Mitchell in illustrating *The Genera of Birds* (1844–9), a book written by George Robert Gray (1808–72), who was then the head of the ornithological section at the British Museum.

England's booming book market provided ample opportunities for Wolf and his talents were keenly sought out. Returning British adventurers such as David Livingstone, Alfred Russel Wallace and Henry Walter Bates, all had Wolf produce work for them. The American zoologist Daniel Giraud Elliot had him contribute illustrations to his acclaimed monograph of the pheasant. Even Charles Darwin commissioned him to produce some illustrations that he used in his study of animal expressions.

Despite his popularity as an illustrator, he continued to pursue his passion for art and developed a reputation for being one of the foremost naturalist painters of his day. Much of his success as an illustrator can be attributed to his artistic sensibilities. He would sketch directly from nature, and this gave his work a fidelity that was lacking from other illustrators who simply drew directly from stuffed specimens. From a technical standpoint, his work demonstrates a painstaking desire for accuracy developed through observation, yet this aspiration never leaves the work feeling cold or detached. Instead, he combines his gift for technical realism with a natural talent for capturing the movement and personalities of his subjects.

It did not take long for Wolf's talents to catch the attention of John Gould, who was keen to have him produce illustrations for his *Birds of Great Britain* (1862–73). Wolf's work seemed to capture Gould's imagination in the same way that Edward Lear's work had done twenty years prior. Since his *Birds of Australia*, all of Gould's output had been drawn from specimens, but this new work presented him with an opportunity that could capitalize on Wolf's talent for drawing directly from life. Wolf worked on a freelance basis, so Gould was unable to lure him on to his staff, but the pair worked together on a number of illustrations and Wolf's birds of prey are among some of the finest illustrations in Gould's celebrated work.

Joseph Wolf died in 1899 at the age of seventy-nine. While ornithological illustration would continue into the twentieth – and indeed the twenty-first – century, the Victorian appetite for expensive and lushly illustrated folios was coming to an end. It is perhaps notable that Wolf would be regarded as an artist to naturalists and a naturalist to artists. His work was produced just as the scientific world was separating from the world of art. Science was modernizing and beginning to explore routes of enquiry that went beyond the limitations of representation.

While ornithology had an early preoccupation with species identification and distribution, many ornithologists were now starting to adopt more formal scientific methods to ask broader questions about birds. Within image-making, the distinction between art and illustration was becoming even more pronounced too. Like Wolf, many have continued to work between these disciplines. The growing popularity of recreational birdwatching in the twentieth century led to many wonderful illustrated books and magazines, while Wolf's work in establishing the genre of wildlife art has seen many generations of artists build upon his inspiring legacy.

GYRFALCON (GREENLAND FORM)
Falco rusticolus

Joseph Wolf, *Traité de Fauconnerie*, 1844–53

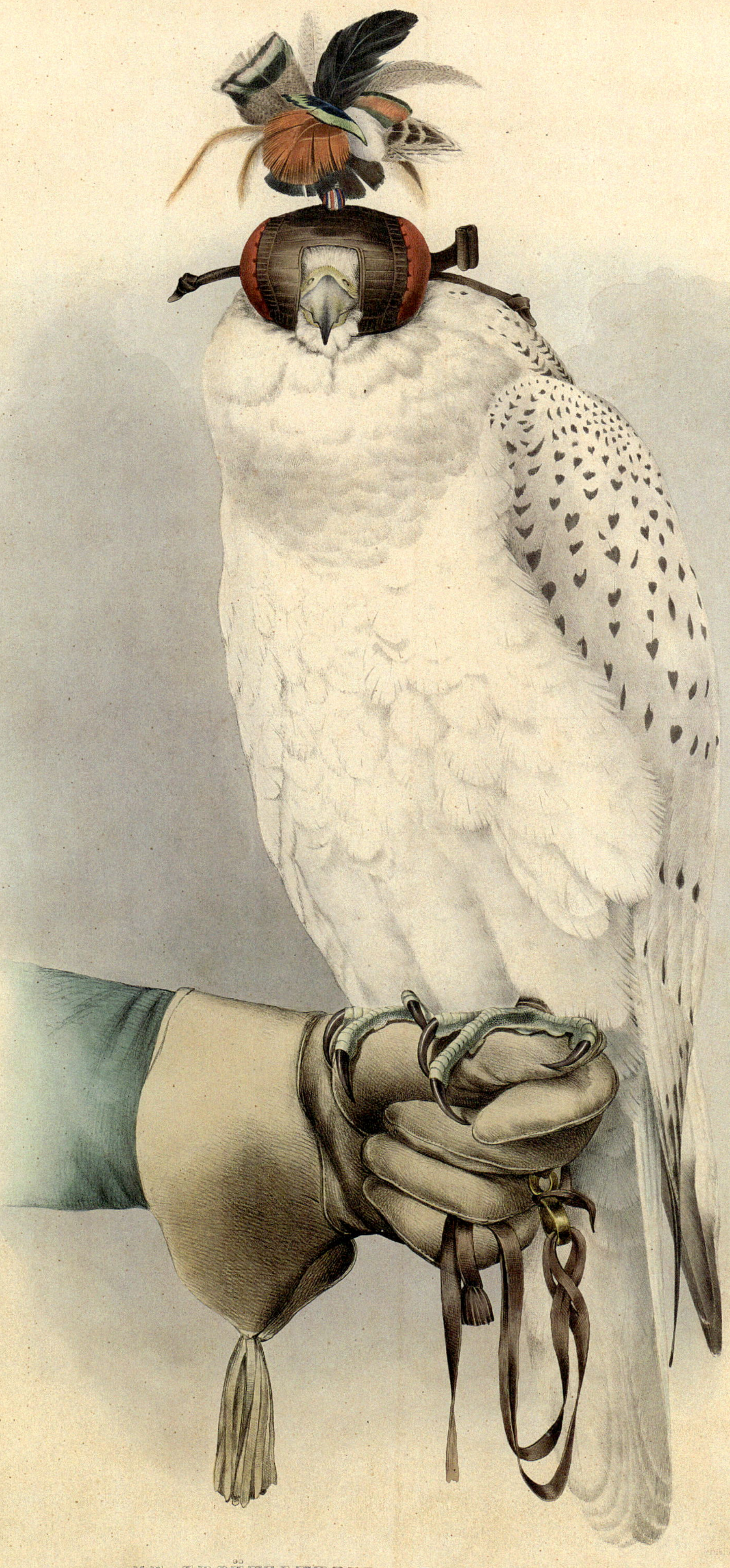

Dessiné par M. WOLF.

Publié chez A. ARNZ et Comp. à LEIDE.

LE GROËNLANDAIS, FAUCON BLANC MUÉ.

PEREGRINE FALCONS
Falco peregrinus

John James Audubon, *The Birds of America*, 1827–38

The peregrine falcon's name is in reference to its migratory habits. The word 'peregrine' means wanderer and North American peregrines have one of the longest migrations of any North American bird. In Audubon's account of this bird, he notes how the French and Spanish Louisiana called this bird the *Mangeur de poulet* (chicken eater), and in other regions it had inaccurate names such as hen hawk, chicken hawk, and pigeon hawk. Ironically, Audubon refers to these birds as great-footed hawks, but peregrines are most certainly falcons.

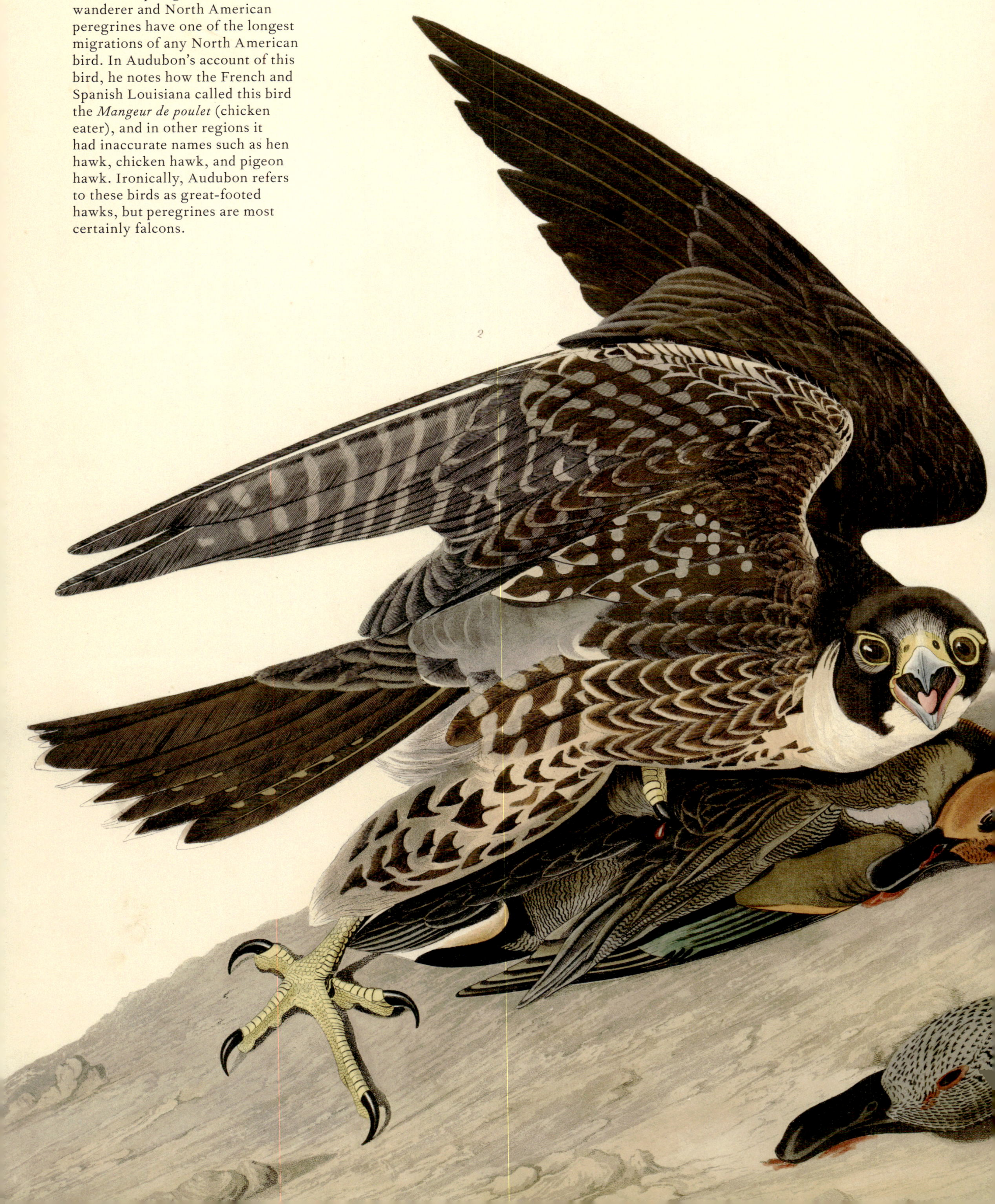

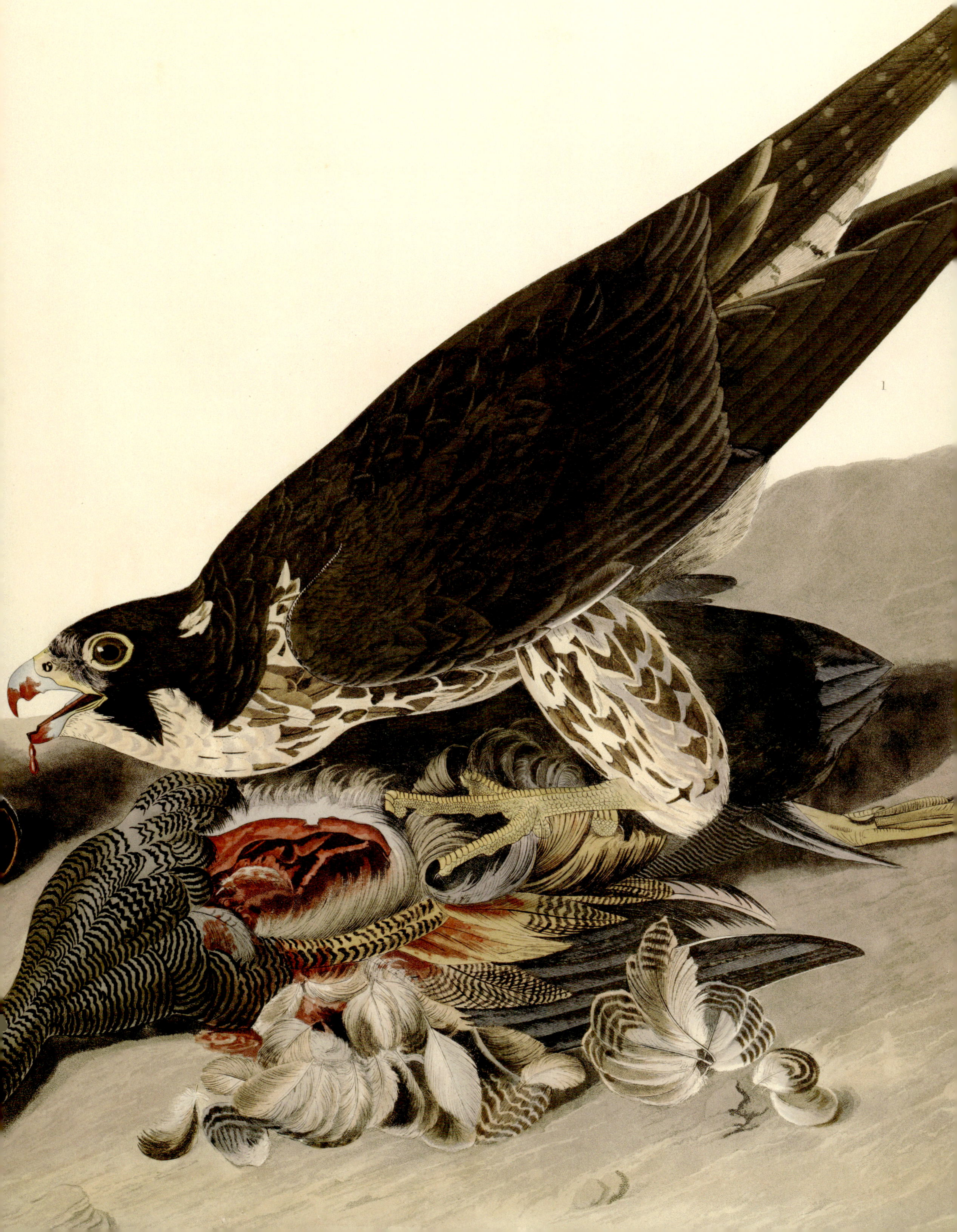
1

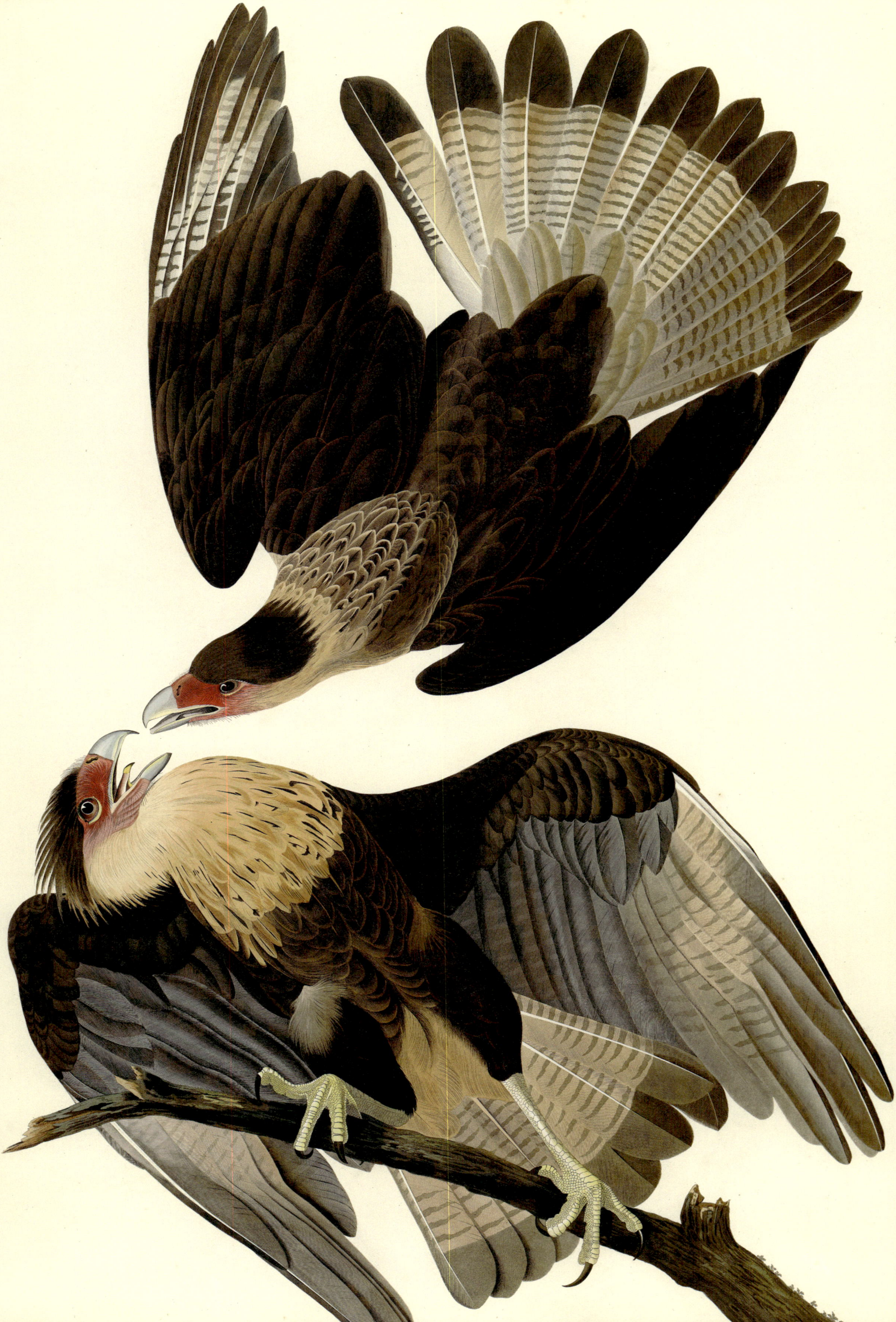

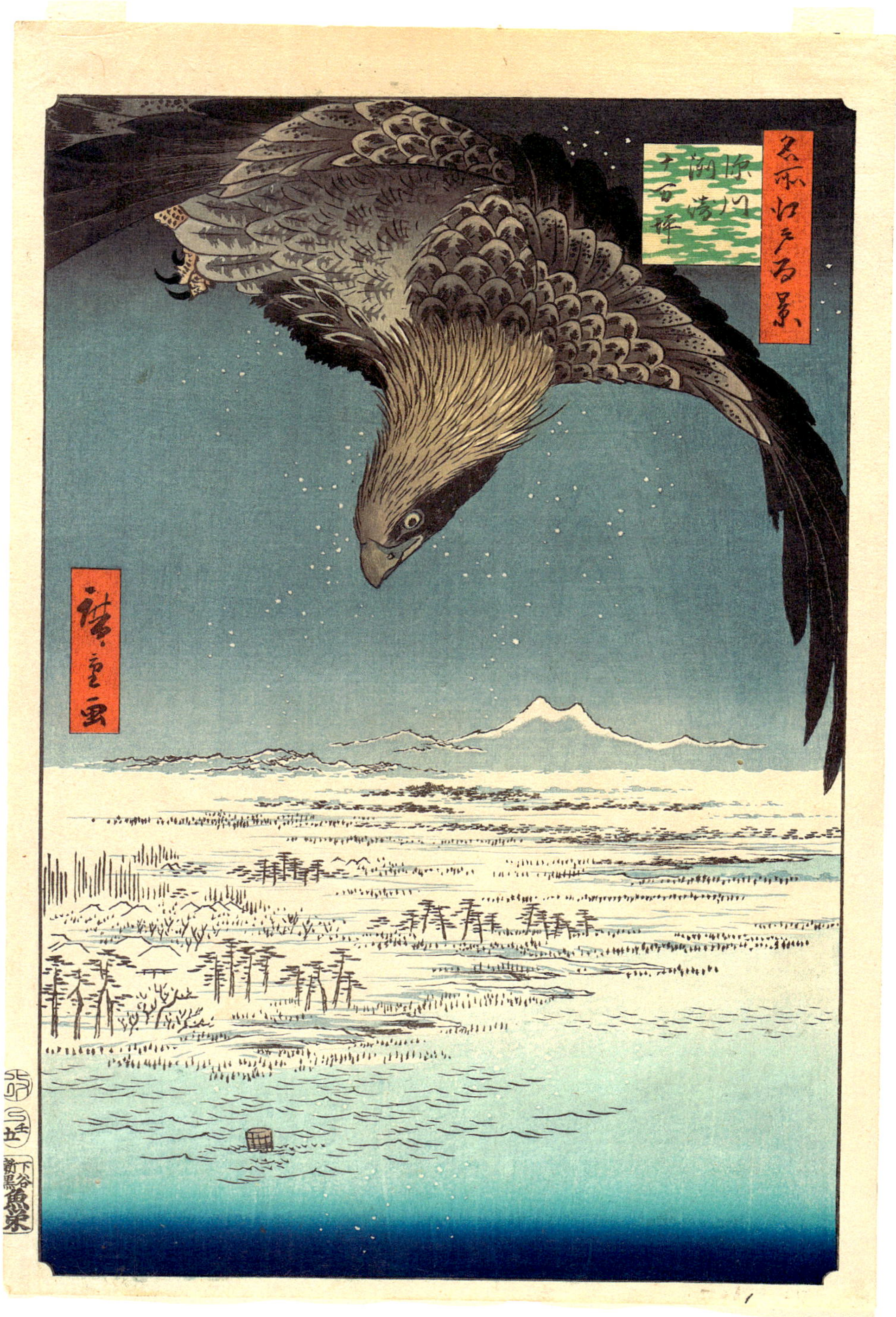

OPPOSITE:
CRESTED CARACARAS
Caracara cheriway

John James Audubon,
The Birds of America, 1827–38

Audubon identifies this bird as the 'Brasilian [sic] caracara eagle'. Today they are typically called crested caracaras. Found in northern parts of South America, as well as much of Central America, Mexico and Cuba, the bird is also seen in the south of the United States. Often found in Florida, Audubon first encountered this bird during the winter of 1831. In tribute to his account of the bird, it is occasionally referred to as Audubon's crested caracara (*Polyborus audubonii*).

JŪMANTSUBO PLAIN AT SUSAKI, FUKAGAWA
Utagawa Hiroshige, No. 107 from
One Hundred Famous Views of Edo, 1856

Hiroshige's series *One Hundred Famous Views of Edo* is his tribute to the Japanese capital. In this scene we see a hawk swooping over the desolate snowy marshlands of Susaki near Fukagawa. This area was once a vast marsh that was renowned for its size. In the background of this print, one can see the snow-capped form of Mount Tsukuba. This is one of the most famous mountains in Japan and is particularly well-known for its distinctive double peaks.

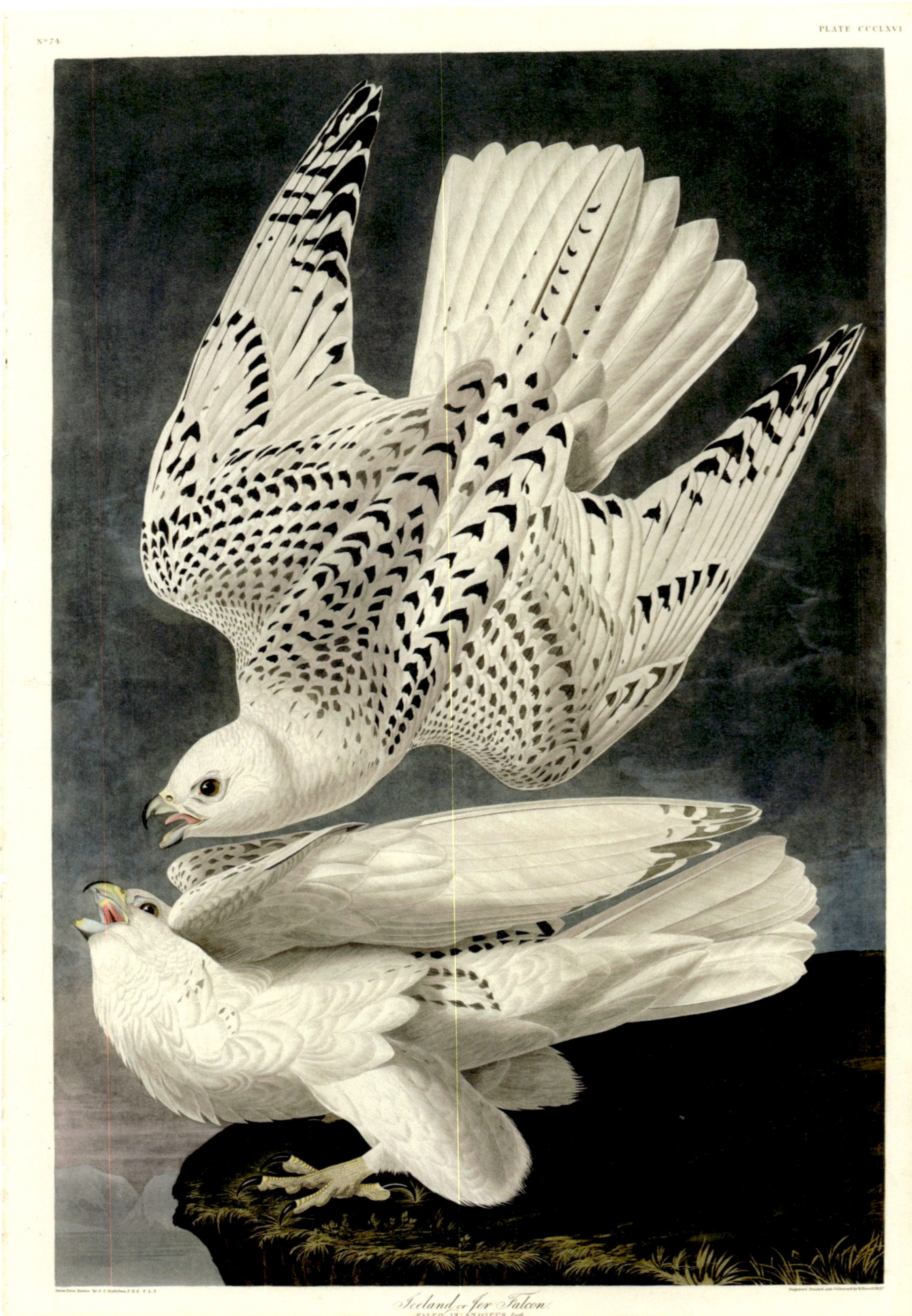

GYRFALCONS
Falco rusticolus

John James Audubon, *The Birds of America*, 1827–38

The gyrfalcon is the largest of the falcon species. This formidable bird is found in the high Arctic, where it typically hunts large ground birds such as ptarmigan and grouse. Celebrated as a hunting bird, this species will also hunt waterfowl and seabirds. Depending on location, the bird's plumage typically varies in colour, with Audubon's depiction showing these birds in striking white. Other gyrfalcons can be seen in shades from grey to brown to black.

WESTERN OSPREY
Pandion haliaetus

John James Audubon, *The Birds of America*, 1827–38

Often found along coastlines, lakes and rivers, the osprey is an impressive bird. Occasionally called the sea hawk, river hawk or fish hawk, these birds are experts at catching fish. Often seen hovering above the water, they wait patiently for the right moment to strike. When the moment arrives, they move in a split second, plunging feet-first into the water and grabbing hold of their prey in their sharp talons. The western osprey can be found distributed all across the world, which is highly unusual for a land-based bird.

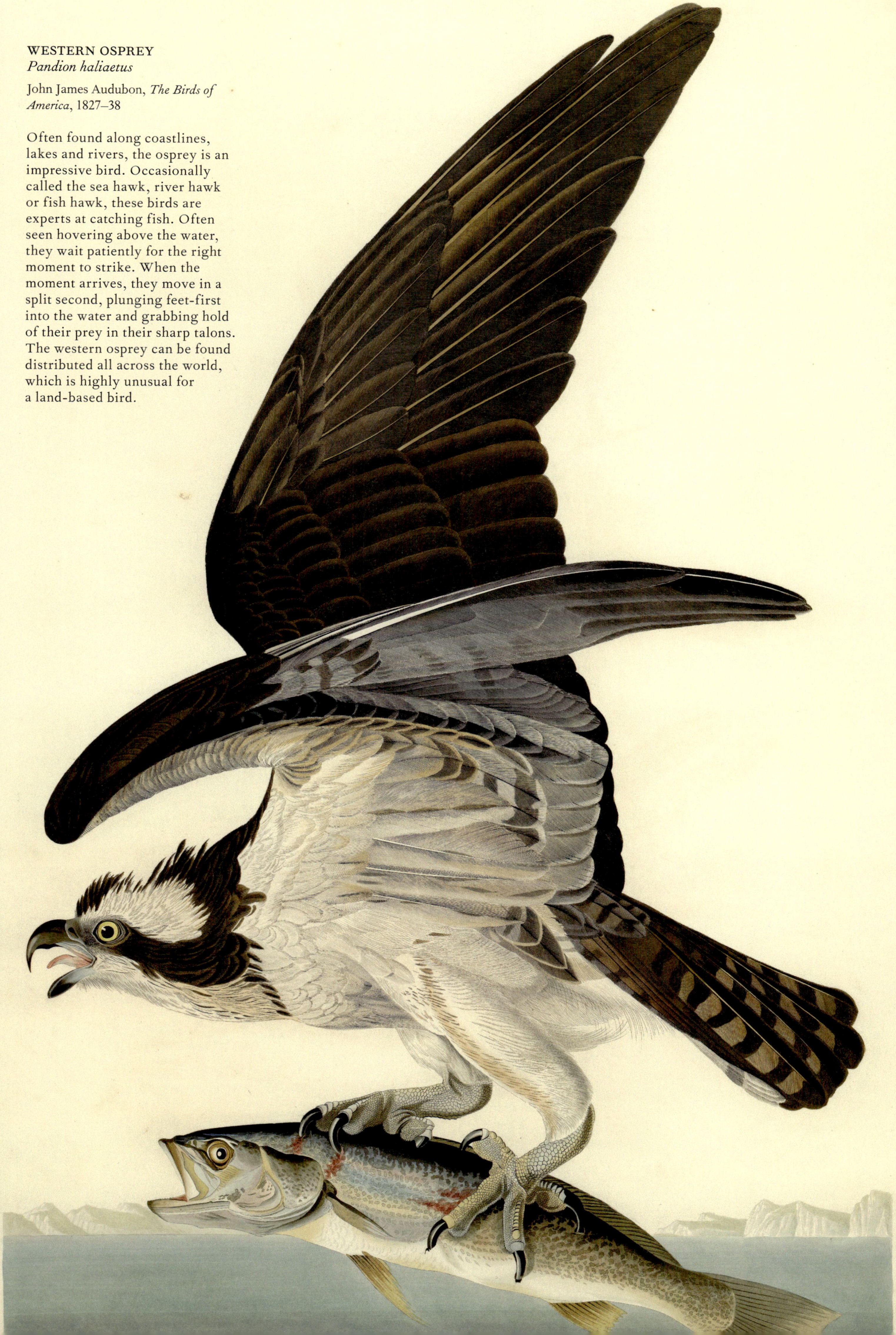

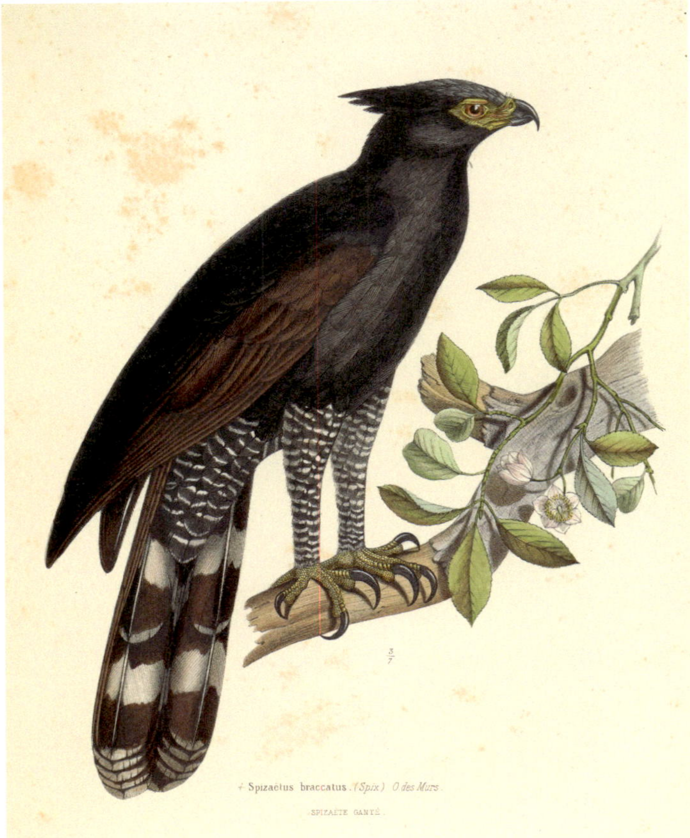

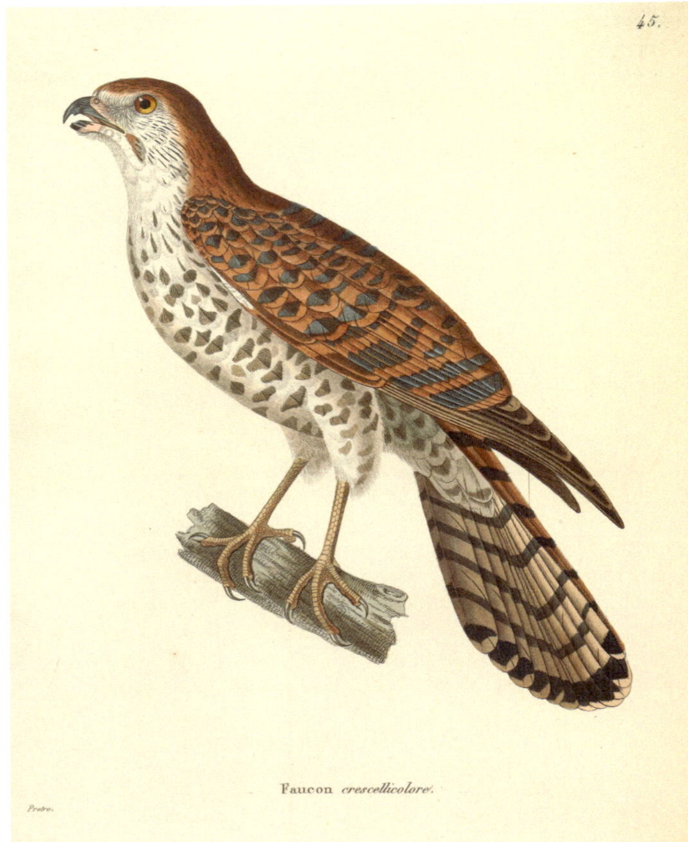

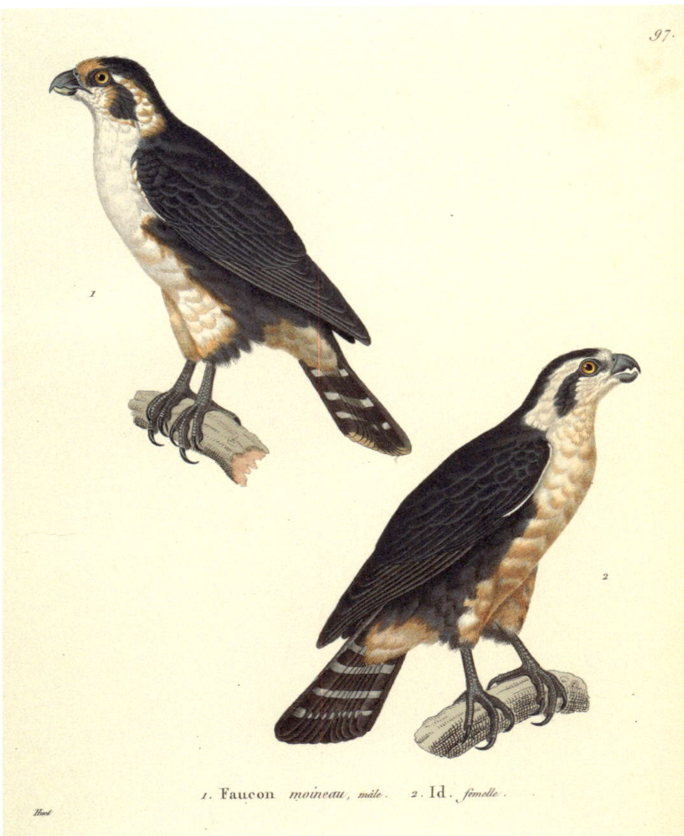

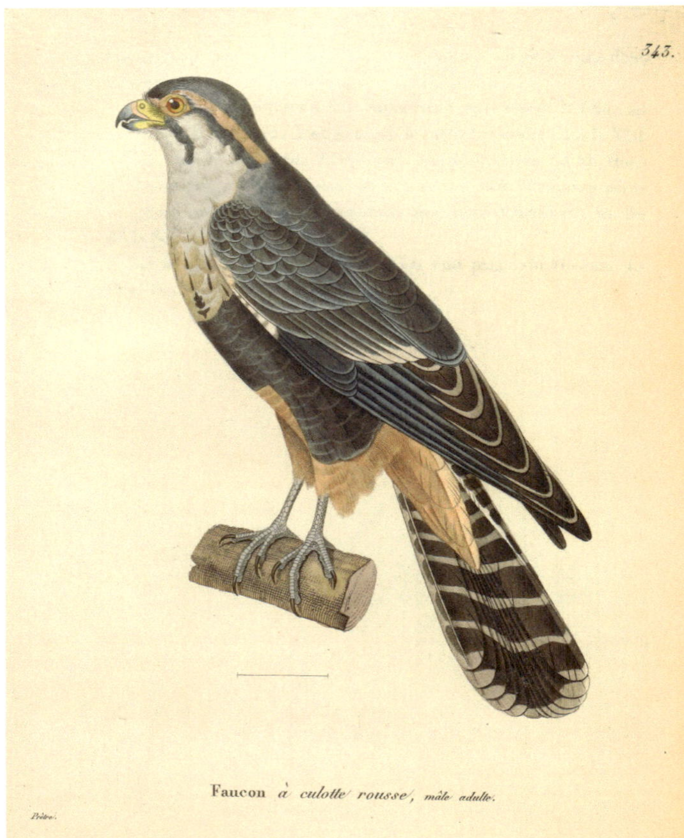

TOP LEFT:
BLACK HAWK-EAGLE
Spizaetus tyrannus

Paul Louis Oudart, *Iconographie ornithologique*, 1845–9

TOP RIGHT:
MAURITIUS KESTREL
Falco punctatus

Jean-Gabriel Prêtre, *Nouveau recueil de planches coloriées d'oiseaux*, 1820–39

BOTTOM LEFT:
COLLARED FALCONETS
Microhierax caerulescens

Nicolas Huet II, *Nouveau recueil de planches coloriées d'oiseaux*, 1820–39

BOTTOM RIGHT:
APLOMADO FALCON
Falco femoralis

Jean-Gabriel Prêtre, *Nouveau recueil de planches coloriées d'oiseaux*, 1820–39

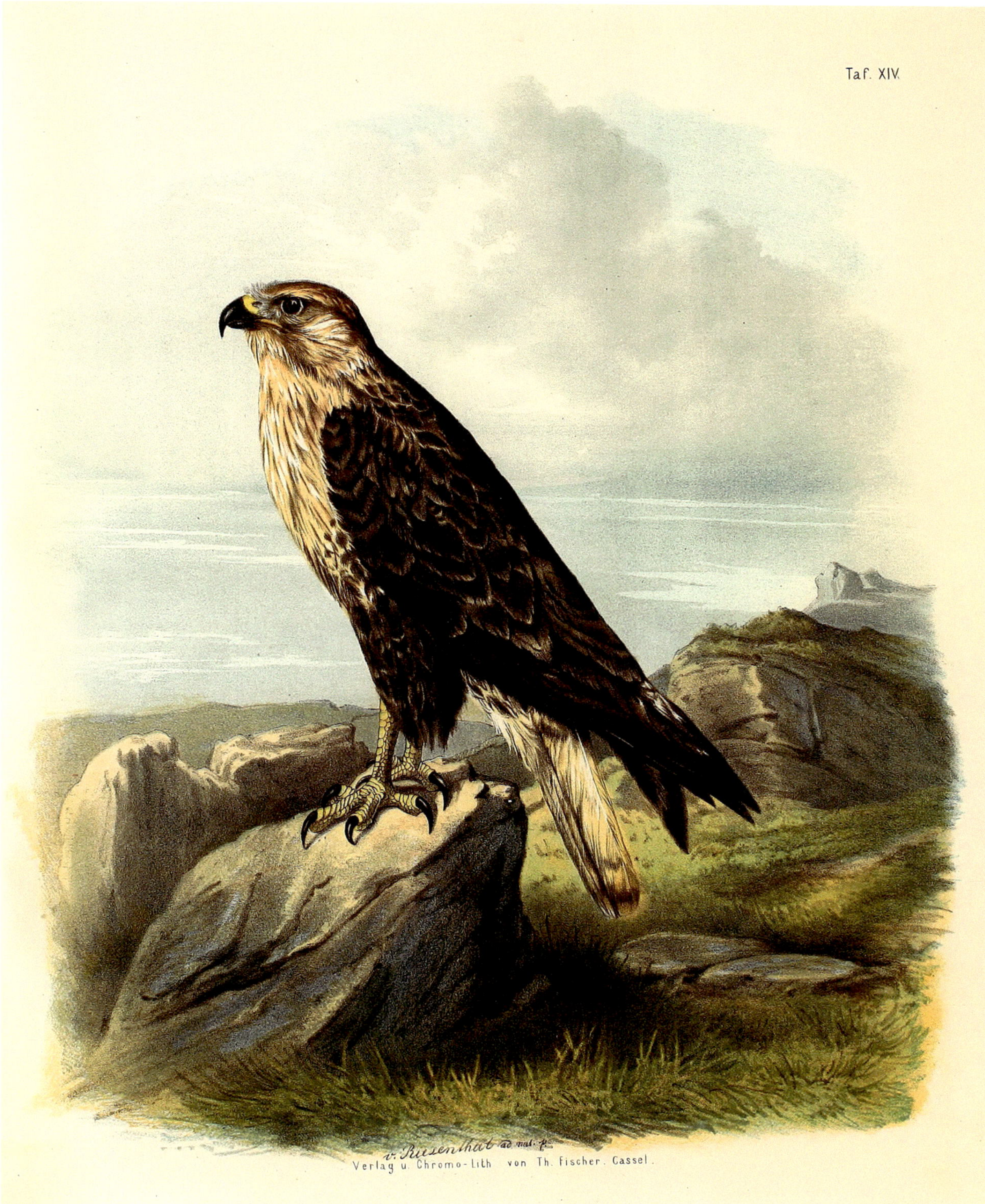

LONG-LEGGED BUZZARD
Buteo rufinus

Oskar von Riesenthal, *Die Raubvögel Deutschlands und des angrenzenden Mitteleuropas*, 1876

Buzzards have amazing eyesight and can survey great distances without ever leaving their perch. This depiction by Oskar von Riesenthal shows a long-legged buzzard. One of the largest species of the genus, it can be found in the dry open plains of North Africa and the Middle East as well as parts of eastern Europe and central Asia. Von Riesenthal was a German naturalist, writer and forester who wrote many books on ornithology, hunting and natural preservation.

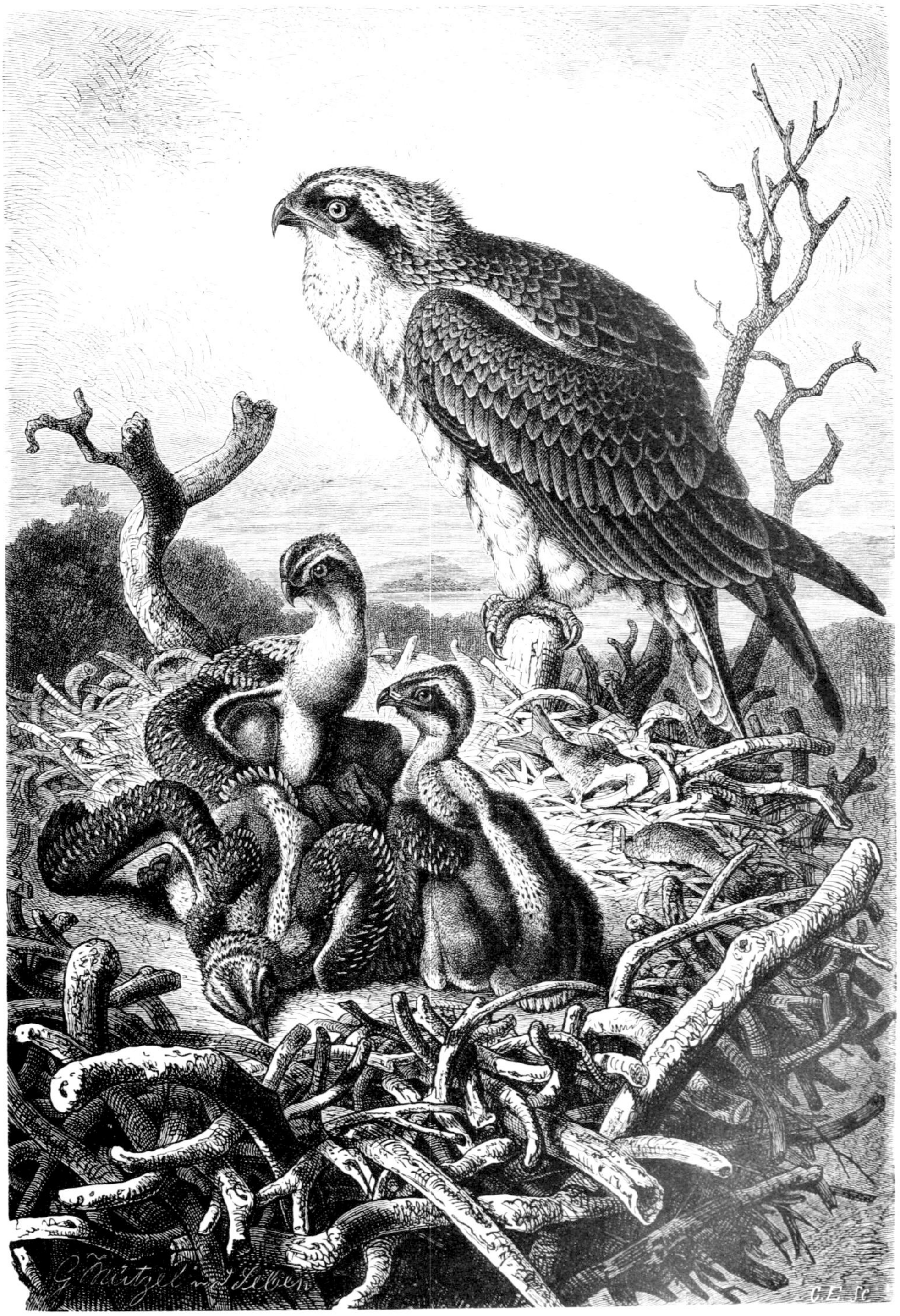

WESTERN OSPREY
Pandion haliaetus

Gustav Mützel, *The Royal Natural History*, 1893–6

Gustav Mützel was a German painter and natural history illustrator. He specialized in birds, often creating work for the German Ornithological Society. Mützel is perhaps best known for the illustrations he produced for the second edition of *Brehms Tierleben*. Published in various editions from 1863, it was a much celebrated and highly successful multi-volume zoological work written by the German zoologist Alfred Brehm and many others.

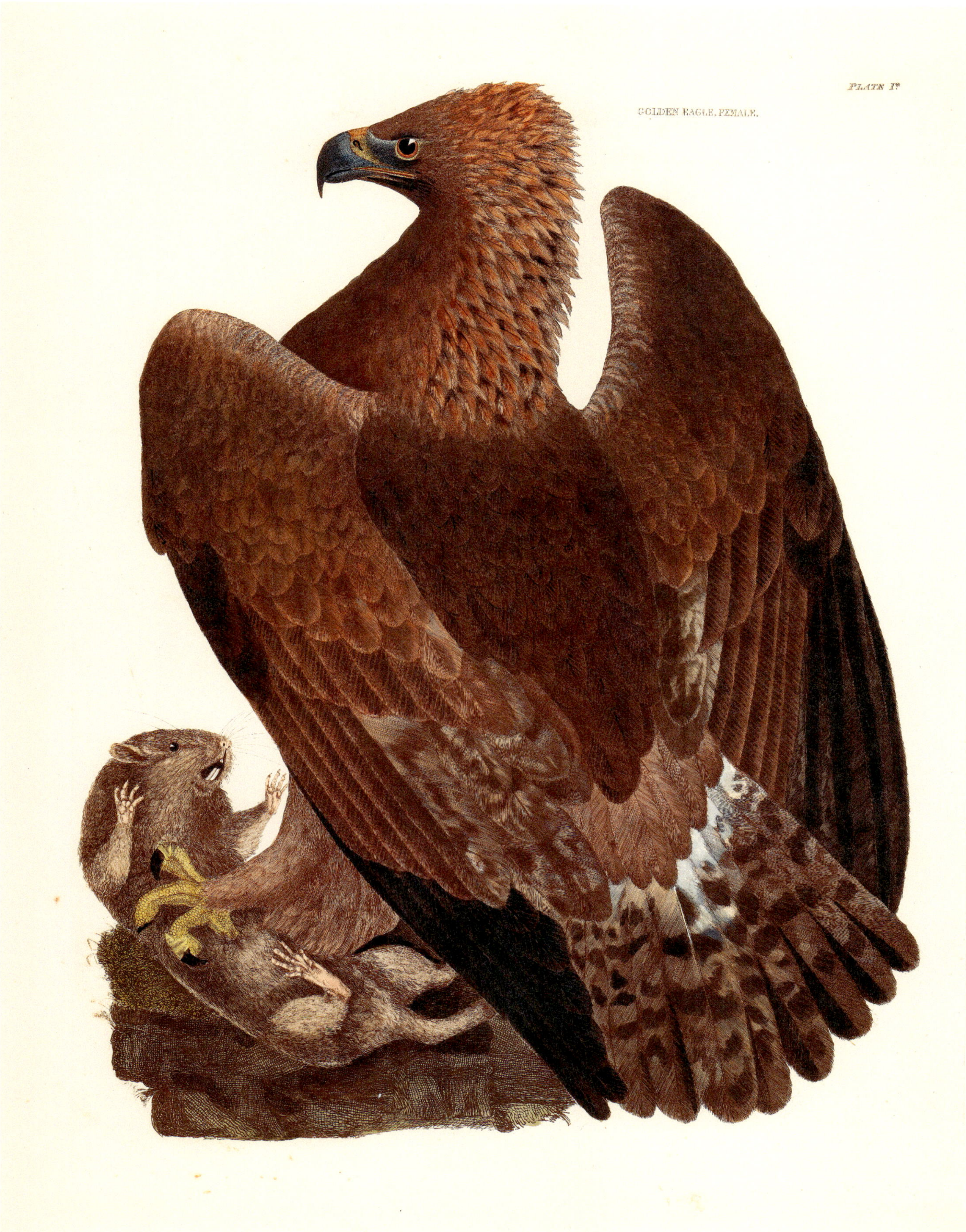

GOLDEN EAGLE
Aquila chrysaetos

Prideaux John Selby, *Illustrations of British Ornithology*, 1819–34

Selby was particularly gifted at capturing the dramatic nature of birds of prey, and was so fond of the golden eagle that he chose it to be the first bird to appear in his *Illustrations of British Ornithology*. This figure, he notes, displays the bird's 'remarkable manner when it has pounced, and exults over its prey.' Selby kept two golden eagles in his possession and drew this work from life. It demonstrates his keen ability to capture the tone and texture of the bird's feathers.

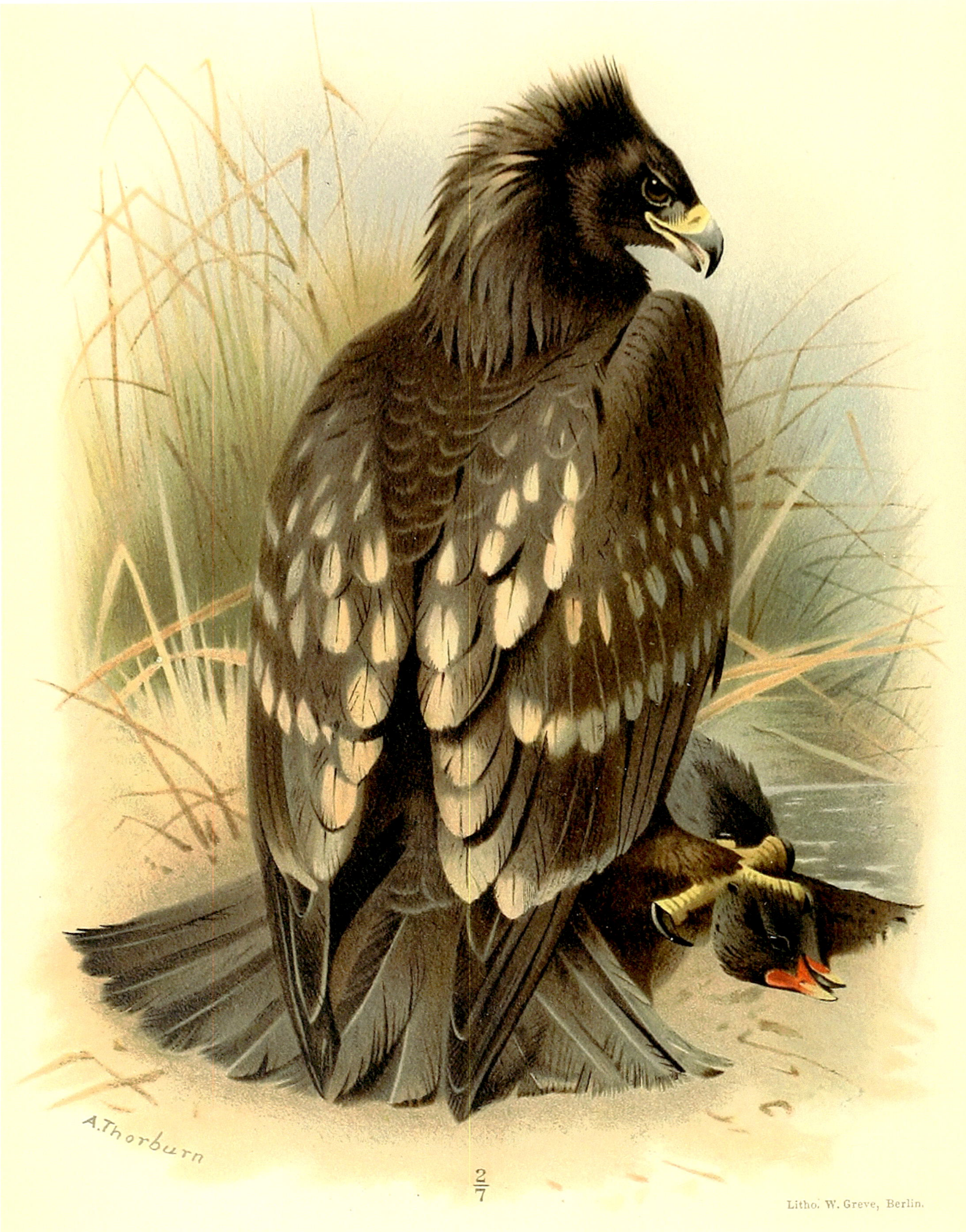

SPOTTED EAGLE
Clanga clanga

Archibald Thorburn, *Coloured Figures of the Birds of the British Islands*, 1885–97

In the world of birds, the eagle is considered to be avian royalty. These birds are praised for their majesty and power, appearing as prominent symbols in everything from religion and folklore to myths and heraldry. This bird is identified as a spotted eagle and is said to have been drawn from a living specimen that was captured near the English town of Colchester on 29 October 1891. This is noteworthy, as this species is exceptionally rare in Britain, tending instead to breed in eastern Europe and eastwards across Eurasia.

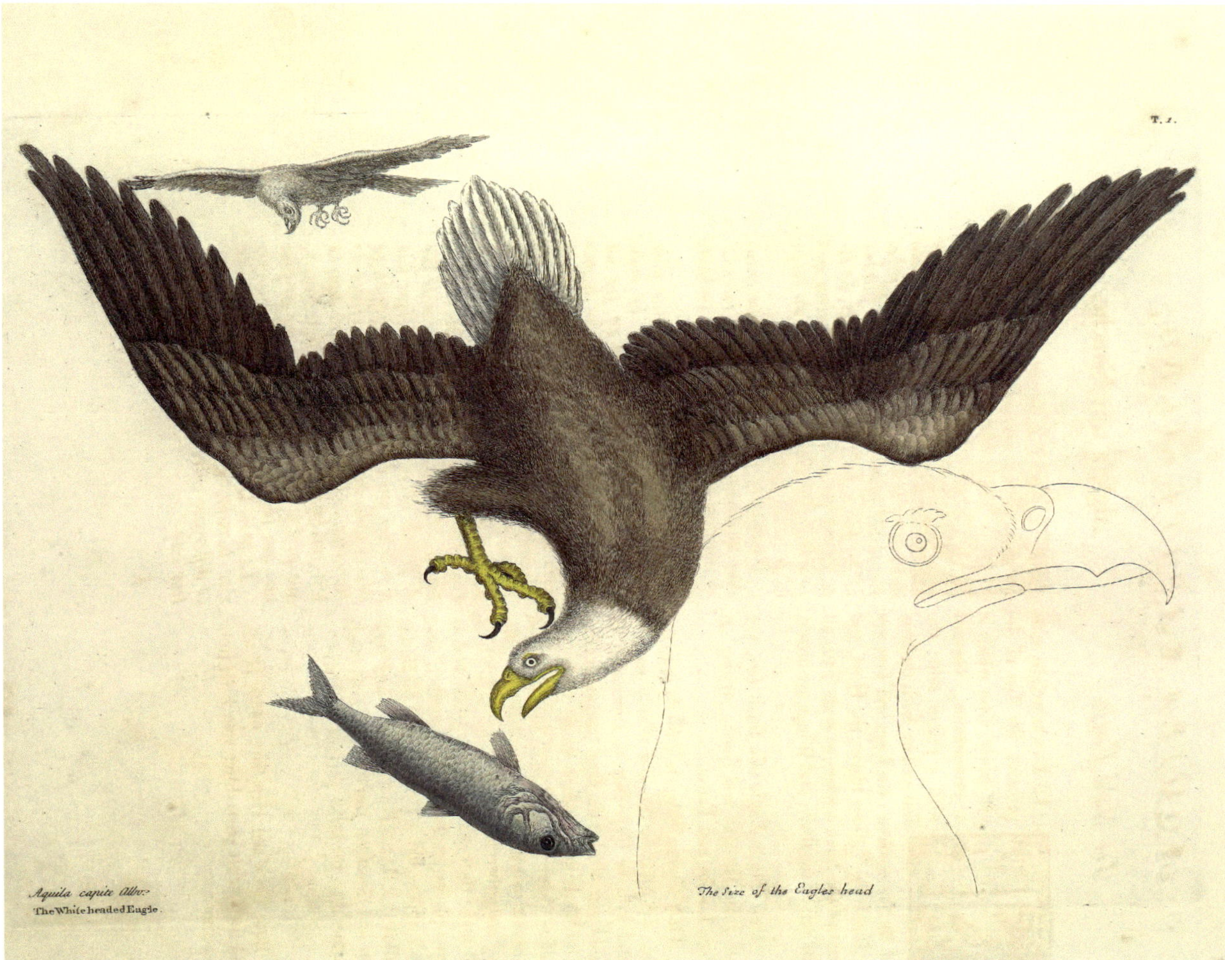

BALD EAGLE
Haliaeetus leucocephalus

Mark Catesby, *The Natural History of Carolina, Florida and the Bahama Islands*, 1729–47

Catesby's bald eagle is arguably the most striking image in his *Natural History* and he chose it as the first plate that would appear in his work. Published between 1729 and 1749, Catesby's illustrations typically lack drama, but his eagle is relatively animated. Despite being 'small of size', Catesby noted that the bird had 'great strength and spirit'. Thirty-three years after Catesby finished his *Natural History* the bird was adopted as the national emblem of the United States of America.

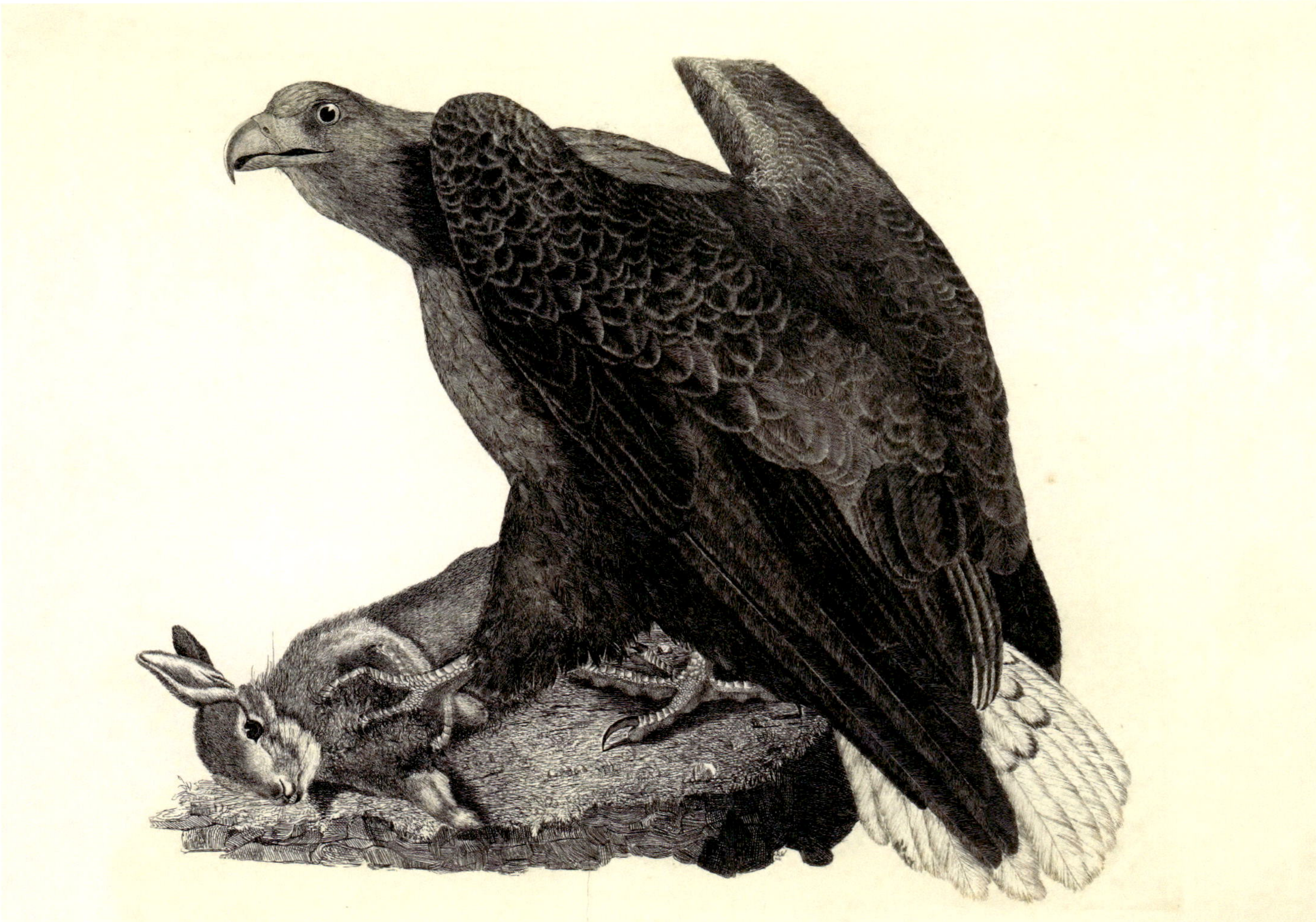

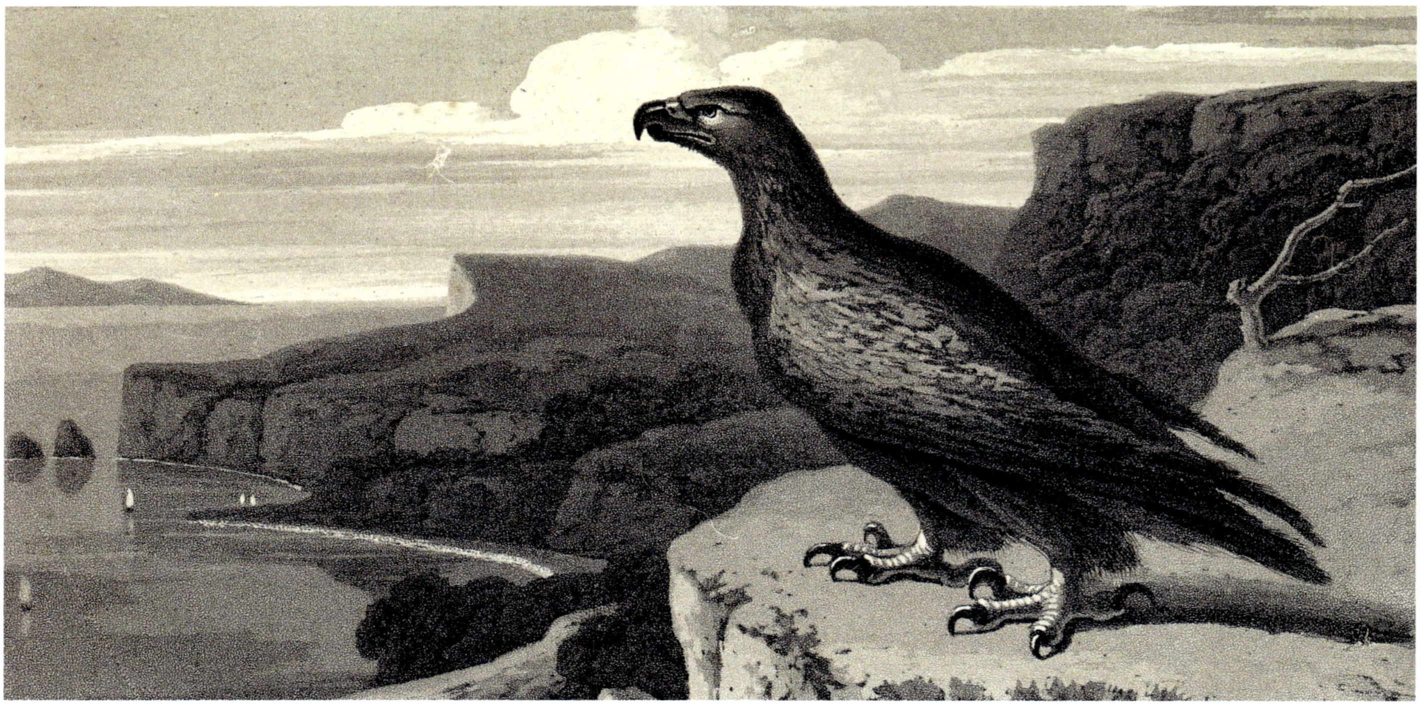

TOP:
WHITE-TAILED EAGLE
Haliaeetus albicilla

Prideaux John Selby, *Illustrations of British Ornithology*, 1819–34

BOTTOM:
GOLDEN EAGLE
Aquila chrysaetos

William Daniell, *Zoography*, 1807

Eagles are often adopted as national birds. The golden eagle, for example, is the national bird of Mexico, Afghanistan, Germany and Scotland. In Poland, the national bird is the white-tailed eagle.

Occasionally, birds will even appear on the national flag, with eagles being the most common. Eagles can be seen on several flags, including those of Egypt, Kazakhstan, Mexico, Moldova, Zambia and Zimbabwe.

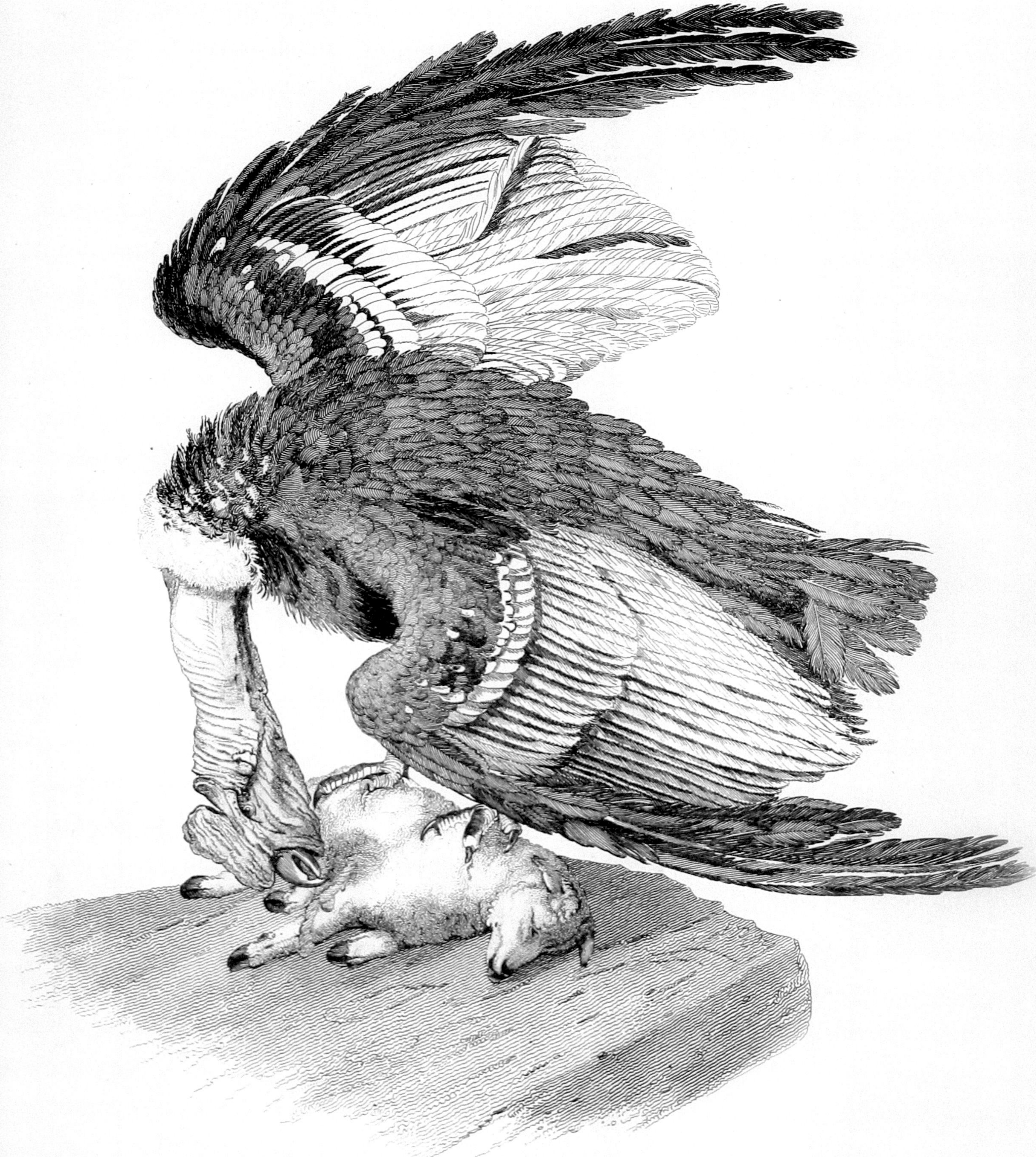

ANDEAN CONDOR
Vultur gryphus

H. Smith, *The Wonders of the Animal Kingdom*, 1830

The vultures of the New World (*Cathartidae*) are not closely related to Old World vultures (part of *Accipitridae*). The condor is a name given to two species of New World vulture: the California condor (*Gymnogyps californianus*), which is found in North America, and the Andean condor (*Vultur gryphus*), which is found in South America. Both are massive birds and are among the largest in the world that can fly.

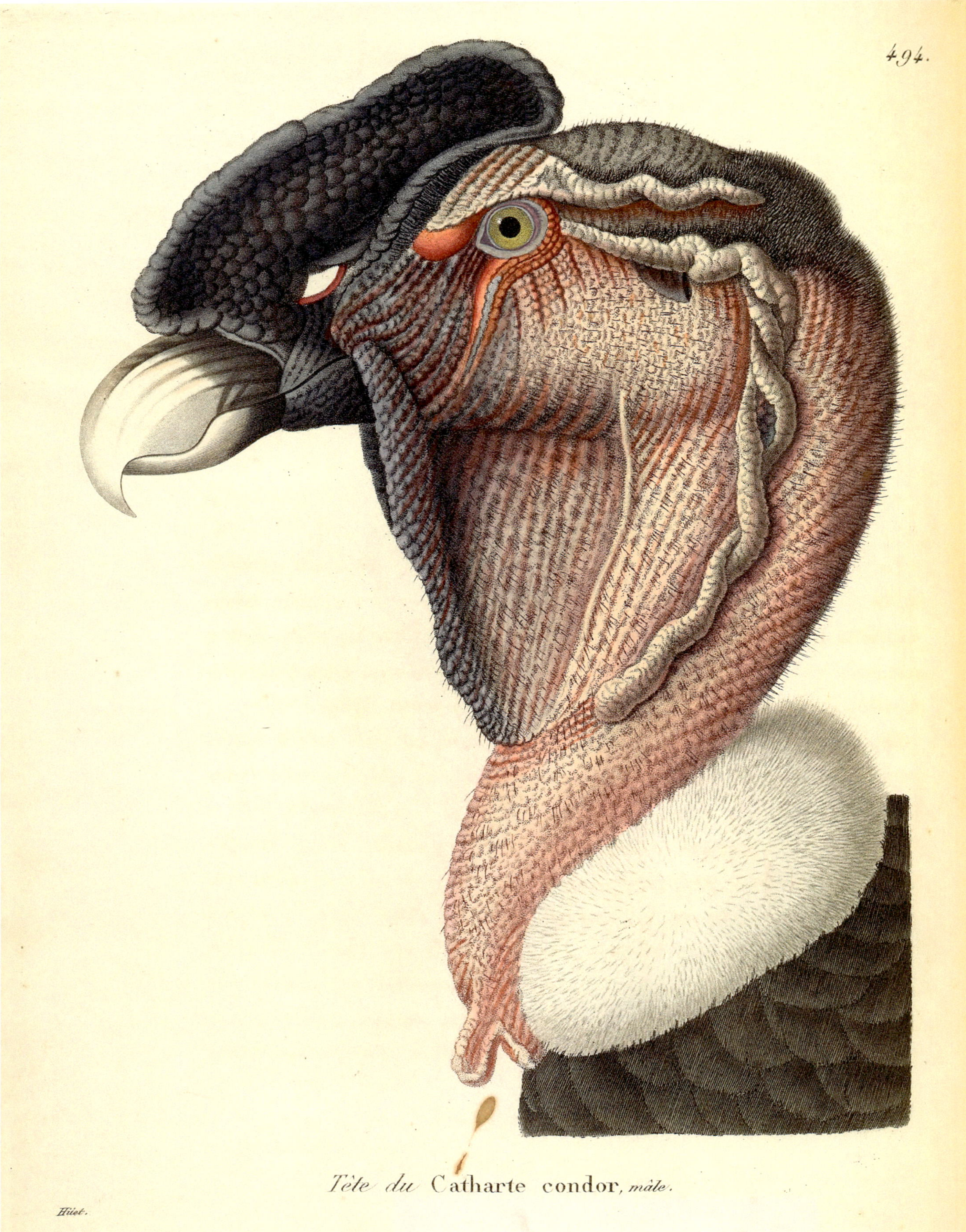

ANDEAN CONDOR
Vultur gryphus

Nicolas Huet II, *Nouveau recueil de planches coloriées d'oiseaux*, 1820–39

The Andean condor inhabits the Andean mountains and is the national bird of many South American countries, including Bolivia, Chile and Colombia. In the Andean regions, the bird is an important figure in myths, folklore and identity, and the bird appears on everything from stamps and banknotes to coats of arms and even the flag of Ecuador. This depiction by Nicolas Huet II shows the bird's distinctive near-featherless head and neck, which is crowned with a dark fleshy comb known as a caruncle.

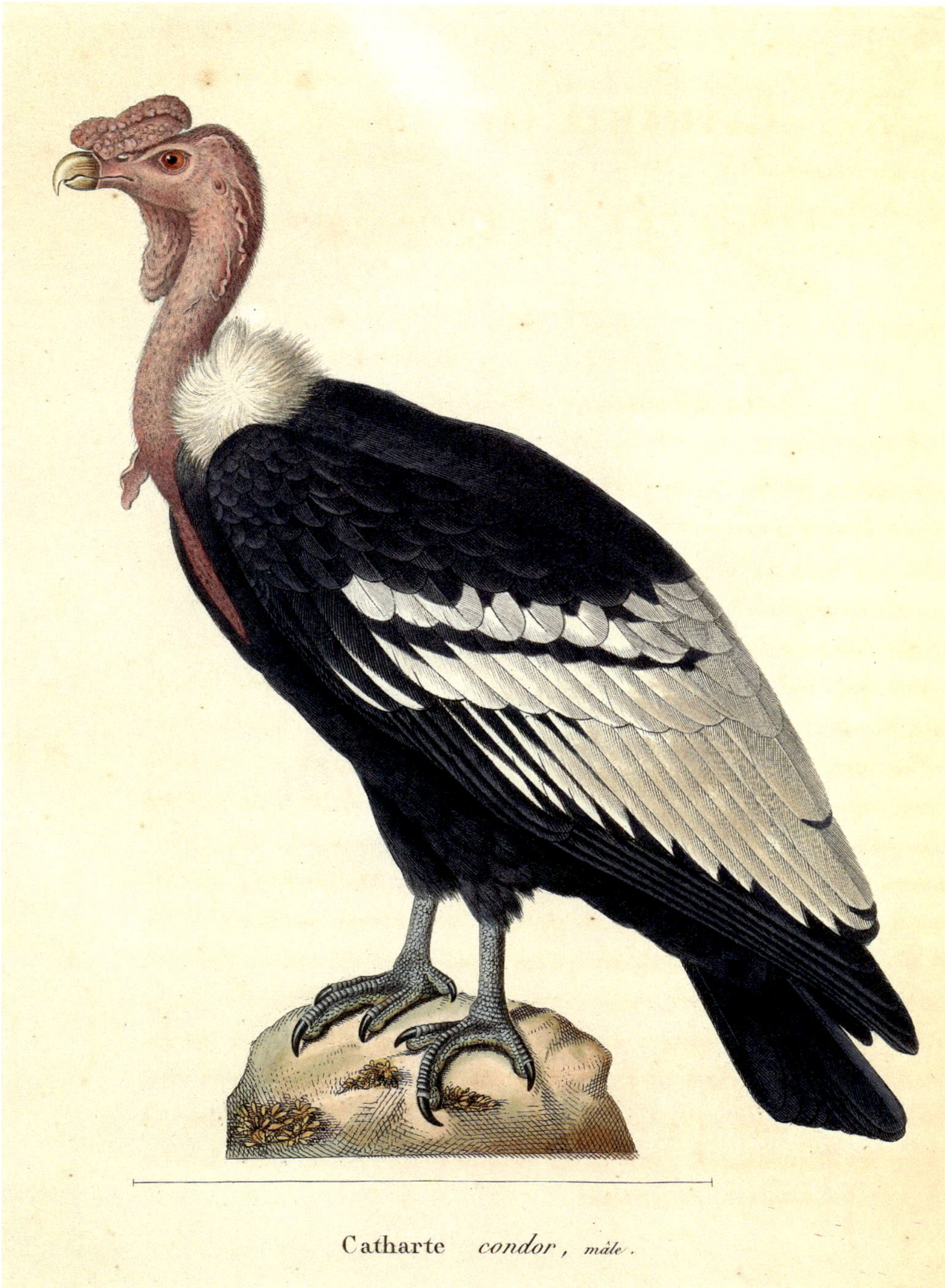

ANDEAN CONDOR
Vultur gryphus

Nicolas Huet II, *Nouveau recueil de planches coloriées d'oiseaux*, 1820–39

In this work by Huet, we see the full scale of the bird. The condor has a wingspan of over three metres and fully grown adults can be as heavy as 15 kilograms. For this reason, the bird tends to live in windy areas, where steady winds can allow this massive bird to glide through the air. The condor is a scavenger, so it keeps a watchful eye for carrion, preferring to feed on the carcasses of larger animals such as llamas, alpacas and rheas. Those that live in coastal areas will often feed on seals, fish and even whale carcasses.

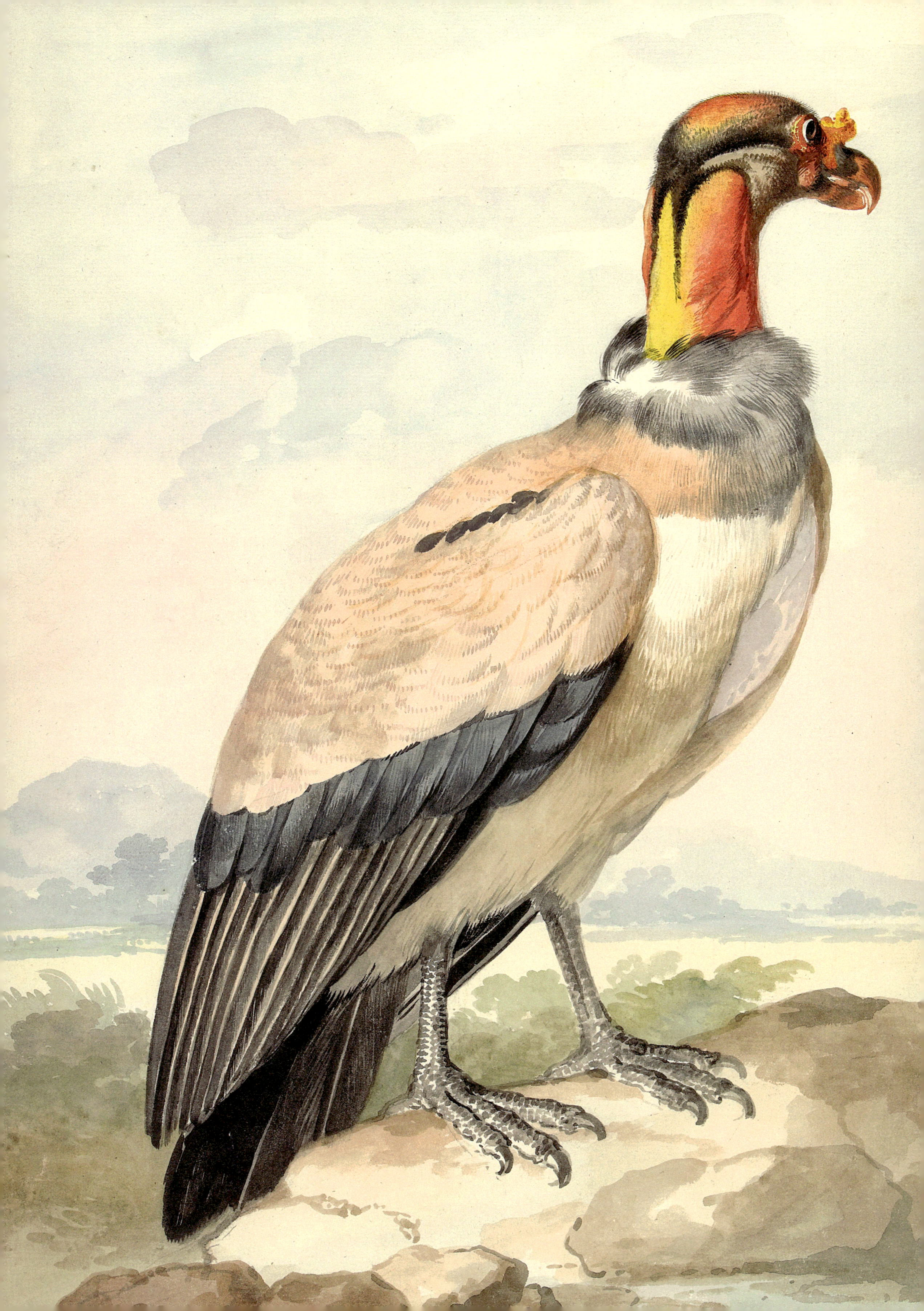

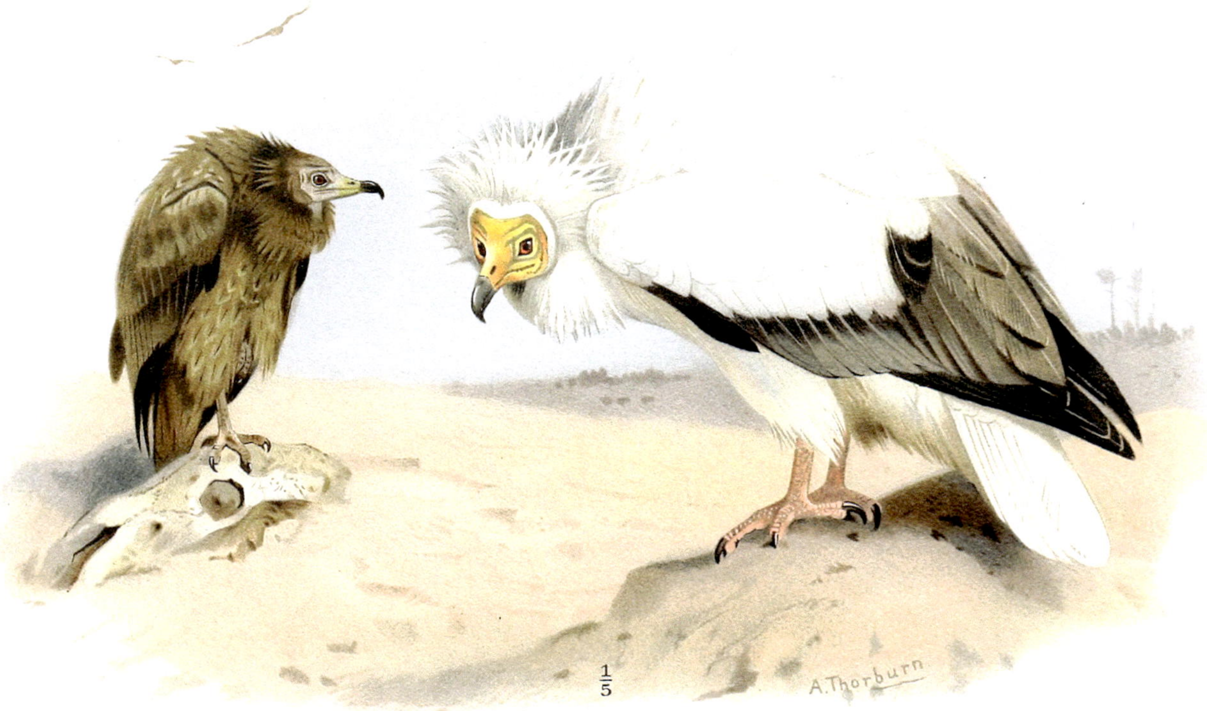

OPPOSITE:

KING VULTURE

Sarcoramphus papa

Aert Schouman, drawing, 1758

Most New World vultures are black, but the king vulture stands distinguished with its colourful head and neck that boasts shades of red, orange and yellow. Excluding the condor, the king vulture is the largest of the New World vultures.

EGYPTIAN VULTURE

Neophron percnopterus

Archibald Thorburn, *Coloured Figures of the Birds of the British Islands*, 1885–97

Old World vultures – such as these Egyptian vultures – belong to the same family as the eagle, yet in terms of human perception, they could not be further apart. While the eagle is seen as majestic and powerful, the vulture tends to be thought of with disgust and contempt. It is their scavenging nature that has earned them such a reputation, yet their impressive ability to digest rotting carcases serves an essential role in the cleaning of our ecosystems.

Index

Page numbers in *italics* refer to illustrations

Bibliography

Adams, H. G., Adams. H. B. *The Smaller British Birds: With Descriptions of Their Nests, Eggs, Habits*. London: Gibbins, 1894

Albin, Eleazar. *A Natural History of English Insects*. London: Published by the Author, 1720

Albin, Eleazar. *A Natural History of Birds: Illustrated With a Hundred and One Copper Plates, Curiously Engraven From the Life*. London: Printed for the Author by William Innys, 1731–38

Albin, Eleazar. *A Natural History of Spiders, and Other Curious Insects*. London: Printed by John Tilly, 1736

Albin, Eleazar. *A Natural History of English Songbirds*. London: Printed by A. Bettesworth and C. Hitch; and S. Birt, 1737

Albin, Eleazar. *The History of Esculent Fish*. London: Printed for E. Jeffery; R. Faulder; J. Cuthell and J Deighton; J. Walker; Hamilton and Co., 1794

Audubon, John James. *The Birds of America; from Original Drawings by John James Audubon*. London: R. Havell & R. Havell Jr.; Edinburgh: W.H. Lizars, Published by the Author, 1827–38

Audubon, John James. *The Birds of America: From Drawings Made in the United States and Their Territories* (octavo). New York: J.B. Chevalier, 1840–44

Audubon, John James, and Bachman, John. *The Viviparous Quadrupeds of North America*. New York: Published by the Author, 1845–48

Bairei, Kôno. *Bairei kacho gafu (Bairei's Album of Flowers and Birds)*. Osaka, 1893

Beebe, William. *A Monograph of the Pheasants*. London: Published under the auspices of the New York Zoological Society by Witherby & Co., 1918–1922

Belon, Pierre. *L'histoire de la nature des oyseaux: avec leurs descriptions, et naïfs portraits retirez du naturel (The History of Birds in Nature of Birds: With Descriptions & Portraits Taken From Nature)*. Paris: Gilles Corrozet, 1555

Bewick, Thomas, *A History of British Birds*. London: John van Voorst, 1797–1804

Broinowski, Gracius J.. *The Birds of Australia Comprising Three Hundred Full–page Illustrations with a Descriptive Account of the Life and Characteristic Habits of over Seven Hundred Species*. Melbourne: C. Stuart & co., 1890–91

Brown, Peter. *New Illustrations of Zoology, Containing Fifty Coloured Plates of New, Curious, and Non-descript Birds, With a Few Quadrupeds, Reptiles and Insects*. London: B. White, 1776

Brown, Thomas. *Illustrations of the American ornithology of Alexander Wilson and Charles Lucian Bonaparte: with the addition of numerous recently discovered species and representations of the whole sylvae of North America*. Edinburgh: Frazer & Co., 1835

Buffon, Georges Louis Leclerc, Comte de. *Histoire naturelle, générale et particuliére, avec la description du Cabinet du roy (Natural history, general and particular, with description by the Royal Cabinet)*. Paris: Imprimerie nationale, 1749–1804

Buffon, Georges Louis Leclerc, Comte de. *Histoire Naturelle des Oiseaux (Natural History of Birds)*. Paris: De l'imprimerie royale, 1770–83

Buller, Walter Lawry, Sir. *A History of the Birds of New Zealand*. London: John Van Voorst 1872–73 (1st edition), 1887–88 (2nd edition)

Cassin, John. *Illustrations of the Birds of California, Texas, Oregon, British and Russian America: Intended to Contain Descriptions and Figures of All North American Birds Not Given by Former American Authors, and a General Synopsis of North American Ornithology*. Philadelphia: J. B. Lippincott & Co. 1853–56, this edition 1862

Catesby, Mark. *The Natural History of Carolina, Florida and the Bahama Islands: Containing the Figures of Birds, Beasts, Fishes, Serpents, Insects, and Plants*. London: Published by the Author, 1729–47

Chapman, Frank M. *The Economic Value of Birds to the State*. Albany: J.B. Lyon Company, Printers, 1903

Darwin, Charles. *Journal of Researches Into the Geology and Natural History of the Various Countries Visited by H.M.S. Beagle*. London: J. Murray, 1845

Darwin, Charles. *The Expression of the Emotions in Man and Animals*. London: John Murray, 1872

Davenport, Charles B. *Inheritance of Characteristics in Domestic Fowl – Carnegie Institution of Washington publication No. 121*. Washington: Carnegie Institution of Washington, 1902

Des Murs, Marc Athanase Parfait Œillet. *Iconographie ornithologique: nouveau recueil général de planches peintes d'oiseaux (Ornithological Iconography: New General Collection of Painted Boards of Birds)*. Paris: Friedrich Klincksieck, 1845-49

Diderot, Denis, and d'Alembert Jean Le Rond (eds). *Encyclopédie (Encyclopedia)*, Paris: André le Breton; Michel-Antoine David; Laurent Durand; and Antoine-Claude Briasson, 1751–72

Diggles, Silvester. *Companion to Gould's Handbook; Or, Synopsis of the Birds of Australia. Containing Nearly One-Third of the Whole, or about 220 Examples, for the Most Part from the Original Drawings*. Brisbane: Thorne & Greenwell, 1877

Dresser, Henry Eeles. *A History of the Birds of Europe: Including All the Species Inhabiting the Western Palaearctic Region*. London: Published by the Author 1871–81

Edwards, George. *A Natural History of Uncommon*

Birds: And of Some Other Rare and Undescribed Animals, Quadrupedes, Fishes, Reptiles, Insects, Et Cetera. London: Printed for the author, at the College of Physicians in Warwick-Lane, 1743–51

Edwards, George. *Gleanings of Natural History: Exhibiting Figures of Quadrupeds, Birds, Insects, Plants Et cetera, Most of Which Have Not, till Now, Been Either Figured or Described: With Descriptions of Seventy Different Subjects.* London: Printed for the author at the Royal College of Physicians, 1758–64

Elliot, Daniel Giraud. *A Monograph of the Phasianidae, or Family of the Pheasants.* New York: Published by the Author, 1870–72

Elliot, Daniel Giraud. *A Monograph of the Paradiseidae or Birds of Paradise.* London: Published by the Author, 1873

Elliot, Daniel Giraud. *A Monograph of the Bucerotidæ, or Family of the Hornbills.* London: Printed by Taylor and Francis, Published by the Author, 1877–82

Fitzinger, Leopold Joseph. *Bilder-atlas zur Wissenschaftlich-populären Naturgeschichte der Vögel in ihren sämmtlichen Hauptformen (Picture Atlas of the Scientific and Popular Natural History of Birds in All their Major Forms).* Vienna: KK Hof und Staatsdruckerei, 1860–64

Frisch, Johann Leonhard. *Vorstellung der Vögel Deutschlandes und beyläufig auch einiger Fremden: nach ihren Eigenschaften beschrieben (German Birds and Some Foreigner: Described According to Their Characteristics).* Berlin: Printed by Friedr. Wilhelm Birnstiel, 1733–63

Gessner, Conrad. *Historia animalium (History of Animals).* Zurich, C. Froschauer, 1551–58

Godman, Frederick Du Cane and Salvin, Osbert (eds). *Biologia Centrali-Americana: Zoology, Botany and Archaeology.* London: Published for the editors by R. H. Porter, 1879–1915

Gould, John. *A Century of Birds from the Himalaya Mountains.* London: 1830–32

Gould, John. *The Birds of Europe.* London: Printed by R. and J. E. Taylor, Published by the Author, 1832–37

Gould, John. *A Monograph of the Ramphastidae, or Family of Toucans. London: Published by the Author,* 1833–35

Gould, John. *A Synopsis of the Birds of Australia and the Adjacent Islands*, Published by the Author, 1837–38

Gould, John. *The Birds of Australia.* London: Printed by R. and J. E. Taylor, Published by the Author, 1840–48

Gould, John. *A Monograph of the Macropodidae, or Family of Kangaroos.* Published by the Author, 1841–42

Gould, John, *A Monograph of the Odontophorinæ, or, Partridges of America.* London: Printed by Richard and John E. Taylor, 1844–50

Gould, John. *The Mammals of Australia.* London: Printed by Taylor and Francis, Published by the Author, 1845–63

Gould, John. *A Monograph of the Trochilidæ, or Family of Humming-Birds*, London: Printed by Taylor and Francis, 1849–61

Gould, John. *The Birds of Asia.* London: Printed by Taylor and Francis, Published by the Author, 1850–83

Gould, John. *Supplement to the Birds of Australia* London: Printed by Taylor and Francis. Published by the Author, 1851–69

Gould, John. *The Birds of Great Britain.* London: Printed by Taylor and Francis, Published by the Author, 1862–73

Gould, John. *The Birds of New Guinea and the Adjacent Papuan Islands: Including Many New Species Recently Discovered in Australia.* London: Henry Sotheran & Co., 1875–88

Graves, George. *British Ornithology British Ornithology: Being the History, with a Coloured Representation, of Every Known Species of British Birds.* London: Printed by W and S Graves, Published by the Author 1811–21

Greene, W. T. *Parrots in Captivity.* London: George Bell and Sons, 1884–88

Gray, George Robert. *The Genera of Birds: Comprising Their Generic Characters, a Notice of the Habits of Each Genus, and an Extensive List of Species Referred to Their Several Genera.* London: Longman, Brown, Green, and Longmans, 1844–49

Huish, Robert. *The Wonders of the Animal Kingdom: Exhibiting Delineations of the Most Distinguished Wild Animals in the Various Menageries of This Country, London: T. Kelly, 1830.*

Jardine, William, and Selby, Prideaux John. *Illustrations of Ornithology.* Edinburgh: W.H. Lizars, 1825–43

Jardine, William. *The Naturalist's Library.* Edinburgh: W. H. Lizars, 1833–43

Johnson, Samuel. *Dictionary of the English Language.* London: Printed by W. Strahan. Published by J. and P. Knapton; T. and T. Longman; C. Hitch and L. Hawes; A. Millar; and R.
and J. Dodsley, 1755

Jones, Howard. *Illustrations of the Nests and Eggs of Birds of Ohio.* Ohio: s.n.,1886

Jones, Thomas Rymer. *Cassell's Book of Birds: From the Text of Dr. Brehm.* London [etc.]: Cassell, Petter & Galpin, ca. 1875

Kashū, Numata. *Shūchō gafu (Picture Album of Various Birds).* 1885

Le Vaillant, François. *Histoire naturelle des Oiseaux d'Afrique (Natural History of African Birds).* Paris: J.J. Fuchs, libraire, 1796–1812

Le Vaillant, François. *Histoire naturelle des perroquets (Natural History of Parrots).* Paris: Levrault, 1801–05

Le Vaillant, François. *Histoire naturelle des oiseaux de paradis et des rolliers, suivie de celle des toucans et des barbus (Natural History of Birds of Paradise and Rollers, followed by that of toucans and Barbus)*. Paris: Denné le jeune, 1806

Lear, Edward. *Illustrations of the Family of Psittacidœ, or Parrots: The Greater Part of Them Species Hitherto Unfigured, Containing Forty-Two Lithographic Plates, Drawn From Life, and on Stone*. London: Published by the Author, 1832

Lilford, Thomas Littleton Powys, Baron. *Coloured Figures of the Birds of the British Islands*. London: R. H. Porter, 1885–97

Linnaeus, Carl. Systema naturae per regna tria naturae (*The System of Nature by Means of the Three Kingdoms of the Nature*). Stockholm: Laurentius Salvius, 1758 (10th edition)

Lydekker, Richard (ed.). *The Royal Natural History*. London & New York: Frederick Warne & Co., 1893–96

Nozeman, Cornelis. *Nederlandsche vogelen; vogens hunne huishouding, aert (Dutch Birds; According to their Households and Characteristics)*. Amsterdam: J.C. Sepp, 1770–1829

Manetti, Saverio. *Ornithologia methodice digesta, atque iconibus aeneis ad vivum illuminatis ornata (Natural History of the Birds, Treated Systematically and Adorned With Copperplate Engraving Illustrations, in Miniature and Life-Size)*. Florence: In Aedibus Mouchianis, 1767–76

Oort, E. D. van. *Ornithologia neerlandica: De vogels van Nederland (Dutch Ornithology: The Birds of the Netherlands)*. The Hague: M. Nijhoff, 1922–35

Phillips, John C. *A Natural History of the Ducks*. Boston: Houghton Mifflin Company, 1922–26

Pope, Alexander. *Upland Game Birds and Water Fowl of the United States*. New York: Scribner, Armstrong, 1877–78

Richardson, John, Sir. *The Zoology of the Voyage of the H.M.S. Erebus & Terror, Under the Command of Captain Sir James Clark Ross, During the Years 1839 to 1843*. London: E. W. Janson, 1844–75

Riesenthal, Oskar von. *Die Raubvögel Deutschlands und des angrenzenden Mitteleuropas; Darstellung und Beschreibung der in Deutschland und den benachbarten Ländern von Mitteleuropa vorkommenden Raubvögel (The Birds of Prey in Germany and Neighboring Central Europe; Representation and Description of the Birds of Prey Found in Germany and the Neighboring Countries of Central Europe)*, Kassel: Verlag von Theodor Fischer, 1876

Rothschild, Lionel Walter Rothschild, Baron. *A Monograph of the Genus Casuarius*. London: Zoological Society of London, 1900

Rothschild, Lionel Walter Rothschild, Baron. *Extinct Birds: An Attempt to Unite in One Volume a Short Account of Those Birds Which Have Become Extinct in Historical Times: That Is, Within the Last Six or Seven Hundred Years: To Which Are Added a Few Which Still Exist, but Are on the Verge of Extinction*. London: Hutchinson & Co., 1907

Rothschild, Lionel Walter Rothschild, Baron (ed.). *Novitates zoologicae*: A Journal of Zoology in Connection with the Tring Museum, vol. 9. Tring: Zoological Museum, 1902

Rowley, George Dawson. *Ornithological Miscellany*. London: Trübner and Co., Bernard Quaritch, R.H. Porter, 1876–78

Sagra, Ramón de la. *Historia física política y natural de la isla de Cuba, botánica (Political and Natural History of the Island of Cuba and its Botany)*. Paris: A. Bertrand, 1838–61

Rüppell, Eduard. *Systematische Uebersicht der Vögel Nord-Ost-Afrika's: nebst Abbildung und Beschreibung von fünfzig Theils unbekannten, Theils noch nicht bildlich dargestellten Arten (Systematic Overview of the Birds of North-East Africa: Together With Illustration and Description of Fifty Species, Some of Which Are Unknown, Some of Which Have Not Yet Been Illustrated)*. Frankfurt: S Schmerber'sche Buchhandlung, 1845

Schlegel, Hermann. *Traité de fauconnerie (Treaty of Falconry)*. Leiden and Düsseldorf: Arnz & Comp., 1844–53

Selby, Prideaux John. *Illustrations of British Ornithology*. Edinburgh: W.H. Lizars; London: Longman, Rees, Orme, Brown, Green and Longman; Dublin: W. Curry Jun. & Co. 1819–34

Selby, Prideaux John. *The Natural History of Pigeons*. Edinburgh: W.H. Lizars, 1835

Selby, Prideaux John, and Turner, William, jr., Rev. *Natural History of Parrots*. Edinburgh, W. H. Lizars, 1836

Sclater, Philip Lutley. *Exotic Ornithology: Containing Figures and Descriptions of New or Rare Species of American Birds*. London: Bernard Quaritch, 1869

Shaw, George. *Museum Leverianum, Containing Select Specimens From the Museum of the Late Sir Ashton Lever, Kt., With Descriptions in Latin and English*. London: Impensis Jacobi Parkinson, 1792–96

Skottowe, Thomas. *Select Specimens From Nature of the Birds & Animals of New South Wales, Collected and Arranged by Thomas Skottowe Esqr*, 1813

Smith, Andrew, Sir. *Illustrations of the Zoology of South Africa, Consisting Chiefly of Figures and Descriptions of the Objects of Natural History Collected During an Expedition Into the Interior of South Africa, in the Years 1834, 1835, and 1836; Fitted Out by 'The Cape of Good Hope Association for Exploring Central Africa'*. London: Smith, Elder and Co., 1838–49

Studer, Jacob Henry. *The Birds of North America: One Hundred and Nineteen Artistic Colored Plates Representing the Different Species and Varieties Drawn and Colored from Nature*. New York: Published under the auspices of the Natural Science Association of America, 1903

Strickland, H. E., and Melville, Alexander Gordon. *The Dodo and Its Kindred; Or, the History, Affinities, and Osteology of the Dodo, Solitaire, and Other Extinct Birds of the Islands Mauritius, Rodriguez and Bourbon*. London: Reeve, Benham, and Reeve, 1848

Swaysland, W. *Familiar Wild Birds*. London, New York: Cassell, 1883

Swainson, William. *Zoological Illustrations, or, Original Figures and Descriptions of New, Rare, or Interesting Animals*. London: R. and A. Taylor, 1820–23

Temminck, C.J. *Nouveau recueil de planches coloriées d'oiseaux : pour servir de suite et de complément aux planches enluminées de Buffon (New Collection of Colored Plates of Birds: To Serve as a Continuation and Complement to the Illuminated Plates by Buffon)*. Strasbourg: Legras Imbert & Comp., 1820–39

Thomson, James. *Arcana naturæ, ou Recueil d'histoire naturelle (Secrets of Nature, or a Collection of Natural History)*. Paris: Bureau du Trésorier de la Société entomologique de France, 1859

Thornton, Robert John. *New Illustration of the Sexual System of Carolus Von Linnaeus: and the Temple of Flora, or Garden of Nature*. London: Printed by T. Bensley, 1799–1810

Turner, William. *Avium praecipuarum, quarum apud Plinium et Aristotelem mentio est, brevis et succincta historia (The Principal Birds of Aristotle and Pliny: A Short and Succinct History)*. Cologne: Johann Gymnich, 1544

Wagler, Johann Georg. *Monographia Psittacorum*. Munich: George Jaquet. 1835

White, John. *Journal of a Voyage to New South Wales: With Sixty-Five Plates of Nondescript Animals, Birds, Lizards, Serpents, Curious Cones of Trees and Other Natural Productions*. London: Printed for J. Debrett, 1790

Wilson, Alexander. *American Ornithology: Or, The Natural History of the Birds of the United States*. New York, Harrison Hall, 1808–14 (1st edition), 1824 (2nd edition)

Wood, William. *Zoography; or, The Beauties of Nature Displayed. In Select Descriptions From the Animal, and Vegetable, With Additions From the Mineral Kingdom. Systematically Arranged*. London: Printed for Cadell and Davies, 1807

Willughby, Francis, and Ray, John. *The Ornithology of Francis Willughby*. London: Printed by A.C. for John Martyn, 1678

Wright, Magnus von, and Wright, Wilhelm von. *Svenska Fåglar Efter Naturen Och Pa Sten Ritade (Swedish Birds Drawn from Nature And On Stone)*. Stockholm: Förlaget Svenska Fåglar, 1927–29

Image credits

Unless stated otherwise, the images used in this book are considered to be out of copyright and in the Public Domain. The author and publisher wish to acknowledge the generosity of the organizations listed below who have made images available without charge.

T = Top; B = Bottom; L = left; R = right

HTTPS://ARCHIVE.ORG : McGill University Library: 1, 46B, 63, 79; University of Pittsburgh Library: 200, 201; University of Toronto: 13.

BIODIVERSITY HERITAGE LIBRARY HTTPS://WWW.BIODIVERSITYLIBRARY.ORG : 8, 9, 15, 18, 22, 24T, 24B, 25, 27, 28, 29, 30-31, 32T, 32BL, 33, 34, 35, 36TL, 36TR, 36BL, 37, 38TL, 38TR, 38BL, 38BR, 39, 42T, 42B, 43, 45, 46T, 47, 48, 50, 51, 52, 53, 54-55, 56, 60TL, 60TR, 60B, 61, 66, 67, 69, 70-71, 73, 74, 77, 78, 82, 83, 84, 85T, 86T, 86B, 87, 88, 89, 90, 91, 93, 95, 96, 100, 101, 103, 104, 105, 106BR, 107, 108, 110, 111, 112T, 112B, 113, 114, 115, 116-117, 118, 119, 120, 122, 125T, 125B, 127, 132, 133, 135, 138, 139, 141, 142, 145, 146TL, 146TR, 146B, 148, 149, 150, 152, 153, 154, 155, 156TL, 156TR, 156BL, 156BR, 157, 158T, 158B, 159, 160, 161, 163, 166, 170, 172, 173, 174B, 175, 176, 177, 179, 180, 182, 183, 184 185, 186, 188, 189, 191, 192, 193, 194, 195, 197, 203, 204, 205T, 205B, 206-207, 209, 210, 211, 213, 216T, 216B, 217, 219, 220, 224, 227, 229L, 229R, 230, 231TL, 231TR, 231BL, 231BR, 232, 233, 235, 236, 237, 238, 239, 240TL, 240TR, 240BL, 242, 243TL, 243TR, 243BL, 243BR, 244, 245, 246, 247, 248, 249, 250TL, 250TR, 251, 252, 253L, 253R, 254, 255, 256, 257, 258, 260, 261, 262, 263, 264, 265, 269, 273, 274, 275T, 275B, 276-277, 278L, 278R, 279, 280, 281, 281, 282, 283, 284T, 284BL, 284BR, 285, 286, 287T, 287B, 288, 289, 290TL, 290R, 290BL, 291, 293, 294, 295, 296L, 296R, 298, 299, 304, 306L, 306R, 307, 308, 309, 310, 311, 312TL, 312TR, 312BL, 312BR, 313, 315, 316, 317TL, 317TR, 317BL, 317BR, 318, 319, 320, 321, 322, 323, 332TL, 332TR, 332BL, 332BR, 333, 334,335, 336, 337, 338T, 338B, 339, 340, 341, 343.

BROOKLYN MUSEUM, New York/Gift of Anna Ferris: 76.

HTTPS://COMMONS.WIKIMEDIA.ORG : 26T, 62, 75, 85B.

HEINRICH HEINE UNIVERSITÄTS UND LANDESBIBLIOTHEK, DÜSSELDORF/ HTTPS://NBN-RESOLVING.DE/URN:NBN:DE:HBZ:061:2-26-P0109-4 : 325.

LIBRARY OF CONGRESS, Washington, DC: 65, 126T,164.

THE LIBRARY OF THE NATURFORSCHENDE GESELLSCHAFT ZÜRICH (ZB Zürich): 128, 134.

METROPOLITAN MUSEUM OF ART, NEW YORK/Bequest of Mrs. Charles Wrightsman: 11; H. O. Havemeyer Collection, Bequest of Mrs. H. O. Havemeyer, 1929: 123; The Howard Mansfield Collection, Purchase, Rogers Fund, 1936: 144R, 329; Purchase, Joseph Pulitzer Bequest, 1918: 144L; Purchase, Rogers Fund and The Kevorkian Foundation Gift, 1955: 171.

MINNEAPOLIS INSTITUTE OF ART. The Minnich Collection/The Ethel Morrison Van Derlip Fund, 1966: 7, 80, 121, 143, 162, 165, 335.

MITCHELL LIBRARY, State Library of New South Wales, Australia: 212, 218.

MUSEUMS VICTORIA COLLECTIONS, Australia: 259.

https://www.rawpixel.com/category/53/public-domain: 26B, 64, 106TL, 106TR, 106BL, 174T.

RIJKSMUSEUM, Amsterdam, Netherlands: 94, 109, 297.

RIJKSMUSEUM, Amsterdam, Netherlands/Purchased with the support of the F.G. Waller-Fonds: 40-41, 181; Gift of M.J. van Zijverden, Warder: 68; Gift of J. Perrée, Eindhoven: 124, 214; M.E. van den Brink Bequest, Velp: 136-137, 147; Gift of Jonkheer B.W.F. van Riemsdijk: 187; Gift of A.H. Beels van Heemstede-van Loon: 202; Gift of Mrs Brandt, Amsterdam: 225, 228; On loan from the City of Amsterdam: 342.

UNIVERSITY OF PITTSBURGH: 17, 23, 49, 72, 81, 126B, 151, 178, 190, 196, 199, 208, 215, 241, 272, 300-301, 305, 314, 328, 330, 331, 326-327, 352.

Acknowledgements

I would like to thank the institutions and libraries that work tirelessly to share their incredible collections with the world. This book would simply not be possible without their hard work, generosity and expertise. A huge thank you also to the team at Laurence King: Zara Larcombe who approached me with the idea; editors Sara Goldsmith and Felicity Maunder; designer Alex Coco; copy editor Laura Gladwin; production manager Simon Walsh; image researcher Peter Kent and ornithology consultant Dominic Couzens. Finally, I would like to thank my parents, Adrian and Rita, for all their support, encouragement and generosity. I dedicate this book to them.

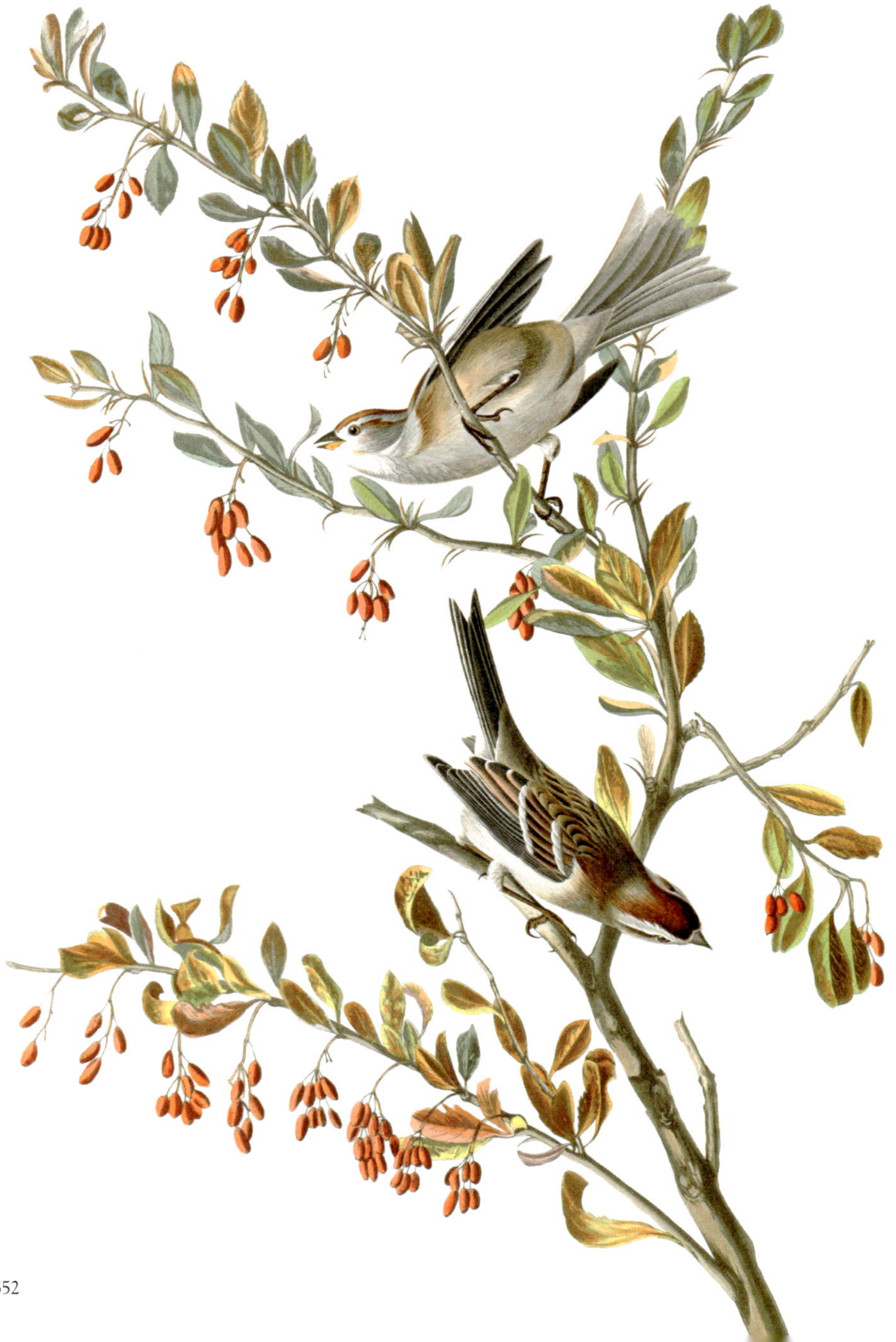